Textbook of
LIMNOLOGY
Fifth Edition

D0073675

Textbook of
LIMNOLOGY
Fifth Edition

Gerald A. Cole
late of Arizona State University

Paul E. Weihe
Central College

WAVELAND
PRESS, INC.
Long Grove, Illinois

For information about this book, contact:
Waveland Press, Inc.
4180 IL Route 83, Suite 101
Long Grove, IL 60047-9580
(847) 634-0081
info@waveland.com
www.waveland.com

10-digit ISBN 1-4786-2307-1
13-digit ISBN 978-1-4786-2307-6

Printed in the United States of America

7   6   5   4   3   2

Contents

Preface

I am honored to present this revision of the *Textbook of Limnology*. The Fourth Edition has served my students well, and I myself was introduced to the subject with a previous edition in Dr. Robert Neely's Limnology class years ago.

In the two decades since the last edition appeared, both the subject and the world around us have changed. Our understanding of ecological theory is richer, our ability to collect and analyze data is much more advanced, and the contribution we make to important conversations about the Earth in the age of the *Anthropocene* is proving critical. And yet, it all still begins with an inquisitive look at tiny organisms floating in a pond, or insects crawling under rocks in a stream, or the distinctive smell underfoot as we muck about in a marsh. We are drawn to the water.

I gratefully acknowledge the support of Central College for a sabbatical while writing this revision. I appreciate the encouragement, advice, and expertise of colleagues and friends, including Russ Benedict, Anya Butt, Al Hibbard, Mark Johnson, Viktor Martisovits, Alexey Pronin, Kristin Siewert, and Jon Witt.

This Fifth Edition is respectfully dedicated to the memory of Dr. Gerald Cole, who in every edition wrote this text first and foremost for the student. I hope this edition will continue that legacy.

Paul E. Weihe

Preface to the Fourth Edition

Since the appearance of the Third Edition of the *Textbook of Limnology* in 1983, there have been many changes in limnologic procedures. Much more experimental work is being done, both in laboratory and natural settings. Instrumentation has advanced markedly, and the modern uses of light microscopy are amazing. With the help of various fluorescent dye molecules, fluorescence microscopy has become a remarkable research aid. The dyes evoke autofluorescence of cells and tiny organisms. Thus, much has been learned about the abundance of bacteria and protists in both marine and inland waters. Further study of their roles in nutrient cycles led to the *microbial loop* concept, which must be included in any modern treatment of limnology or oceanography.

Over the years most limnologists have shown no interest in fishes and other vertebrate predators. This has changed and now the "downward" effects of upper trophic levels on those below them is emphasized; the words "trophic cascade" and "cascading effects" are new and represent another important concept in aquatic studies. Today more attention is directed toward the interactions of organisms in food webs. Moreover, the PEG model that explains seasonal dynamics of lacustrine planktonic events has been recommended to me as an influential and significant concept since the Third Edition.

I had planned to add several new acronyms to the glossary, but soon realized that this might not be worthwhile. Authors save lines of print by using acronyms that are defined early in their papers; however, there seems to have been a trend toward multi-authorship in recent years. Thus, page space saved by using countless acronyms may be cancelled out by extra lines in citation sections!

A high percentage of the literature cited in the Third Edition could be omitted and replaced by more recent references. Despite this, there is historic value in many of the early publications and I have retained citations to papers that were influential decades ago.

The deaths of several prominent limnologists since publication of the Third Edition have saddened us all. The loss of that super-ecologist G. E. Hutchinson is especially worthy of note. His ideas opened the doors for a multitude of research projects and new concepts.

At the outset of the revision process, Waveland editors requested comments from professors who use my book for their classes. I am grateful for their valuable responses. Moreover, many authors have sent me reprints over the years and I appreciate their input. Finally, I am happy to acknowledge the literature and suggestions about updating the chapter on lotic waters that I received from James V. Ward.

Gerald A. Cole

Introduction

Limnology Defined

The most frequent question asked the student of limnology is, "What is limnology?" It was first defined by F. A. Forel (1892), a Swiss professor, who has been called the Father of Limnology. Translated, Forel's definition was "the oceanography of lakes." Forel's pioneer investigations were focused on Le Léman (Lake Geneva); although Alpine lakes are nowhere near an ocean, early limnologists found the techniques and theory of oceanography to be useful inland as well.

The term "limnology" is derived from the Greek word *limne* meaning pool, marsh, or lake; the science arose specifically from lake investigation, however. As time passed, limnology became the science of inland waters, concerned with all the factors that influence living populations within those waters. It now includes study of running water (lotic habitats) as well as of standing water (lentic habitats). It embodies the largest of lakes to the smallest of ponds and embraces ephemeral waters and wetlands as well as permanent lakes that have existed for millions of years. It is incorrect to limit it to the study of fresh waters; for in arid regions extremely saline pools and lakes are found, and these fall within the realm of limnology. Despite the variety of ecosystem types and geographic locations considered, limnologists study a surprisingly small amount of water.

Most of the earth's water is out of sight, locked in the rocks of the primary lithosphere, but of the more visible or accessible supply, about 97.3% is in the ocean basins (Table 1-1). Another 2.19% is represented at this time by ice in glaciers and the polar caps. That leaves a little more than one half of a percent

Table 1-1 An approximate partitioning of the earth's water supply*

	Volume (10^3 km^3)	Percent
Oceans	1,310,302.1	97.3
Ice	29,491.9	2.19
Groundwater	6,733.3	0.5
Soil moisture	74.1	0.005
Atmospheric water vapor	13.5	0.001
Inland freshwater lakes	126.0	0.009
Inland saline lakes	104.0	0.008
Rivers	1.3	0.0001

*Means from various previous estimates including Williams, 1986.

1

to be shared by the ground water, soil moisture, atmospheric water vapor, and the lakes and rivers that are the concern of limnologists. A little less than 0.02% of the world's water supply is found in inland lakes, both fresh and saline, and the amount present at any time in river channels is about one half a percent of that!

Limnology: An Interdisciplinary Science

Geology

The origin of lake basins, their resultant morphology, and subsequent modification of shape are the results of geologic processes. Because erosion and sedimentation are within the scope of geology, it is apparent that the birth, life, and death of a lake, as its basin fills, are geologic functions.

The substratum on which a water-filled depression lies or from which it receives its soluble salts and other nutrients is dependent on its geologic legacy. This heritage of nutrients available to the aquatic habitat via the weathering of soils within its drainage area is the **edaphic** (soil-related) factor. **Allochthonous materials** are produced outside the lake basin, and **autochthonous** from within.

The so-called trophic nature of a body of water (referring basically to its soluble nutrients and resultant biotic productivity) is the result of interaction of at least three important factors. First is the edaphic factor, which determines whether a lake is rich and productive or comparatively sterile. Second are the morphologic features, best defined as dimensions of the basin. These first two factors are of a geologic nature. The third is the climatic element, a meteorologic-geographic matter. There is a range of climates from severe to favorable for growth and productivity, with many meteorologic ramifications. These are fairly obvious, including extreme temperature differences, duration of growing season, solar radiation, precipitation, evaporation rates, and wind.

Aquatic systems in which these three main factors are working in harmony may exhibit the extremes of a trophic scale. Thus, a remarkably poor lake, exhibiting what will be defined later as **oligotrophy**, has soil, basin shape, and climate working in concert. Many lakes fit on the scale somewhere between oligotrophy and marked **eutrophy**, characterized by water rich in nutrients and high biotic productivity, and are a result of edaphic, morphologic, and climatic features working in unison.

Also within the province of geology is **paleolimnology**, the study of lacustrine (L. *lacus,* lake) sediments and relics preserved in them. The nature of the sediments implies something of past conditions in the lake as well as its responses to external influences, which involve chemistry as well as structure. In some instances, laminated annual layers, called **varves**, can be discerned and counted. The chitinous remains of aquatic arthropods, calcareous molluscan shells, siliceous structures from diatoms and sponges, and other resistant structures tell something about the communities that occupied the basin in the past and what changes have occurred in the biota. The study of these sediments has become increasingly valuable as we attempt to understand global climate change (Cohen, 2003).

Closely related to paleolimnology is **palynology**, the study of spores and pollen grains. These plant structures remain preserved in lake and bog deposits, where they imply much about neighboring plant communities and climates of the past. Sometimes fossilized pollen from aquatic plants are present, and from them information about conditions in the lake itself can be inferred.

Physics, Mathematics, and Computer Science

The first concern of physical limnology is the very nature of the water molecule. Water is of course a mixture of hydrogen, deuterium, and tritium isotopes combined with isotopes of oxygen, rather than a single type of molecule. Water's high specific heat and the nonlinear relation of density and viscosity to temperature are unique properties. They play important roles in the penetration, absorption, and distribution of light and heat and in the resultant density stratification in lakes.

In addition, the various movements of water—eddies, currents, and waves—come under the heading of physical limnology; we are indebted to those limnologists well grounded in mathematics for our knowledge of these topics. Here we discover meteorologic overtones in the synthesis called limnology. Meteorologists have applied techniques proved useful in investigating aerial conditions to the study of lacustrine currents.

Physicists' special contribution to limnology has been the development of electronic apparatus to facilitate rapid, accurate measurements. In the United States the late L. V. Whitney's efforts were especially valuable; his electric subsurface thermometers, conductivity meters, and photometers represented milestones in the journey of North American limnology and were produced early in the age of modern electronics.

Limnologists have benefited from widespread, cheap computing and network capabilities. Probes and sondes are deployed to measure and store data for later download, or even continuous remote transmission. A limnologist today might have river discharge readings or lake chemistry loaded automatically to her smart phone. Even better, the Internet has allowed widespread collaboration between "citizen scientists" sending environmental observations to a researcher half a world away, greatly increasing the number and type of observations available.

Chemistry

Analysis and study of the chemical constituents in natural waters is a big part of limnology. Much of chemical limnology has been inorganic, but more is being learned continuously about the importance of complex organic compounds in the dynamics of fresh water. It is apparent that water chemistry is closely allied to geology and to the biology of aquatic habitats. The relationship of physical limnology to the distribution of chemical compounds within inland waters is discussed subsequently.

Biology

The organisms of inland waters range from bacteria to mammals. Bacteriologists, botanists, and zoologists are, therefore, members of the limnologic fraternity. Their discipline is aquatic biology, the study of aquatic species and populations. This definition rests on shaky grounds for when biologists concern themselves with environmental factors that affect the species or community, they have ventured into limnology.

The concept of biologic productivity, as well as the search for factors that make one lake more productive than its neighbors or than a distant counterpart, is a major and unifying theme of limnology. The basis for biologic productivity is photosynthesis or primary production. The plant physiologist, who is interested in photosynthetic mechanisms and rates, and the phycologist, who is interested in algal nutrients and growth, have become major contributors to the science of inland waters.

These photosynthesizers are the base of aquatic food chains which are better understood in recent years. Limnologists were key in developing the Trophic Cascade concept, with whole-lake food chain manipulation. Addition or subtraction of predators (such as stocking large piscivorous fish) allow us to observe and control populations farther down the chain.

An Overview of Aquatic Ecosystems

Nearly 50 years before the term "ecosystem" was first used, Stephen A. Forbes (1887) delivered a talk that, although published in an obscure journal, subsequently became a classic paper, "The Lake as a Microcosm." Forbes' concept of a little lake world emphasized the interacting biotic and abiotic processes in the system; today we would also recognize important external influences on the lake. Forel's first volume on Lake Geneva (1892) dealt with environmental factors rather than with the lacustrine biota. Since then geographic-physical-chemical studies have been termed *forelian* limnology. The work performed during the first 40 years of the 20th century by E. A. Birge, C. Juday, and their students at the University of Wisconsin marked the onset of modern American limnology and made conditions in Wisconsin lakes a touchstone for later studies in other regions. Birge, one of the first Americans to work seriously with the microcrustaceans known as Cladocera, was led from a biologic study of their spatial and seasonal distribution in Lake Mendota to a study of the physical and chemical reasons accounting for puzzling fluctuations of cladoceran populations—in other words, to forelian limnology.

Limnologists have always been on the cutting edge of ecosystem science. The first measurement of energy flow through ecosystems (trophic dynamics) was conducted by Raymond Lindeman (1941, 1942a, 1942b) on Cedar Bog Lake in Minnesota. Biogeochemistry was advanced through the classic *Treatise on Limnology* (Hutchinson, 1957). We are now attempting to understand and predict rapid, catastrophic changes in ecosystems; such "regime shifts" in lakes are described by Carpenter and associates (2011) but have implications for many ecosystems.

Lake Regions

Benthic Zones and Benthos

The adjective *benthic* applies to bottom regions. The word *benthos* designates the community of bottom-dwelling organisms.

Littoral zone. The peripheral shallows are subject to fluctuating temperatures and erosion of shore materials through wave action and the grinding of ice. The result is a bottom region of relatively coarse sediment, especially evident near unprotected shores. These shallows are usually well lighted and inhabited by rooted aquatic plants extending to some lakeward depth and contributing fragments to the littoral sediments as well as providing support and sustenance for other organisms. This is the **littoral zone**, the shore region (Fig. 1-1).

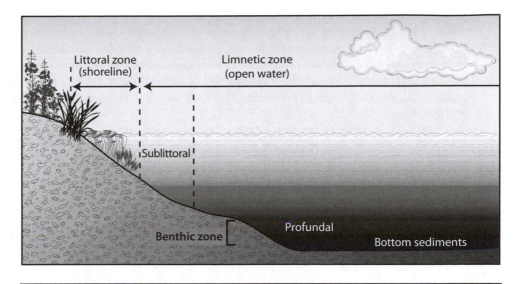

Figure 1-1 Lake zones based on lake bottom (benthic) characteristics: near the shoreline the littoral region contains rooted vegetation; the open-water or limnetic zone lies beyond. In deeper lakes, stratification causes the limnetic zone to form layers, with a profundal zone lying beneath the lowest layer.

The zone from shoreline to the depth where "weeds" disappear is often defined as the littoral region. Sometimes wave action is so extreme that large aquatic angiosperms are absent and only algae, attached and streaming in the currents or growing as benthic mats, are present, their outer margin fixing the lakeward limits of the littoral area.

The littoral benthos is composed of many species and taxonomic groups. Great diversity and high annual production set this community apart from the benthos in deeper areas. This is a result of the abundance and variety of habitats and ecologic niches in near-shore regions as compared with the deeps.

Sublittoral zone. The sublittoral region extends lakeward from the littoral. Its sediments are finer grained. Although dimly lighted and lacking a benthic macroflora, it is usually well oxygenated. The sublittoral fauna contains fewer species than the littoral assemblage; this is a result of the reduced number of niches.

In some lakes the old shells of gastropods and pelecypods that inhabited the littoral zone are found accumulated at sublittoral depths. These *shell zones* are thought to mark the place where interactions of wind, wave, shoreline configuration, currents, and countercurrents have carried and dropped the calcareous remains of the littoral molluscan fauna. There is some evidence that clams retreating to deep water for the winter season die in the sublittoral region and contribute to the shell zone. Whatever the case, parts of the **sublittoral zone** serve as the cemetery of some littoral species.

Profundal zone. The **profundal zone** can be defined in lakes deep enough to exhibit summer temperature stratification. Under such conditions a deep, cold region is formed where currents are at a minimum and where light is much reduced. The temperature is nearly uniform throughout this region, and under some conditions oxygen is scarce or depleted, although gases such as methane and CO_2 are abundant. The hydrogen ion concentration is high (pH is low) because of the presence of carbonic acid, and this stratum of water is characterized by decay rather than by production of organic matter. This is the hypolimnion, the "below lake" stratum that overlies the profundal zone.

Profundal sediments are fine particles largely made up of material produced within the lake and whatever matter from outside is small enough to be blown in or carried by currents to the deeps before settling. The profundal benthos has attracted the lion's share of attention from limnologists. Commonly, lake researchers move directly to the deeper parts, passing over the rich littoral biota to dredge up the impoverished profundal fauna. The benthos of the profundal zone tells something of conditions in the hypolimnetic water and, ultimately, a great deal about the nature of the lake. Moreover, dredge samples of profundal ooze are easily washed through sieves, leaving the animals exposed.

Open-water zones. Beyond the pond weeds (a common term for the marginal vegetation) is the open water. This is the limnetic or **pelagic zone**, a region in the lake where shore and bottom have lessened influence. It is the habitat of the plankton, an assemblage of tiny free-floating, drifting, or swimming plants and animals representing many taxa. In the upper windswept and well-illuminated layers photosynthesis prevails in daylight hours because phototrophic algae, the primary producers, fix inorganic carbon to manufacture organic compounds. It is called the **trophogenic zone**, where synthesis of organic carbon compounds occurs. Farther shoreward the littoral macrophytes and benthic algae, bathed by the same upper stratum, are part of the trophogenic layer. They are of less relative importance in primary productivity except in shallow lakes.

The limnetic trophogenic zone is the layer extending from the surface to the depth where photosynthesis no longer occurs. The trophogenic zone overlaps the *epilimnion*—the "above," or "upper," lake, at the surface its lower boundary is independent of the lower boundary of the epilimnion. Ideally, the sunny epilimnion is mixed throughout by the wind, many of its algal producers being moved throughout it as a result. The limnetic or plankton community includes the plant members (phytoplankton) and the animal species (zooplankton).

Below the trophogenic layer is a darker **tropholytic zone**, a region where respiration and decomposition predominate; organic compounds are broken down rather than synthesized. Ideally, between the two regions a compensation depth marks the place where photosynthetic processes are matched by respiratory events. In stratified lakes much of the tropholytic zone lies within the hypolimnion, but the two are not necessarily synonymous. The trophogenic-tropholytic boundary is a function of light penetration.

Productivity

Productivity of photosynthetic organisms is the base of every food chain, and it powers every ecosystem. A continuum spans the very clear, least productive *oligotro-*

phic systems to the *eutrophic* systems which are very green with algae or macrophytes. This trophic state or status is often obvious and, when extremely productive, results in nuisance conditions.

Thoreau (1854) described lakes with different productivities. In the *Walden* chapter "The Ponds" he described White and Walden Ponds as "great crystals on the surface of the earth . . ." with "their remarkable transparencies, their hues of blue and green, their lack of muck," but "not very fertile in fish." These characteristics contrast with Flint's Pond, "more fertile in fish . . ." but "comparatively shallow," "not remarkably pure," and having a "sedgy shore." His words about the farmer who "ruthlessly laid bare" the shores of this lake are especially significant. Thoreau had recognized, in the first examples, oligotrophic lakes, and in the second, a eutrophic lake.

Further lake studies in Europe showed there were contrasting phytoplankton types that could be related to water chemistry. Cyanobacteria (sometimes referred to as "blue-green algae"), for example, were characteristic of eutrophic waters in the lowlands. Oligotrophic waters from mountainous regions produced a different limnetic flora. As in any ecosystem, the availability of nutrients will be a key factor determining the producers present and their abundance. In recent years, we have learned more about the effects of nutrient ratios in water (**stoichiometry**) and how human activities change these dynamics (Elser et al., 2009). However, it is helpful to bear in mind that the concepts of eutrophic and oligotrophic refer to the chemical conditions (nutrient abundance) of water, as well as the growth of the plants and algae.

Thienemann (1925), summarizing much of his earlier study of a lake district in Germany's Eifel highlands, proposed broader interpretations of eutrophy and oligotrophy. He found that conditions in the deep lakes differed from those of the shallow lakes, even though they lay in the same climatic and edaphic region. The summer oxygen supply in the hypolimnion of a deep basin was always abundant, and the bottom fauna was diverse, including many species incapable of withstanding low oxygen tensions. By contrast, the dissolved oxygen became critical in the shallower Eifel lakes, often disappearing completely as a result of decay. Only hardy species were found in the profundal benthos in such lakes.

The mention of the gas oxygen prompts discussion of an important limnologic concept. Gas exchanges accompany the metabolic processes of organisms when organic substances are built up or broken down. The rate at which oxygen is consumed, for example, or the total amount that disappears is a measure of the organic matter mineralized through bacterial action or dissipated by respiration of aquatic organisms. Increases in carbon dioxide follow oxygen's decrease, so that, typically, there is an inverse relationship between these gases. The production of oxygen as a by-product of photosynthesis, quantitatively assayed in the epilimnion, is marked by a concurrent decrease in CO_2. The rate of either process is a measure of the remarkable chemical reaction of photosynthesis, whereby new organic matter is formed from mineral substances. All subsequent echelons of production in a body of water stem from this new material and from other organic matter that may have been blown or washed into the lake. Herbivores and carnivores—heterotrophic rather than autotrophic organisms—build their bodies and owe their existence to this primary organic material. They consume and destroy it, spending a part of the chemical energy stock available to them. Their corpses, dead algal cells, and the organic juices leaking from both are eventually mineralized by bacteria, with attendant consumption of O_2 and elaboration of CO_2. This disappearance of oxygen in the tropholytic zone, then, mirrors the basic production in upper layers (strata).

Thienemann (1925, 1927) was an outstanding student of the Chironomidae; the profundal representatives of this fly family were especially engaging to him. The adult chironomids or midges superficially resemble mosquitoes but do not sting. Their larvae are quite different, however. The species of limnologic interest are wormlike and live in the sediments. One group, bright red and commonly called "blood worms," belongs to the genus *Chironomus* and tolerates low oxygen tensions. These were the characteristic midge larvae of Thienemann's shallow lakes, much as they had been found to typify eutrophic waters elsewhere. Other species, including those of the genus *Tanytarsus*, are intolerant of low oxygen levels. They occur in profundal deposits of the deeper Eifel lakes, along with a varied collection of other types of animals, in contrast to the taxonomically meager benthos of the shallow lakes.

The oxygen supply and the nature of the benthos became important indicators to Thienemann as he categorized the two kinds of lakes: shallow, *Chironomus* eutrophic lakes and deeper, well-oxygenated, *Tanytarsus* oligotrophic lakes. The former includes the bizarre phantom larva (glassworm), *Chaoborus*, a member of another fly family, although often called a midge. It is present in lakes where oxygen's lack is too severe even for *Chironomus* but is found less often in the bottom fauna of oligotrophic lakes.

The types of biota, oxygen distribution, lake dimensions, water transparency, and sediments all relate to plant and algal production. These features are used to contrast an oligotrophic from a eutrophic system (Table 1-2).

Table 1-2   Features contrasting oligotrophic and eutrophic lakes—factors contributing to or resulting from the two types

Oligotrophy	Eutrophy
Deep and steep-banked	Shallow, broad littoral zone
Epilimnion volume relatively small compared with hypolimnion	Epilimnion/hypolimnion ratio greater
Blue or green water; very transparent	Green to yellow or brownish green; limited transparency
Water poor in plant nutrients and Ca^{2+}	Plant nutrients and Ca^{2+} abundant
Sediments low in organic matter	Profundal sediments an organic **copropel**
Oxygen abundant at all levels, at all times	Oxygen depleted in summer hypolimnion
Littoral plants limited, often rosette type	Littoral plants abundant
Phytoplankton quantitatively poor	Abundant phytoplankton
Water blooms of cyanobacteria lacking	Water blooms common
Profundal bottom fauna diverse; intolerant of low oxygen tensions	Profundal benthos poor in species; survive low oxygen
Profundal benthos quantitatively poor	Profundal benthic biomass great
Tanytarsus-type midge larvae in profundal benthos; *Chaoborus* usually lacking	*Chironomus,* the profundal midge larva; *Chaoborus* present
Deep-water salmonid and coregonid fishes	No stenothermal fish in hypolimnion

All of these factors interact at any particular site, making each unique. The morphology of the basin and its setting are important. For example, a deep eutrophic lake with a cool, oxygen-rich hypolimnion may have a benthic fauna more like an oligotrophic system, even if nutrients are abundant.

Bog Lakes and Dystrophy

Northern waters, protected from wind and poorly drained, may become bog lakes. They are fringed by floating mats of vegetation growing inward to encroach on the open water. Cedar Bog Lake (site of Lindeman's famous study), is a senescent lake, nearly obliterated by the marginal mat. It occupies a depression in soils moderately supplied with calcium, and its waters are circumneutral in pH and fairly clear. Farther northeast in Minnesota, northern Wisconsin, and the Upper Peninsula of Michigan are some glacial soils derived from Precambrian rocks.

The Canadian Shield is deficient in calcium, and bog lakes there show features defined as **dystrophy**. The stained waters are acid and brown and low in electrolytes with reduced transparency caused by colloidal and dissolved humus material. A striking example is the brown water of the Tahquamenon River draining bogs in Michigan's Upper Peninsula: it looks much like root beer as it cascades over the falls near the town of Paradise.

The importance of calcium shortage in bringing about dystrophy was demonstrated by Hasler and co-workers (1951) in a brown-stained lake of the Wisconsin Highlands. Adding commercial hydrated lime, mostly $Ca(OH)_2$, raised the pH from 5.4 to 7.1 and resulted in clearing the water and increasing vertical light penetration. Dystrophy seems to be linked to the blocking of bacterial action by calcium's scarcity; organic matter, then, does not decay rapidly and is not recycled in a normal fashion. Stained, murky waters, characterizing dystrophy, are not restricted to northern bogs; many southern waters in North America and elsewhere appear darkly colored although eutrophic.

Dystrophic lakes probably belong to the oligotrophic series even though, except for some in deep basins, the oxygen in lower strata is much reduced or lacking. Furthermore, the benthos is taxonomically poor, consisting of a *Chironomus* fauna reminiscent of eutrophy. Despite this, the scantiness of the plankton and benthic crops, the lack of dissolved nutrients, and the nature of the algal flora in part suggest oligotrophy.

Primary Productivity and Lake Classification

It is not always easy to place a lake into a definite class. Some seem to be mosaics of what we assume to be eutrophic as well as oligotrophic characteristics. Rodhe (1969) approached the problem directly. In an **autotrophic lake**—one in which the organic compounds produced are from photosynthesis rather than imported from the outside—the trophic status can be related to the primary productivity. Going directly to the source of biologic productivity, we can ask how much inorganic carbon is fixed, and at what rate, by the phototrophic producers. Rodhe suggests rates of 7 to 25 g of carbon per year during the open season for every square meter of lake surface in examples of oligotrophy; 75 to 250 g of carbon per square meter per year in natural eutrophic lakes; and 350 to 700 g of carbon per square meter per year in polluted lakes. This leaves us with **mesotrophy** lying somewhere between annual primary production of 25 to 75 g of carbon fixed and converted to organic molecules by the lacustrine green plants under an average square meter of lake surface. Rodhe's definition of ranges and limits are not applicable everywhere, but his approach frees us from some of the difficulties in pigeonholing certain perplexing aquatic ecosystems.

The following general, simplified equation shows the bare essentials of primary production:

$$CO_2 + H_2O \rightarrow (CH_2O) + O_2$$

| inorganic carbon source | hydrogen donor | solar energy | carbohydrate unit | by-product |

The equation shows a simple inorganic carbon molecule being converted to a high-energy, chemically reduced organic compound. Hexose sugar is often portrayed as the latter, as follows:

$$6\,CO_2 + 6\,H_2O \xrightarrow{\text{solar energy}} C_6H_{12}O_6 + 6O_2$$

In each equation there is a plant pigment that traps some of the sun's energy to power the conversion; chlorophyll a is typical for the chemistry shown. A most important factor in Rodhe's categorization, and not shown above, is the nutrient supply. The two equations apply to both nutrient-poor oligotrophic waters and nutrient-rich eutrophic bodies of water; the rate of productivity depends to a great extent on the plant nutrients that are available.

Desert waters, where solar radiation is intense, growing seasons are long, and nutrients are condensed, often surpass the limits of Rodhe's polluted waters in carbon fixation per year. Carpelan (1957) discussed primary productivity amounting to 1.42 kg of carbon per square meter per year in some California brine pools. The productivity of Arizona's Montezuma Well is about 798 g of carbon per square meter per year, although no pollution is involved. Similarly, enriched waste lagoons may annually fix about 3.0 kg of carbon for every square meter of surface. A record high for annual production may have been found by Talling and associates (1973) in two Ethiopian lakes, rich in sodium carbonate—perhaps near 5.0 kg of carbon per square meter. Similarly, the Kenyan Lake Nakuru, also a concentrated soda lake spreading over about 62 km^2, may be one of the world's most productive lacustrine ecosystems. From the data of Melack and Kilham (1974), one can estimate an annual primary productivity in the neighborhood of 3.9 kg of carbon per square meter.

Extremely low rates of carbon fixation are also known to fall below Rodhe's oligotrophic boundary. Goldman and others (1967) reported about 0.014 g of carbon per square meter for an average summer day (February) in an Antarctic lake. Likewise, the pristine mountain lake, Waldo, in Oregon, fixes about 0.03 g of carbon per square meter during an average summer day (Malueg et al., 1972).

Researchers have attempted to relate productivity (which can be challenging to measure in the field) to other parameters; it is also possible that any one measurement fails to provide a complete picture; trophic status is related to factors beyond photosynthetic output. Carlson (1977) developed a widely-used index incorporating transparency, chlorophyll, and total phosphorus concentration. A modern approach to determining trophic status in streams based on similar principles was recently developed by Dodds (2006).

Limnology and Society

Modern limnology was born of international research on Alpine lakes in Europe, and has since spread across the globe. Today, limnologists from many nations work on every continent, helping society answer fundamental questions about science and how to effectively manage our aquatic resources (Fig. 1-2). Recent years have even seen limnologists studying "extremophiles" (organisms surviving in harsh environments) to hypothesize about extraterrestrial life; it is believed that life outside Earth may resemble organisms of habitats like boiling geothermal pools, or perennially dark lake bottoms covered by permanent ice (Cavicchioli, 2002).

Here on Earth, humans exert an influence on water as never before: our alterations of water flow now rival those of nature (Meybeck, 2003), even causing earthquakes documented at over 90 locations worldwide (Gupta, 2002). We continue to

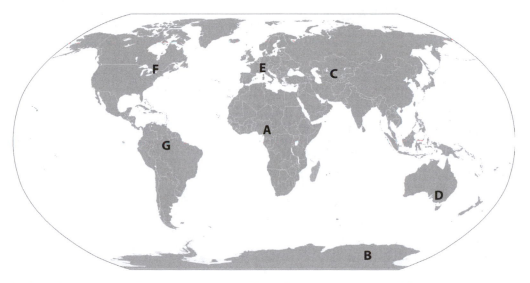

A. Africa: Limnologists help defuse Nyos and similar "killer lakes" (Ch. 13).

B. Antarctica: Lake Vostok has a diversity of life under four kilometers of ice (Ch. 2).

C. Asia: The Aral Sea, once the fourth-largest lake on Earth, lost its inputs of river water; it is now about one-tenth of its 1960 surface area (Ch. 8).

D. Australia: Residents of the driest inhabited continent balance human and natural water requirements in the Murray-Darling Basin (Ch. 7).

E. Europe: Well-studied Alpine lakes stand as sentinels of climate change (Ch. 5).

F. North America: After a surprising recovery from being declared "dead," Lake Erie is once again threatened by cultural eutrophication (Ch. 15).

G. South America: The first bridge across the Amazon River opened only in 2011; the river's exact origin and length are still under investigation (Ch. 8).

Figure 1-2 Important sites of limnology research providing "case studies."

pollute our waters, and the effects are seen worldwide rather than in isolation at one waterbody by one effluent (Elser et al., 2009).

The environmental challenge of our times, global climate change, is significantly influenced by the world's lakes, rivers, and especially wetlands (Bridgham et al., 2013; Crawford et al., 2014). Bogs, swamps and similar peat-building systems contain perhaps a third of the world's soil carbon. Limnologists are considering the dynamics of a potential "positive feedback" in which carbon releases to the atmosphere may change the climate, leading to yet more carbon release.

The next three chapters consider biological limnology. In Chapter 2 we look at the biology of important groups of organisms (taxa). In Chapter 3 we discuss groups of organisms interacting to form ecological communities; Chapter 4 considers how the biota interact with the abiotic environment to form ecosystems.

2

Survey of Limnological Biodiversity

In recent years, biologists have come to realize that the single-celled, prokaryotic organisms known as the *Archaea* are less like the Eubacteria than we are like a mushroom or a palm tree. The differences in cell structure and genetics are so profound that it has caused us to redraw the Tree of Life. Limnologists encounter examples from each of these newly-recognized domains (Archaea, Eubacteria, and Eukarya) in the waters they study.

Molecular biology has added to the ongoing use of traditional methods such as anatomical study to refine our understanding of the relationships between organisms; and of course, we continue to discover and name new species. Biologists are free to disagree about how to assign organisms to taxa and how to name the groups; no doubt this will continue for years to come. In this book, we will use the most common nomenclature and avoid consideration about taxonomic rank and phylogenetic history.

Life in inland waters presents challenges in almost all cases, however. Evidence shows that some major taxonomic groups of the inland aquatic biota invaded from the sea, others from land, and a minority seem to have arisen in fresh water itself. Formidable barriers must be surmounted in adapting from life in a marine or terrestrial environment to the habitats afforded by lakes, ponds, and streams. For this reason entire groups of animals on the marine or terrestrial class levels, or even those ranked as phyla, have no representatives in fresh water. Other marine and terrestrial groups have contributed disproportionately to the limnologic biota. Although the array of tiny plants and animals to be found floating, swimming, and gliding about in collections taken from freshwater habitats may seem staggering at first, the inland waters are taxonomically impoverished when compared with the sea. Groups such as the Cephalopoda, Ctenophora, and Polyplacophora, for example, have no freshwater representatives, and from a total of about 5,000 species of the Porifera, less than 300 are found in fresh waters (Reiswig, 2006).

Physical Factors

Water levels in freshwater environments range from stable conditions throughout the year to such extremes as Australia's tremendous Kati Thanda (Lake Eyre), often dry for decades between its irregular fillings. Some temporary ponds have a regular

13

Case Study: Lake Vostok

Recent study of extreme environments reveals whole ecological communities in unlikely places with surprising diversity. The legendary biodiversity of tropical rainforests and coral reefs is impressive, but a single ice core from Lake Vostok (Antarctica) revealed 3,500 species thriving in complete darkness under pressure with 3,700 meters of ice pressing down from above (Shtarkman et al., 2013). About 94% of the sampled organisms were Eubacteria.

seasonal rhythm, a dry phase alternating with a wet phase. The liquid phase appearing each spring is certain in most northern depressions that hold snowmelt. The basins of temporary waters vary, from those whose sediments are moist even during the empty phase to those that are powdery dry between fillings.

Oceans and inland waters differ in behavior and temperature range. Even the largest lakes, characterized by stable water levels, show no significant response to the moon. The tidal regime has resulted in marine communities adapted to conditions alien to inland lakes: within the shallows alternately bared and covered by tides, complex zonation of plants and animals occurs, and their biologic clocks operate on a twice-a-day schedule.

The crashing of ocean surf against the shore can be matched, to some extent, in windswept lakes that are subject to a long fetch, but the continuous unidirectional flow of rivers and brooks is not seen in marine habitats. Moreover, the annual variation of temperatures in surface waters of the ocean is usually less than 10° C, with the exception of the North Atlantic and North Pacific, where cold continental air masses may depress temperature as much as 18° from that of the warmest time of year. This range is small compared with the annual variation of temperature in inland waters, where 25° C would not represent an unusual range for lake surfaces in temperate regions.

Chemical Factors

The pH of the well-buffered seas varies little, from 8.1 to 8.3. However, all extremes are known in inland waters: alkaline lakes may go above pH 10 or 11, and the acid waters of some northern bogs sometimes fall below pH 4.0.

Marine life is adapted to waters with a **salinity** scarcely varying from 3.5% and composed largely of NaCl. Freshwater life has involved survival in and adaptation to dilute water, best described as a calcium bicarbonate type subject to much variation. Even when inland waters approach or surpass the sea in salt content, the proportion of major ions may be much different from, and bear little resemblance to, that found in sea brine.

Adaptations to Fresh Water

The major challenge to fresh water biota is osmoregulation. Marine organisms are characterized by blood and other body fluids that are isotonic to seawater; moving to hypotonic waters brings up the problem of the organism maintaining a stable internal environment against the constant threat of osmotic water diluting its fluids and leaching out the salts. Few marine plants or animals can survive dilute water. At the opposite extreme are the hypersaline waters found in some desert regions. Here the salinity may be seven times greater than the sea's brine, and only a few forms can colonize such habitats. Most of these have freshwater ancestors rather than immediate marine antecedents. The freshwater plants and animals show greater general adaptability and ecologic versatility than their marine relatives.

A further factor in an organism's successful freshwater existence is its adaptation to prolonged desiccation. As a result, resistant stages and mechanisms for enduring dry or other unfavorable periods are common in inland water forms but are lacking in their close marine relatives. The gemmules of freshwater sponges and the statoblasts of inland bryozoans are well-known examples of such mechanisms for endurance. These structures are analogous to the seeds of plants in that they encase and protect the organism during dormancy, however the spread of the organism is strictly an asexual reproduction by means of cloning.

The immigrants from land that are found in fresh water have overcome osmotic problems, but the switch from aerial to aquatic respiration may have been more of a hurdle. In many instances the structures that were used to guard against desiccation in terrestrial existence serve as barriers in the water, thereby reducing the area of unprotected membranes through which environmental water enters the rather concentrated body fluids. The abundance of oxygen mixed with other gases of the air, however, is far greater than the concentration of oxygen in water; except for some ingenious arthropods that carry bubbles of atmospheric gas underwater with them, most aquatic animals of terrestrial origin have devised new respiratory structures that function in an aqueous environment.

Many freshwater animals lack a larval stage, thus differing from their close relatives in the sea. It has been theorized that such stages are weak links in the life cycle, especially in lotic environments, or that nutrients in dilute inland waters are so scarce that diminutive larvae could not fend for themselves. Nature's answer has been to pack salts and yolk into the egg where embryologic development includes larval stages, and to set free the young, which resemble miniature adults, when they are ready to cope with conditions in the freshwater habitat. Similarly, the motile medusa stage of the hydrozoans was eliminated in the highly evolved freshwater *Hydra*. The unusual and sporadic occurrence of one of the freshwater jellyfish, the medusa of *Craspedacusta,* is always exciting to zoologists, although marine medusae are common.

Doing away with delicate motile stages, whether adult or larval, may be of adaptive significance for many species, but there are a few freshwater larvae that seem to be faring well. For example, the nauplius of copepods and some branchiopod crustaceans is well developed in inland waters. The spectacular spread of the Asiatic clam *Corbicula* in North America and *Dreissena* in the Old World (and more recently in the North American Great Lakes) implies that there has been no disadvantage in retaining the *veliger,* a typical marine-mollusc larva.

The presence or absence of the nauplius larva (see Figs. 2-6, *F* and 2-8, *F*) is especially interesting. In the marine world many diverse groups of the Crustacea produce this larva, which links such unlikely forms as the barnacles with other, more conventional Crustacea. It is the simplest larva that hatches from a crustacean egg, bearing three pairs of jointed swimming appendages that are destined to become the first and second antennae and the mandibles. The first pair of appendages is unbranched, but the following two are characteristically Y shaped or biramous (two branched) and appear one pair per segment. With subsequent molts, additional segments bearing appendages appear. These stages are the metanaupliar instars, although often simply designated as later nauplius in-stars. Many freshwater crustaceans have relegated the nauplius-metanauplius stages to their embryology.

The Prokaryotes

Prokaryotic organisms have a structural organization that sets them off from the larger, more familiar eukaryotes. Prokaryotes are simple cells (or groups of cells) lacking nuclear membranes, the endoplasmic reticulum, chloroplasts, and mitochondria. Cell division occurs via binary fission rather than mitosis. In prokaryotes that possess flagella (cyanobacteria, the so-called "blue-green algae," have none), these locomotor organelles are distinct from the complex flagella and cilia of the plant and animal eukaryotes. The flagella and cilia of the newer, more complex species, however, have a similar basic structure.

The prokaryotes are remarkably adaptable, existing abundantly in the soil, the sea, and fresh water. They are the oldest organisms on earth and have evolved tremendous metabolic diversity. Because the early earth had a reduction atmosphere, the first prokaryotes were forced to rely on anaerobic energy supplies, and today many are still obligate anaerobes. These include at least the simplest fermenters, the photosynthetic microbes (other than the cyanobacteria), and the methane formers. Later developments in cell structure and physiology allowed prokaryotes to exploit nearly every habitat on Earth, including diverse aquatic habitats. Much later, the eukaryotes arose from prokaryotic ancestors; cell structure and biochemistry reveal a common ancestry.

Prokaryotes are essential to the functioning of ecosystems. Cyanobacteria are important primary producers in many waters; Eubacteria are both important decomposers and a major food item in trophic dynamics and energy transfers. Much attention has also been focused on them as mediators in cycles of such elements as hydrogen, sulfur, carbon, iron, manganese, nitrogen, and phosphorus. Despite this recognition, systematics and diversity of prokaryotes, especially in aquatic habitats, have only recently been studied effectively.

Despite centuries of ever-improving microscopy, microbiology lacked two important abilities necessary to study microbes in limnological and other natural environments. First, we couldn't obtain pure cultures. Food and medical microbiology utilized aseptic technique to isolate and grow individual species or strains of prokaryotes for individual study. Unfortunately, microbes in the wild cannot be isolated for study because they cannot be grown using established techniques. Second, we couldn't understand the functioning of microbes. Cells with similarities in appearance may actually be quite unrelated and have very different ecology, and this is understood only through understanding of their physiology and molecular biology.

An exciting series of developments now allow us to overcome these difficulties and explore environmental microbiology, opening a whole new area of study to the limnologist. It is now possible to sample waters and use **metagenomics** to characterize the entire microbial community present. The "genetic fingerprints" of the organisms are obtained by rapidly copying (amplifying) sequences of genetic material (DNA or RNA). Specific genes can be identified, and sequences "expressed" to determine the functioning of the genes, and the physiological ecology of the microbes themselves (Logue et al., 2008, Debroas et al., 2011). We can then describe new species and understand their ecological roles, such as oxidation of methane and similar single-carbon compounds (Kalyuzhnaya et al., 2008). These techniques have proven robust and valid (Lyautey et al., 2005).

Archaea

In recent years we have discovered that these organisms are neither "bacteria" nor limited to extreme environments; we have come to understand that Archaea are fundamentally different in anatomy, physiology, and evolutionary history (Woese, 2004). Archaea were initially studied for their remarkable ability to thrive in the most stressful habitats. One group is thermophilic, some forms functioning at 110° C (the maximum temperature for Eubacteria is 90° C). Archaea are found in environments such as thermal hot springs (Barns et al., 1994). Another limnologically important group in this domain includes organisms that produce methane from the digestion of organic matter in anaerobic conditions (Liu and Whitman, 2008). These organisms break down biomass and release substantial amounts of methane, a major "greenhouse gas."

Metagenomics has revealed many more types of Archaea, suggesting that we have only begun to characterize this large and diverse group. Auguet and others (2009) studied 67 sources from around the Earth and identified 13 lineages with freshwater habitats showing great diversity. Although nomenclature and phylogeny are still under development, study of the Archaea is certain to produce greater insights into aquatic ecology and the limits of life.

Cyanobacteria

Formerly known as "blue-green algae" (Fig. 2-1 on the following page), cyanobacteria derive their name from the pigment phycocyanin. Another pigment, phycoerythrin, is red and, mixed with the phycocyanin and two or three other pigments, gives various blue-green colors ranging from red through violet to black. They are photosynthetic, as are many other prokaryotes, but uniquely use chlorophyll a. Apparently the invention of chlorophyll a was a major evolutionary step, and it appears as the dominant photosynthetic pigment in all higher plants. Cyanobacteria are ancient: fossils preserved in rocks known as stromatolites date from 3.5 billion years ago (Whitton and Potts, 2012).

According to Stanier (1973), all the cyanobacteria that have been studied can use light as an energy source and CO_2 as a carbon source. Thus, they are primarily photoautotrophs. Some can assimilate organic substances such as acetate and amino acids, and some are organotrophs, being able to grow on a few sugars in the dark. Some can carry out vital ecological transfomations—fixing nitrogen, for example. Cyanobacteria are closely studied now because of the unpleasant growths they form in lakes where eutrophication has been brought about by domestic, agricultural, and industrial pollution (see Chapter 15). In such cases, accelerated growth of cyanobacteria results

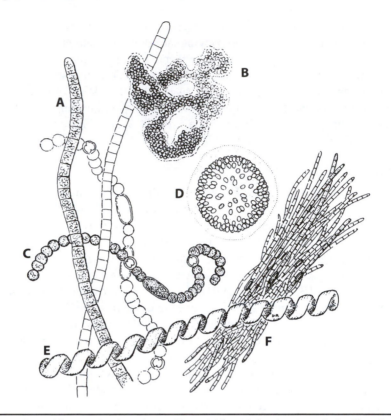

Figure 2-1 Some blue-green algae (cyanobacteria) from inland waters. **A**, *Oscillatoria*; **B**, *Microcystis aeruginosa*; **C**, *Anabaena*; **D**, *Coelosphaerium*; **E**, *Spirulina*; **F**, *Aphanizomenonflos-aquae*.

in noxious "water blooms" of such forms as *Microcystis* and *Anabaena*. Often their respiratory demands surpass their daylight oxygen production, and upon death their decay promotes further deoxygenation.

A few lakes have been plagued by races or strains of cyanobacteria that produce dangerous toxins. Six genera have been implicated in fish kills and in the deaths of horses, cattle, swine, dogs, chickens, squirrels, and other animals: *Microcystis, Nodularia, Cóelosphaerium, Gloeotrichia, Anabaena,* and *Aphanizomenon.* Putrid masses of these organisms accumulate in the shallows or at the strand line, where they are often unsightly and possibly dangerous. International concern has resulted in attention to the problem and intense study (Anderson et al., 2002).

Many experiments have been performed over the years to elucidate the value of cyanobacteria to grazing zooplankters, and yet the question remains open (Agrawal and Agrawal, 2011); however, it is clear that many species produce potent toxins effective on grazers. Arnold (1971) reviewed the literature and reported on the effects of seven species as a diet for *Daphnia pulex.* (He also compared the cyanobacteria with green algae as food sources.) Arnold concluded that cladocerans would survive only at low levels of abundance if their food mixture were dominated by cyanobacteria.

This has been the opinion for many years—the cyanobacteria have been considered weed species. But we must not forget that the cyanobacteria of some African soda lakes are noted for the abundant animal life they support. Lake Nakuru has the highest known biomass of lacustrine primary consumers: the herbivores amount to about 19 g/m² on a dry weight basis, and the lesser flamingo (*Phoeniconaias minor*) is the predominant consumer (Vareschi, 1978). The phototrophic base for this unusual situation is mostly a cyanobacterium, *Spirulina platensis,* which has a mean biomass (dry weight) of at least 400 g/m² in Nakuru. The beak of this bird is remarkably adapted for straining out the coiled filaments of *Spirulina* (Jenkin, 1957), and it can even feed on rotifers when cyanobacterial densities are low (Vareschi, 1978). The latter author reported 1.5 million flamingos congregated at the lake in July, 1973; he estimated the birds at that time consumed 93% of the daily production of *Spirulina,* amounting to about 103 metric tons on a dry weight basis. The important food fish *Tilapia nilotica* in Lake Turkana (formerly Lake Rudolf) consumes and digests *Spirulina* and *Anabaenopsis,* two genera typical of concentrated sodium carbonate water. Furthermore, in the deeps of freshwater lakes elements of the bottom fauna, such as *Chironomus plumosus,* feed on blue-greens that reach the profundal ooze (Jónasson and Kristiansen, 1967; Iovino and Bradley, 1969).

In European lakes, and to a lesser extent in North America, the reddish filamentous *Oscillatoria rubescens* has gained a reputation as an indicator organism. Its initial appearance in lakes seems to announce the approach of cultural eutrophication (Hasler, 1947; Edmondson, 1991).

Eubacteria

Microbiologists traditionally group Eubacteria based on cell shape, cell-wall chemistry (indicated through Gram staining), aerobe/anaerobe, or spore formation. Sigee (2005) provides a summary of 18 groups common in limnology. They have a diverse ecology and physiology, and many of the organisms have strains or related species in terrestrial or marine environments. Because of their small size compared to cyanobacteria, they are not usually encountered in limnology class, and yet these organisms are quite important ecologically.

Eubacteria are by far the most abundant organisms in water, easily encountered in the millions to hundreds of millions in a typical mL sample. Most are free-living decomposers, digesting detritus and releasing nutrients in the process. They are also an important part of the food chain. A study of the eutrophic Lake Mendota in Wisconsin (Pedrós-Alió and Brock, 1982) found high rates of Eubacterial biomass (89–205 g C/m²/yr); this was about 50% of the primary production by photosynthesizers. This was a significant food source for free-swimming animals.

A minority of Eubacteria is harmful to humans, but such pathogens are occasionally found in lakes and streams, often because of contamination with human waste. Use of the waters for swimming or drinking can prove a safety hazard, so public health officials are careful to sample water in suspect areas and post warnings or close beaches as needed. Fortunately, sunlight inactivates the cells quite effectively (Whitman et al., 2004).

The Eukaryotes

The familiar organisms of the biology student are the Eukaryotes. Even the smaller unicellular species are normally visible with the naked eye. All of these organisms have well-developed organelles covered by membranes, and in general a more elaborate cellular anatomy. The relationships between the groups is an area of active research, and the taxonomy will certainly be revised extensively in years to come. Here, we group the photosynthetic primary producers (Plants and Algae) together, and then survey the consumers (Animals and Protozoa).

The **Protists** (sometimes called Protoctists) also include Euglenoids, with a variety of plant-like and animal-like organisms. The *Euglena* are of particular interest to biologists: they employ several different pigments, can propel by a crawling motion or with a flagellum, and can be autotrophic photosynthesizers or heterotrophic grazers. Despite their curious biology, these organisms are not especially numerous or influential (Sigee, 2005).

Algae and Plants

The green algae (Chlorophyta) are diverse (more than 8,000 freshwater species), widespread in inland waters, and often abundant. These algae are of interest since such an organism was the likely ancestor of terrestrial plants. This Phylum includes such familiar genera as *Spirogyra, Volvox, Chlorella,* and *Ulothrix.* The Desmidiaceae (a family of green algae) are especially interesting because typically they are found in acid bogs, in very dilute water low in electrolytes, and in oligotrophic lakes (Fig. 2-2, *B, D, F, G,* and *H*).

The Chlorophytes may be useful indicators of overall functioning of the ecosystem, although complicated by the occurrence of physiologic races and taxonomic problems. Brook (1965) classified desmids on the basis of their occurrence in oligotrophic, mesotrophic, or eutrophic waters. In Brook's opinion, 27 of 46 species were oligotrophic indicators; one genus, *Cosmarium* (Fig. 2-2, *H*)**,** had more eutrophic species than oligotrophic, while the beautiful *Micrasterias* (Fig. 2-2, *F*) and *Xanthidium* (Fig. 2-2, *G*) typified oligotrophic waters.

The Pyrrophyta (dinoflagellates), are extremely abundant in the sea; fresh water has only about 200 species. These algae possess both chlorophyll *a* and chlorophyll *c* and also golden-brown pigments. The armored *Ceratium hirundinella* (Fig. 2-3, *E* on p. 22) is a spectacular and easily identified form that is very common in plankton collections. Some Pyrrophyta produce toxins; of particular note is *Pfiesteria*, which causes lesions on fish and is a potent neurotoxin to humans (Grattan et al., 1998); outbreaks are linked to cultural eutrophication of rivers (Heisler et al., 2008).

Two groups of golden brown algae containing flagellated forms are the Cryptophyta, or cryptomonads, and the Chrysophyceae, a family of the Chrysophyta (Fig. 2-3). The cryptomonads include many small, colorless, as well as pigmented, species. The Chrysophyceae is a family containing interesting colonial flagellates such as *Dinobryon* (Fig. 2-3, *D*), *Synura* (Fig. 2-3, *G*), and *Uroglenopsis,* and scaled and spiked unicellular forms typified by *Mallomonas* (Fig. 2-3, *A*). The most important group of the Chrysophyta are the diatoms (Fig. 2-3, *C, F, H,* and *I*), the Bacillariophyceae of many authors. There are perhaps 2,000 freshwater species of these golden algae.

Because they use silica to build their frustules (glassy shells), this compound is an essential nutrient to them. Diatom frustules appeared in lake varves during early Tertiary times and therefore serve as a natural marker useful in paleolimnological studies (Bradbury, 1988). The diatoms compose the great "pastures" of the sea, carrying on most of the world's photosynthesis.

There are other important large plant-like species that are considered to be algae and closely related to the chlorophyceans. These are the Charophyta, including *Chara* and *Nitella*. Some species of *Chara* owe their name "stonewort" to the encrusting calcareous coverings they form on their "stems" in hard-water lakes. *Nitella* is a delicate plant found in acidic and dilute waters; it is not calcified.

Macrophytes, the species of large, vascular plants found in inland waters, are nearly all terrestrial flowering plants that have adapted to the unique challenges of aquatic growth: less gas available when dissolved in water than in air; reduced light in the water column; lack of wind or insect pollination under water; and the threat of

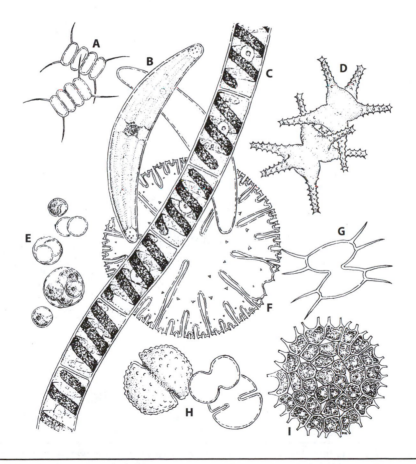

Figure 2-2 Some green algae (Chlorophyta) from inland waters. **A**, *Scenedesmus*; **B**, *Closterium*; **C**, *Spirogyra*; **D**, *Staurastrum*; **E**, *Chlorella*; **F**, *Micrasterias*; **G**, *Xanthidium*; **H**, *Cosmarium*; **I**, *Pediastrum*. **B, D, F, G,** and **H** belong to the Desmidiaceae.

damage from water movements. Although wetland species (see Chapter 7) face only some of these problems, or only intermittently, truly aquatic species must develop effective strategies to face these daily challenges. Nearly all of these are flowering plants (Anthophyta/Magnoliophyta). In North America roughly 20 families of the monocotyledons and 30 of the dicotyledons have aquatic representatives. The freshwater monocotyledonous species, however, outnumber the dicotyledons (Hutchinson, 1975). The mosses (Bryophyta) and ferns (Pteridophyta) have contributed relatively few species to aquatic environments; the ferns especially are poorly represented and are considered curiosities. Several species of moss and the pteridophyte *Isoetes* extend into deep water of transparent lakes.

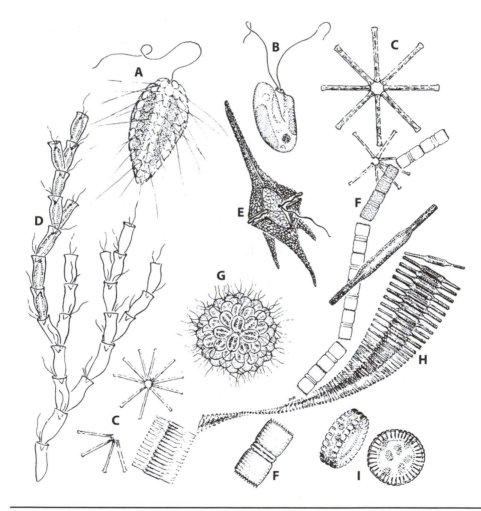

Figure 2-3   Representatives of the golden brown phyla in inland waters. **A**, *Mallomonas*, Chrysophyceae; **B**, *Cryptomonas*, Cryptophyta; **C**, *Asterionella*, Bacillariophyceae; **D**, *Dinobryon*, Chrysophyceae; **E**, *Ceratium hirundinella*, Pyrrophyta; **F**, *Melosira*, Bacillariophyceae; **G**, *Synura*, Chrysophyceae; **H**, *Fragilaria*, Bacillariophyceae; **I**, *Cyclotella*, Bacillariophyceae.

Floating-leaved plants such as water lilies (*Nymphaea*) have contact with the atmosphere, allowing direct gas exchange for respiration and photosynthesis. In most species, the stomata are located on the top surface of the leaves, the opposite of the usual terrestrial anatomy (Dunn et al., 1965). Gas exchange occurs through these openings to the air, and in fact the plant actively pumps air from one leaf, down through the petiole, and the circulating through the plant body to oxygenate the buried portions (Richards et al., 2012).

Submersed plants remain under the surface, although some, such as many pondweeds (*Potamogeton*) grow flowers emerging above the water surface on an aerial spike (Crow and Hellquist, 2006). One unique adaptation to gas exchange in water is to utilize dissolved bicarbonate ions instead of gaseous carbon dioxide (Osmond et al., 1981); this overcomes the limitation caused by low CO_2 to support photosynthesis.

Protozoa and Animals

Many students encounter Protozoans, tiny animal-like Protists, during a "pond study" or a biology lesson as children. The familiar *Paramecium* or *Amoeba* is common in still, shallow water in many parts of the world. The classification of even these familiar animal-like organisms has been reconsidered in light of recent work, and the process is still ongoing. At present, it is not possible to definitively classify Protozoans or state the relationship of various Protista to other kingdoms. Still, a helpful review by Corliss (2002) provides an excellent overview of their basic biology, diversity, and notes on ecology.

For some groups of freshwater animals there are many zoologist specialists. Other groups are relatively neglected because of taxonomic difficulties, or perhaps because of a general lack of appeal. A small number of phyla or classes are of limnologic importance. The rotifers, annelids, crustaceans, insects, molluscs, and fishes are the major taxa in the eyes of many limnologists. The reasons for the interest in fishes are obvious. Rotifers and crustaceans are well represented in the lake plankton; annelid worms and insects make up a great part of the bottom fauna, and the latter are especially well represented in lotic environments. Molluscs are abundant and diverse in streams, but they are found in most aquatic settings.

Because some taxa are nearly restricted to fresh waters, they are of particular interest. There is hardly a body of inland water that lacks a rotifer fauna, although the group is poorly represented in the seas. Insects belong to the terrestrial fauna, but there are many in all sorts of freshwater habitats whereas almost none exist in the oceans.

The discussion here is limited to relatively few groups. For details not presented here at least four books are recommended. Volume 2 of Hutchinson's *Treatise on Limnology* (1967) deals with geographically broad aspects of lake biology. Thorp and Covich (2010) and the updated version of Pennak (Smith 2001) are authoritative texts on North American aquatic invertebrates. A worldwide review of freshwater fish is available from Berra (2007).

Ciliates. The most common and familiar of protozoans are members of the phylum Ciliophora (Fig. 2-4 on the next page). We now realize that the thousands already described are only a fraction of the total diversity (Šlapeta et al., 2005). They exist mainly as free-living filter feeders of prokaryotes and are in turn an important food for predators

(Sigee, 2005). Motile forms live in the water column as free-swimming plankton, but many are attached to a larger substrate in the water column or in the benthos below.

The peritrich ciliates include several genera, most of which are epizoic or *aufwuchs* forms. The telotroch stage of *Vorticella* (Fig. 2-4, *A*) occasionally swarms throughout the lake and shows up in plankton collections. A colonial, branched peritrich, *Zoothamnium limneticum* (Fig. 2-4, *D*), is said to be a euplankter (true plankton form), feeding on Eubacteria and nanoplankton algae. In general, there are many ciliates in the plankton habitat about which we know little, because they become distorted and unrecognizable when preserved and because they are not easily retained in plankton nets.

The tintinnid ciliates are much like *Difflugia* in that they build tests (a type of hard shell) of foreign particles around themselves. Indeed, *Codonella cratera* (Fig. 2-4, *B*), a common limnoplankter, was originally described as a species of *Difflugia* on the basis of its test. There are about 1,000 tintinnids, but most are marine. These ciliates are true plankters, grazing on small diatoms and chrysomonads. Another genus, *Tintinnidium*, is probably less abundant than *Codonella* in inland plankton communities.

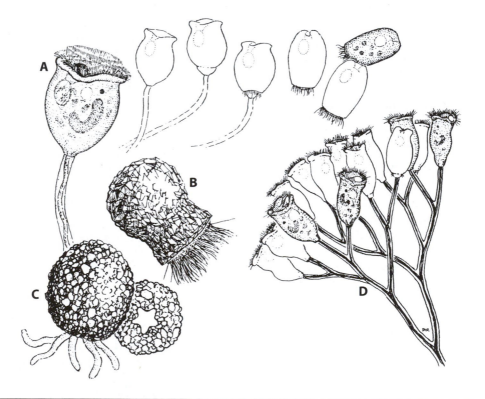

Figure 2-4   Some protozoans occasionally found in the freshwater plankton. **A**, *Vorticella*, a peritrich ciliate, its free-swimming telotroch stage shown at right; **B**, *Codonella cratera*, a tintinnid ciliate; **C**, *Difflugia lobostoma*, Testacea, Rhizopoda; **D**, *Zoothamnium*, a peritrich ciliate.

An amoeba-like ciliate is *Difflugia* (Fig. 2-4, *C*) whose taxonomy is in doubt and may contain more than one genus. The *Difflugia* build tests of tiny sand grains and are able to float with the aid of gas vacuoles. We owe much of our knowledge of these shelled amoeboid forms to Schönborn (1962), who studied the biology of *Difflugia limnetica*. In early summer this testacean rises from littoral sediments to become limnetic (pelagic); in October it sinks to the sediments and forms a winter cyst. During the spring months it is a member of the littoral **microbenthos**. (This cycle is rather typical of plankton organisms, both plant and animal.) The distinction between benthos and plankton communities is hazy because of the movements between the two. Quite often plankters cannot be found in collections from the open water, because they are resting in some sort of cyst or diapausing form in the bottom deposits.

Sponges. These simple and ancient animals belong to the phylum Porifera. Of at least 15,000 species only about 200, belonging to six families, are found in fresh water; the rest are marine (Manconi and Pronzato, 2008). They attach to varied substrates in both lentic and lotic environments. Both gemmules and siliceous sponge skeletal elements called *spicules* are useful in studies by both the taxonomist and the paleolimnologist.

Rotifers. The phylum Rotifera (Fig. 2-5 on the next page) consists of small (most under 0.5 mm body length), pseudocoelomate animals, characterized by a ciliated "corona" at the anterior end. The cilia serve for locomotion and for spinning food into the mouth, below which an internal mastax apparatus, consisting of trophi or jaws, is located. There are about 2,000 species of rotifers known, nearly all restricted to fresh water (Wallace and Snell, 2010), although a few are marine; some are even found in wet terrestrial habitats such as moss beds or inside pitcher plants, and in a surprising variety of temporary sites (Walsh et al., 2014).

Some rotifers belong to the plankton community and will be discussed in more detail later. Most of the rest are benthic, especially littoral forms. Their feeding spectrum is broad and is based on the shape and size of the mastax, corona, and certain appendages. The mastax alone can be placed in at least eight categories, making for specialized and size-selective feeding. As is true for other small organisms, their metabolism is high. Among the algal feeders an individual's daily intake of food may be four to five times its dry weight. Other filterers feed on detritus and prokaryotes. Among the species belonging to a single genus are both herbivorous and carnivorous species. Even obligate parasitic rotifers are known. Certain species have bizarre food habits: feeding on dead microcrustaceans and worms; specializing in cropping the young of other rotifer species; and reaching between the valves of small cladocerans to devour the soft body within.

Reproduction in the rotifers is varied, although some generalizations can be made. In one group classified as the bdelloid (leechlike) rotifers, all reproduction is done by parthenogenetic females; to date no bdelloid males have been discovered. The females produce diploid eggs that are termed amictic; these develop without fertilization. Boschetti and others (2012) recently explained the mystery of how these organisms have survived tens of millions of years in challenging environments without sexual reproduction: they use horizontal gene transfer to incorporate DNA from the environment and express it themselves. Interestingly, these genes come from quite unrelated organisms, even including prokaryotes.

Most rotifers, however, probably produce males although many have never been observed. The males are small and very short lived; they do not feed, and they die soon after fertilizing a female. The male is produced by a haploid, so-called mictic egg that is not fertilized. If he fertilizes a mictic egg, it becomes a diploid "resting egg," a zygote that will develop and hatch later. The resting egg becomes a female, and she eventually produces amictic eggs that start several generations of other parthenogenetic females. Eventually (and the reasons are not well understood), mictic females

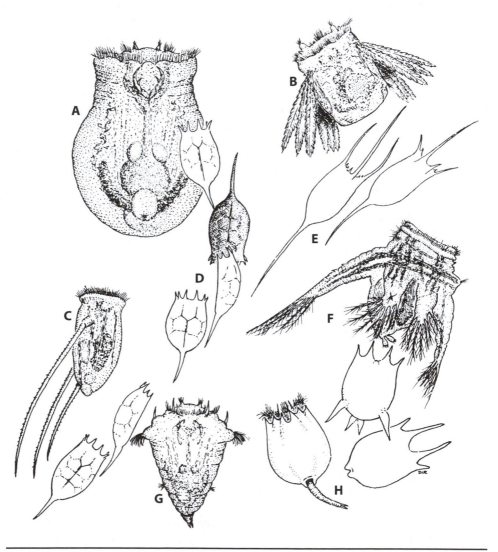

Figure 2-5   Some common planktonic rotifers. **A**, *Asplanchna sieboldi*; **B**, *Polyarthra*; **C**, *Filinia*; **D**, *Keratella cochlearis*; **E**, *Kellicottia*; **F**, *Hexarthra*; **G**, *Synchaeta*; **H**, *Brachionus plicatilis, B. calyciflorus* (two at right).

develop and they produce haploid mictic eggs that may eventually become males, if unfertilized, or resting eggs, if fertilized.

Rotifer taxonomy has been hindered by many factors, including seasonal variations in form and complicated life cycles. In some species the female emerging from the resting egg is so different from later generations that it was not recognized immediately as conspecific! It is apparent now that some species are not cosmopolitan; some of these may be limited in their distribution by the type of algal resources available to them. Other species seem to be more widespread, but there is still much to be learned about these tiny aquatic animals.

Annelida. The segmented worms are a large and diverse phylum found in marine, terrestrial, and freshwater habitats worldwide. The taxonomic relationships are undergoing active revision (Zrzavý et al., 2009), but it is understood that several classes should now be combined into the Clitellata (Marotta et al., 2008), with at least one more class which will contain the former Polychaetes and some others.

The leeches, Hirudinea, are predominantly a freshwater group, although a few species are marine or terrestrial. In North American lakes and streams at least 76 species are known (Govedich et al., 2010). They are not taxonomically easy and have been neglected by most invertebrate zoologists. Although they are commonly and incorrectly called bloodsuckers, only a few are ectoparasites, not all of which feed on warm-blooded animals. Many species are scavengers; others are active predators. In temperate North America they frequent littoral areas but have been reported from deeper benthic zones in Central American lakes. The leeches are rarely taken into account in limnologic studies.

At the other end of the spectrum are the inland polychaete worms. The class Polychaeta is predominantly marine and composed of thousands of species. A polychaete worm is much more the typical annelid than is the large earthworm dissected in beginning zoology laboratories. There are eight species known from North American waters, and most occur near the sea coasts. An exception is *Manayunkia speciosa*, a small tube builder found in some of the Great Lakes, in Lake Champlain, and reported from other widely scattered locations. In general, however, the polychaetes have been unsuccessful in crossing osmotic barriers from the sea and are considered curiosities in fresh water.

The Oligochaeta command the most attention of limnologists. The oligochaetes, sometimes described as aquatic earthworms, are found everywhere in the bottom deposits of ponds, lakes, and streams. A diverse worm fauna composed of three or four families is found in littoral areas, but in poorly oxygenated profundal sediments the Tubificidae predominate. Often tubificid worms are abundantly present in eutrophic profundal sediments or polluted reaches of streams, although species diversity is low. Also, in unpolluted situations tubificid individuals sometimes make up more than 95% of profundal benthic communities (Edmonds and Ward, 1979). It is no coincidence that R. O. Brinkhurst, a leading authority on the oligochaeta, also wrote a book entitled *The Benthos of Lakes* (Brinkhurst, 1974).

Crustacea. Most of the world's crustaceans are marine. Those that are found in inland waters, however, are very important members of aquatic communities. There are various ways to classify the Crustacea. There are at least six classes and a bewildering array of subclasses and superorders, as carcinologists treat the crustaceans as either a distinct phylum or simply a class or subphylum of the Arthropoda. Martin and Davis (2001) provide a comprehensive treatment of the entire Subphylum.

The class Cephalocarida was only recently discovered (Sanders, 1955). Howard L. Sanders was a graduate student at Yale when he collected puzzling specimens from the bottom sediments of Long Island Sound. Eventually he realized that the find represented far more than just a new species or genus. Sanders named the animal *Hutchinsoniella* (Fig. 2-6, *A*) in honor of Professor G. E. Hutchinson. The remarkable feature of *Hutchinsoniella* and the other, subsequently discovered Cephalocarida is that they link the long-extinct trilobites and some enigmatic fossil crustaceans with the modern living Crustacea. Furthermore, from a cephalocaridan ancestor it would be easy to derive at least the modern Branchiopoda and Malacostraca, two other crustacean classes. The Cephalocarida are entirely marine and widespread bottom dwellers.

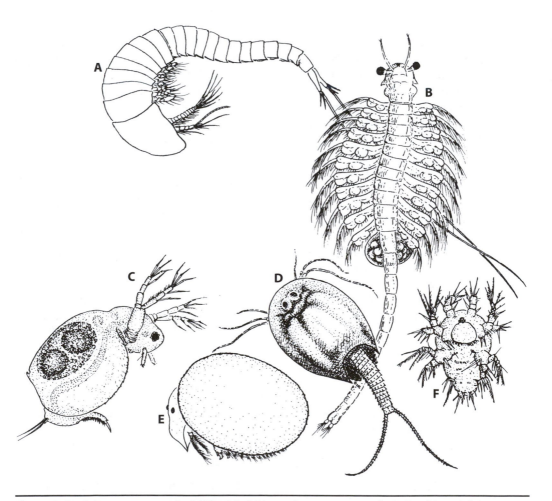

Figure 2-6 Some crustaceans, including inhabitants of temporary ponds. **A**, *Hutchinsoniella,* marine; **B**, female *Artemia sp.,* from hypersaline waters; **C**, *Moina wierzejskii,* Cladocera; **D**, *Triops,* Notostraca; **E**, *Lynceus,* Laevicaudata; **F**, crustacean metanauplius larva.

BRANCHIOPODA. Except for the cladocerans, all the branchiopods (Fig. 2-6, *B, C, D, E*) are confined to inland waters. They are found especially in temporary pools but may occur in permanent habitats where fishes and other large predators are absent. Their permanent habitats include some hypersaline lakes and shallow northern lakes where prolonged ice and snow covers lead to anoxic conditions. Some Alaskan lakes are this shallow type, where the summer swimmer encounters a rich "temporary pond" fauna in what appears to be a respectable body of water. The occurrence of these branchiopods is understandable, since fish are wanting. However, the presence of *Lepidurus arcticus* (Notostraca) in permanent Norwegian lakes, where it serves as an item in trout diet, and of the conchostracan *Cyclestheria hislopi* in Lake Victoria (Hartland-Rowe, 1972), is baffling when one believes the branchiopods cannot tolerate predation.

The best-known Branchiopods are the Anostraca. They are cosmopolitan animals, occurring in all types of temporary ponds, or as in the case of *Artemia*, in hypersaline waters (including temporary, salty ponds). Most feed by filtering suspended particles or by scraping organic matter and algae from surfaces. The giant of the group, *Branchinecta gigas*, preys on other anostracans in alkaline ponds in western North America.

The Notostraca are widespread in continents of the world but are lacking in the eastern half of North America. They have biting mouthparts and chew on plants or prey on other animals. Only two genera, *Triops* (Fig. 2-6, *D*) and *Lepidurus,* are known; they are termed the tadpole shrimps. Fossil notostracans extend back to the upper Carboniferous. They are now restricted, or nearly so, to temporary ponds.

CLADOCERA. Dodson and others (2010) provide an excellent description of the Cladocera (Fig. 2-6, *C* and Fig. 2-7) and describe the ongoing taxonomic revisions. Cladocerans arose relatively late, the oldest known fossil being from Oligocene strata.

The typical cladoceran is 1.5 mm long and has a distinct head and a body covered by an unhinged bivalved carapace. It has a single compound eye and a smaller eyespot, or ocellus. The first antennae, or antennules, are small, although larger in some males. The second antennae are large, branched swimming structures, the typical biramous crustacean appendage. Most species have five or six pairs of legs beneath the carapace. There is a posterior extension of the abdomen to form the postabdomen or abreptor, presumably formed from fusion because it terminates in a pair of claws; features of the postabdomen are very important in taxonomic determinations. There is a dorsal heart and a more posterior dorsal brood chamber beneath the carapace roof and over the upper part of the body. Here the eggs are held during development.

Typically the life cycle includes many generations of females parthenogenetically producing more females from diploid "eggs." There comes a time when some eggs develop into males, and meanwhile other females are producing haploid sexual eggs that require fertilization. The diploid males produce haploid gametes. The carapace surrounding the sexual eggs secretes a dark, sometimes sculptured case, the *ephippium*, which is shed at a subsequent molt. The ephippial eggs are the resistant, diapausing stage so characteristic of freshwater animals. They will hatch later as parthenogenic females. There are no larval stages.

Keys prepared for identification of Cladocera (Brooks and Dodson, 1965) usually feature female structures, because males appear so seldom. An exception is made in the genus *Moina* (Fig. 2-6, *C*), commonly found in temporary puddles and in short-lived sewage beds subjected to frequent drying. Here the male's first antennae and ephippial structures are used to distinguish species (Goulden, 1968).

Planktonic Cladocerans may be more familiar because of their active movement and frequent appearance in net samples. In northern lakes the Holarctic *Leptodora* occurs (Fig. 2-7, *A*). This is a transparent crustacean, the largest of the cladocerans, and is noteworthy on several accounts. It rises at night to prey on the other zooplankton. At this time it is easily collected, but during the day it is not readily found. It has a greatly reduced carapace, consisting of no more than a bubble-like dorsal brood pouch. It is parthenogenic during the summer, but males appear in late fall, and fertilized eggs serve as the overwintering stage. At the vernal overturn, these sexual eggs hatch to release a naupliar stage, quite unlike the other cladocerans.

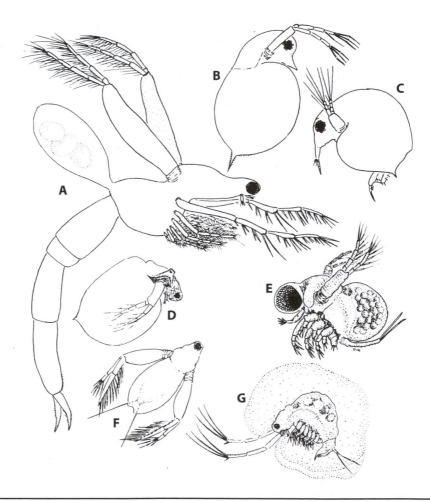

Figure 2-7 Some planktonic cladocerans. **A**, *Leptodora kindtii*; **B**, *Daphnia rosea*; **C**, *Bosmina longirostris*; **D**, *Ceriodaphnia lacustris*; **E**, *Polyphemus pediculus*; **F**, *Diaphanosoma*; **G**, *Holopedium gibberum*.

COPEPODA. The Copepoda are a large subclass of the phylum Crustacea containing more than 14,000 known species (Reid and Williamson, 2010). Most are in the sea, but a few have come to fresh water where there are three principal free-living orders: the Harpacticoida, the Cyclopoida, and the Calanoida (Fig. 2-8). Two parasitic orders occur also.

The harpacticoids (Fig. 2-8, *D*) are mostly littoral microbenthic forms or groundwater species. The cyclopoids (Fig. 2-8, *B*) are found in a variety of habitats, from caves, wells, and small puddles to the open waters of the largest lakes. Most of them are littoral, and only a few are typically limnetic. The calanoids, however, are all open-water plankters, if we overlook some marine species that are benthic.

Freshwater copepods are small, and it would be unusual to find one as long as 5 mm. Their first antennae are conspicuous and longer than the second pair (the reverse

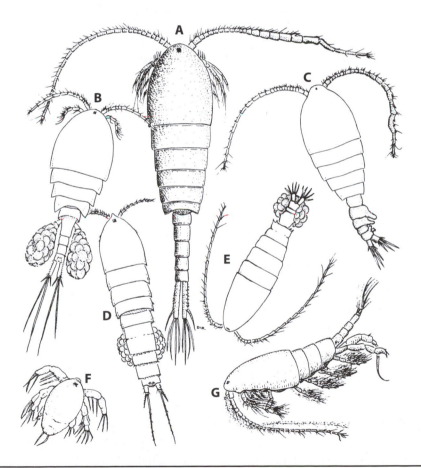

Figure 2-8 Some copepods of inland waters. **A**, *Limnocalanus macrurus,* Centropagidae, Calanoida, male; **B**, *Eucyclops serrulatus,* Cyclopoida, female; **C**, *Epischura lacustris,* Temoridae, Calanoida, male; **D**, *Canthocamptus,* Harpacticoida, female; **E**, *Diaptomus siciloides,* Diaptomidae, Calanoida, female; **F**, nauplius larva of *Diaptomus;* **G**, *Senecella calanoides,* Pseudocalanidae, Calanoida, male.

of the cladoceran plan). They are not covered by a carapace, although the cephalotho-rax segments are fused dorsally. Their mouthparts include a maxilliped (an ancestral thoracic appendage), and they lack abdominal appendages. The last abdominal seg-ment bears two cylindrical rami tipped with setae (Fig. 2-8).

The Copepoda brought the typical nauplius larva of the marine crustacean to fresh waters. In cyclopoids there are usually five naupliar stages followed by six *copepo-did* stages, which give the distinct appearance of a copepod. The sixth copepodid stage is the adult. The calanoids differ by having six distinct nauplii before the first copepo-did form.

The common calanoid in North America, *Diaptomus* (Fig. 2-8, *E*), is divided into subgenera. The Europeans first raised these to generic rank, and it is now being done in North America. Studies of diaptomid zoogeography profit from the splitting of the genus. The genus *Diaptomus,* being widespread, tells us little; but when the subgenera are raised to generic level, we see interesting patterns of distribution from continent to continent.

Diacyclops bicuspidatus was the first known of many cyclopoids that perform the trick of leaving the plankton and becoming microbenthic. This mechanism occurs at the fourth copepodid stage in *D. bicuspidatus.* Then the entire population moves to the deeps and burrows into anaerobic sediments, diapausing for months before returning to the upper waters to mature and mate, completing its life cycle within a month or two. **Diapause** is a regular feature of some cyclopoid life cycles and may serve to pro-tect them from predators; in some lakes *D. bicuspidatus* is diapausing during the sum-mer, while the larger *Mesocyclops edax is* in the plankton above. Elgmork, who has worked for many years on problems of copepod encystment, reviewed the state of our knowledge (Elgmork, 1980).

OSTRACODA.   The Ostracoda are tiny crustaceans found in marine and inland waters. They have unsegmented bodies enclosed by two hinged valves. Thus they resemble tiny clams. Their thoracic appendages number only three pairs, and the abdomen is represented by a limbless structure, the furca. The typical two pairs of crustacean antennae are present and serve in swimming activity. There is an excellent fossil record of the ostracods (called ostracodes by the paleontologist); most fossils and living species are marine. The North American freshwater ostracods have been studied by relatively few workers, and we still have much to learn about their distribu-tion and taxonomy. A review of the group can be found in Smith and Delorme (2010).

Most freshwater ostracods are bottom dwellers, although some appear occasion-ally in plankton samples. One truly planktonic species, *Cypria petenensis*, is endemic to Laguna de Petén and eight nearby smaller lakes in Guatemala (Deevey et al., 1980). A few species occur in temporary ponds, and there are semi-terrestrial forms known from the moist humus of tropical forest floors. In North America a species of *Scottia* creeps about in damp moss near water. All members of one family, the Entocytheri-dae, are found living commensally within the gill chambers of crayfish.

Some copepod predation involves cannibalism. The smaller male cyclopoid is in danger of being devoured by the female, and both adults may eat individuals of the small younger stages. In these instances the sex ratios toward the end of a cohort's life are skewed in favor of the females. This may explain a phenomenon sometimes observed in populations of large diaptomids in temporary ponds in the American West. At the beginning of the pond's existence the sex ratio is 1:1, but toward the closing days the females far outnumber the males.

The effects of copepod predation explain in part the phenomenon of **cyclomorphosis**. In rotifers, limnetic dinoflagellates, and cladocerans there is a type of polymorphism in which members of a species vary seasonally in body form. The environmental cues that trigger the changes and the adaptive significance of these changes were sought for many years. In most instances predation was not taken into account. Work done on changes in *Bosmina* morphology as a result of predation pressure from *Epischura* and *Heterocope* has thrown light on one reason for the so-called cyclomorphosis (Kerfoot, 1977; O'Brien and Schmidt, 1979). Different survival rates of contrasting morphs in the presence of copepod predators accounts for spatial and temporal separation of *Bosmina* clones. Some of the clones seem to be reproductively isolated also, with apparent incipient speciation caused in part by effective copepod predators.

MALACOSTRACA. A large group within the Crustacea, these organisms are described by Covich and others (2010). These are represented best in marine habitats, but some of the Peracarida and Decapoda have invaded inland waters. Of special interest to limnologists are three orders of the Peracarida—the Mysida, the Isopoda, and the Amphipoda.

The Mysida, a large marine group, has perhaps 70 species occurring in fresh water. The species of special interest is the lacustrine possum shrimp, *Mysis relicta* (Fig. 2-9, *B* on the following page). *Mysis* occurs normally in northern oligotrophic lakes, but it has spread elsewhere to serve as a forage item for sports fishes. *Mysis* is an effective predator, and it has had profound effects on native zooplankton populations in some of the lakes to which it has been introduced. *Mysis* is discussed in later pages.

The Isopoda and Amphipoda, the two other orders of the Peracarida, have also entered fresh water. They have sessile compound eyes quite in contrast to the stalked eye of *Mysis*. Most Isopoda are flattened dorsoventrally rather than compressed laterally, as are the amphipods. The Australian phreaticoid isopods are an exception to this rule.

In North America the aquatic isopods are mostly lotic, although *Caecidotea forbesi* is a temporary-pond inhabitant, and there are occasional lacustrine occurrences of *Lirceus* and *Caecidotea*. Stream biologists deal especially with isopods, finding them feeding in leaf packets and beneath the stones of small stream beds. There are far more lake-dwelling asellids in the Old World.

The Amphipoda are common in all types of aquatic habitats except for anaerobic hypolimnion waters. They will be discussed in more detail than the isopods in later pages. They are known as scuds, side-swimmers, and erroneously as freshwater shrimps. They are important members of the stream bottom fauna, the lake littoral zone, and rarely as members of limnetic communities. *Hyalella* (Fig. 2-9, *F*) occurs from South America to the Canadian timberline. There are probably many undescribed species masquerading as the ubiquitous *H. azteca,* described originally from Mexican specimens. *Hyalella* is the commonest malacostracan in the littoral plants of glacial lakes all across the continent and in waters farther south, west of the Great Plains. *Crangonyx* is a species-rich genus in the southeastern part of the United States; a few species are found in Canada, but except for some Washington–British Columbia occurrences, they are absent in the West. In Europe and Asia there are many species of *Gammarus* and closely related genera. North America has only about 16 freshwater *Gammarus* species and one of those (*G. lacustris*) is shared with Eurasia. *Gammarus* appears in the scientific literature of streams where it feeds on fallen leaves and their associated fungi.

The decapod crustaceans include freshwater shrimp and crabs, which are not well-studied, and also the familiar crayfish. The North American crayfish belong to the family Astacidae. Now more than 400 species have been described from many types of inland waters. Some forms are semiterrestrial, digging tunnels down to the

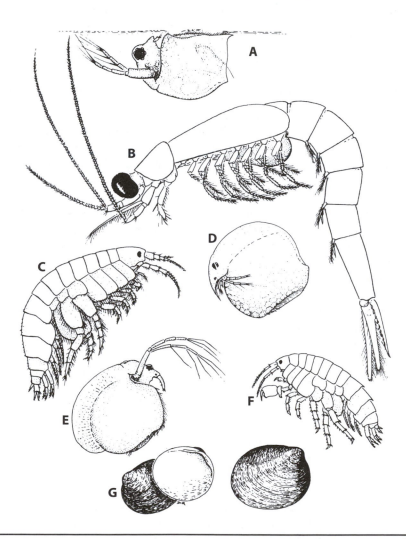

Figure 2-9   Miscellaneous lacustrine species of special interest. **A**, *Scapholeberis kingi,* female; a hyponeustonic cladoceran. **B**, *Mysis relicta,* female, a nektonic glaciomarine relict from northern lakes; **C**, *Diporeia,* female, a benthic glaciomarine relict amphipod; **D**, *Chydorus sphaericus,* a common chydorid cladoceran; **E**, *Bunops* sp., a macrothricid cladoceran; **F**, *Hyalella azteca,* male, a New World amphipod; **G**, *Pisidium conventus,* a cold-water pisidiid clam.

water table in low-lying fields; the mouths of these burrows are marked by chimneys of mud pellets.

The decapods have moved three pairs of thoracic appendages into the head region to serve as mouthparts. Thus, they have three maxillipeds and only five pairs of pereiopods, the so-called walking legs. Their third maxilliped is homologous to the second pereiopod of the Peracarida. The decapod "head" or cephalothorax is covered by a carapace as seen in the crayfish and the marine lobster. The developing eggs are carried beneath the abdomen and are secured by abdominal appendages rather than being held by a thoracic marsupium.

Other decapods in North American freshwaters include the glass (or grass) shrimps, *Palaemonetes* (at least nine inland species), and the less common river shrimps, *Macrobrachium*. The crayfish, especially, have many unpigmented, blind, **troglobitic** species. Speleologists and zoologists interested in commensal ostracods and branchiobdellid annelids of crayfish are adding to our knowledge of these decapods.

In some tropical regions true decapod shrimps are found and harvested in fresh water. It seems bizarre to the zoologist farther north that shrimps and even decapod crabs occur far inland in lakes and rocky streams of Honduras and Mexico, for example. The most unusual case is seen in Lake Tanganyika, the largest of the African rift lakes, where 15 shrimp species and ten crab species occur. Fourteen of the shrimps and eight of the crabs are believed to be endemic to the lake (Coulter, 1991).

Foreigners who have been in Sweden during the summer are amazed at the fervor with which the populace greets the crayfish season and its culinary delights. It is not so seasonal a ritual, nor so joyous, in Louisiana where "crawdads" are eaten throughout the year and where the oppression of a long, harsh winter is unknown.

Hexapoda (Insects). The insects comprise more than 75% of all the described animal species and are an extremely successful group. Perhaps 60–80% of all aquatic animals are insects, with diversity estimates of at least 100,000 species, and perhaps twice that number, being aquatic or having aquatic larval stages (Dijkstra et al., 2014). The limnological insects are described in the DeWalt et al. chapter of Thorp and Covich (2010).

Of the 10 orders listed in Table 2-1, only five (Fig. 2-10) have almost all their species with at least one aquatic stage: the Ephemeroptera, mayflies; Odonata, dragonflies

---

Table 2-1 Aquatic and Semiaquatic insects

Order	Common Name	Active Aquatic Stages
Ephemeroptera	Mayflies	Naiad
Odonata	Dragonflies, damselflies	Naiad
Plecoptera	Stoneflies	Naiad
Trichoptera	Caddisflies	Larva, pupa
Megaloptera	Dobsonflies, alderflies	Larva
Heteroptera	True bugs	Nymph, adult
Neuroptera	Spongeflies	Larva
Lepidoptera	Moths	Larva
Coleoptera	Beetles	Larva, adult
Diptera	True flies	Larva

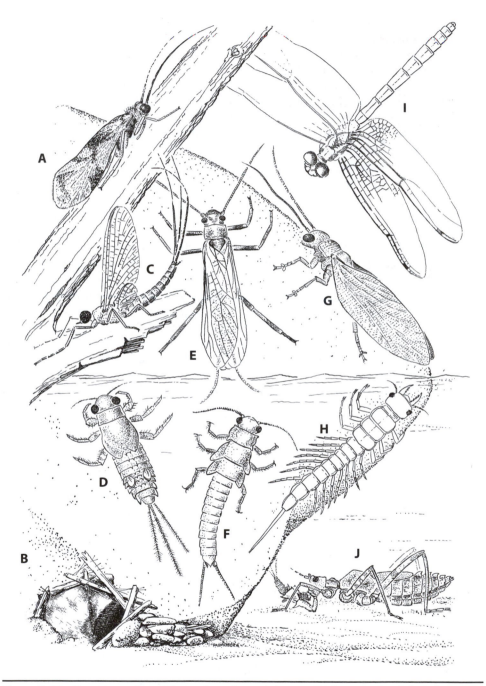

Figure 2-10 Representatives of five orders of aquatic insects. Trichoptera: **A**, adult *Hydropsyche;* **B**, case and net of larva. Ephemeroptera: **C**, adult *Ephemerella;* **D**, naiad or nymph of same genus. Plecoptera: **E**, adult *Isoperla;* **F**, naiad or nymph, *Pteronarcys*. Neuroptera: **G**, adult *Sialis;* **H**, larva, same genus. Odonata: **I**, adult *Macromia;* **J**, naiad or nymph, same species, feeding on *Ephemerella* naiad.

and damselflies; Plecoptera, stoneflies; Trichoptera, the caddisflies; and Megaloptera, the dobsonflies. This gives the wrong impression concerning some of the remaining five insect orders that are aquatic or semi-aquatic. The true flies (Diptera) especially are extremely important in aquatic habitats (Fig. 2-11). Their abundance and diversity in terrestrial environments far outshadow the freshwater situation, but there are at least 27 fly families with aquatic members. Of these, some are primarily aquatic or have tremendous representation in freshwater and inland saline bodies of water. For example, in North American waters the Chironomidae (midges) are represented by about 142 genera made up of well over 1,000 species described and named to date. In some lakes and especially in rivers there may be 100 to 200 species of midges, occupying all types of microhabitats and playing a variety of trophic roles (Coffman, 1978).

Similarly, the orders Heteroptera (true bugs) and Coleoptera (beetles) have some families that are primarily aquatic and are important members of some ecosystems. Their land membership, however, is relatively far more numerous.

It has been understood for decades that aquatic animals can act as indicators of water quality and biodiversity (Resh and Unzicker, 1975). Insects in particular are

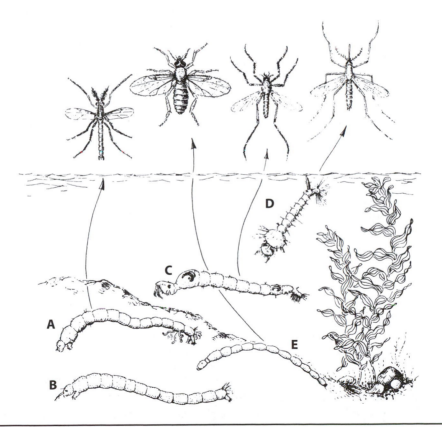

Figure 2-11 Some aquatic flies from different families: the adults superficially similar, the larvae quite different. **A**, *Chironomus*, Chironomidae; **B**, *Tanytarsus*, Chironomidae; **C**, *Chaoborus*, Chaoboridae; **D**, *Culex*, Culicidae; **E**, *Culicoides*, Ceratopogonidae. The pondweed is *Potamogeton richardsonii*.

known to be useful sentinels of changes due to pollution such as acidification (Bell, 1971), or metal contamination (Warnick and Bell, 1969). These factors have led to the development of indices of biological integrity, widely adopted as an approach to assess the overall health of a water body. These indices have been developed for Belgium (Gabriels et al., 2010), Kenya (Raburu et al., 2009), and Mexico (Weigel et al., 2002), among many others. Each index can be tailored to local landforms and biota, but the general approach is to determine the sensitivity of the various taxa to low oxygen, siltation, toxins or other degradation, and create a weighted scoring system assigning numbers based on sensitivity. Users count organisms and multiply by the weighted quality factor.

Mollusca.   The mollusc phylum includes a tremendous number of species in marine and inland waters. Many snails have become terrestrial, but the spectacular squids, octopi, and cuttlefish remain only in the world's seas.

The phylum in North America has been reorganized, using molecular and other modern data sources to now encompass two classes of Bivalves (Cummings and Graf, 2010) and nine classes of gastropods (Brown and Lydeard, 2010).

BIVALVIA.   Molluscs with opposing shells are widespread in aquatic habitats worldwide; estimated numbers of species range wildly, from 8,000 to 30,000 (Bieler and Mikkelsen, 2006). These animals possess gills for effective respiration in oxygenated water and a filter-feeding apparatus. The rock-hard shell (test) is hardened with carbonates and pulled tightly by strong muscles.

The bivalved molluscs assigned to the Unionoidea are typical of the river fauna. Although two genera, *Anodonta* and *Margaritifera*, have some holarctic members, the others occur only in North America. Exceptionally rich faunas are found in streams of the Mississippi drainage, and many forms are restricted to some particular southeastern rivers. The unique clam fauna of the Cumberland River system in Tennessee and Kentucky contains many endemic species threatened by impoundment, flooding, and silting of the streams. Mussels are found in the shallows of lakes as well as in streams, but most species and the most diverse communities are present in medium-sized to large rivers.

Mussels feed on tiny particles that they bring in by way of ciliary currents through a posterior incurrent siphon. They are dioecious and produce a unique larva, the *glochidium.* The glochidium parasitizes a fish, continuing its development in the gills or surface flesh of the host. Eventually, it develops into a juvenile clam and drops from the fish host to start its independent life as part of the bottom fauna.

The fish host species is unknown for about 50% of the North American mussels. Some glochidia have been found to develop directly without the fish, and at least one unionid clam uses a salamander as the host form. In general, however, the unionacean mussel is not present in waters from which fish are excluded, quite in contrast to the pisidiid fingernail clams (discussed below) and the Asiatic *Corbicula*, neither of which has a parasitic stage in its life cycle.

One North American unionid clam, *Lampsilis cariosa*, is noted for the unusual adaptation that has come about in the naturally frilly siphons and posterior mantle margins that typify its genus. A female, half buried in the substratum with her posterior end protruding, presents a spectacle that is irresistible to potential hosts. The mantle-siphon structure resembles a little undulating fish, and the real fish that comes to investigate and tug at the lure receives a discharge of glochidia which may be drawn into its gill chambers.

The tables are turned by *Rhodeus*, a European cyprinid fish known as the bitterling. The female has an elongate ovipositor with which she deposits her eggs in the incurrent siphon of a mussel. Further development occurs in the mantle chamber of the clam.

The fingernail clams are a group of little bivalved freshwater organisms with a long history of marine ancestry. They are known as the Sphaeriidae, after the genus *Sphaerium*. Related is *Pisidium,* the tiny pea clam (Fig. 2-9, *G*). A third genus is *Musculium*, its size intermediate between *Sphaerium* and *Pisidium.*

The story of the zebra mussel (*Dreissena polymorpha*) in North America is told well by van Driesche and van Driesche (2000). So named for the alternating light-and-dark stripes on the shell, this mollusc is only the size of a thumbnail but thoroughly filters a liter of water every day, removing food necessary to support the diverse Unionid taxa and also interfering with their sperm transfer. Tens of thousands of individuals in a square meter affix to rocks, human-built structures, or even other molluscs. Up to 70% of other mussels die in a single year. The zebra mussel has spread surprisingly quickly throughout the United States and Canada since its arrival to the Great Lakes area in ship ballast in 1988.

GASTROPODA.   The snails and slugs of rivers and lakes have been studied more intensively than the clams because they concern the parasitologist. Gastropods serve as intermediate hosts for innumerable flukes, a number of which use humans and their domestic animals as definitive hosts. The discipline known as medical malacology focuses on snails because of this.

In inland waters the gastropods are mostly members of two distinctive groups. Caenogastropods are gill bearers and can be identified by the presence of an operculum "trap door," a discoid structure that blocks the shell aperture when the animal has drawn completely into its shell. Some members, especially the species of *Campeloma* and *Viviparus*, are ovoviviparous, eventually giving birth to many tiny, shelled offspring. Pulmonates have no operculum to protect soft tissues and breathe using lungs instead of gills. All are monoecious, and some individuals fertilize their own eggs, although outcrossing is preferred when possible.

Species of the lacustrine genus *Valvata* occur throughout lakes of the Northern Hemisphere. Two other Caenogastropods, *Elimia* and *Pleurocera*, are typically river forms, although some "pond snails" (*Lymnaea*) may be found in streams. One family, the Hydrobiidae (mud snails), contains many tiny forms including the widespread *Amnicola*. By contrast, there are many endemic species of hydrobiids in isolated, islandlike springs in arid parts of North America. Endemism is also a feature of snails in the world's ancient lakes: there are more than 40 endemic prosobranchs in the Siberian Lake Baikal.

Despite branchial respiration, *Pomatiopsis*, consisting of a few species in eastern North America, occupies habitats intermediate between aquatic and terrestrial settings. It is found abundantly in the vicinity of water on moist soil, sometimes beneath mats of dead rushes and cattails. *Pomatiopsis* has attracted attention because it is similar to the oriental *Oncomelania,* the intermediate host of *Schistosoma japonicum,* a blood fluke of the human.

The subclass Pulmonata includes many land snails in addition to some common aquatic species. The group is characterized by (1) the adult lacking an operculum, (2) the absence of gills, and (3) possession of a vascular chamber that serves as a lung. The land snails and the pulmonates from temporary ponds draw into their shells and secrete mucus, which hardens to form an epiphragm that seals the shell aperture in times of stress.

Pisces

Nelson (2006) provides a concise yet thorough catalog of fish diversity. The freshwater fishes make up a heterogeneous group. Phylogenetically, they range from the Chondrichthyes (if we include the occurrences of sharks, sawfish, and rays in some lakes and streams of Central and South America) through many families of the Osteichthyes. Physiologically they can be categorized according to their osmotic abilities— their tolerance to seawater or, conversely, to dilute water. One euryhaline group of fishes enters fresh water irregularly, and their inland occurrences are rare and sporadic. Two examples, the mullets (Mugilidae) and a few of the sea basses (Serranidae), do very well in fresh water. If transplanted from the sea, they survive and reproduce in dilute water.

At the other extreme are those fishes that are found only in fresh water and cannot endure seawater to any degree. Their inability to deal with the osmotic hazards of the sea, along with their fossil record, implies they have had no recent marine background or ancestry. There is no evidence that they have dispersed except by way of continental connections; the sea has always been an insurmountable barrier to them. This unique division of fishes includes the following: the paddlefishes, Polyodontidae (Fig. 2-12, *A*); the minnows, Cyprinidae (Fig. 2-12, *B, E*; Fig. 2-13 *D;* Fig. 8-4, *C, D, E*); the sunfishes, Centrarchidae (Fig. 4-1, *A*); the perches, pike-perches, and darters, Percidae (Fig. 2-11, *A*); the pikes, Esocidae (Fig. 2-13, *C*); the suckers, Catostomidae (Fig. 2-12, *D;* Fig. 8-4, *A, B*); and most of the catfishes, Ictaluridae (Fig. 2-12, *F*). These are the true freshwater fishes, but there are many others in the gradient that extends from them to the euryhaline mullets and sea basses.

On the basis of their size and ability to move at will against turbulent forces, the fishes are classified as elements of the **nekton** community. But because of their diverse feeding habits, they could be assigned to other aquatic communities as well. As "top carnivores," four piscivorous fishes are shown in Figure 2-13. These elongated, cylindrical predators seem closely related, although they belong to four different families. Other piscivores, not depicted, are the black basses, species of *Micropterus*; they are more elongate than their fellow family members, the smaller sunfish, also of the Centrarchidae (Fig. 4-1, *A*). Other feeding habits are described in the legend to Figure 2-12, a figure that portrays fishes that take their food from various communities: the zooplankton and phytoplankton; the macrophytes (and their epiphytic algae) of the littoral community; the zoobenthos and phytobenthos; and dead plant and animal detritus. In this sense, a fish could be assigned to each community in a lake or stream ecosystem. Moreover, in the life of any one species there are shifts in feeding habits during growth and maturation that link it to several communities. The aquatic communities mentioned in this paragraph will be described in detail in the following chapter.

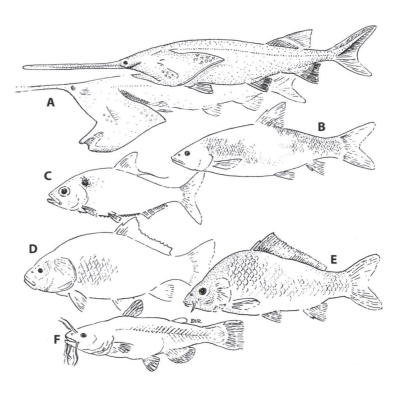

Figure 2-12 Freshwater fishes from ponds, lakes, and large rivers. **A**, Paddlefish, *Polyodon spathula,* zooplankton feeder of the Mississippi Valley; **B**, grass carp, *Ctenopharyngodon idella,* voracious consumer of aquatic vegetation, introduced to North America and Europe from Asia; **C**, thread-fin shad, *Dorosoma petenense;* adults are planktivorous. **D**, bigmouth buffalo, *Ictiobus cyprinellus,* efficient strainer of small bottom fauna in deeper pools of large rivers, impoundments, and natural lakes; **E**, carp, *Cyprinus carpio,* an omnivorous feeder introduced in North America from Asia via Europe; **F**, black bullhead, *Ictalurus melas,* a bottom scavenger feeding on a variety of plant and animal material.

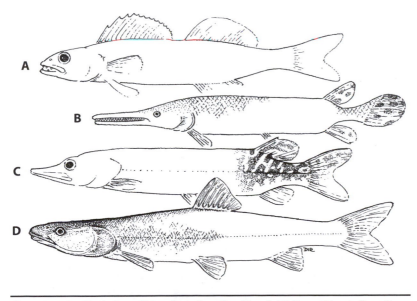

Figure 2-13 Piscivorous fishes from four families. **A**, Walleye, *Stizostedion vitreum* (Percidae), 75 cm, 25 kg; **B**, short-nosed gar, *Lepisosteus platostomus* (Lepisosteidae), 60 cm, 17 kg; **C**, northern pike, *Esox lucius* (Esocidae), 75 cm, 25 kg; **D**, Colorado squaw fish, *Ptychocheilus lucius* (Cyprinidae), 200 cm, 40 kg.

3

The Limnetic Communities

Community ecology studies the interactions of various species within their habitats. We consider the limnetic communities of lakes and ponds here; wetlands are described in Chapter 7 while streams are the subject of Chapter 8.

General ecological principles are presented first, then we survey the various specific communities individually. The biology (classification, anatomy, ecology) of the individual organisms were presented in the preceding chapter. The communities of a typical lake are diagrammed in Figure 1-1.

Community Ecology in Limnology

The important ecological dynamics found in marine or terrestrial habitats are also found in limnetic communities, although the specific organisms and aspects of the functioning differ. Key concepts in community ecology were, in fact, developed or refined in limnology. For example, the Competitive Exclusion Principle was first demonstrated by Gause (1934) using *Paramecium*. G. E. Hutchinson pioneered niche theory in biodiversity by asking "why are there so many kinds of animals?" in pools and other habitats (Hutchinson, 1959). Brooks and Dodson (1965) demonstrated the interactions of competition and predation in structuring plankton communities. Three of the most important ecological phenomena are discussed in this section: population dynamics, diversity, and succession. They all have seen notable research, and unique applications, in limnology.

Population Dynamics

All organisms are evolved to employ life-history strategies, maximizing either fast growth and production of many offspring, or persistence and competitive ability, known respectively as *r*- and *K*-selection. This concept came from symbols in the logistic equation that expresses the numerical growth of a population to an upper plateau. Several similar equations exist, but the following illustrates the point:

$$\frac{d\mathrm{N}}{dt} = r\mathrm{N}\left(\frac{K - \mathrm{N}}{K}\right)$$

The intrinsic reproductive ability is r; this is the maximum rate of increase of which members of a population are capable. K symbolizes the upper numerical limits of the population, and this is largely an environmentally controlled and density-dependent plateau or asymptote; it is the carrying capacity of the environment. Out of the equation came r-selection to describe species that have evolved to make the most of unpredictable, uncertain environments. They are often short-lived and reproduce at an early age. They produce a great many offspring, investing little energy in the individual egg and offspring, which are small and grow rapidly. They are hardly affected by interspecific and intraspecific competition. Catastrophic mortality may strike the populations at any time, but the sheer number of offspring produced ensures survival of the population.

K-selection involves the opposite of most phrases that describe r-selection, but there is a continuum between the two types. The work of Belk (1977b) on fairy shrimps from two kinds of temporary ponds illustrates a continuum even among forms that we might describe as r-selected. Belk contrasted the irregularity and uncertainty of the rain ponds of the Arizona Sonoran Desert with the environmental certainty of the vernal snowmelt ponds on the high-altitude Kaibab Plateau, also in Arizona. The former fill with winter and summer rains, although remaining dry some seasons and years. On the Kaibab Plateau deep winter snows are predictable, as are the vernal snowmelt ponds. The ponds are long lived and sometimes, depending on the summer rains, even persist until the following winter.

Four species of fairy shrimps (*Anostraca*) representing three genera were found in the Kaibab ponds. Six different species, also representing three genera, occurred in the short-lived uncertain desert ponds. The desert females produced four to five times as many eggs with volumes 0.42 times the bulk of those produced by the Kaibab females. Relatively speaking, the Kaibab species were farther along a gradient toward being K-selected. Another factor entered into the selection for more voluminous eggs and larger hatchlings in the snowmelt ponds: a predaceous cyclopoid copepod was part of their environment. Belk showed experimentally that the predator had very little success in capturing the Kaibab larvae but easily captured and consumed the smaller, more r-selected desert rain-pond larvae.

Diversity

Superficially, diversity would appear to be a simple concept, but this is not true. So many technical, conceptual, and semantic problems had accumulated on the subject by the end of the 1960s that it was suggested that much of the literature on species diversity had little biologic importance (Hurlbert, 1971). Many workers have contributed to the topic of species diversity, and it is discussed now in all modern ecology texts. In the new millennium, ecologists still debate the proper measurement of diversity and application of an index to real-world species (Buckland et al., 2005). Species diversity has been discussed in relation to evolutionary time, ecologic time, year-to-year climatic stability or predictability, habitat heterogeneity, various gradients such as latitude and altitude, productivity of habitats and stability of primary productivity, degree of competition among species, predation, and both natural and anthropogenic disturbance, among many.

The term diversity has two main components: *species richness,* which refers to the number of species in a sample, community, or habitat; and *evenness* with which individuals are distributed numerically among species. The maximum equitability would occur if all the species in the sample were represented equally—evenness in the best sense of the word. Although there may be more than 24 species diversity indices, the best ones express *heterogeneity* by combining species richness and evenness.

A measure widely used by ecologists is the Shannon-Weaver (1949) index, H', which accounts for both ingredients of heterogeneity:

$$H' = -\Sigma p_i \log p_i$$

In this equation, p_i refers to the proportion or frequency of the ith species, where $i = 1, 2, 3, \ldots n$. The product of the frequency and log frequency for every species in the collection or community is calculated, and the sum of these products becomes the diversity index. Referring to species alone is too narrow an approach, for p_i can be considered any important value in the community examined, being numerical, biomass, productivity, energy flow, and so on.

Often diversity is measured in *bits* by using logarithms to the base 2 in the equation, reflecting the information-theory origin of the Shannon-Weaver index and, incidentally, criticized as not being applicable to ecologic matters (Goodman, 1975). It is permissible, however, to express the index in terms of $\log_e$, or with common logarithms ($\log_{10}$). Whatever the case, consistency should be the rule in making comparisons, and it should be stated clearly which log base was used in the calculations.

The maximum diversity (H'_{max}) obtainable would be found if all species were equally abundant. The relative value of the actual diversity index is then H'/H'_{max}. If H' were found by using $\log_e$, the denominator of this ratio would be the natural logarithm of the total number of species that are present in the sample ($\log_e S$).

The notion of the effective, apparent, or functional number of species as an index of diversity is favored by many ecologists. High equitability would result in a greater number of effective species than would a situation where only one or two forms were extremely abundant in relation to all others in the community. One way of arriving at this theoretical number of functional species and, hence, this index of diversity is from $1 / \sum p^2_i$. This equation is the reciprocal of an index proposed by Simpson (1949). The Simpson index D_s expresses the probability that two individuals selected at random from a sample will belong to the same species; it varies inversely with heterogeneity.

In Table 3-1 on the next page, some diversity indices are shown for macrobenthic animals in an Italian lake; they were calculated from the data of Nocentini (1966) and Bonomi and Ruggiu (1966). Certainly other indices could have been selected and the data could have been treated in greater detail (Alatalo and Alatalo, 1977). Although the contrasts among animal populations in littoral, sublittoral, and profundal depths are indicated clearly, they are less than what would have been revealed in some highly eutrophic lake; stress from anaerobic conditions would have been accented in the profundal region.

Succession

Probably the most influential theory in ecology states that communities undergo *succession,* a predictable series of communities (**seres**) changing over time towards a stable "climax." In a geologic time frame, all aquatic environments are predicted to

Table 3-1   Data concerning diversity of macrobenthos from three benthic zones in Lago di Mergozzo, Italy, based on mean annual numbers collected at each depth

	Littoral (0–5 m)	Sublittoral (15–20 m)	Profundal (50 m)
Mean number of species	112	60	11
Simpson's diversity index	0.07	0.12	0.67
Effective species	13.9	8.6	1.5
Shannon-Weaver index, H'	3.58	2.64	0.77
H' / H' max	0.76	0.64	0.32

fill with eroded soil from surrounding land and biomass produced by the biota within. The eventual terrestrial climax is the termination of this **hydrarch succession**.

Johnson and Miyanishi (2008) reviewed studies on succession conducted over a century and found little evidence to support hydrarch succession as a natural process. Instead, they agree with van der Valk and Davis (1978) that shallow water systems do not develop linearly toward a terrestrial endpoint, but rather have recurring cycles of regeneration, alternating between open water and dense vegetation.

Paleolimnologic evidence implies that past stable lake stages were in equilibrium with their climatic and edaphic surroundings. These lakes were oligotrophic as well as eutrophic in nature. Instead of a natural process of filling, increased erosion in the drainage basin (often due to forest clearing and farming) upset the equilibrium, and the lakes rapidly changed to states nearer the eutrophic end of the spectrum. A similar upset occurs when fertilizers are added, but at present there is no satisfactory evidence that a natural, gradual accumulation of nutrients dictates an inexorable trend from oligotrophy to eutrophy. A lake brought to eutrophy by addition of nutrients may revert to its original condition if the nutrient supply is withheld. This has been documented in a series of papers on Lake Washington summarized by Edmondson (1991). Furthermore, many old, now shallow lakes still have oligotrophic water (in the sense of Weber, 1907) and scanty plankton populations despite their filled-in condition.

Limnologists also recognize a second form of succession. Variously termed *algal* or **phytoplankton succession**, this phenomenon is unique to aquatic environments (both freshwater and marine), and has little in common with traditional hydrarch succession, or to classical terrestrial succession. Instead of a permanent change (at least until a disturbance), phytoplankton succession is seasonal. Every year a series of algal assemblages occurs, changing over the months of the growing season, and repeating the following year.

Algal succession was described in more than 30 African lakes since 1900, in a review by Talling (1986). Although the majority of these lakes are located in the tropical region, they still experience seasonal changes based on water inputs (especially associated with rainy seasons) or circulation patterns based on light and temperature. One pattern noted is that following a profound mixing, diatoms bloom; later water is warm and stagnant and is dominated by cyanobacteria.

Lewis (1978) characterized the changes in environmental conditions across the seasons in Lake Lanao in the Philippines, observing increasing light and cell sinking rate together with decreased nutrients. These caused the characteristic progression of algae:

Diatoms → Chlorophytes → Cyanobacteria → Dinoflagellates

This was also noted in Lake Valencia, Venezuela (Lewis, 1986). Communities differ quantitatively as well as qualitatively: it is interesting to note that the cyanobacterial community dwarfed all others; of all algae sampled, about two thirds by volume and over 80% by number were cyanobacteria.

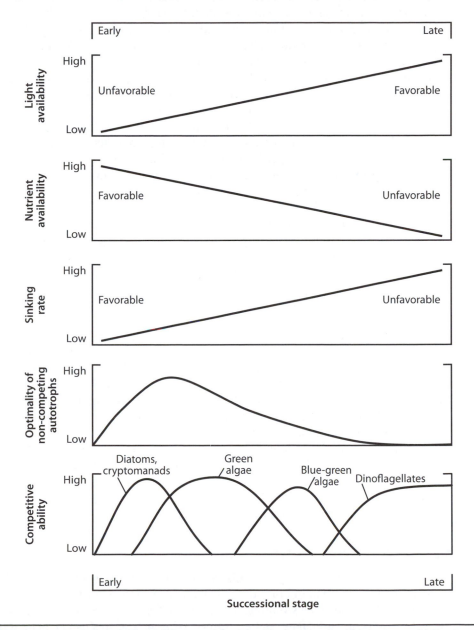

Figure 3-1 Environmental conditions and resulting algal succession in Lake Lanao, Philippines. Adapted from Lewis, 1978.

Another important paper that appeared in 1986 was concerned with algal succession in Lake Constance (Bodensee), an elongate lake bordered by Austria, Germany, and Switzerland (Sommer et al., 1986). The paper consists of 24 statements explaining the plankton events throughout a typical year in Lake Constance, although some data from other lakes are cited. The statements are known as the PEG model, an acronym for the international Plankton Ecology Group, and they are based on the concept that normal planktonic events can be explained; such events are neither chaotic nor random. Starting with the winter's end and spring's arrival, the PEG model takes us to the next winter, when reduced light energy brings about a decrease in primary productivity and algal biomass hits its annual low point. There is a resultant decrease in herbivorous plankters, deprived of food and experiencing low water temperatures.

The above discussion of the PEG model touches on a tiny part of the 24 statements in the 38-page PEG paper. The statements also deal with the complex interactions of various nutrients; P, N, and silica depletion and availability; algal size, growth rates, edibility and the effects of grazing and the light regimen on their populations; the generation spans of zooplankters, diapause "strategies," and the effects of size upon feeding abilities and protection from predators (including fish). Various workers have tested the PEG sequences in other lakes, finding agreement in many instances along with occasional discrepancies.

One more type of succession is important in limnology: succession in biofilms. Biofilms are thin coatings of microbes (bacteria, fungi, and algae) adhering to surfaces in lakes and streams. These are important drivers of ecological processes, particularly detrital decomposition and nutrient cycling (Peter et al., 2011).

The general process of biofilm succession (Fig. 3-2) was described by Jackson and colleagues (2001) for both wetland and lake environments. Initial colonization appears random, although some bacteria are better-suited to a particular situation than others. Characteristic changes proceed as better competitors emerge, lowering diversity. The successful species then fill the habitat, increasing chemical and structural diversity and leading to availability of new niches. Sekar et al. (2002) found an interaction between heterotrophic bacteria and algae in biofilm succession. In the light, biofilms developed a more productive and complex community with a progression from chlorophytes to diatoms to cyanobacteria, although chemistry rather than light shaped the community structure.

Biofilms are discussed later in this chapter, and in stream ecology (Chapter 8). Interestingly, the formation and ecology of biofilm is recognized as important in a variety of contexts in addition to natural habitats, most prominently in dentistry (O'Toole et al., 2000) where many of the same genetic and physiological, and successional phenomena are recognized.

The Plankton Community

The plankton community is a mixed group of tiny plants and animals floating, drifting, or feebly swimming in the water mass. The name is owed to Hensen (1887), who applied it to what we would now call **seston**. Seston includes the living plankton and the nonliving particulate matter, termed **tripton**. The freshwater plankton lacks many elements that are abundant in the sea, where nearly every phylum is represented.

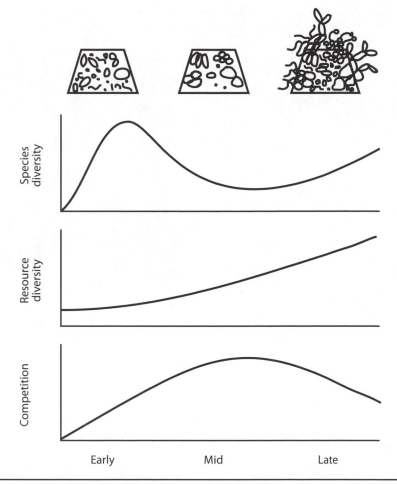

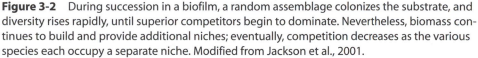

Figure 3-2 During succession in a biofilm, a random assemblage colonizes the substrate, and diversity rises rapidly, until superior competitors begin to dominate. Nevertheless, biomass continues to build and provide additional niches; eventually, competition decreases as the various species each occupy a separate niche. Modified from Jackson et al., 2001.

The individual plant, animal, or bacterium in the plankton community is called a *plankter*. The plant plankters compose the **phytoplankton**, and the animal plankters are the **zooplankton**. There are all sorts of schemes for classifying the plankton further; a division based on size is addressed later in this section.

The complex plankton community comprises primary producers, herbivores, carnivores, detritivores, and decomposer organisms. Thus, plankton consist of prokaryotes, plants, and animals. Of these, the primary producers are the basis for the planktonic food web and for food energy in other aquatic communities. They are the photosynthetic algae and cyanobacteria, joined occasionally by other photosynthetic bacteria. Although their standing stock at any time is low, the rate at which they fix carbon is high. Thus, an annual production of a kilogram (dry mass) per square meter

Figure 3-3 The plankton community is commonly sampled using a Wisconsin net (left) or similar apparatus, towed behind a motor boat (horizontal sample) or pulled up through the water column after having first been lowered to the bottom (vertical sample). On the right is an Ekman dredge used to obtain a sample of the benthos at the bottom of the lake; a small weight ("messenger") sent down the tow rope trips the mechanism, causing the jaws to spring together.

of lake surface could come from a rapidly reproducing population with a mass of no more than 5% of this at any single sampling time during the year.

The Phytoplankton

Plankters have been the subject of many studies on adaptations for flotation. Oil droplets, gas bubbles, gelatinous envelopes, and water-filled and saccoid bodies are adaptations for reducing weight or specific gravity. Horns, spines, setae, and elongate sticklike bodies are some of the structures that increase total surface area and resistance to sinking. The weight-reducing factors and the increases in specific surface area are thought to function in response to viscosity of the medium, which is usually dependent on temperature. On this basis, flotation devices should be needed most in warm and least in cold water. There have been many studies on the effectiveness of these devices; for example, Conway and Trainor (1972) showed that *Scenedesmus* with bristles and spines has greater buoyancy than spineless types of this green alga. But observations of natural habitats at low and high temperatures often reveal some inadequacy in the scheme. Sometimes the structures needed for buoyancy are absent in habitats in which we believe they would be necessary.

Some devices that once were believed to function primarily in flotation may serve in other capacities. Certainly, rigid projections, large size, gelatinous coating, and colonial organization protect phytoplankton against grazing by zooplankters, especially the filter feeders. There are instances, however, where grazing enhances the growth of an algal species. The gelatinous sheath that surrounds the cells of the planktonic chlorophyte *Sphaerocystis* is partially disrupted when ingested by the herbivorous cladoceran *Daphnia*. Despite this, many cells pass unharmed through the crustacean's gut and subsequently show heightened growth rates. They have taken up nutrients during their intestinal journey, and this is followed by increased rates of photosynthesis and cell division (Porter, 1976).

One theoretical consideration of cell shape in phytoplankters had been largely overlooked. It may be difficult for us to comprehend the watery world of the plankter, since we exist at an interface where there is an abrupt density difference of 27,000 times (earth and atmosphere). Hutchinson (1970a) tells us that cellular properties interacting with water turbulence to determine sinking rates may be as important as the familiar skeletal and muscular structures interacting with gravity. The structure of an organism depends on where it lives. Adaptations for sinking have been almost completely ignored. Munk and Riley (1952) theorized that a phytoplankton cell will absorb nutrients by diffusion and forced convection, the current relative to the cell as it sinks. A plankton cell such as a diatom absorbs nutrients as it sinks, which is the only way it can move from the impoverished envelope of water immediately surrounding it. We can appreciate the security enjoyed by plants rooted in lotic environments constantly washed by water containing fresh nutrients, in contrast to the nourishment of phytoplankters.

Size, shape, and density determine the rate of sinking for algal cells. As they sink, the cells divide, some of the lighter daughter cells standing a good chance of being swept to upper layers again by turbulence. Munk and Riley's work shows that phytoplankton adaptations include response to three problems: flotation, absorption, and herbivory. In large diatoms the first two are the most important; in smaller forms the grazers present the greatest threat. It becomes evident from Munk and Riley's computations and subsequent conclusions that phytoplankters are not just a group of cells floating in a willy-nilly manner. Bolstering this concept is the experimental work of Anderson and Sweeney (1977), who found that a species of marine diatom regulated its buoyancy by selective ion regulation. Short-term physiologic changes controlled the cell density and hence the settling rate. Probably ionic regulation is less important as a flotation mechanism in the phytoplankters of dilute freshwater habitats. Some cyanobacteria in fresh waters, however, exhibit diel vertical migrations that are based on turgidity and flaccidity of intracellular gas vacuoles. Klemer and associates (1982) showed this phenomenon in *Oscillatoria* relates to nutrient shortages. Relative gas vacuolation increases with inorganic carbon limitation, and the cells float to the surface where atmospheric diffusion alleviates the carbon shortage. Similarly, nitrogen limitation decreases vacuole turgidity, and the cyanobacteria sink to form subsurface populations.

Although the lacustrine phytoplankton is taxonomically diverse, it can be divided into two main groups on the basis of size. The **net plankton** are retained by a no. 25 plankton net with openings (mesh size) approximating 50 μm; smaller algae, protozoans, and bacteria not captured by the net are subdivided further according to size. At this time much of the net plankton would be classified as **microplankton**, from 20 to 200 μm. The plankters that slip through the openings of a fine silk net make up the **nanoplankton**, 2–20 μm; the **picoplankton**, 0.2–2 μm; and the **femtoplankton**, 0.02–0.2 μm. Since the invention of membrane filters, water that has been strained through the plankton net can be subsequently filtered by using membrane or fiberglass filters that stop all the tiny cells.

Before the size-related nomenclature just discussed was put forth, a simpler system prevailed. The organisms other than the net plankters were grouped as the nanoplankton, a word derived from the Greek word *nanos,* meaning dwarf. For the sake of simplicity, in the following pages nanoplankton-nanoplankters will refer to all organisms not retained by the net composed of no. 25 silk.

It is obvious now that the cyanobacteria are not the only bacteria in the plankton community; there is another bacterial plankton segment of non-photosynthetic forms. More than 300 million per milliliter are present in some lake waters, where they convert *dissolved organic matter* (**DOM**) to *particulate organic matter* (**POM**). These bacteria play important roles in a nutrient cycle termed the *microbial loop,* the subject of the following section.

The minute chlorophyllous forms that make up the nanoplankton exhibit high metabolic rates. As a result, more than 75% of the mean annual production in some lakes is owed to these tiny organisms. The net phytoplankton composed of the larger cyanobacteria, diatoms, and green algae are photosynthetically inefficient by comparison. In a study of natural phytoplankton, Ostrofsky and Peairs (1981) found that the most efficient photosynthesizers were the nanoplankters with one exception—*Dinobryon*. The large, arboreal branching of the *Dinobryon* colony (Fig. 2-3, *D*) makes it a member of the net plankton. The individual cells, however, are tiny, with a favorable (large) surface/volume ratio. They would slip through the plankton net's mesh.

Within both the net phytoplankton and the nanoplankton, individual species may be doing far more than their share of the autotrophic work. Stull (1971) found that the uptake of ^{14}C by phytoplankton was often out of proportion to numerical dominance in Castle Lake, California. Using autoradiography, she showed that species that took up most of the carbon were sometimes greatly outnumbered by inactive species.

Using the same radiographic technique six years later, Paerl and Mackenzie (1977) demonstrated how important time of day is in studies of aquatic organisms. During the early daylight hours the nanoplankters in a New Zealand lake showed high photosynthetic activity as compared with the net plankters. The net plankters contributed increasingly more to carbon fixation during the afternoon and early evening, showing the errors that would have resulted if conclusions had been drawn on data from one time of day only.

According to Findenegg (1971), the differences in photosynthetic activity of phytoplankton in 30 alpine lakes were due largely to the varied composition of the algal communities from lake to lake. Using an *activity coefficient,* the ratio of carbon assimilated in photosynthesis to carbon content in the algae, he found that chlorophyceans were above the mean and that the diatom populations were somewhat below average. The activity coefficient of the small flagellate *Cryptomonas* (Fig. 2-3, *B*) was 10 times higher than that of *Oscillatoria rubescens,* the cyanobacterium that made up the least active community.

On the basis of existing reports in the literature, Williams (1969) generalized that the phytoplankton of oligotrophic lakes includes such members as the desmid *Staurastrum* (Fig. 2-2, *D*), the chrysophyte *Dinobryon* (Fig. 2-3, *D*), noted for its intolerance of anything but low phosphate concentrations, and the diatoms *Tabellaria* and *Cyclotella* (Fig. 2-3, *I*). The eutrophic lake, however, has a different group of diatoms and a mixed crop of cyanobacteria that may eventually replace them. The filamentous diatom *Melosira* (Fig. 2-3, *F*) and *Stephanodiscus*, a form similar to *Cyclotella,* appear first. Then with agricultural runoff or increased erosion due to stripping forest cover, the beautiful *Asterionella* (Fig. 2-3, *C*) appears, perhaps in response to a need for soil substances. Later another diatom, *Fragilaria* (Fig. 2-3, *H*), shows up, especially if sewage enters the lake. Various paleolimnologic studies have shown that this may be a good generalization about the sequence of organisms.

The fossil diatom assemblages found in cores of sediment from a group of Minnesota and South Dakota lakes show a sequence somewhat similar to that outlined earlier (Bradbury, 1975). Land clearing and agricultural activity in the region during the late nineteenth century were marked by an increase in ragweed pollen (*Ambrosia*) in the cores; diatom stratigraphy showed changes that could be correlated with these disturbances in the drainage basins. It is important, however, to note from Bradbury's report and from a paper dealing with phytoplankton of the Laurentian Great Lakes (Stoermer, 1978) that generic designations alone do not always provide precise information. Stoermer showed that there are almost as many species of *Cyclotella* in the most disturbed and polluted parts of the Great Lakes as there are in the least modified and most oligotrophic regions. The species, not the genus, serves to indicate water quality.

The Microbial Loop

With the development of techniques in environmental microbiology, enormous masses of nonphotosynthetic bacteria have been found in the euphotic zones of marine and inland waters. Although there were pioneer workers decades ago, population estimates were hampered by lack of the technology enjoyed today. It was known that alkaline and sewage-polluted waters contained the greatest number of bacteria, approaching 4×10^8 cells per ml. Oligotrophic waters have smaller populations, and we know now that there are obligate oligotrophs that are absent from eutrophic waters where other bacterial species thrive. Formerly it was believed that most aquatic bacteria were benthic. Now it is known that bacteria make up more than 50% of combined pico- and nanoplankton. They play important roles in a cycle of energy not detected in earlier studies. The cycle has been named the **microbial loop**, and the acronym DOC (for dissolved organic carbon) has gained popularity. The history of the microbial loop concept goes back at least as far as a 1974 paper written by oceanographer L. R. Pomeroy, although Azam and others (1983) coined the name.

The microbial loop, or web, starts with photosynthetic organisms releasing in solution a significant portion of the organic compounds they produce. The colorless heterotrophic bacteria, members of the picoplankton, flourish at even low concentrations of these nutrients, increasing their numbers at a tremendous rate. In turn these prokaryotes are grazed efficiently by a host of small ciliates and flagellates best described, on the basis of size, as nanoplankters. Most are **phagotrophic**, and they are extremely successful in clearing the bacteria from a large volume of water. They, in turn, are victims of larger ciliates and microzooplankters such as rotifers. Surprisingly, the large crustacean net plankters, *Daphnia* and *Ceriodaphnia,* also prey upon organisms sized from one to 50 μm, including larger members of the picoplankton as well as nanoplankters and some of the microplankton. Some copepods feed upon rotifers, and thus some "loop" energy can pass to another level of the euphotic-zone community and into the conventional food web.

The complex microbial loop is completed when the ingested organisms are mineralized and excreted as N- and P-containing labile carbon compounds that serve as plant nutrients. Thus, there is a retention and recycling of important nutrients and energy in the upper waters rather than a great loss via decomposition to the deep water and sediments. In productive lakes, especially, the planktonic bacterial biomass represents a huge sink of nutrients that had been unsuspected in earlier years. Two rewarding and clarifying discussions of the microbial loop are those authored by Stockner and Porter (1988) and Berninger and associates (1991). The 1991 paper discusses quantitatively the elements of the loop in 108 bodies of water throughout the world. The idea has remained influential for decades, helping provide a framework for understanding carbon cycling especially (Fenchel, 2008)

The Zooplankton

The zooplankton of inland waters derives its members mainly from the Protozoa, the Rotatoria, the Cladocerans, and the Copepoda. In addition, there are occasional minor elements contributed by ostracod crustaceans, water mites (arachnids), larval molluscs such as *Dreissena,* mysid crustaceans (Fig. 2-9, *B*), and the larva of the insect *Chaoborus* (Fig. 2-11, *C*).

We now have the ability to track zooplankters to better understand their fundamental ecology. Genin and colleagues (2005) observed the movements of more than 375,000 individuals in the open ocean and discovered that these organisms swim against strong currents successfully and in unison. Many phytoplankton cells pass through zooplankters' digestive tracts unharmed and, in fact, have benefited by the nutrients gained en route (Porter, 1976); the grazing zooplankters release ammonia and phosphate, and these nutrients are simultaneously taken up by phytoplankton cells, a recycling that may amount to 50% or more of the herbivores' total phosphorus content each day (Lehman, 1980). Still, some vertebrate plankton predators take the larger, pigmented zooplankton species selectively, leaving a population of smaller and more transparent species (Hrbácek, 1962; Brooks and Dodson, 1965; Nilsson and Pejler, 1973). Invertebrate plankton predators select smaller prey items, thus giving the advantage to larger zooplankters or those protected by spines and other outgrowths (O'Brien and Schmidt, 1979; various papers by Dodson and Kerfoot reviewed in Zaret, 1980). Despite these advances, we are still learning new insights into zooplankton. A recent paper, for example, found that the bacterium causing cholera (*Vibrio cholerae*) can "hijack" *Daphnia*, controlling its behavior (Nihongi et al., 2011)!

The Nekton Community

Animals substantially larger than plankton organisms and capable of efficient, directed locomotion make up the nekton community. Most of them are fishes and a few of the larger invertebrates. By contrast, in the sea many phyla make up the nekton. In a sense, they tie the other communities together, contributing to the detritus pool and feeding on many sources: the large plants and the algae of the littoral zone; the algae of the plankton; the animal representatives of the littoral, planktonic, and benthic communities, and other members of their own nekton community.

Size is the fundamental difference between the nekton and zooplankton, and it gives the former the power to migrate and move independently of currents against which smaller swimmers would be powerless. This allows nektonic species to move to different areas to spawn or forage, their impact being felt in several areas of an aquatic system.

In lakes and rivers the nekton is impoverished compared with the sea: squids and whales are lacking, for example. Fishes and the larger crustaceans are the nekton community of limnology unless we count loons, grebes, and cormorants temporarily foraging beneath the surface, and the seals of Lake Baikal.

The nekton is entirely animal and therefore made up of relatively large consumers, some of which occupy the top trophic level in aquatic ecosystems. The larger, slow-growing animals of the nekton, quite in contrast to the zooplankton, may have a mean standing stock (mean biomass) hardly surpassed by its annual production. Smaller nektonic forms, however, may have annual productions three or four times the mean biomass.

Invertebrate Nekton Elements

The strong swimmer, *Mysis relicta*, is classified as nektonic by some authors; others might not agree and there is, therefore, a hazy distinction at times between plankton and nekton members. It is about as good an example of a freshwater invertebrate nekton organism as can be presented, however. *Mysis* is not the top predator in nekton communities, being a preferred food item for fishes. For this reason it has been introduced to lakes far from its natural range. A series of papers have been published on its impact on zooplankton populations in Lake Tahoe (see Goldman, 1981), and a description of the cascading effects in Flathead Lake is found in Chapter 4.

Most freshwater amphipods are stream dwellers or benthic forms in the shallow weed beds of lakes. A "species flock" of endemic species in Lake Baikal is equal to about one-third of all the freshwater forms, yet only one of the Baikal endemic amphipods is truly nektonic; this is *Macrohectopus branickii*, a large transparent form whose elongated body has been described as "mysidiform." *Macrohectopus* feeds on both phytoplankton and zooplankton. Its transparent body probably is a response to fish predation; it is always limnetic and serves as an important forage item for some fishes.

Two other freshwater amphipods, *Gammarus lacustris* in North America and *G. pulex* in Tibet, keep to the margins of lakes when fish are present. In the absence of fish they become pelagic and serve as top consumers, feeding on zooplankters (Uéno, 1934).

In Montezuma Well, Arizona, where no fish occur, an endemic amphipod *Hyalella montezuma* is nektonic (Cole and Watkins, 1977). It spends the daylight hours in the open water because it falls prey to insects visually hunting in the peripheral weed beds. After dark a predaceous leech rises to the upper layers of water to hunt *Hyalella,* using mechanoreception to trace the swimming individuals. The amphipod escapes by moving laterally into the pondweeds during the night hours. These movements are minor when compared with those of Lake Baikal's *Macrohectopus*, but typically the common *Hyalella azteca* (Fig. 2-9, *F*) remains in littoral vegetation. An undetermined species of *Hyalella* does not venture into the open water with *H. montezuma,* but remains in the vegetation close to the shoreline of Montezuma Well. *H. montezuma* differs from other members of the genus by having mouthparts adapted for filter feeding. Laboratory experiments with radioactive algae revealed its efficiency in this method of gathering food (Blinn and Johnson, 1982).

Except for some benthic species that occasionally swim up into the water column at night (*Pontoporeia* and *Anisogammarus*), the nektonic way of life is very rare in freshwater amphipods. The phenomenon is common in the sea, where many species are pelagic.

A situation in the arachnids analogous to the *Hyalella montezuma* case seems to be the occurrence of an endemic water mite (*Piona limnetica*) in the plankton of Lake Madden, Panama. The predaceous mite is truly limnetic and is found only in the waters of that artificially created lake (Gliwicz and Biesiadka, 1975). Pelagic water mites elsewhere may be more common than generally known. For example, Riessen (1980) reported on the daily vertical migration of pelagic water mites in a Quebec lake. Such animals would lie very close to the boundary between large plankton species and the nekton forms.

Vertebrate Nekton Elements

The vertebrate components of the nekton species are found in all parts of the lake or stream. The planktivorous species such as the alewife, *Alosa pseudoharengus*, and the paddlefish, *Polyodon* (Fig. 2-12, *A*), seem to fit the nekton definition better when compared with forms such as the grass carp, *Ctenopharyngodon* (Fig. 2-12, *B*) and the bullhead, *Ictalurus* (Fig. 2-12, *F*). The former is a littoral form that eats pondweeds most of its life; the latter is associated with the zoobenthos and bottom detritus. Fishes belong to every aquatic community and may be more omnivorous than generally is supposed. For example, the coregonid fishes (whitefishes) are classed as planktivores because of their closely spaced gill rakers and pelagic haunts. In eutrophic Lake Itasca, Minnesota, *Coregonus artedi* feeds to a marked extent on the benthic larvae of *Chironomus plumosus* (Fig. 2-11, *A*).

In the past, researchers of zooplankton glossed over the effects of predation on populations. Emphasis was placed on competition between closely related species—niche partitioning, character displacement, and competitive exclusion. A question typically asked was "Why are calanoid copepods large in temporary ponds?" rather than "Why are they smaller in permanent ponds?" With the appearance of the papers by Hrbácek (1962) and Brooks and Dodson (1965) the element of predation in fish-containing permanent waters began to be considered. Those authors showed how vertebrate predators modify the composition of zooplankton populations. Large pigmented species are replaced by smaller transparent forms when planktivorous fish are introduced to a body of water. Light and all factors contributing to visibility—hemoglobin and carotenoid content, body size, and conspicuous eye spots—are factors in fish predation. One explanation for the phenomenon of cyclomorphosis in microcrustaceans may be explained by fish predation. In 1972 Zaret discussed the appearance of two morphs of the cladoceran *Ceriodaphnia cornuta*. One is small-eyed and reproductively inferior; it survives in the presence of planktivorous fish. The other occurs in parts of the lake where fish are scarce and it is not endangered by the visual cue offered by its large black eye. The two morphs of *Ceriodaphnia* co-occur temporally but are separated spatially in the lake. Seasonal predation pressure from fishes in other lakes could account for some of the so-called cyclomorphosis seen in zooplankton.

Fish are not the only vertebrate planktivores that are size selective and, as a result, modify zooplankton communities. Salamanders also are visual hunters, selecting the larger and more conspicuous forms. More remarkable is the case of the sandpiper-like bird, the red phalarope (*Phalaropus fulicarius*). It takes freshwater zooplankters from the upper layers of shallow arctic ponds, preferring the larger *Daphnia,* chydorid cladocerans, and larval fairy shrimps (Dodson and Egger, 1980).

You have just learned some examples of *cascading effects,* the consequences of predation, not only on prey populations but indirectly on other communities as well. This subject is discussed further in the next chapter.

The Neuston

The **neuston** is the community of flora and fauna associated with surface tension. The difference in density between the water and the air above is on the order of 1,000 times. This great density difference produces a film with which small organisms are associated.

Another thin organic film on the upper surface of the water is said to contain lipoproteins. This layer and the material collecting in it support the animals of the *epineuston.* Two abundant taxa in this assemblage are small arachnids, the Acari, and the insect order Collembola. The collembolans, or springtails, have a markedly hydrophobic cuticle fitting them for such a habitat. The water striders, consisting of two hemipteran families, the Veliidae and Gerridae, are among the most conspicuous of the epineuston. These are all essentially terrestrial organisms, or at least aerial rather than aquatic.

The conspicuous green covering on some ponds is made up of the duckweeds, members of the Lemnaceae. These tiny angiosperms float with rootlets hanging in the water below the surface.

The dusty or oily appearance of some pond surfaces is due to an epineustic microflora. Floating chrysophyceans, euglenophytes, and chlorophyceans contribute to this appearance.

The *hyponeuston* is the community living at and under the surface film. A host of algae and protozoans can be found here. They provide food for other organisms, such as mosquitoes, which in turn are preyed on by cyprinodont top-minnows such as *Gambusia*.

Two crustaceans are of special interest here. One is the rare ostracod *Notodromas*; the other is *Scapholeberis*, a cladoceran (Fig. 2-9, *A*). Both are remarkably adapted for moving smoothly upside down along the underside of the film.

The Benthic and Littoral Communities

The Benthic Community

The benthos is composed of bottom-dwelling organisms and has the obvious major categorical division between phytobenthos and zoobenthos. Further categories are based on the lake region—littoral, sublittoral, and profundal—and on the size of the organisms. The study of benthic animals has generally involved the use of sieves in separating organisms from the sediments. Sieves have various openings, and many sizes have been used in benthic investigations; in many instances the size has not been revealed. Somewhat arbitrarily we can say that screens with mesh openings from 1.0 to 0.425 mm (diameter) are used commonly to retain the so-called **macrobenthic** species. **Meiobenthic** forms pass through these coarse sieves but are stopped by a mesh size of 0.050 to 0.045 mm. The **microbenthic biota** is made up largely of protozoans and any rotifers, nematode worms, or gastrotrichs that are small enough to pass through the 0.050 to 0.045 mm spaces. Some workers have gone directly to the size or mass of the organisms to categorize the benthic groups; thus the meiofauna is made up of metazoans with a wet mass less than 10^{-4} g. The ciliates range from 10^{-10} to 10^{-6} g.

The Phytobenthos—Littoral Plants

The **phytobenthos** includes the aquatic macrophytes and the bottom-dwelling algae. The zonation seen in aquatic plants of the littoral zone is caused by more than one factor. Wave action and the physical and chemical nature of the substrate are obviously important. Spence and Chrystal (1970a, b) found a nice correlation between depth distribution and the inherent photosynthetic ability of some species of *Potamogeton*. Deep-water species, such as *P. praelongus* and *P. obtusifolius,* are more shade tolerant and carry on photosynthesis at light intensities too dim for the maximum efficiency of a shallow-water form, such as *P. polygonifolius*. *P. obtusifolius* owes much of its success to a low respiratory rate, coupled with photosynthetic efficiency in dim light.

Pressure limits the downslope distribution of the tracheophytes. It is significant, therefore, that the greatest depth achieved by a vascular plant is in Lake Titicaca, 3,809 m above sea level. There, where a species of *Potamogeton* grows to a depth of slightly more than 11 m, pressure from the rarified atmosphere is 275 mm of mercury less than pressure at sea level. This is equivalent to 3.7 m of water pressure (Hutchinson, 1975). Charophytes, mosses, and some benthic algae achieve the greatest depths in lakes.

The littoral macrophytes support epiphytic algae and attached animals such as bryozoans. Snails and other organisms crawl over the weeds, consuming the epiphytic plants. Berg (1950) and McGaha (1952) showed that the pondweeds are extremely important to immature insects of various orders. Leaf miners and channelers devour the plant tissue, while other insects are adapted for living in rolled-up leaves, where they trap planktonic forms. These authors showed that the weeds are far more than clinging, hiding, and resting places for animals, an idea that had been put forth more than 30 years earlier. Nevertheless, the relative importance of the algal community growing on the macrophytes must not be underemphasized. The annual productivity of these tiny plants far surpasses their average biomass, and probably they are more important as a direct source of food for grazing invertebrates than are the large plants to which they are attached. Moreover, where the epiphytic algae are concerned, the macrophytes may serve as little more than supporting structures. Cattaneo and Kalff (1979) compared the primary production of algae growing on artificial and natural aquatic plants (*Potamogeton richardsonii*, Fig. 2-11) and found that despite some evidence for limited nutrient transfer from the living plants, the substrate is nearly neutral with respect to growth and primary production of the attached algae.

Some years ago it was found that the macrophytes not only take up inorganic nutrients from the water and sediments but also release quantities of dissolved organic compounds that play a part in the lake's economy (Wetzel, 1969; Allen, 1971).

The spongy tissues of many aquatic plants store metabolic gases (Hartman and Brown, 1967), letting them out at times that only serve to confuse the student of photosynthesis. Sometimes oxygen released to surrounding lake water at night was produced hours earlier in the daylight; the diel oxygen curve in the adjacent lake water is anomalous as a result of this delay.

Most submersed aquatic plants are extremely susceptible to desiccation soon after being taken from the water. On the other hand, the floating leaves of water lilies are waxy coated and have unusual resistance to desiccation.

The highest known rates of primary production in natural communities come from beds of emergent reeds and rushes. Such plants stand rooted in fertile soils with available water resources unknown to terrestrial plants and are exposed to solar radiation far in excess of that energy available to the submersed phototrophs deeper in the lake.

The Zoobenthos

The littoral **zoobenthos** is extremely varied when compared with that of deeper regions, obviously a reflection of the abundant microhabitats. Protozoans, sponges, coelenterates, rotifers, nematodes, bryozoans, decapods, ostracods, cladocerans, copepods, pelecypods, gastropods, insects, and leeches are abundant in the shallows.

A unique feature of the littoral benthos is its nearness to the plankton. Littoral bottom samples often include planktonic forms, which serve as a source of nourishment for littoral animals. However, these plankton organisms are available to the benthos of deeper regions only after they have died, settled, and undergone some mineralization.

The meio-microbenthic fauna is composed of small animals that are permanent bottom dwellers, the very young stages of macrobenthic species, or benthic stages of limnetic forms. The **meio-microbenthos** is an important segment of lake production despite the diminutive nature of its members. In certain Polish lakes this fauna contributed up to 50% of the total benthic production (Kajak and Ryback, 1966). Evans and Stewart (1977) found that the micro-crustaceans in the shallows of Lake Michigan far surpassed larger bottom-dwelling animals numerically; Babitski (1980) reported that the microbenthic forms were much more abundant than the macrobenthic species and that they made up as much as 50% of the total biomass in some Russian lakes. Their annual productivity amounted to quadrupling the average standing crop. Anderson and De Henau (1980) studied the meiobenthos (passing through a 0.425 mesh, but retained on a 0.045 mm mesh) in nine Canadian lakes. They found that one third of the total benthic biomass was meiobenthic, and they estimated macrobenthic productivity was equaled or surpassed by meiobenthic productivity. Fenchel (1978), while discussing carbon flow in meio-microbenthic populations, reminds us of the spectacular increase in metabolic rate as body mass decreases, a phenomenon accounting for the relative importance of the microscopic animals in benthic productivity.

Early investigation of the meio-microbenthos in eutrophic lakes of the United States were published by G. M. Moore (1939) and Cole (1955). Their research involved some interesting burrowing and creeping invertebrates. One group of these littoral microcrustaceans, the chydorid cladocerans, is especially noteworthy.

Most arthropods of an aquatic habitat leave scanty remains after death, but certain cladocerans are exceptions. The head shields and carapaces of *Bosmina* are especially resistant, while *Daphnia*, *Diaphanosoma*, and the copepods that are part of the limnoplankton leave little in the way of chitinous remains. The most remarkable example of resistance comes from a cladoceran family, the Chydoridae. They are typical of the littoral region, but following death, parts of their bodies are swept lakeward. The result is that a sample of surficial sediment taken offshore, perhaps near the center of a small lake, contains the head shields, carapaces, and perhaps postabdomens of all the chydorid species living in the lake.

Today the chydorid crustaceans are the most-studied family of cladocerans. We are indebted to David G. Frey for giving impetus to this branch of research by pointing out methods by which sedimentary chydorid relics can be identified as to specific rank (Frey, 1959). Since the late 1950s he and his students have shed light on some taxonomic and ecologic riddles in the group. Fryer (1968), supplementing the work performed by the Frey school, related the chydorid's anatomy to its ecologic role. It is now possible to gain more paleolimnologic insight from core samples than ever before, although knowledge of the biology of the chydorids has only recently started to be gathered, and in North America their taxonomy needs work (Frey, 1982).

When we move to the sublittoral region, we find that species diversity drops sharply. Some unionid clams, ostracod crustaceans, copepods, and cladocerans from the littoral region are there, but very few are typically sublittoral dwellers.

In the eutrophic profundal zone the benthos becomes more impoverished. It is distinguished by a few hardy forms that can tolerate low levels of oxygen. A typical macroscopic assemblage would include bright red midge larvae of the genus *Chironomus*; a few species of oligochaete worms, *Limnodrilus* being a common genus; tiny pisidiid clams belonging to the genus *Pisidium*; and the phantom larva of the dipteran *Chaoborus*. There are a few microscopic forms present, although they are often overlooked. If conditions are not too severe, some ciliates (*Loxodes, Coleps, Rhagadostorna,* and *Metopus*) and other protozoans (the rhizopod *Pelomyxa* and flagellates such as *Bodo, Chilomonas, Monas,* and *Phacotus*) may be found in or near the sediments. The rotifer *Rotaria rotatoria* and several nematodes are especially hardy in low-oxygen conditions.

The profundal zone is most interesting from an overall limnologic point of view. Beneath hypolimnetic waters the dark environment can impose stress to which organisms in shallower depths are not subjected. The profundal sediments in a eutrophic lake lie below water that is quite different from the water overlying sublittoral and littoral deposits. Generally, the water is colder, lower in dissolved oxygen and pH, and higher in carbon dioxide, methane, bicarbonate, organic compounds, phosphorus, and nitrogen compounds, including ammonia. The environment is apt to be a chemically reducing one, the redox potential being low; this means there is an excess of reducing substances over oxidizing ones. The reducing substances in hypolimnetic waters and in the organic ooze compete with the animals for oxygen.

In some respects the eutrophic profundal benthos resembles the fauna of grossly polluted waters or of hypersaline desert lakes. Comparatively few species can survive such conditions, and there is little diversity. Low diversities mean "monotony"—a great number of individuals, but only one or two species represented. This is what would characterize a polluted environment. Below the sewage outfall in a stream, diversity is much lower than farther downstream where unfavorable conditions have ameliorated somewhat and more species have colonized.

The oligotrophic profundal benthos is more diverse than the bottom fauna of eutrophic lakes. A varied group of animals live in the firm, oxygenated sediment-water interface at the bottom of oligotrophic lakes. Most of these animals are intolerant of low oxygen tensions, such as found in eutrophic lakes. Midge larvae belonging to the genera *Orthocladius*, *Tanytarsus*, *Calopsectra*, and *Sergentia* are forerunners of a monotonous *Chironomus plumosus* type of fauna, which prevails if eutropy occurs and oxygen is at a premium in the summer hypolimnion.

Tubificid worms of the Oligochaeta are represented by several species in aerated, unpolluted sediments. According to Brinkhurst (1966), the occurrence of *Limnodrilus hoffmeisteri* nearly alone and in great abundance would indicate pollution. A prime example of the loss of diversity with pollution and oxygen reduction comes from the western basin of Lake Erie. Formerly, the sensitive mayfly *Hexagenia*, along with many other high-oxygen-requiring forms, occurred in great numbers. In a hopeful example of lake regeneration after pollution control, the insect reappeared in the mid-1990s after a 40-year absence (Bridgeman et al., 2006).

The cold oligotrophic lakes to the north have a special fauna consisting of at least four widely separated taxa. There is a little assemblage of bottom forms that have a similar Holarctic geographic distribution and may have migrated via proglacial lakes, marginal to Pleistocene ice. These species are found today in northern oligotrophic lakes. One of these, *Mysis relicta*, has already been mentioned. Others include the pontoporeiid amphipods (Fig. 2-9, *C*), which can be found in brackish waters of the Baltic Sea and many Eurasian freshwater lakes. Members of this family have been collected from more than 50 oligotrophic lakes in glaciated North America. Bousfield (1989) discussed their distributions and the theories concerning these so-called glacial relicts. In the same paper he divided the genus *Pontoporeia* into three genera, including two new ones.

A cold-stenothermal clam, *Pisidium conventus* (Fig. 2-9, *G*) has a geographic distribution much like the pontoporeiid amphipods, although probably restricted to freshwater lakes. Also, a comparison of the rich North American *Pisidium* fauna with that of Europe suggests that it moved in the opposite direction from the amphipods. The occurrence of *Mysis*, *Pisidium conventus*, and pontoporeiids in oligotrophic Green Lake, the deepest of Wisconsin's inland lakes, is typical today.

Some benthic species typical of the eutrophic profundal zone occur also in polluted habitats, where oxygen is also in short supply. If anaerobic spells are lengthy, however, even the hardy species show signs of stress. Chironomid larvae reduce the time spent in feeding and increase the respiratory undulations, finally ceasing even that and becoming anoxybiont when the oxygen falls to 0.5 mg/liter. *Pisidium* closes its valves tightly and becomes dormant during prolonged anaerobic periods. *Chaoborus* is not so hard pressed, since it is able to move up to the aerated epilimnion each night. In meromictic lakes the inability of the average benthic animal to survive anaerobiosis and the reduced substances associated with the lack of oxygen is evident. In Fayetteville Green Lake, New York, no macroscopic animals were collected below 20 m, although nearly 300 dredge samples were raised from that isobath to the maximum depth of 59 m (Eggleton, 1956). Twenty meters was the level where O_2 disappeared and H_2S became evident.

Animals in the profundal benthos differ in their feeding behavior, but much of their food comes from the phytoplankton; spurts of growth often follow the arrival of planktonic algae at the bottom. Jónasson and Kristiansen (1967) showed how feeding behavior varies with the season and with the taxonomy of phytoplankton in Denmark's eutrophic Lake Esrom. Diatoms are distributed to the bottom in spring and autumn when the lake circulates completely. Most green algae are rapidly circulated to the deeps, but blue-greens sink slowly, reaching the bottom after having undergone some decay. The ooze-dwelling animals of the profundal zone rely heavily on the phytoplankton and are essentially primary consumers. Some, especially the tubificid worms, feed on bacteria. In Lake Esrom there are three primary consumers: the tubificid worm *Ilyodrilus*, *Chironomus anthracinus*, and *Pisidium*. Two other forms are carnivores: *Procladius* (a chironomid) and the planktonic larvae of *Chaoborus*.

Procladius is not a permanent member of the Esrom profundal benthos, because it is absent during the summer anaerobic period. Even *Chaoborus* is not especially tolerant, but it can escape anoxia by nocturnal vertical migration. The diel migrations, however, vary seasonally and with respect to the four larval instars within the lake species, *C. flavicans* (Goldspink and Scott, 1971).

Hemoglobin occurs in some benthic animals. The bright red species of *Chironomus* are especially characteristic; even the tiny burrowing cladoceran *Ilyocryptus* of the littoral and sub-littoral microbenthos is red. Hemoglobin permits the animals to quickly gain trace amounts of oxygen from the environment; this helps, but only for a matter of minutes. The resistant species of *Chironomus*, *C. plumosus*, and *C. anthracinus* may be able to excrete some end products of anaerobic respiration into the water (Walshe, 1947, 1950).

A great many studies of bottom fauna have quantified profundal benthos on the basis of standing crop in weights, numbers, or volumes. It is not uncommon to find macrobenthic animals present in numbers of $10,000/m^2$, and communities five times that density are not rare. *Chaoborus* has been collected in dredge samples in such numbers that populations over $90,000/m^2$ can be inferred.

A mixed sample of bottom fauna is probably 85% to 90% water, but there are taxonomic differences. The average water content of *Chironomus plumosus* from Lake Itasca, Minnesota, is 88.2% of the fresh live weight, ranging from 85.4% to 91.8%. A standing stock of profundal benthic animals averaging 4 g/m^2 (dry mass) is considered high. Most lakes would perhaps have only one tenth this value. In North America, Last Mountain Lake in Saskatchewan may hold the record: Rawson and Moore (1944) reported 8.6 g/m^2 for the mean dry weight of their samples.

Estimation of standing crop on the basis of a few samples can be biased, for it depends on when and where the collections were made. Commonly there is an upward migration of midge larvae to form a summer concentration zone at the boundary of the sublittoral and profundal zone (Eggleton, 1931; Deevey, 1941). Also, it is not unusual for predation and general mortality to take a toll of 1% each day, so that June collections gather more animals than August samples.

Production differs from numbers or masses of the biota. It is expressed as a rate, annual productivity per unit area, for example. Brooks and Deevey (1963) presented some estimated annual production rates ranging from 0.6 g/m^2 for Lake Nipigon, Ontario, to 14.8 g/m^2 for Beloye, Russian Federation. Two of the highest rates from North America were 7.2 and 5.5 g/m^2 per year for Linsley Pond and Lake Mendota, respectively. Since then some figures derived from a Lake Manitoba study imply productivity of about 7.5 g/m^2 (Tudorancea et al., 1979). Annual production amounting to a dry mass of 10 g/m^2 is rare, but there are some unusual exceptions. Some single-species productivities reported by Waters (1977) include 40.5 g/m^2 per year for a chironomid (*Glyptotendipes*) in a Scottish lake and his own finding of 27.1 g/m^2 for *Gammarus pseudolimnaeus*, a benthic amphipod in a Minnesota stream (Waters and Hokenstrom, 1980).

The Periphyton (Aufwuchs) and Biofilms

The microbenthic flora blends with another community, the **periphyton**, a word used originally to describe the organisms attached to artificially submerged objects. It has been broadened to include the entire sessile community of organisms. If used strictly, it would refer to organisms on plant stems, leaves, and perhaps submersed sticks. The German term *aufwuchs* enjoyed popularity for years following the Frey and Fry translation of Ruttner's work (1953), but "periphyton" seems to have come back and is widely used.

The periphyton includes both flora and fauna. Most people have experienced the treacherous slipperiness of stones in streams. A film of diatoms attached by gelatinous stalks accounts for most of this. Living with these algae are other encrusting types, bacteria and fungi and also such animals as the stalked peritrichous ciliates *Vorticella* and the branched *Carchesium*. *Vorticella* often abounds in the open-water plankton of lakes while in its telotroch stage, a motile wandering phase in the life of this protozoan. Similarly, the typical attached form of *Vorticella* is gathered in plankton collections because it rides about on strands of cyanobacteria.

Periphyton assemblages on stones (**epilithic** organisms) can hardly be distinguished from microbenthic assemblages unless the strict sessile nature of the organisms is observed. **Epipelic** algae are those on sediment surfaces and come closer to being defined as microphytobenthos. **Epiphytic** organisms are the aufwuchs of plant leaves and stems, hardly different, if at all, from the narrow definition of periphyton. **Epizoic** periphyton communities dwell on animals and are spectacularly demonstrated by the "mossy" backs of some freshwater turtles. A unique community of such epizoic periphyton as the green alga *Basicladia* and a group of suctorian ciliates make up the major part of this movable aufwuchs.

Studying the role of attached algae in lacustrine productivity, Wetzel (1964) discussed the confusing history of the terminology referring to these forms. Probably one can now speak of epizoic, epiphytic, epilithic (the lithophyton), and epipelic periphyton as four kinds of aufwuchs. Care should be used to designate periphyton algae as opposed to periphyton animals.

The early techniques for studying the periphyton were reviewed by Sládecková (1962). The production of both primary (algal) and secondary (animal) periphyton has been estimated in various ways. A common technique is based on submerging racks of glass plates (microscope slides are often used) for a period of time and measuring the organic matter that accrues. Castenholz (1960) measured the increment of organic material in Washington's Grand Coulee chain of lakes in this manner. The maximum daily rate, amounting to a dry weight of 1,043 mg/m^2, was in saline Soap Lake. Wetzel (1964) used carbon-14 methodology to determine annual primary production in Borax Lake, California. He found that 731.5 mg of carbon per square meter were fixed on an average day by the periphyton algae, dominated by the rare chlorophycean *Ctenocladus,* a saline-water form. This represented 69% of the carbon fixed by the autotrophs in the entire lake. The planktonic algae took second place (23.6%) in production rate, and the macrophytes fixed 7% of the annual carbon. Such relatively high littoral production by periphyton algae is probably found only in shallow lakes. The mean depth of astatic Borax Lake is usually less than 1 m, despite its area of 40 ha. Naiman (1976) reported some remarkable periphyton production from a shallow thermal stream in the Mojave Desert of California. An unshaded benthic mat of cyanobacteria was responsible for the mean fixation of 3.3 g of carbon per square meter each day.

In recent years we have come to understand the importance of not just the larger periphyton, but also the key contribution of the thin (thickness of dozens to a few hundred microns) layers of microbes adhering to the aquatic surfaces (Costerton et al., 1987; O'Toole et al., 2000). These are important drivers of many biogeochemical processes through primary production, release of digestive enzymes causing decomposition, and mediating interactions between the macrobiota. Although sensitive themselves to nutrients from the water (Romani et al., 2004), they also cause the release of stored nutrients, for example from dead and decaying cattail (*Typha*) litter: both the fungi (Kuehn et al., 2011) and the interaction of bacteria, fungi and algae (Francoeur et al., 2006) accelerate the decay and subsequent nutrient release, even when the plants are emergent above the water surface.

Groundwater (Interstitial) Communities

The microbes, algae, protozoans and metazoans living in the watery spaces among sand grains make up a community called **psammon** (from Greek *psammos,* meaning sand), also widely known as **interstitial biota**. The tiny organisms in the sand interstices below the lake or sea water compose the **hydropsammon**. There is horizontal and vertical zonation, and the word *biotope* is used to describe these habitats. The bio part of the name is obvious; the tope comes from the Greek word *topos,* meaning place. The interstitial populations below or adjacent to streams make up the *hyporheos* (Greek *hypo,* below; *rhein,* to flow). Much of the recent advances in interstitials has been on the rheic biota (Williams and Fulthorpe, 2003; Boulton, 2007).

A valuable paper concerning the published work on interstitial habitats is that of Boulton and associates (1992), although its emphasis is on stream and streamside communities. In the manuscript the authors discuss the hyporheos of a Sonoran Desert stream in Arizona. They list four subsurface biotypes: the epigean hyporheic, from 0 to 50 cm beneath the stream waters; the phreatic, from a depth of 50 cm to the bedrock; the parafluvial, beneath the surface lateral to a stream; and the dry-channel hyporheic topotype that is specific to temporary stream beds when the overlying water has disappeared.

Schmid-Araya (1998) reviewed the literature on Rotifers of interstitial sediments, both lentic and lotic. These animals vary markedly over the space of a few centimeters. Overall, she found remarkable diversity within this specialized niche.

A host of tiny animals in the psammon may be entirely absent from the microbenthos of the littoral region only a few meters away. This community has been neglected in North America, although a few workers have published on the psammon. The first detailed study was that of Pennak (1940), writing about the sandy beaches of Wisconsin lakes. His description of the biota to be found there is enlightening. He wrote that if a sample of 10 cm^3 containing 2 or 3 ml of water is collected about 1.5 m from the water's edge, it will contain 4 million bacteria, 8,000 protozoans, 400 rotifers, 40 copepods, and 20 tardigrades, along with a few other species. The copepods will be largely harpacticoids.

The biota of the interstitial water of beach sand often contains species previously unknown, and some remarkable discoveries have been made in this area. For example, Pennak and Zinn (1943) found a small, elongate crustacean in the marine psammon at a beach at Woods Hole, Massachusetts. These shores had been trodden for 70 years or so by countless zoologists, including students, researchers, and all-around collectors.

At first glance, the animal living in the interstitial water seemed to be a copepod. However, further study showed that it represented a new, previously unknown, primitive order of the Crustacea. It could not be assigned to any known group. Since then, other Mystacocarida, as Pennak and Zinn named the group, have been found along the Mediterranean and other coasts of Europe, Africa, and western South America.

The Detritus Community

Following primary production all ensuing transfers, transformations, and degradations of energy must be termed secondary production, the production of heterotrophs supported by the net energy synthesized by autotrophic producers. There are two main routes by which energy flows from the primary producers of ecosystems, the first of which having been emphasized far beyond the other. It involves grazing of green organisms by the herbivores; the energy of photosynthesis is passed on directly to the consumers. The second pathway involves energy flow from dead material. At any particular time there is 10 to 100 times as much dead material present in an aquatic habitat as there is living tissue.

The term **seston** has been used in aquatic ecology for decades. It is defined as particulate material suspended in the water and includes both living plankters, their corpses, and other nonliving fragments. Net seston and nanoseston are determined by particle size, the former being retained by the plankton net and the smaller nanoseston slipping through its meshes. The nonliving fragments have been called **tripton**, but now the term **detritus** has replaced tripton. As the notion of its importance has grown, the meaning of the word has been broadened; detritus includes more than particulate dead matter. It also comprises dissolved carbon-containing substances excreted by living plants and animals, and the soluble organic material leaking from decaying plant and animal tissues as they are digested by bacterial enzymes. The energy in this material is not lost to food webs, as will be discussed later.

The origins of detritus are diverse. Its autochthonous plant components include cells and tissues that were ungrazed, the undigested plant material in animal feces, and an array of dissolved substances released by littoral macrophytes and phytoplankters. Detrital components of animal origin include bits of decaying tissue or whole bodies, feces with their associated microbial populations, and the soluble organic compounds secreted and excreted by the living fauna and released from decaying parts.

Allochthonous detritus imported from the catchment area is a product of primary and secondary production elsewhere, but it is utilized by the aquatic biota. Leaf litter is a common example of particulate detritus derived from terrestrial ecosystems.

All the particulate and dissolved organic material mentioned above have their accompanying bacterial or fungal associates. The enzymatic versatility of these organisms makes the energy from the detrital pool available to the less talented animal detritivores. Bits of detritus, apparently very resistant to digestion, may pass through the guts of consumers many times. The animals are gaining energy, however, from ingesting the associated microbes that were able to break down the chemical-bond energy within the detritus and to pass it on within their cells; eventually most of the refractory detritus has been transformed and degraded to CO_2. The mycelia of fungi and bacterial bodies are part of the detrital community, but their role is dual: they are both detritivorous and an integral part of the detritus.

Detritus is broadly categorized as particulate (POM) or dissolved (DOM) organic matter. To emphasize the carbon contained in detritus, the acronyms POC and DOC are often used.

Dissolved Organic Material

Bacterial action and autolysis affect the decay of dead plants and animals, which release soluble organic compounds. Also, newly formed photosynthates, especially glycolates, are leaked directly into the water by phytoplankton, and back in 1969 Wetzel reported the excretion of DOM from the submersed macrophyte *Najas*. Soon thereafter the littoral community of lakes was shown to be an important major source of soluble organic substances because of the leaks from the emergent *Scirpus* (bulrush) as well as *Najas* (Allen, 1971). Meanwhile the zooplankters and other animals are excreting dissolved organic nitrogen and phosphorus as well as inorganic compounds of these two elements (Lehman, 1980).

Dissolved organic matter can be converted to particulate material by two main mechanisms. The first method might be termed agitation-aggregation. Decades ago Baylor and Sutcliffe (1963) found that air bubbling through filtered seawater brought about formation of tiny platelike aggregates of organic composition. The same year Riley (1963) went a step further to show that these amorphous organic aggregates are colonized by bacteria and algae and that they increase from 5 μm to several millimeters in length. Hynes (1969) reported that leaves falling in brooks lose weight as soluble material leaches from them and that this substance rapidly becomes particulate. The particles soon aggregate to form clumps 10 times larger than the original. Johnson and Cooke (1979) published results of studies on air bubbles dissolving in seawater; residual particles of about one tenth the diameter of the original bubbles remain as the bubbles dissolve. They either disperse or clump to form larger particles. The phenomenon comes about because such dissolved stuff as fatty acids, proteins, humic acids, and sterols adhere to bubble surfaces. Thus, the great store of dissolved organic material in the seas and in inland waters is made available to small filter-feeding animals and is passed on when they are eaten by predators.

The second manner whereby DOM becomes POM involves microbes and algal cells. Bacteria can take in DOM and reproduce rapidly in the dilute culture that is lake, river, or ocean water. Attached to inorganic flecks and organic aggregates or floating free in the water, the bacteria divide and redivide, incorporating DOM into their cells and daughter cells.

In addition to the typical prokaryotic bacteria, a few of the cyanobacteria and eukaryotic algae can convert DOM to POM. This is because some algae are able to use dissolved organic substrates as carbon and energy sources. Decades ago Saunders (1972) used autoradiography to show that *Oscillatoria aghardii* cells take up acetate and glucose labeled with ^{14}C. Eppley and MaciasR (1963) showed that the flagellated green cell *Chlamydomonas mundana*, common in anaerobic sewage lagoons, uses light to absorb acetate. Its photosynthesis and release of O_2 are feeble, not compensating for its O_2 consumption. *C. mundana* is more like a photosynthetic bacterium than an alga, and it perpetuates rather than alleviates anaerobiosis. Another alga that grows best in a dissolved organic substrate is *Ochromonas malkamensis*. It is remarkable because it contains only chlorophyll *a*, lacking chlorophyll *c* that is typical of the other chrysophyceans. It has been suggested that lack of the accessory pigment reduces its photosynthesizing abilities and brings about its need for DOM. Some species of *Scenedesmus* and *Chlorella* are common in sewage water because they require unusual amounts of phosphorus; they consume CO_2 and release O_2 abundantly, being good photosynthesizers. They are not equivalent to the sewage forms that depend on DOM.

Particulate Organic Material

Particulate organic matter is defined on the basis of what is retained by a membrane filter. The filter pore size is about 0.45 μm and, therefore, POM by definition is composed of particles with diameters greater than 0.45 μm. DOM is the dissolved detritus that can be drawn through the filter. By convention, 0.5 μm is often cited as the lowest range for POM, but there are individual published papers on the subject where both 0.45 and 0.5 μm are designated the dividing line between POM and DOM! POM is categorized further as coarse (**CPOM**), with a diameter greater than 1 mm, and as fine (**FPOM**), which is less than 1 mm across. A further division is sometimes made: ultrafine particles (**UPOM**) are recognized as those with diameters between 0.45 (or 0.5) and 50 μm. With woody debris—twigs and bark, for example, another classification prevails: fine material lies between 10 and 100 mm; coarser detritus (10 cm) consists of boles, branches, and root masses.

The Detritivores

Bacteria constitute a major portion of the diet of most ciliated protozoans. The ciliates are competitors with bactivorous invertebrates, for example, certain rotifers. On the other hand, the ciliates are acceptable prey for many other invertebrates, because the particles that were originally bacterial cells become part of a larger protozoan body—in the neighborhood of 0.5 mm. Some are consumed by filter feeders in the open water. Others are preyed on by benthic forms and periphytic species in the littoral region. The oligochaete worm *Chaetogaster langi* actively searches for and preys on ciliates. Also, cladocera, copepoda, dragonfly naiads, mosquito larvae, and even small fish have been observed to feed on these protozoans (Taylor, 1980). Although a great deal is yet to be learned, it appears that the ciliated protozoans are important in detritus food webs and are significant conduits in the flow of energy to top consumers in aquatic ecosystems. From DOM to the production of bits large enough to be consumed by macroscopic forms, the ciliates are an essential part of the pathway.

Larger detritivores, feeding on organic particles, consume some ciliates as well as the associated bacteria and fungi. This is because a few ciliates may be grazing on the bacterial cells attached to the detritus.

The macroinvertebrates that consume detritus and its associated microflora and fauna compact the material and send it on its way as fecal particles. Some of this material sinks relatively rapidly to the sediments, where it becomes a source of energy for the zoobenthos. Nonliving seston that sinks more slowly may be degraded, perhaps consumed several times, before it reaches the benthic detritivores. Much of its mass as well as its nutritive value may have been lost in the water column.

The detritus community is discussed in Chapter 8 in connection with the lotic ecosystem. Although stream biologists especially have been concerned with the processing and utilization of detritus, this information applies to lentic communities also. Pomeroy (1980), who reviewed our knowledge of detritus as a food source in aquatic ecosystems, calculated that 1,000 kcal of energy in dead plant material might yield eventually 10 to 30 kcal of energy in the biomass of top predators.

Glacial Relicts

A cluster of animals with their closest relatives in ocean habitats are lumped as glaciomarine relict species (Fig. 2-9). They occur in deep, cold oligotrophic lakes distributed in a rough circle from northern Germany and the Scandinavian countries to Siberia, Canada, and the northern United States. The Baltic Sea Basin may have been a center in the transition of these animals from salt to fresh water. One theory states that, where heavy Pleistocene glaciation depressed land that was inundated by the sea and subject to later up-warping, some species were stranded in "relict" lakes. Other notions are that some of these species were living in dilute waters during preglacial times and that proglacial lakes, in contact with ice and marginal to it, provided a migratory path. As a result, there are many species with marine affinities living in the deeps of northern lakes. Holmquist (1962) rejected the word "relict" for these organisms because most are not detached remnants of existing marine populations past or present. A marine relict, she pointed out, is exemplified nicely by *Zostera* (eelgrass), a widespread littoral marine angiosperm that exists in the Caspian Sea. The Caspian was formerly part of the oceans, but was isolated through uplifting. Väinölä and others (2008) provide an excellent survey of amphipods across the world, explaining the relict distributions and evidence for the phylogenetics and distribution observed in this diverse group. Some 1870 species are found, mainly in cooler waters but distributed across every continent, and in a variety of surface and subterranean habitats.

The relict concept probably is overstressed with regard to such copepods as *Limnocalanus macrurus* and *Senecella calanoides*, although they represent marine families. Their present distribution, nearly circumpolar, and their restriction to northern oligotrophic lakes implies, however, spreading via proglacial lakes. *Eurytemora* is different. It seems to be actively invading fresh water even today. It first appeared in Lake Erie in 1961 (Engel, 1962) and since then has spread to all the other Great Lakes (Watson, 1974).

Mysis relicta (Fig. 2-9, *B*) belongs to a large marine order, the Mysidacea. Although not a copepod, *Mysis* must be discussed with the glaciomarine species. It is believed to have spread from the Baltic region via proglacial lakes. Today it is considered both a deep-water nektonic species and a member of the profundal benthos in northern oligotrophic lakes.

Modern cladistics analysis of *Mysis* (Audzijonyte et al., 2005, 2008), confirms the relict distribution. Genetic evidence points to a single ancestor that colonized the inland lakes from the ocean perhaps 3 million years ago.

Impermanent Habitats

The temporary pond could be considered the aquatic habitat most unlike the oceans. At least the dry phase of an ephemeral pool is a phenomenon unmatched in the sea.

Most temporary waters are likely to be found in arid and semiarid regions, where evaporative loss exceeds the water income. This does not mean that humid regions lack them. A great number of vernal ponds fill with snowmelt in the temperate regions, and in polar climates there is a physiologic "arid" period in winter when the water is present only as ice.

Hartland-Rowe (1972) elaborated on some older classifications of temporary waters, pointing out that they are markedly **astatic** (their surface levels fluctuate). He defined two categories: *seasonally static* waters, which dry up annually, and *perennially astatic* pools, whose levels rise and fall but do not dry up every year. The seasonally astatic pools comprise the cold, clear spring ponds of the North and the transitory waters of the desert, including those that fill with warm summer rains, those that fill with winter precipitation, and those that fill during both seasons. The perennially astatic waters range from depressions that always contain some water, but whose marginal flats are both exposed and inundated with some periodicity during the year, to such temporary waters as Australia's great Kati Thanda (Lake Eyre), which has no regularity to its cycle of wet and dry phases and is often empty for decades.

Temporary ponds reflect the immediate climate and meteorologic regime far more than do ponds deep enough to be permanent. The biota of transitory waters are adjusted to a set of unusual environmental factors of which the dry phase commands the most attention. Mechanisms to survive the dry phase include resistant cysts or diapausing eggs (usually shelled embryos), resistant diapausing immature or adult stages, or escape by flight or burrowing.

When in the watery phase, the biota is subjected to widely varying physiochemical characteristics. The oxygen tension often falls to low levels; salinity varies and becomes unusually high in some pools toward the end of the wet phase; and temperatures are extremely variable even during short periods.

The temporary pond can be likened to a refugium for some species because there are few predators present, although some African depressions that are dry for 8 months each year support cyprinodont fishes of the genus *Nothobranchius.* Usually the predators in impermanent water are hemipterans and coleopterans that fly in to feed on the abundant resources offered by the ponds. In woodland pools there are often many insect and even salamander predators.

Restriction to temporary waters for some groups can be a matter of need for drying of the fertilized egg at some stage. This is controversial; it is true for certain species, but evidence is conflicting for others.

Adaptation to ephemeral waters may be a kind of retreat to a haven where competitors are excluded. Proctor (1957) showed that *Haematococcus,* a small green flagellate characteristically found in bird baths and funeral urns, readily survives drying in temporary waters but is inhibited by the presence of substances produced by less talented algae in permanent, larger ponds.

Temporary pond faunas often include species that develop more rapidly than relatives in permanent waters. There is a race for life in many ephemeral pools. Rzóska (1961), who studied the short-lived temporary pools in the Sudan, described life in such habitats as a race against time, often culminating in death for thousands of animals in the last few hours.

The adaptations to life in temporary waters and the stresses with which the biota must cope are numerous (Belk and Cole, 1975). There are different ways of coping with the temporary habitat, and therefore the life cycles of even closely related animals in some ephemeral ponds are not identical. Populations of the fingernail clam, *Sphaerium occidentale,* survive the dry state as adults and juveniles. By contrast, *Musculium partumeium* and *M. securis* complete reproductive activity and their adult stages before the pond disappears, and we find only juveniles surviving the dry phase by burrowing into the substratum (McKee and Mackie, 1981; Way et al., 1981).

The branchiopod *Lynceus brachyurus* (Fig. 2-6, *E*) shows adaptations for survival in relatively long-lived woodland pools that contrast with other species from smaller puddles and from short-lived desert ponds. Insect predators are abundant in the forest pools, and the morphology of *Lynceus* may be an adaptation to their presence. It is spherical rather than laterally compressed, as are many other conchostracans, and its shiny, hard valves are extremely smooth with no concentric ridges. As a result of this structure *Lynceus* is a speedy swimmer, and even when captured by an insect, it often pops out of the predator's mandibles and escapes (D. M. Kubly, personal communication, 1979). This theme of special adaptations to life in temporary and variable habitats is repeatedly considered in the discussion of wetlands (Chapter 7).

4

Ecosystems, Energy, and Production

The Community Concept and Ecosystems

In Chapter 3 we considered the interacting biota of ecological communities. But these biota also interact with the nonliving (abiotic) environment, so in this chapter we examine ecosystem ecology. Ecological interactions may be thought of in physical terms: organisms absorb, transform, and eventually release chemicals and energy in multiple forms.

Although the concept of the **ecosystem** has at times been criticized as simplistic, thinking of living systems as "machine-like" (O'Neill, 2001), yet the idea still has value as a way of organizing ecological complexity (Pickett and Cadenasso, 2002). It is useful to consider a water body as a system with interacting parts, even though it is open—usually the system receives sunlight, exchanges gases with the atmosphere, absorbs runoff from land, and organisms come and go. Forbes' "little world" is hardly a world alone unto itself; it is part of a biosphere.

Food Chains and Ecomorphs

Like any system, an ecosystem has components with specific functioning: producers such as plants capture energy for biological use, herbivores are primary consumers that eat the producers, while secondary and tertiary consumers are higher on the *grazing* food chain. As described in the previous chapter, a detrital community uses the corpses and waste of that grazing food chain as food, recycling the chemicals in the process.

Each of the *trophic levels* (links in the food chain) contains a whole collection of species with different niches, or particular approaches to using the environment. Ecologists have found similar niches in equivalent ecosystems (temperate-zone ponds, tropical lakes, etc.) worldwide. The organisms occupying each niche are ecomorphs and often share a remarkable resemblance, even if they are unrelated taxa located in geographically separate sites (Harvey and Partridge, 1998).

Figure 4-1 shows the effects of convergent evolution of unrelated fish, each occupying a similar niche. In North America there are many species of fishes belonging to the sunfish and bass family, the Centrarchidae. Many are deep bodied and laterally compressed and appear nearly circular when observed from the side. Various feeding

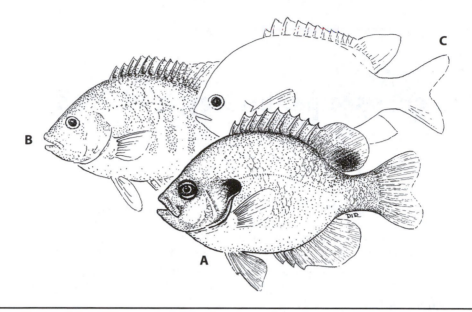

Figure 4-1   Ecomorphs of sunfish-shaped fish from different families. **A**, Bluegill sunfish, *Lepomis macrochirus* (Centrarchidae), North America; **B**, convict cichlid, *Cichlasoma nigrofasciatum* (Cichlidae), South America; **C**, Panamic sergeant major, *Abudefduf troscheli* (Pomacentridae), marine, Gulf of California.

habits are associated with the different species. Some feed on insects and microcrustacea; others are mollusc feeders; some are piscivorous; and many are omnivorous. In Mexico and Central America the centrarchids drop out and representatives of another family appear. They belong to the Cichlidae, a diverse group remarkably resembling the centrarchids. All the trophic roles played by centrarchid species farther north are matched by the cichlids. An excellent analysis of cichlid ecomorphology in Belize was conducted by Cochran-Biederman and Winemiller (2010), who described a spread of anatomical and behavioral responses in just one stream, as these fish occupied the available niches. A more famous example comes from Africa in the remarkable piscine fauna of Lake Tanganyika. At least 172 cichlid species are present and 97% of them are endemic. In that huge lake they play many ecological roles (Coulter, 1991).

The ecomorph concept is also known in other species. Moen and colleagues (2013) sampled frogs from 44 sites across three continents, finding excellent evidence for convergent ecomorphs. Wellborn and others (2005) found both life history and allozyme diversification in the amphipod *Hyalella azteca* that agreed with previous work on the taxon. The authors point to "the evolution of similar phenotypic solutions to comparable ecological challenges" in the group.

Biogeochemical Aspects of Ecosystems

Early efforts to understand the dynamics of an ecosystem focused on the chemical nature of organisms with respect to their environment. Later the quantitative roles of organisms in major chemical cycles became an important goal in ecologic research.

This was essentially a biogeochemical approach. Originally biogeochemistry described the study of organisms in relation to mineral formation, but it expanded to range from soil science and geochemistry to ecology, where it explores the interactions among the biosphere, atmosphere, lithosphere, and hydrosphere (Likens, 1981). Also, the use of isotopic tracers has helped quantify some relationships between living and nonliving components of ecosystems.

The total mass of the biosphere is much less than that of the atmosphere, hydrosphere, and the crustal portion of the lithosphere—1 million times smaller than the lithosphere, for example. The biosphere is in dynamic equilibrium with these components of the physical environment, and the geochemical importance of organisms must not be underestimated. Although relatively common atoms compose the bodies of organisms, a few essential elements are so scarce that demand surpasses supply at times, and growth is thereby limited.

The chemistry of aquatic plants and animals is much like their surroundings, although some atoms are remarkably concentrated in the organisms. Redfield (1958) reviewed data showing that the relative composition of phosphorus and nitrogen in marine plankters is nearly identical to the ratio in seawater. The relative abundance in the organisms is 1:16, almost matching the mean of 1:15 in seawater samples.

The term *cycle* is applicable to both biogeochemistry and ecology. The chemical parts of an ecosystem can enter, leave, and reenter living systems repeatedly. Sometimes there may be a loss, as when silicon locked up in diatom frustules is sedimented and buried by subsequent deposition of other material, but this is only temporary on geologic time scales. Nutrients and other minerals are continually recycled through biotic communities.

Energy Flow in Ecosystems

All systems, including ecosystems, rely on energy input to operate. Limnetic systems use several energy forms: illumination, heat, and kinetic (water movement) are each discussed in Chapters 9–11. More important to ecological function is the energy flow from light into the organisms in food chains. We live on a solar-powered planet.

All energy flows obey the laws of thermodynamics; the first two laws are critical to ecosystem function, while the third law relates to the nature of "absolute zero," –273° C. The first law states that the amount of energy in the universe is a constant: no "new" energy is ever created—energy simply changes form. The second law refers to disorder in a system (entropy): entropy increases in any physical process or chemical reaction, so useful energy is less available (often lost as heat). The combined effect of the first and second laws on ecosystems is that only a fraction of energy is utilized, severely limiting system productivity and the length of food chains.

Teal (1962) measured the flow of energy in a Georgia salt marsh (Fig. 4-2 on the following page) and the study illustrates energy flow through ecosystems. Primary producers (plants and algae) were able to capture 6.1% of sunlight for use in photosynthesis, a rate comparable to very efficient crop yields and much higher than typical 1–3% efficiencies in natural systems (Gebhardt, 1986). Of the energy in the salt marsh cordgrass (*Spartina*), only about 4.6% is consumed by the herbivorous insects. It is obvious why food chain length is limited. More on efficiency of energy transfer appears later in the chapter.

The Nobel prize-winning physicist Schrödinger (1945) wrote a book that is still fascinating and relevant decades later. He underlined the improbable state of organization—the living system—that characterizes the organism. He asked how an organism avoids decay and postpones that state of entropy we call death. His answer was that we continuously feed on negative entropy, the reciprocal of disorder. We take in orderliness in the form of complicated organic compounds, using much of their contained energy. In ecosystems the primary producers have their most important source of negative entropy in solar radiation. The sunlight that supports the green plants is the orderliness that ultimately maintains the entire ecosystem, allowing it to stave off the final decay of entropy.

In fact, all living things, and ecosystems generally, organize to use available energy efficiently and prosper. Nearly a century ago, Lotka (1922) suggested what would later be proposed as a potential fourth law of thermodynamics: the *maximum power principle*. Since all life requires energy and life evolves, Lotka reasoned that both species and the ecosystems they live in evolve (self-organize) over time to maximize

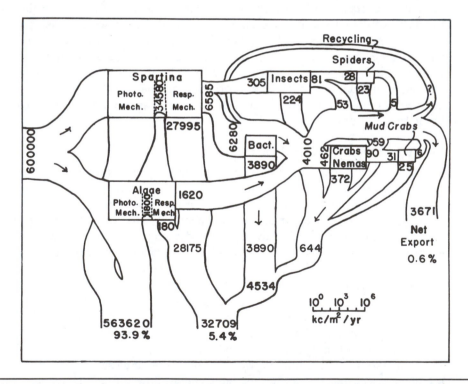

Figure 4-2 Energy flow through a Georgia salt marsh. Numbers are kilocalories or percentages of flows. Note that less than 40,000 out of 600,000 incoming kcal of solar energy (= 6.1%) is captured by the *Spartina* plants and algae (34580 and 1800 kcal, respectively), and only about 8,000 kcal (6585 plus 1620) then flow to subsequent trophic levels. The mechanisms of plant biochemistry are complex and relatively inefficient; photosynthesis and respiration mechanisms within the *Spartina* plant body use 27995 of the 34580 kcal absorbed by the plant, representing 81% of the energy obtained from light. From Teal (1962). Reprinted with permission.

efficient energy use. The concept was championed for decades by the late Howard T. Odum (1956a, 1995), and still is useful in ecology, for example to understand competition dynamics (DeLong, 2008). The appeal of the maximum power principle is that it synthesizes ecology, evolution, and thermodynamics.

The trend to seek a common denominator in ecosystem ecology has led to increased use of the calorie. Solar radiation, plant and animal tissue, respiration, and other metabolic output can all be quantified as calories. Improved apparatus for direct calorimetry of very small samples has led to better results than those derived from earlier attempts with wet oxidants such as $K_2Cr_2O_7$. From the many determinations summarized and listed by Cummins and Wuycheck (1971) it has become obvious that the caloric content of all tissue is not alike; there are important systematic, as well as ecologic, differences. The 1971 list permitted a new precision in converting mass to energy equivalents.

The grand mean of aquatic primary producers is 4,639 cal/g of ash-free dry weight. The diatoms with 5,310 cal/g surpass the cyanobacteria (4,882 cal/g), which in turn exceed the chlorophytes that average 4,780 cal/g. The invertebrate consumers, averaging 5,470 cal/g, have a greater energy content than the plants upon which they depend. In all organisms the energy in 1 g is less if it is dry weight including ash; for this reason ash-free dry weight is the best base for conversion to calories.

Community Metabolism

There is a technique by which communities can be appraised in a manner analogous to quantifying an organism's metabolism. The community metabolism is measured for a period of time. The gross or total photosynthesis (P) is determined during the daylight hours and compared with total respiration (R) during a 24-hour period. From this a **P/R ratio** can be established without knowing specific rates of photosynthesis and respiration for individual organisms in the system. The methods of measuring are addressed later. It is sufficient to say here that oxygen increment during daylight hours reflects net photosynthesis; it is assumed that some oxygen produced is consumed by respiration before being assayed. Gross primary production is proportional to the total oxygen evolved. Since respiration dominates at night while oxygen is consumed, the nocturnal consumption of oxygen (when darkness is 12 hours) is doubled to represent total diel respiration (R). Half this daily respiration is added to the net photosynthesis, the sum being gross primary production (P).

The P/R ratio varies from day to day, season to season, or year to year, but its long-term mean tells something of the ecosystem's nature. In North America Odum (1956b) is usually credited with developing the idea of a P/R proportion to define a community type. Winberg (1972) suggested that this concept was expressed in Russia 20 years earlier in terms of autotrophic or photosynthetic processes in relation to all those that are the opposite—heterotrophic phenomena. Total respiration is a measure of all the energy used by organisms (the combined metabolism of the producers plus the consumers) in the ecosystem. When the ratio is unity, all organic material produced by autotrophic organisms is destroyed by destructive processes, and a dynamic balance is achieved.

Oligotrophic lakes typically have P/R ratios not differing significantly from 1.0 (Fig. 4-3, *A*). These communities are in one kind of equilibrium: the solar energy

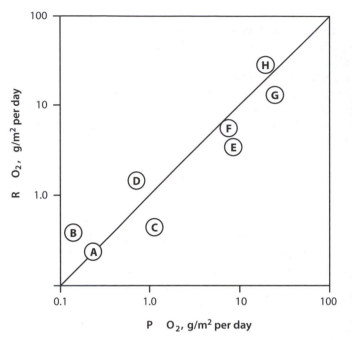

Figure 4-3   Log-log plot of P vs. R. On the diagonal line, P/R = 1.0. Above the diagonal indicates heterotrophy; below the line, autotrophy. **A**, unpolluted portion of Lake Päijänne, Finland (Tuunainen et al., 1972); **B**, grossly polluted portion of Lake Päijänne; **C**, enriched, eutrophic portion of Lake Päijänne; **D**, Root Spring Massachusetts (Teal, 1957); **E**, Severson Lake, Minnesota (Comita, 1972); **F**, Montezuma Well, Arizona (unpublished data); **G**, ponds of tertiary-treated waste water, Arizona (Foster, 1973); **H**, sewage lagoons in the Dakotas. Adapted from Bartsch and Allum, 1957.

received and converted to highly organized organic molecules, negative entropy, is approximately matched by conversions to entropy, although small amounts of organic material are stored in the bottom deposits. An ecosystem such as Silver Springs, Florida, must then be considered a community in dynamic equilibrium, for a balance is achieved between the organic matter produced and that destroyed or leaked downstream each year (Odum, 1957). Likewise, unpublished studies on Montezuma Well, Arizona, show that it is in a similar balance: its photosynthesis is nearly matched by respiration (Fig. 4-3, *F*).

In eutrophic lakes the P/R ratio is substantially greater than 1.0 (Fig. 4-3, *C* and *E*). Autotrophic organisms produce much more organic material than can be accounted for by respiration. There is a balance, however, because the excess organic substances are stored as sediment, awaiting some future time to be oxidized. Communities with P/R > 1.0 are designated autotrophic; photosynthesis of organic molecules surpasses their destruction by heterotrophic processes. Many fertilized fish ponds and enriched sewage lagoons are good examples of this type (Fig. 4-3, *G*). Their primary productivity is high, and only a part is consumed by heterotrophic events, termed respiration here. Thus Foster (1973) found that Arizona ponds enriched by treated sewage water converted 2.80 kcal/cm^2 of the annual solar radiation to organic molecules via photosynthesis. Of that primary production, 1.44 kcal was destroyed in community respiration; the storage in the sediments was about 1.3 kcal/cm^2, accounting for most of the rest. The **gross production** amounted to the fixation of about 2.5 kg/m^2 of carbon.

In such highly productive ecosystems, temporary halts in the input of solar energy stops the synthesis of organic molecules, and the degradation of stored energy assumes predominance. Thus, during cloudy periods and during winter when ice and snow cover shallow eutrophic lakes, the P/R ratio often falls below unity.

Aquatic ecosystems to which allochthonous organic molecules are added in substantial quantities have P/R ratios significantly below 1.0 (Fig. 4-3, *B, D*, and *H*). The structured potential energy of those molecules entering community metabolism to be degraded did not come from autochthonous primary production. As a result, respiration, in the broadest sense, surpasses photosynthesis, and the community can be termed heterotrophic. The heterotrophic events surmount the phototrophic action of green plants, just as they do in shallow, darkened, ice-covered lakes.

Saprobity, the total of all the energy consumption processes, is another classification of **heterotrophy**. A sequence of saprobic conditions from oligosaprobic to mesosaprobic through polysaprobic corresponds roughly to a spectrum from pure oligotrophic to extremely polluted conditions. There is an α and β level for each division so that six categories represent the trend. The saprobic scheme is usually employed in discussions of polluted waters and their saprobionts, the indicator species ranging from the most tolerant polysaprobic species to the least tolerant (Caspers and Karbe, 1967).

Dystrophic lakes receiving humic substances from external sources are usually heterotrophic. Winberg (1963), summarizing work done by himself and other Russian limnologists, pointed out that dystrophic Lake Glubokoye, surrounded by bogs and with heavily stained water, has a low annual productivity and, despite some variation, typical heterotrophic P/R values. One year the mean ratio seemed to be 1.04, but usually the figure was much less. It was 0.82 the following year, for example.

Macan and Worthington (1951) referred to Cedar Bog Lake, Minnesota, as dystrophic despite Lindeman's (1941) statement to the contrary. The quotient of the primary production and the sum of respiration from most trophic levels gives a value of about 3.7. The overall P/R ratio is lower than this, for neither bacterial respiration nor days of winter anaerobiosis are included. Nevertheless, Cedar Bog is autotrophic, producing more than it respires each year. Its community metabolism is not typical of dystrophy.

Within a single Finnish lake, Päijänne, a series of areas from oligotrophic to polluted can be discerned (Tuunainen et al., 1972). Päijänne has a maximum length of 120 km and has been polluted at sites along the irregular shoreline by pulp mill influents and by water entering from a nearby polluted lake. Twenty years before the report, Lake Päijänne was oligotrophic, but by 1970 changes were evident. Some areas were still relatively clean with P/R values averaging near unity (Fig. 4-3, *A*); another area showed the eutrophicating effects of sewage: P/R = 2.42 (Fig. 4-3, *C*); and others reflected the organic load and perhaps toxicity, imposed by pulp mill wastes: P/R = 0.04 to 0.51 (Fig. 4-3, *B*). Meriläinen and others (2001), however, reported that cessation of mill effluent resulted over time in less brown staining of the water, more normal water chemistry, and recovery of biodiversity and productivity.

A similar relationship is seen sequentially in streams where pollution and recovery are involved in a given stretch. Immediately below the source of pollution, heterotrophic processes predominate (P/R < 1.0), the pollution being so great that there is no photosynthesis. Organic matter decomposes while flowing downstream, and gradually photosynthetic algae appear and begin to function. In a region best termed the early recovery zone, the nutrient-rich algae produce organic material in excess of its decay to create an autotrophic segment. In healthier reaches farther downstream, both primary production and respiration lessen, and the P/R figure approaches unity, perhaps much like conditions just above the pollution source.

Teal (1957) published results of work on Root Spring, a small **limnocrene** (a pooled spring) in an article that has become a classic in the history of investigations of community metabolism and ecosystem energetics. Root Spring, lying but a few kilometers from Walden Pond in Concord, Massachusetts, is supplied by cold water from adjacent banks of glacial drift. As is typical of small springs and brooks, there is no plankton, but benthic algae contribute to the input of energy via photosynthesis. However, the major source of organic input for the little ecosystem (diameter about 2 m!) is leaf litter from nearby apple trees and other plant debris that falls into the pool, accounting for 76% of the annual energy supporting the spring community. As a result Root Spring is heterotrophic, with a P/R figure of about 0.30 (Fig. 4-3, *D*).

Steinböck (1958) proposed **allotrophy** as a term to describe conditions where the main energy source comes from outside the lake. Allotrophic lakes are, of necessity, heterotrophic on the basis of the P/R ratio. Windblown organic material is especially significant in some desert basins. Hutchinson (1937) coined **anemotrophy** to describe certain allotrophic, arid land pools that receive their primary energy-rich carbon compounds via the wind.

Some pools from arid and semiarid regions are persistently turbid with fine-grained suspended material, which serves as a trophic base for the fauna. Wind is a key factor here also, but much of the fine detritus is autochthonous, stirred from the basin floor itself. Light penetration is very low, and phytoplankton is absent. Having seen this type of ecosystem in Africa and Nevada, Hutchinson (1937) offered the word **argillotrophy** to describe the situation where a fauna, characterized by low diversity, is dependent on suspended clayey detritus rather than on algae. Argillotrophic systems were treated in detail by Daborn (1975), who based much of his discussion on his studies of Fleeinghorse Lake, Alberta, Canada.

Trophic-Dynamic Ecology

Lindeman's posthumous paper (1942b) marked a turning point in ecology. The article was an outgrowth of his doctoral dissertation based on research at Cedar Bog Lake and of his association (after completion of the Minnesota work) with the Yale group, where an unpublished manuscript by G. E. Hutchinson was especially influential (Cook, 1977). Lindeman's paper (published after his untimely death at the age of 26) has become one of the most cited works in the literature of ecology. His effort seems to have crystallized earlier studies of energy flow through ecosystems and has stimulated further investigation of the dynamic approach to the study of communities. In retrospect there were flaws in his conclusions where he oversimplified and used terminology that is no longer in favor; but after Lindeman, the study of energy transfer in ecosystems came into its own.

The Problem of Efficiency

Measurements of energy passage in communities are notoriously difficult, yet subsequent studies agree with Lindeman's opinion that energy conversions from trophic level to trophic level are rather inefficient in ecosystems. From the energy available to a trophic rank, only a small part is incorporated into individual bodies; thus some energy is lost. Such low efficiencies of conversion can be inferred from the second law of thermodynamics. Lindeman (1942b) was concerned with progressive efficiency, the ratios of energy used by successive trophic levels. This is expressed as a percentage by:

$$\text{Efficiency} = \frac{\lambda}{\lambda - 1} \times 100$$

The formula applies to different types of efficiencies; Lindeman used it to compare the degree to which one level (λ_n) used the energy resources available to it from the previous level (λ_{n-1}). Following Lindeman's concern for progressive efficiency, there was an eruption of efficiencies—growth efficiencies of individual species determined from laboratory feeding experiments, and a host of other types. Twenty of these categories of efficiency from earlier authors are presented in a list by Kozlovsky (1968); still other types have been suggested.

The efficiency concept may be worth pursuing, but there are many inherent problems in it. The first trophic level, phototrophic plants using radiant energy, is easier to analyze than the subsequent categories where omnivory obscures precise grouping and there is not always strict dependence on energy from the λ_{n-1} level. Furthermore, as Darnell (1961) observed, there are ontogenetic changes in food habits of fish. In Lake Pontchartrain, Louisiana, he found that a fish may be, successively, a detritus feeder, an herbivore, and a carnivore as it ages, thus shifting from category to category through time.

Primary Production

All production within an ecosystem stems from the energy in organic substances that autotrophic organisms create from inorganic raw materials. In some instances the organic energy was produced earlier and stored, as in leaf litter on a forest floor, or was imported as allochthonous energy produced elsewhere by autotrophs. With this in mind, the energy approach to studying ecosystems has the goal of balancing the whole energy budget and assumes that first, there is an increase in organic compounds during a period of time and, except for import, only primary production can bring this about; second, there is a decrease in that stock as it is required by organisms in aerobic environments, fermented in anaerobic surroundings, or oxidized chemically without the assistance of organisms; and third, any difference between addition and subsequent subtraction of chemical energy can be accounted for by either storage or some sort of export from the system.

Primary Producers (Autotrophs)

There are two main sorts of autotrophs or primary producers, both of which rely on inorganic substances as electron donors. The use of inorganic reductants is termed **lithotrophy**. Using two prefixes, primary producers can be categorized as (1) **chemolithotrophic** forms (chemoautotrophs or chemosynthetic bacteria) and (2) **photolithotrophic** types (**photoautotrophs** or photosynthetic forms). For growth they both require a source of energy and usable inorganic carbon. The chemolithotrophs derive their energy from inorganic chemical bonds. They oxidize compounds of sulfur, nitrogen, and ferrous iron; a few can oxidize hydrogen. The phototrophs use solar radiation as an energy source. They have been divided into two groups by Broda (1975): the photosynthetic bacteria (anaerobes) and the photosynthetic "plants" (aerobes). He termed the aerobes' type of photolithotrophy *photophytotrophy* or simply **phytotrophy**, making this division on the basis of the electron donor, which is usually H_2S in the photosynthetic bacteria and H_2O in organisms that possess chlorophyll *a*. As CO_2 is assimilated and reduced to organic molecules, the by-products are S and O_2, respectively. The division is phylogenetically hazy because the phototrophs include prokaryotic cyanobacteria as well as eukaryotic photosynthesizers. Furthermore, the discovery that at least one strain of *Oscillatoria limnetica* can fix CO_2 facultatively in an anaerobic environment, using sulfide as the sole reductant, is significant; this is typical of other bacteria but not of the phytotrophs (Cohen et al., 1975).

A most remarkable community far from the light was discovered at some sea-floor hydrothermal vents in the Galapagos region (see Karl et al., 1980, for a review). There in the east Pacific, associated with sea-floor spreading centers, hydrogen sulfide in the warm water arising from the vents supports a tremendous population of sulfur-oxidizing bacteria. They, in turn, support a community of clams and worms. This is one of the best examples of chemolithotrophy known to date.

When the materials used by bacteria to acquire energy come from outside the lake ecosystem, **chemosynthesis** is slightly analogous to photosynthesis taking place within the lake; both require energy resources from outside. It cannot be equated with photosynthesis, however, when the energy source for such chemotrophs comes from decomposition of organic material produced originally by phototrophic organisms. In such instances, chemosynthetic bacteria decrease the total energy in the ecosystem rather than adding to it. They are playing a part in the use and degradation of the energy of primary production, typical of what is termed secondary production. Some limnologists believe this generally holds and that all kinds of chemotrophic biosynthesis should be designated secondary production.

The most important lacustrine producers are aerobic phototrophs; photosynthesis creates the bulk of the new organic compounds. Littoral macrophytes (with associated epiphytic algae), benthic algae, and phytoplankters convert radiant energy to potential chemical energy. Except for shallow lakes, the phytoplankters are the most important producers.

Williams (1972) suggested that photosynthetic bacteria other than cyanobacteria may make substantial contributions to primary production in certain saline inland waters. This was borne out later by data from a saline beach pond on the Sinai coast, where more than 82% of the annual primary productivity comes from the activity of phototrophic bacteria (Cohen et al., 1977). This may occur elsewhere in meromictic lakes where aerobic-anaerobic boundaries lie near the surface. Here green and purple sulfur bacteria reduce carbon dioxide to organic molecules, using light energy for the process. Since the electron donor is something other than water, bacterial photosynthesis differs from the typical green plant type. The various types of photosynthesis may also differ in other ways: oxygen is not always evolved; different wavelengths of light are used; different pigments function; and photosynthates vary. However, all photosynthesis depends on radiation for energy and converts stable inorganic compounds to energy-rich, but unstable, complex organic molecules—negative entropy in the sense of Schrödinger (1945). Rabinowitch (1948) expressed this conversion in this manner for green plants: "In endlessly repeated cycles the atoms of carbon, oxygen, and hydrogen come from the atmosphere into the biosphere . . . . After a tour of duty which may last seconds or millions of years in the unstable organic world, they return to the stable equilibrium of inorganic nature."

The production of photosynthetic organisms must be defined carefully, because the word "productivity" has been used with different meanings. For some authors it goes without saying that production is net production or effective production, the growth increment. Perhaps this has been true more of Russian scientists, who have treated gross production as a superfluous concept (Winberg, 1971). Others feel that gross productivity is of considerable interest even though it includes a metabolic loss. In any case, partitioning the various contributions of the plankton, and separating gross and net productivity is difficult in practice.

Biomass

The Eltonian pyramid is usually an excellent generalization to portray relationships among an ecosystem's trophic levels, whether constructed in terms of numbers, mass, or energy. The concept was first presented in a book written by Elton (1927) and has become ingrained in the field of ecology. The base of the pyramid is made up of the autotrophs that synthesize their tissues from inorganic compounds; typically these are green plants. The next layer is composed of heterotrophs, the herbivores that require the autotrophs for food. Above this are predaceous heterotrophs that feed on the herbivores. Secondary carnivores make up the next higher level, and so on. Because much energy is expended in metabolism at each level, the mass of living material produced decreases at each higher level. On an annual basis, at least, the trophic structure of an ecosystem takes on the form of a pyramid (Fig. 4-4, P on the next page). (In a pyramid constructed on an energy basis, solar radiation would form the broad base in the best tradition of the ecosystem concept.)

From our experience with terrestrial ecosystems we would expect a random sample to yield data showing that plant biomass surpasses that of the herbivores, which in turn is greater than the carnivore biomass. The sample of primary producers might exceed the others by 1,000 times, and this could apply also to their mean annual biomass. The situation in the plankton community is quite different (Fig. 4-4, B); the producers differ from their terrestrial counterparts, annual production (P) exceeding the mean biomass (B)—in some instances by a hundred times. Relevant ratios are shown in Figure 4-4, P/B: the P/B figure for the phytoplankton exceeds that of the carnivores by about 13.5 times. The pyramids from Krivoïe do not represent the entire lake ecosystem. If the nekton and benthos were included, the contrast between the typical terrestrial ecosystem and the aquatic ecosystem would be emphasized far more. With the addition of carnivorous fish and bottom-dwelling invertebrates, both herbivorous and carnivorous, an inverted pyramid would result; Figure 4-4 B would appear top-heavy and mushroom shaped.

Blueweiss and associates (1978) discussed and quantified the relationships between life-history events and body size. Tiny organisms generally exhibit low-standing crops but high productivity. Thus, the mean biomass or standing crop of phyto-plankton (Fig. 4-4, *B*) does not reflect annual production. Biomass, or the sampled standing crop or stock, must not be confused with productivity; biomass is static, whereas productivity involves rate. The average sampled biomass of the planktonic algae is far less than the annual production (Fig. 4-4, *B*, *P*). The contrast is not as great in the larger carnivores, where sampled biomass serves as a closer approximation to

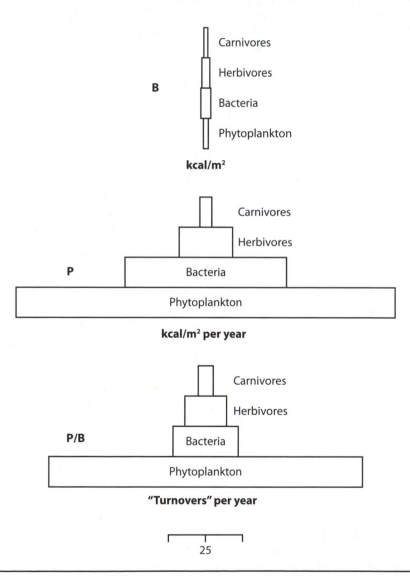

Figure 4-4   Eltonian pyramids from the plankton community of oligotrophic Russian Lake Krivoie. **B**, Mean annual biomass; **P**, annual production; **P/B**, turnover number, quotient of annual production divided by mean annual biomass (data from Winberg, 1972). Probably the bacteria are underestimated in these pyramids.

annual production; larger species usually have fewer generations per year and produce new tissue at a slower rate.

There are several methods of expressing biomass. Of these, wet weight requires no special treatment. Dry weight is found following oven drying at 80° to 105° C until a constant mass is achieved. If a dry sample is burned in a muffle furnace at about 500° C, only the ash content remains. Above 500° C some CO_2 may be driven from $MgCO_3$ and $CaCO_3$, leaving MgO and CaO, and also some potassium and sodium may be lost, thereby reducing accuracy. The difference between dry weight and ash weight ("loss on ignition") is an approximation of organic content, termed ash-free dry weight. Some authors have presented biomass in terms of ash-free wet weight, a description of the unlikely occurrence of a wet, organic plant or animal body lacking salts.

A further way of determining phytoplankton biomass is to extract and measure the chlorophyll from membrane or glass-filtered cells, using acetone as a solvent. The mass of the ash-free phytoplankton substance from which it came may have been some 35 times greater.

Assimilation Numbers

After being warned of the error inherent in considering biomass the equivalent of production, the student must be aware of an important exception. Knowing the mass of chlorophyll is very close to knowing primary production. This applies especially to chlorophyll *a*. Its abundance tells a great deal, and because it is lively stuff, it is possible to establish empirically its relation to photosynthetic production.

Two researchers (Ryther and Yentsch, 1957), working at the Woods Hole Oceanographic Institute on Cape Cod, experimentally demonstrated something that had been suspected by Ryther and some earlier authors: a fairly constant relationship exists between chlorophyll and photosynthesis at any given light intensity. Thus, primary production studies, accompanied by collections and assays of chlorophyll *a*, permit calculation of an **assimilation number**. The milligrams of C fixed per m^3 during an average hour of daylight is divided by the mg of chlorophyll *a* in the same volume:

$$\frac{\text{mg C per hour}}{\text{mg chlorophyll } a} = \text{assimilation number}$$

Similarly, the average hourly photosynthetic rate beneath a square meter of lake surface could be divided by the chlorophyll *a* present beneath the same area. The quotient would be an assimilation number expressed in terms of unit area. In the same manner the oxygen released in the presence of a given mass of chlorophyll *a* can be calculated. This yields a number almost three times greater than the carbon result because the molecular weight of O_2 is 32 compared with carbon's atomic weight of 12.

After a mean assimilation number has been established from research expeditions to some body of water, the mass of chlorophyll *a* within the illuminated portion is all that needs to be determined for future primary production assays. (However, a radical change in the composition of the algal community would negate this.) Multiplying the biomass of chlorophyll *a* by the assimilation number supplies a reasonable estimate of photosynthetic rate, precluding the need for tedious oxygen or carbon procedures. Thus McConnell (1963), studying a small Arizona impoundment named Pena Blanca Lake, settled on 2.04 as an oxygen-chlorophyll assimilation number. Following this, he limited his field investigations of primary production there to determining the lake depth where about 1% of the incident solar radiation remained and to quantifying the chlorophyll in the stratum bounded by the surface and that depth.

The determination and use of an assimilation number make up one of four principal methods of measuring primary productivity in aquatic ecosystems. The others, to be treated later, are assessing oxygen released during a period of time, assaying radiocarbon taken up, and finding the changes in pH caused by the removal of carbon dioxide from the water by photosynthesizing organisms.

Nutrients and temperature (controlling enzyme activity) affect photosynthetic rates directly, as do age and adaptation of the cells. Furthermore, as phytoplankton populations increase, shading limits photosynthesis. Talling and associates (1973) suggested that an upper limit of gross production equals no more than 17.8 g of carbon per square meter per day, or the release of about 47 g of oxygen. This rate of production must be extremely rare because when chlorophyll concentrations roughly exceed 70 mg/m^3, most of the incident light is absorbed within 1 m of water, limiting the photosynthetic capacity of the community. The figure put forth by Talling and his coauthors came from a tropical lake unusually high in salts and the raw stuff of photosynthesis, inorganic carbon.

Efficiency of Primary Production

The efficiency with which green plants convert radiant energy to potential chemical energy is low. An exact figure does not apply to all chlorophyllous communities, but 1% to 3% is often quoted. It depends, in part, on the manner by which efficiency is calculated. The general equation would be:

$$\text{Efficiency} = \frac{\lambda_n}{\lambda_{n-1} \times 100}$$

where λ_{n-1} represents solar radiation and λ_n equals the energy in the primary producers. There are several ways in which the sunlight in the denominator could be treated: total incident sunlight, the visible radiation useful for photosynthesis (from about 380 to 720 nm), light penetrating to the community after correcting for reflectance and backscattering, light absorbed by the plant cells, and light absorbed by specific photosynthetic pigments. Each of these leads to a different index of efficiency for the plants.

Similarly, the numerator could be either the energy fixed during photosynthesis (gross production) or this energy minus respiration (net production). There are then 10 possible producer efficiencies, a bewildering array. Lindeman (1942b) compared solar radiation with gross primary production in Cedar Bog Lake, coming up with an efficiency of 0.1%. Had he corrected for the fact that only about one half the incident wavelengths are useful in photosynthesis, the efficiency would have been doubled. Further corrections for reflectance (conservatively, 5%) and the diminishing radiation, after it strikes the lake surface and passes through water, would raise the efficiency of gross production even more.

Generalizations about the efficiencies of plants in capturing solar energy, calculated as the ratio of gross production in relation to total radiation per unit area, reveal that terrestrial communities often do a better job. Aquatic phototrophs are distributed downward from the surface, and as a result their energy resources are diminished. The plants are less efficient than the animals that ultimately depend on them, yet another aspect invites attention. Ratios of energy expended in relation to gross production show that many terrestrial as well as aquatic primary producers respire only about 15% to 25% of the energy they fix. This indicates that their inefficiency in capturing the energy available is compensated for by their ability to convert it to effective biomass.

We now recognize new urgency in accurate accounting for photosynthesizer efficiency: our assumptions and methods affect the function of ecosystem models embedded in climate change predictions. So, we tease out various factors including storage versus biomass increase, provision for feeding symbionts (like N-fixers), and respiration (Cannell and Thornley, 2000). Melis (2009) calculated an ideal maximum for energy use efficiency in plants: 45% of light is absorbed, 30% of light energy is chemically converted, and a mere 8–10% of light energy converted to biomass. The fact that less than a third of sunlight is effectively captured in photosynthesis seems low; however, it is still better than almost all conversion efficiencies of photovoltaic panels (Green et al., 2012).

Organotrophy

Organotrophy differs from lithotrophy and should not come under the heading of primary productivity. Chemoorganotrophic bacteria, for example, derive their energy from fermentation of, or oxidation of, a variety of organic compounds; they do not create organic molecules from inorganic sources, the hallmark of primary production. The organotrophs are heterotrophic rather than autotrophic.

One group of motile microorganisms, called the nonsulfur purple bacteria, is considered to be an ancient group; these organisms are mostly anaerobic, but a few have become facultatively aerobic. Generally, however, they use a wide range of organic substrates as electron donors under anaerobic conditions in the light. They do not qualify, therefore, as autotrophs. They are photoorganotrophic bacteria.

It would be easy to classify organisms on a trophic scale if they did not cross functional boundaries. Unfortunately, not all lithotrophic forms are obligatory users of inorganic energy and inorganic carbon supplies. There are well-known examples of photosynthetic algae that occasionally behave as heterotrophic organisms. These are algae that can derive carbon from various organic molecules, glucose and acetate for example. **Mixotrophy** is the ability to use both organic and inorganic carbon sources.

Secondary Production

The secondary producers cover a varied spectrum of organisms that rely on energy ultimately made available to them by primary producers. These include bacteria and fungi, plus the animal members of the community, including herbivores, omnivores, and predators. These are the true heterotrophs, ranging from ciliates to the sharks found in some freshwater lakes. Modes of feeding are so diverse among heterotrophs that assignment to discrete trophic levels is difficult, although we tend to categorize these animals as primary, secondary, tertiary, and quaternary consumers.

From examining trophic relations of the animals in an ecosystem, it is obvious why food "chain" gave way to food "web" in the parlance of biologists, the term web designating more complexity. Taxonomic categories often include animals of contrasting feeding habits, assignable to different trophic levels. The calanoid copepod *Diaptomus* (Fig. 2-8, *E*) feeds on limnetic phytoplankton in a manner reminiscent of its marine relative *Calanus*, which it closely resembles. Another calanoid, *Epischura* (Fig. 2-8, *C*), often present in the limnoplankton, bears a superficial likeness to *Diaptomus* but preys on it. To confound the issue further, some *Diaptomus* species in small mountain ponds attain unusual sizes and, as R. Stewart Anderson (1967) discovered, have abandoned the herbivorous way of life to feed on smaller congeneric forms with which they coexist.

In addition, feeding habits change throughout life in many species, and there are marked seasonal aspects in diet for others. Long-standing generalizations about the niche occupied by a species in an ecosystem often do not hold up. Coregonid fishes, for example, are notably zooplankton feeders in lakes, but they turn to consuming bottom fauna at times.

Estimating Secondary Production of Individual Species

The measurement of secondary production is difficult for a multitude of reasons. There is no hourly rate to be established as in studies of photosynthesis, and standing crops contain no chlorophyll *a* to be converted to dynamic rates by assimilation numbers. Another inconvenience stems from the fact that some aquatic animals are perennial, whereas other species require only a week or so to complete a life cycle.

The simplest measurement, based on the annual take of economically important forms, such as fish and the larger molluscs and crustaceans, is not really simple even when forthright procedures are followed. If a known number of fish is introduced to a small pond early in the year and harvested later, a small part of the annual production can be computed. This quantity is termed yield and gives only a rough approximation of net production. Mortality before capture, which is part of the yearly production, was not taken into account. The portions taken up by predators or removed by bacterial decomposers and scavengers can only be guessed without additional data. Further complications arise from reproduction and the appearance of a new generation of individuals, many of whom died during the year. To all of these variables could be added metabolic loss, another part of gross production.

Nearly a century ago, the Danish marine biologist Boysen-Jensen, working with the bivalve *Solen* and other benthic invertebrates, developed a growth equation still used today (Gray and Elliot, 2009). His technique was based on the yearly difference between numbers and individual masses of a given age class of the clam. It was rationalized that the production in time could be approximated by:

$$P = (N_1 - N_2) \times \frac{w_1 + w_2}{2}$$

N_1 and N_2 are the numbers of animals per unit area at the beginning and end of the year, respectively, and w_1 and w_2 are the corresponding mean weights of the individuals. The product of the average weight and the number of individuals that disappeared represents the production that had been either consumed by members of higher trophic levels or that had perished and been mineralized by microorganisms.

Boysen-Jensen's approach to estimating production, with various refinements, has served subsequent workers. It is especially useful for reckoning production in populations with long life cycles. The problem becomes more difficult when animals have short life cycles and are constantly producing young, as is true for many plankters.

Many modern authors have used the Allen curve method (Fig. 4-5) to estimate annual production. This is especially useful when members of one generation or cohort are being studied through the year, as when Allen (1951) studied trout in a New Zealand stream. Through the course of a year in the life of a population, regular sampling yields information on the numbers of individuals and the average mass of each. Preferably, numbers per square meter and grams dry weight are determined. When the numbers are plotted against individual mass, a curve results that shows growth on one axis and mortality on the other. Any segment of the curve from time T to T has an area beneath it equal to net production in grams dry weight per square meter. This does not take into account production of gametes and the losses during seasonal nonfeeding periods, and emigration may not be distinguishable from mortality as numbers decline through time.

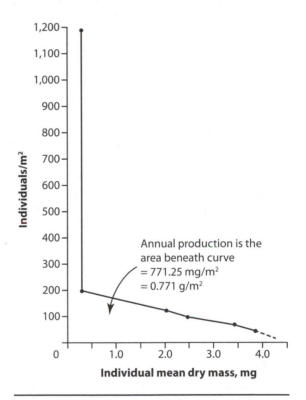

Figure 4-5　Allen curve showing annual production of the chironomid larva, *Tanytarsus*, in Sugerloaf Lake, Michigan. Estimations from the data of Anderson and Hooper, 1956.

Edmondson (1960) published a scheme for estimating production rate in planktonic rotifers. The problem in working with such animals is that samples taken at intervals could contain identical numbers yet tell nothing about the number of intervening generations and the population's production. Similarly, the ratio of eggs carried to females in the samples reveals nothing of the birthrate nor how often eggs were produced. Edmondson studied the rotifers in the laboratory, subjecting females carrying newly extruded eggs to different temperatures and noting the varying times it took the eggs to hatch. With information gathered on effects of temperature on duration of embryonic development, the lake temperature at time of collecting became a significant datum. The number of eggs produced per female per day, B, was calculated by $B = E/D$, where E stood for the number of eggs per female in the plankton sample and D was the duration of the embryonic stage. Thus, if 40% of the females carried single eggs, E would be 0.4. If the temperature were such that the value of D was 4 days, production would be 0.1 eggs per female per day. At some higher temperature, implying complete embryonic development in half a day, the same egg/female ratio (E) would represent eight times the daily production of the former case.

Edmondson went further to estimate an instantaneous birthrate, $b' = \ln(B + 1)$, and a coefficient of population increase, r', from counting the populations, N, on two successive dates, t_0 and t_1:

$$r' = \frac{\ln N_t - \ln N_0}{t}$$

From these calculations the death rate, d', was estimated as:

$$d' = b' - r'$$

With rotifer biomass data on hand, the production could be fixed as the product of mortality and mass.

Edmondson's rotifer technique can be applied to other plankters that carry their eggs until hatching and hold them after being killed and preserved; most copepods are especially suitable for such investigation. Actually, the method had its beginnings with Elster (1954) and Eichhorn (1957), who worked with calanoid copepods and showed the dependency of embryonic duration on temperature.

As data accumulate, it may become possible simply to apply the concept of the van't Hoff temperature coefficient, the Q_{10} value denoting the degree to which a process is accelerated for each 10° C increase. There are remarkable similarities in the temperature-development curves among some groups of related species, although divergences can be expected because of special adaptations to cold water or to transitory ponds (Fig. 4-6).

Estimates of secondary production have grown more sophisticated and reliable, thanks to better chemical analyses, microscopy, and instrumentation for data acquisition and analysis (Avila et al., 2012; Berggren et al., 2010; Brand and Miserendino, 2011; Tagliapietra et al., 2007).

Estimating Secondary Production at the Trophic Level

To understand the "big picture" of ecosystem function, we must consider whole trophic levels. Therefore, the best approach to the complexities of community secondary production is probably the study of individual species in the laboratory to gain information that will aid in interpreting quantitative field data. Some important facts to be assembled in such investigations are caloric equivalents of dry weights; respiratory rate as a function of temperature, age, stage of development, and size; and growth efficiencies under various conditions.

Consideration of certain basic assumptions generally shows what knowledge should be sought. First, the food requirement for an individual can be quantified as (1) calories as the sum of energy used in growth, (2) metabolic loss via excretion and respiration, and (3) the energy not used and therefore eliminated with the feces. Second, net production during the life cycle can be quantified as the sum of (1) somatic body

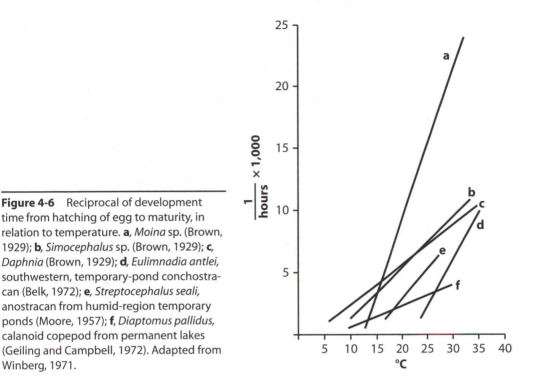

Figure 4-6 Reciprocal of development time from hatching of egg to maturity, in relation to temperature. **a**, *Moina* sp. (Brown, 1929); **b**, *Simocephalus* sp. (Brown, 1929); **c**, *Daphnia* (Brown, 1929); **d**, *Eulimnadia antlei*, southwestern, temporary-pond conchostracan (Belk, 1972); **e**, *Streptocephalus seali*, anostracan from humid-region temporary ponds (Moore, 1957); **f**, *Diaptomus pallidus*, calanoid copepod from permanent lakes (Geiling and Campbell, 1972). Adapted from Winberg, 1971.

growth, (2) reproductive-organ and gamete production, and (3) the energy in molted exoskeletons of many invertebrates.

The items enumerated above can be seen to belong to the various components of the energy budget equation $C = P + R + F + U$, where C is consumption, the total intake of food energy during some specified time; P is the energy content of the biomass and its products, resulting from food digested; R is respiration, the digested energy that is converted to heat and lost in metabolic processes; F is egesta, the undigested portion of the food energy that was taken in; the U is excreta, the energy content of digested material passed from the body. Clearly, P is net production in this equation, which quantitatively describes ecologic transfers of energy. Techniques for measuring the various elements of the equation are important because the results can be applied in ecosystem studies.

Teal (1957) studied individual organisms of the Root Spring fauna both in the laboratory and isolated in bottles submerged in the spring water. From the data acquired, he was able to approximate what was going on at each trophic level in the spring by correlating information derived from field collections. Comita (1964) spent many months studying *Diaptomus siciloides* in the laboratory and came up with an annual energy budget for the species. With these data, he later calculated energy relations of the entire plankton community (1972). Many field samples implied five generations of *Diaptomus* per year. Extrapolation from laboratory data permitted Comita to make judgments about the annual energy flow in the entire lake for this species, the most important primary consumer.

The great amount of work necessary for measuring secondary production in aquatic invertebrates, many with several overlapping generations each year, has inspired a search for applicable generalities. For example, the relationship between production and mean biomass, expressed as the P/B ratio, recalls the assimilation number in primary production. P/B quotients serve to translate biomass to a dynamic growth rate. The P/B value is the number of annual turnovers—how many times the mass or energy is replaced during the year. Studies of freshwater benthic animals show that Waters' (1977) mean value of 4.5 is applicable for **univoltine** species (those with one generation per year). An exceptionally low P/B quotient of 0.62/year was reported for a slow-growing marine clam (Table 4-1) by Hughes (1970). The P/B number is greater in smaller species of a taxonomic group: for example, species of chironomids with diminutive larvae have a higher figure than those with larger larvae. Furthermore, cladoceran P/B ratios surpass those of the copepods, and daily rates for both groups are lower in the cold water of northern lakes, except for some stenothermal species that thrive at such temperatures (Winberg, 1972).

Studies of growth efficiencies have established values for ratios such as **P/C**, production in relation to total food *consumed*, and **P/A**, the production divided by *assimilated* food. The former has been symbolized K_1, equaling gross growth efficiency, and the latter is K_2, the net growth efficiency. P/C may be subject to more variability than P/A; yet Mann (1969) states that P/C figures for entire populations are better for showing magnitude of energy flow in and out of a given trophic level.

Table 4-1   Energy budget for a population of the marine bivalve *Scrobicularia plana* in a tidal mud flat*

		Kcal/m^2	kJ/m^2
C	Food consumed	553.8	2317.6
Pr	Gametes liberated	34.1	142.7
ΔB	Net increase in biomass	23.4	97.9
E	Mortality	13.3	55.7
Pg	Tissue produced, ΔB + E	36.7	153.6
P	Total production, $P_r + P_g$	70.8	296.3
B	Mean biomass	113.5	474.8
R	Respiration	265.3	1110.3
F	Feces	217.7	911.3
U	Urine and exudates (not applicable)	—	—
A	Assimilation, C-F-U	336.1	1406.8

P/B, turnovers/annum	0.62
A/C × 100, assimilation efficiency	61%
P/C × 100, (K,), gross growth efficiency	12.8%
P/A × 100, (K2), net growth efficiency	21%
E/C × 100, ecologic efficiency	2.4%
R/A × 100, loss through metabolism	79%

*Means of populations on upper and lower levels of the flat.

Data from Hughes, 1970.

Biomass and Efficiency of Secondary Producers

The Eltonian pyramid can be developed to model individual numbers, individual sizes, biomass, mean energy content, and annual production rates of successive trophic levels. Ideally, there is a diminution upward from the base, although in the case of individual body sizes, the pyramid might balance on its vertex. In an aquatic pyramid of mean mass, the base might be the dissolved solids in the water, followed by layers representing primary producers, bacteria, and the animals. In agreement with the generalization that biomass differs from production, a pyramid based on biomass may be imperfect, whereas production data provide good building blocks for an ideal pyramid (Fig. 4-4).

A generality that is useful, yet only approximate, is to consider a loss of 99% from solar radiation to primary producers, and from then on, as energy flows from one trophic level to the next, a loss of 90% at each exchange (the so-called "Rule of 10" in food chains). This means that biomass or production of the first carnivore level, should be about 1% of primary production or 0.01% of solar radiation. An approximation of such a condition was shown by McConnell (1963), who found that the harvest of centrarchid fishes from Pena Blanca Lake was about 0.98% of the photosynthesis. This generality about energy flow between trophic levels is **ecologic efficiency**, the P/C ratio for all members of a trophic level collectively. It is only a rough approximation because loss of energy varies from 70% to 95% in transfers from plants to herbivores, and herbivores to carnivores. Nevertheless, the principle that animals are more efficient than plants in using energy available to them still applies. Slobodkin (1972) questions the universality of the 10% loss in energy exchanges while denying the probability that evolution could stabilize ecologic efficiencies; he points out, however, that animal effectiveness in taking prey is subject to evolutionary selection as is the ability to escape your predator.

Lindeman (1942b) believed there were grounds for the notion that efficiency increases from the herbivores to the top carnivores of an ecosystem. This idea has not found support in subsequent research on individual P/C or P/A figures; the carnivorous fly larvae in Lake Manitoba seem to be no more efficient than their prey and other herbivore-detritivores in the lake (Tudorancea et al., 1979).

Another ratio R/P (respiration in relation to gross production), shows that, typically, animals spend a greater part of their energy on respiration than plants do. Furthermore, in many instances it appears that carnivores outdo the herbivores in this respect.

Trophic Cascades

Once researchers began to understand the contribution of environmental factors on primary production ("bottom-up" control) and the significance of predation pressure cascading down food chains ("top-down" control), the search was on to determine the importance of each process in aquatic food chains (Hillebrand, 2002; Caraco et al., 2006). The idea that predators can control populations in the trophic levels below them dates from the landmark "Green World" hypothesis (Hairston et al., 1960). The "top-down" control is widely known as the *trophic cascade* (Paine, 1980), and has been amply demonstrated (see Brett and Goldman, 1996, for a review of over 50 studies). Carpenter (1988) edited an important book dealing with the complex biological interactions in lakes.

The trophic cascade is tested whenever predators are added or removed from an ecosystem, deliberately or inadvertently. For example, over a century ago Flathead Lake in Montana was a pristine mountain ecosystem with native species. Ellis and others (2011) describe how first Kokanee sockeye salmon (*Oncorhynchus nerka*, a high-level predator) and later *Mysis diluviana* (opossum shrimp, a predator of herbivorous zooplankton) changed the trophic dynamics of the ecosystem (Fig. 4-7).

The introduction of *Mysis* influenced the food web dramatically. By preying on ostracod zooplankters, *Mysis* both outcompeted Kokanee for that food resource, and allowed the phytoplankton (prey for the zooplankton) to increase. Cladocerans were reduced by 78% overall, and primary production increased significantly as a result (Fig. 4-8).

Whole-lake experiments allow us to directly test important hypotheses in limnology, including the trophic cascade. In some cases such manipulations are associated

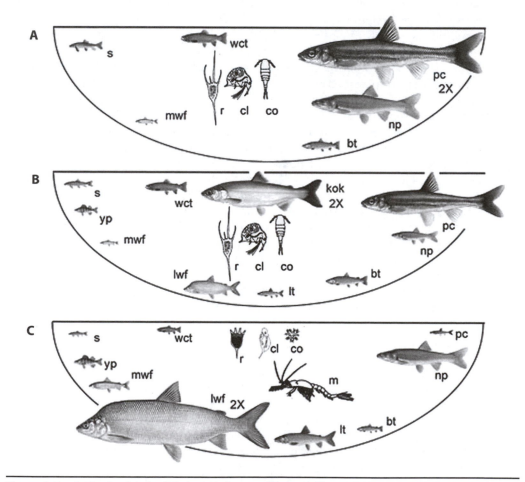

Figure 4-7   A, The food web of Flathead Lake was dominated by native species 120 years ago; **B,** non-native Kokanee salmon [kok] altered this substantially, but **C,** introduction of *Mysis* opossum shrimp [m] had a greater effect, described in Figure 4-8. From Ellis et al. (2011). Reprinted with permission.

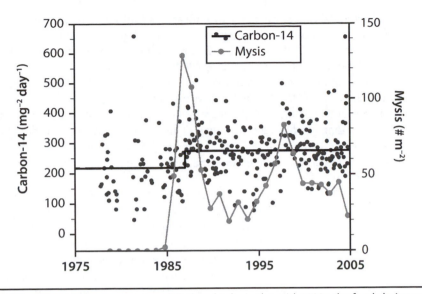

Figure 4-8 After *Mysis* grazers entered Flathead Lake in the mid-1980s, the food chain was permanently altered. Note the marked increase in primary productivity as measured by Carbon-14 uptake. Adapted from Ellis et al. (2011).

with fish stocking for recreation, in other cases they are for research purposes alone. It has become clear that nutrient addition and "bottom-up" effects interact with predatory "top-down" effects (Carpenter and Kitchell, 1996), and specific outcomes cannot be readily predicted. Pace and others (1999) found that nutrient addition did feed a bottom-up increase in biomass, but the effect was mediated at the level of *Daphnia* predators, without increasing higher up the chain. Pace and Cole (1996) found that the lowest level of production (bacterial) benefitted from nutrient addition, but again *Daphnia* controlled further effects. Adding a top-level predator, Pike (*Esox lucius*), to a lake drastically reduced the minnow (*Pimephales neogaeus*) population, with additional effects on lake chemistry as well.

On a different level there are well-known cascading effects and upward consequences when prey and predator interact. Interesting limnological examples of such coevolutionary relationships come from the many endemic species in Africa's ancient Lake Tanganyika. Prosobranch gastropods have developed beautifully sculptured and heavily calcified shells that are unusual for freshwater snails. In response to this protective armor the mollusc-eating fish have evolved strong jaws and massive crushing teeth. More interesting is the presence of ten crab species in the lake and their adaptations to feeding. These decapod crustaceans have developed unusually large calcareous claws that aid in crushing snail shells (Michel et al., 1992).

External Metabolites and Cyclomorphosis

More than four decades ago it was observed that certain planktonic rotifer and cladoceran species assumed different forms at certain seasons. The phenomenon of periodic morphological changes in some populations was termed *cyclomorphosis*. Later it was discovered that some body modifications (tall helmets, tail spines) in prey species are brought about by their predators rather than by seasonal factors. It is likely that temperature or other abiotic factors work together with predation pressure (Yurista, 2000; Laforsch and Tollrian, 2004).

A most remarkable case was worked out by Gilbert (1966) following the observations of Beauchamp (1952). The herbivorous rotifer *Brachionus calyciflorus* responds to some substance produced by its enemy, the carnivorous *Asplanchna,* by reproducing young with long posterolateral spines. The spines effectively hinder *Asplanchna* in attempts to engulf *Brachionus.* (*Asplanchna* has no difficulty in eating the short-spined variety.) This unusual type of embryonic induction occurs before cleavage commences and is ineffective in later stages.

A species of *Daphnia* in the Upper Midwest United States was described in the late nineteenth century. The so-called *D. minnehaha* was the new form, but it was rarely seen again and was forgotten by most workers.

In 1980 Edwards reported the findings from a comparative study of the mandibles of *Daphnia* species founded on scanning electron microscopy (SEM). She could find nothing to separate *D. minnehaha* from *D. pulex.* The next year Krueger and Dodson (1981) described a series of experiments that had shown a water-soluble factor released by the phantom larva, *Chaoborus americanus,* which affects *D. pulex* embryos so that they develop into *D. minnehaha.* The latter is characterized by a spine that makes it less easily eaten by *Chaoborus.*

Since these pioneer research efforts, many more examples of prey organisms' responses to ectocrine substances released by their predators have been published. Perhaps these could be included under the heading of cascading effects of upper trophic levels. Dodson continued his interest in the subject, and in a 1989 paper he stated that predator-induced responses are widespread and ecologically important in freshwater zooplankton populations. More than 20 species of freshwater protozoans, rotifers, and cladocerans are known to respond to predator-produced morphogens, as ectocrine substances inducing body changes have been termed. Ringelberg (1991) described experiments showing that *Daphnia hyalina* alters its phototactic reactions in the presence of a fish ectocrine. Thus, behavioral alteration as well as morphological changes can be induced by external metabolites released by predators. The seasonal connotations of the word *cyclomorphosis* may rarely apply. The structural and behavioral changes seen in many prey species may be almost always explained by predator induction.

5

Lake Origins and Evolution

It is somewhat ironic that the discipline studying ancient history of lakes (*paleolimnology*) is one of the most modern and active disciplines in limnology. It is nevertheless true—we now have better techniques to study lake formation and change, and also a pressing need: our lakes, ponds, and reservoirs are undergoing rapid change. Even more critical is the role lakes play as "sentinels," bearing witness to changes in the land and atmosphere around them (Schindler, 2009; Williamson et al., 2009). Concepts from paleolimnology are included here and in other chapters. A compilation by Cohen (2003) about lake history and evolution is a fine introduction to the field.

We live on a dynamic planet. When we think of events happening in "geologic time" or "at a glacial pace," we may erroneously assume that such processes are all in a distant past or are too slow for us to observe, but nothing could be further from the truth. In fact, today and every day, our continents move around the globe, mountains rise or are eroded down, glaciers scrape and melt, and life modifies the surroundings, including surface waters. And as we shall see later in the chapter, a changing planet affects, and is affected by, humans.

This area is intimately linked with geology, and particularly the branch termed geomorphology. Hutchinson (1957) devoted a chapter to the subject of lake origins, assigning them to 11 major categories; only 2 were not geologic. (The 11 main classifications were fractionated to 76 subdivisions, and 12 of these were divided further.)

Some geologic phenomena that build lakes have been somewhat restricted to certain times, climates, or geographic areas. As a result there are *lake districts* where bodies of similar age and origin are now clustered, or were in the past.

The concave shape of a lake's basin leads inexorably to accumulation of sediments and thus to eventual death of the lake. Erosion and wind bring in allochthonous materials; settling materials formed within the lake itself contribute a portion to the bottom deposits. The trend is toward a terrestrial community where the lake once stood. Some lakes are fleeting features of the landscape, whereas others occupy such deep depressions that millions of years would be required for their filling, or for ongoing forces to preserve the basin, or even for causing it to enlarge.

Glacial Lakes

Background of Glaciology

Important and fundamental contributions to limnology have come from studies of glacial lakes. For this reason, glacial events as lake-forming phenomena are considered first. North American limnology owes much to early workers who investigated glacial lakes such as Lake Mendota, Wisconsin; Cedar Bog Lake, Minnesota (Fig. 5-1); and Linsley Pond, Connecticut.

A short introduction to some geologic terminology is appropriate. We are living in the Holocene epoch, as the recent epoch is generally called. It is the time following the last glacial period, a postglacial span during which our modern soils and landscapes were fashioned. The Holocene's beginning commenced with the ending of the Pleistocene epoch some 10,500 or more years BP (before the present). Together the Holocene and Pleistocene epochs comprise the Quaternary period, which was preceded by the Tertiary period. The Pleistocene was marked by four great ice ages, of which the last, the *Wisconsin*, is of limnologic interest in North America. It corresponds to the North European *Weichsel*, the Polish and Russian *Varsovian*, and the Würm glaciation of the Alps. Perhaps around 17,000 years BP the Wisconsin ice began to dwindle, with deglaciation completed to a great extent 6,000 or 7,000 years later. At the glaciers' maxima about 31.5% of the world's land area was covered.

Figure 5-1　Cedar Bog Lake, Minnesota, the senescent kettle lake made famous by Lindeman (1942). Bog forest and sedge mat occupy most of the former basin. Courtesy of the Cedar Creek Ecosystem Science Reserve and the University of Minnesota.

Today's glaciers, existing in cold regions that receive abundant snowfall or where snow persists, cover about 10% of the land. However, warming continues, and at an accelerated pace from human activities (global climate change). The loss of ice cover on Earth is faster, shrinking glacial and ice sheet coverage (Shepherd et al., 2012).

The ice sheets of Pleistocene glaciation left the upper latitudes of the Northern Hemisphere well endowed with lakes. Immediately beyond the ice margin in periglacial areas, tundra and permafrost prevailed, and local glaciers developed in mountains farther south. Farther from the ice and beyond the permafrost, climatic conditions were such that lakes developed in basins that are now dry or contain saline relicts only.

The Pleistocene was a period of greater precipitation, lower temperature, and marked cloudiness as compared with the present; and most important, it was a time of reduced evaporation—a pluvial age. Pluvial lakes owed their existence to the climatic trends that produced glaciers to the north and at higher altitudes, but the depressions they occupied were products of phenomena other than glacial activity.

Lakes Associated with Existing Glaciers

Certain lakes exist only when in contact with active glaciers. This type of lake is, of course, much scarcer at present than in the past. Pools of water lying in the icy depressions on glaciers or collecting beneath to form bizarre subglacial lakes, and streams dammed by advancing tongues of ice from mountain glaciers, are good examples of such types today. The literature concerned with these lakes stresses their impermanence caused by impounding ice that gives way to release waters with a rush (sometimes in a dangerous manner).

Summer visitors to the Columbia Ice Fields in western Canada may have seen another characteristic type. A pool of water lies at the glacier's extremity in a depression bounded by the ice and a morainic dam below.

Lakes Formed Near Glaciers

Waters associated with active glaciers are perhaps curiosities, but lakes that exist now because of long-gone glaciers are near the heart of limnology. Of these, some lakes formed because of colder past climates or the proximity of glaciers, and indeed this type of lake—the **cryogenic lake**, occupying a *periglacial frost-thaw* basin—is being produced today.

Lozinsky (1909) introduced the term *periglacial* to refer to the cold climate beyond glacial ice sheets. According to Péwé (1969), the modern periglacial zone is marked by permafrost areas, where the ground is permanently frozen, with only a relatively thin upper layer thawing in the summer. Geologists find evidence for former periglacial environments in unusual soil configurations called ice-wedge casts. These indicate the former presence of vertical, wedge-shaped veins of ground ice. Mean annual temperatures from ${}^{-}1.0°$ to ${}^{-}2.0°$ C are a prerequisite for the existence of ice wedges, but their active growth requires temperatures from ${}^{-}6.0°$ to ${}^{-}8.0°$ C.

The physiographic province known as the Arctic Coastal Plain in Alaska covers more than 67,500 km^2, lying adjacent to the Arctic Ocean from 69° to 71° North latitude. In it are tens of thousands of elliptic lakes, oriented on NW-SE axes and, on the average, pointing N 12° W. Their basins are above permafrost, where subzero temperatures are prevalent not far below the soil surface. Limnologists have explained the genesis of these elliptic lakes by localized melting that forms tiny pools. Loss of insulation by damage to plant cover may initiate this process. The long axes of most of these lakes lie at right angles to the modern summer winds. Livingstone (1963a) gives a good account of these Arctic Coastal Plain lakes and of the theories and controversy concerning their orientation.

A district of mostly extinct lakes in the New Jersey coastal plain is composed of hundreds of basins stretching from the Raritan River at New Brunswick, about 8 km south of the greatest advance of the Wisconsin ice, to the tip of Cape May, 65 km farther south. These lakes are shallow, about 3 m, and in many cases are filled with peat. They range up to 2.5 km^2, but most are much less expansive. The presence of many ice-wedge casts in the soils of this part of New Jersey leaves little doubt that these are periglacial frost-thaw basins that formed near the continental ice mass more than 10,000 years ago (Wolfe, 1953). They are irregular in outline and unoriented, quite unlike the Alaskan periglacial lakes that are still forming today.

Pingo is an Eskimo word applying to certain hills rising above the surrounding tundra. A pingo is formed when water under pressure ascends through gaps in the permafrost, freezes, and lifts an ice dome that is covered with alluvium and tundra vegetation. During uplift the alluvial cap ruptures, and the ice, no longer insulated, melts to form a crater that may contain water. Likens and Johnson (1966) described an unusual lake in a pingo near the small Alaskan settlement named Circle. The pingo is 450 m in diameter. The thaw crater in its center is 13.8 m deep and contains 8.8 m of water forming a nearly circular lake, quite different from the oriented lakes of the coastal plain farther north.

Lakes Formed Where Glaciers Existed

Relevant terminology.   Ice that moves slowly down mountain slopes, spreading out over the piedmont regions, perhaps joining other ice masses to form expansive sheets, is not clean. It carries rocky material of many sizes, from boulders to silty particles, and it scours and pushes materials ahead as it advances. When the ice melts, the rocky debris is left behind.

Glacial *outwash* applies to material carried from and beyond the ice as a glacier retreats by melting. A melting, isolated ice mass, stranded by its retreating parent ice sheet, can also produce outwash material. This may be designated as *drift*, a general term for any accumulation of direct or indirect glacial origin, including boulders, sand, and gravel. The word *moraine* designates drift deposited directly by a glacier; outwash, therefore, is not morainic.

Lakes in ground moraine.   Ground moraine is the drift left on a surface over which a glacier has moved. Upon melting, the glacier leaves an expansive region of irregularly surfaced drift, perhaps containing masses of detached ice like raisins in a cookie. Such dead ice blocks may persist for centuries because of the insulation afforded by overlying drift; but they eventually melt, leaving water-filled pits. *Kettle* (or *pit*) *lakes* occupy such depressions. The irregular low spots in the ground moraine contain water if the climate permits. Thus, in recently deglaciated regions such as the Ungava Peninsula of Quebec, much of the northern area is water. Shallow lakes with irregular shorelines are abundant. Erosion subsequent to deglaciation has not altered them much or filled them in. Farther south, in the Great Plains physiographic province, a later stage is seen. In the rolling plains, the "knob and kettle" topography of North Dakota, for example, prairie sloughs represent deeper ground-moraine lakes of the past. Many have been obliterated in the last 10,000 years.

Kettle lakes may be many hundreds of years younger than one might expect. They do not necessarily form immediately after the glacier's recession. Insulated by an overburden of drift, the detached ice mass persists long after the parent glacier has disappeared from the region. A typical plant succession may follow, supported by the drift above the residual ice. Herbaceous species, appearing first, give way to alder thickets, which are followed in turn by forests. As a result, cores of sediment taken from some North American kettle lakes yield spruce needles, twigs, bark, and other evidence of a boreal forest *below* the sediments of lacustrine origin (Florin and Wright, 1969).

Lakes formed by morainic impoundment. Several categories of morainic lakes are formed by ponding rather than by ground moraines. A glacier, moving down a valley, blocks and impounds a tributary, and long after its ice has wasted, a damming lateral moraine is left behind. Lakes resulting from such impoundment are not rare, but a more common type is formed by a terminal moraine, marking the farthest extent of the ice. In many such cases, however, *corrasion* (mechanical erosion) was also involved. Thus, the grinding abrasion of a tongue of ice moving down a river valley converts it from a V-shaped trough to a typical glacial valley, U shaped in cross section.

Fremont Lake (Fig. 5-2) in the Wyoming mountains illustrates the interplay of corrasion and morainic ponding (Rickert and Leopold, 1972). The elongate lake, which must be one of the finest pristine lakes remaining in the United States, is in a trough of granite, scoured by glacial debris and ice and plugged downstream by a terminal moraine. The lake covers 20.6 km^2 and is 185 m deep.

As a more complicated example, the 11 Finger Lakes

Figure 5-2 Fremont Lake, Wyoming, an ultraoligotrophic glacial lake formed by corrasion and morainic damming. Photo by David L. Gaskill, USGS.

of New York state are narrow bodies of water, semiparallel, oriented north-south, and impounded at *both* ends by drift. They were formed in river valleys, gouged out by ice to depths below sea level in the two deepest, Seneca and Cayuga lakes. The southern limits are blocked by complex terminal moraines of a Wisconsin ice advance somewhat older than the moraines at the northern shores.

Lakes formed by hydraulic force on drift.   During temporary decreases in the rate of glacial wasting, recessional moraines are left behind. Where retreats are marked by several rate changes, complex and jumbled morainic hills eventually form. Such morainic mounds and valleys in Itasca State Park, Minnesota, are called the Itasca Moraine. Hundreds of lakes, ponds, and bogs occupy depressions in that region. Many of these existing lakes are elongate, a characteristic that leads to the proposal that melt water, streaming under pressure beneath the ice, washed out deep elongate basins in the drift. This explains the morphology of Lake Itasca itself and a topographic continuation of its southeast arm—a narrow, deep valley, dry now except for a few deeper depressions that contain isolated, well-protected lakes. One of these, Demming Lake, is best known, thanks to the work of Hooper (1951) and Baker and Brook (1971). Hydraulic force, then, is another glacial activity in lake building. Hutchinson (1957) discussed examples of this phenomenon that occurred in Denmark, Germany, and Poland.

Glacial scour lakes.   Although emphasis has been placed on basins in drift depressions and ponded by drift barriers, corrasion as a lake-forming force needs attention. The striking corrasive effect of ice moving down valleys is seen in modern *fjords*. They are restricted to coastal areas of high relief where there was severe Pleistocene glaciation. Towering cliffs rise abruptly above deeply scoured, elongate, narrow passageways. Fjords commonly connect with the sea, but many are lakelike, above sea level, and separated from marine waters by rocky sills at their outer ends. The process of isostatic elevation (uplift after having been depressed by ice) is very important in the genesis of a fjord lake. The fabled Scottish Loch Ness is one of these.

Great Slave Lake in the Northwest Territory of Canada is the deepest lake in North America. Its surface is 150 m above sea level, but its maximum depth is reached about 464 m below sea level. This basin is largely the result of corrasion beneath a great mass of continental ice. A more remarkable example is the Norwegian Mjiisen, a glacial scour lake with its shoreline 395 m above sea level, but having maximum depth at 720 m, a point 325 m below mean sea level.

Familiar examples are the Laurentian Great Lakes of North America, which owe a great part of their origin to glacial corrasion, although tectonic events were also involved. Lake Superior especially, lying in an enormous Precambrian syncline, might be better classified as a lake of mixed origins.

Cirque lakes.   At the very head of a mountain glacier is an abrupt crevasse, the Bergschrund of German authors. It is bounded by the ice and an abrupt, nearly vertical headwall of rock. Daily freezing and thawing of the melt water at the base of the crevasse, and of rain and melted snow trickling down the headwall, shatter bits of the rock from it. In addition, the glacier grinds the rock beneath it, perhaps by rotational slip of the ice. The result is a cirque, typically with a concave floor that meets the headwall sharply above and is bounded below by a lip or threshold of rock. These armchair-shaped hollows can hold cirque lakes. In English-speaking regions **tarn** has become the name for these appealing mountain lakes.

Tectonic Lakes

Tectonism refers to the earth's crustal instability and all its behavior. Warping, faulting, fracturing, buckling, folding, thrusting, and quaking are words that apply to deformation and adjustment of the earth's shell; they describe **tectonic** phenomena. From crustal movements come depressions that can hold water, in some cases making up the most remarkable lakes of the world.

Graben, Fault-Trough, or Rift Lakes

A **graben** is a depressed, usually elongate block of the earth's crust lying between faults and adjacent highlands. Typically the graben or fault-block is untilted, forming the flat bottom of a trough. Its development was preceded by crustal fracturing called faulting (Fig. 5-3). Two parallel faults or rifts isolated a block that sank lower to become the graben. In addition, its depression may have been relatively increased by

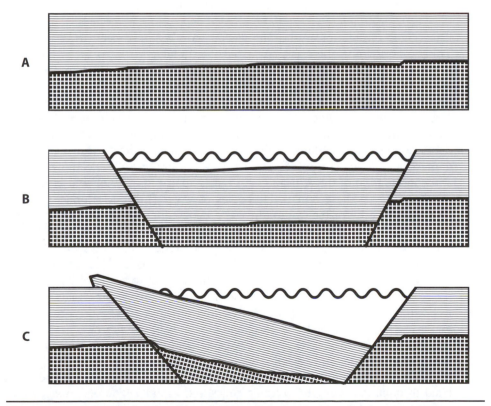

Figure 5-3 Formation of graben lakes. **A,** The rocks before faulting; **B,** lake occupying a symmetrical graben; **C,** lake occupying a graben with a tilted fault-block.

attendant crustal uplift of land on either side. The ideal symmetric graben forms a steep-walled, flat-bottomed lake such as Lake Tahoe, California-Nevada.

The most remarkable graben lakes are two ancient bodies of water, Baikal of Siberia, with an unsurpassed depth of more than 1.6 km, and the second deepest, Tanganyika of East Africa. Both are typically elongate.

A mosaic of crustal plates makes up the surface of our planet. The resulting fracture systems create extensive rifts. At least two of these are of special limnological interest. The great Mid-Atlantic Ridge is an elongate volcanic mountain chain lying beneath the Atlantic Ocean from above the Arctic Circle to at least 60° south of the equator. Rifting has occurred and still proceeds along the crest of the Mid-Atlantic Ridge. Iceland lies atop the Ridge, and some remarkable subarctic lakes are found in the exposed spreading axis. One of these is named Thingvallavatn, and results of ecological studies on the lake and its surrounding catchment area are presented in a remarkable book edited by Pétur Jónasson with contributions by 33 other authors (Jónasson, 1992). The lake lies in an elongate graben, the floor of which is subsiding at the rate of from four to five mm every year. Thus, the lake is expanding and deepening; its present maximum depth is 114 m, the deepest 14 m being below sea level. Thingvallavatn is no older than 10,000 years and illustrates the principle that different interacting geologic processes may create lake basins. Tectonic subsidence and glacial erosion have played roles in forming its basin; four postglacial lava flows have entered the basin, creating a relatively flat-bottomed depression, and one flow impounded the lake. Thus, rifting, volcanism, and glacial scouring have been involved in forming Thingvallavatn, cited here as an example of a rift lake.

The Red Sea trough and its northward continuation, the Jordan-Dead Sea Rift, are part of another fracture system. Extending from the south end of the Red Sea, two enormous rents in the earth's crust make up the African Rift valleys. The eastern rift extends south for 1,500 km into Tanzania, containing many more small lakes than the western rift. Lake Malawi, at least 760 m deep, and Tanganyika, with a maximum depth of more than 1.4 km, are the most remarkable of the African lakes and occupy depressions in the western rift. The results of a symposium concerning the resources and conservation of the African Great Lakes appeared in 1992 and serve as a source for literature on these remarkable bodies of water (Lowe-McConnell et al., 1992). Extremely worthy of recognition here is a book edited by G. W. Coulter (1991) that discusses life in Lake Tanganyika. Most of the other large African lakes are mentioned, and a remarkable list of more than 1,600 references is presented.

The tectonic lakes of the African Rift valleys are not so idealized as Lake Tahoe, for asymmetry is common among them. With respect to sea level, the lowest part of the African Rift valleys (if indeed it belongs to the same system) holds the Dead Sea, its bottom being 793 m below sea level. Utah's Great Salt Lake also occupies a graben and is reminiscent of the Dead Sea, although not below sea level.

The large freshwater lakes in the Nicaraguan Lowland, Nicaragua and Managua, are in an area famed for volcanic activity. At one time they were believed to represent results of Pleistocene and Holocene volcanism, being remnants of a Pacific bay isolated through damming by volcanic cones and ashfalls. The occurrence of sawfish, tarpon, and a shark, erroneously assigned Pacific affinities, strengthened this idea. However, the modern concept is that the lakes occupy a late Tertiary–early Quarternary graben, and the evidence for a marine past is lacking. The disastrous earthquake of 1972 in Nicaragua emphasized that the lakes are in a region of tectonic as well as volcanic activity.

Uplift Lakes, the Result of Epeirogeny

Epeirogeny is the term applied to wide-reaching tectonic events that raise large crustal blocks and sometimes bring about the formation of enormous basins. It is a slow process when compared with orogeny (typical mountain building by folding and thrusting of the earth's crust). The prime example of an epeirogenetic lake is the Caspian Sea, with the greatest area of them all. It appears to have been originally a submarine depression. During the Tertiary it was raised to inland status, cut off from the sea, and subsequently freshened.

The major high-altitude lake in the world is Lake Titicaca, lying across the boundary of Peru and Bolivia high in the Andes more than 3,800 m above the sea. It is fed largely by snowmelt. A rather rapid epeirogeny was one of the tectonic events in the history of the lake. Its basin, a low spot in a great plateau of inland drainage known as the Altiplano, is thought to have been raised on high from altitudes but a few hundred meters above the Pacific during and since the Tertiary.

Floridians are proud of Lake Okeechobee, which, they point out, is the largest body of fresh water entirely within the boundaries of one state. (The dilute nature of the water must be stressed in order to take the Great Salt Lake out of the competition.) Okeechobee occupies a shallow depression in land that was lifted by epeirogeny from below the sea surface. It holds water largely because of the damming effect of an accumulation of plant material.

Earthquake Lakes

Reelfoot Lake, Tennessee-Kentucky, came into being almost overnight in 1811, or so the mythology of the region tells us. A series of earthquakes created lowlands into which Mississippi River water spilled. The wooded, subsided areas received waters rich in nutrients. The largest depression is Reelfoot, noted for its exceptional sports fishery. Much of the basin is now marshland, and because of erosion from adjacent farms, the lake is shoaling rapidly.

Landslide Lakes

Many lakes throughout the world owe their existence to the impoundment of stream valleys by rock slides, mud flows, or other mass movements of soil and rock. The lakes formed by such impoundment are often short lived because a stream obstructed by such a barrier might eventually overflow and destroy it.

Mountain Lake, Virginia, has been the scene of summer biology instruction for many years and the site of some interesting limnologic research. Hutchinson and Pickford (1932) showed that Mountain Lake's morphology overrides its edaphic heritage, displaying many aspects of eutrophy despite its nutrient-impoverished water. Roth and Neff (1964) summarized the pertinent preexisting literature on Mountain Lake and published the results of their own extensive limnologic study there.

Mountain Lake is the only natural lake of consequence in the unglaciated Appalachians. Sometime during the early Holocene a mountain valley was dammed by a landslide of sandstone. Its level may have fluctuated much since then, but now it is about 32 m deep. It has not breached the slide-rock dam, because it is fed by no more than precipitation, seepage, and underwater springs.

In the Victorian highlands of Australia there is but one permanent natural body of water. This is Lake Tali Karng, a landslide lake covering an area of 16.2 ha and reaching a maximum depth of 51 m (Timms, 1974).

Although most landslide lakes are solitary, Martin (1960) studied an entire high-altitude lake district of this type in Costa Rica. The lakes are boglike, in some instances completely filled in. They are found in narrow valleys at altitudes where precipitation is high, but streamflow is not continuous. Landslides are common along the steep-sided valleys, impounding runoff to form numerous small ponds. Many of these have been able to run the course of their life cycles to extinction because streamflow has not been great enough to destroy the landslide dams.

Lakes Formed by Volcanic Phenomena

Geography of Volcanic Lakes

Volcanism has modified and shaped past landscapes in many areas of the world and continues to do so today. The depressions owed to volcanic activity, then, occur in many countries, and their formation is in no way related to climate. To keep basins full, however, a water source is needed, and today there are dead volcanic lake districts in arid zones where lakes flourished in the pluvial climate of the Pleistocene.

Important contributions to limnology have come from studies of volcanic lakes. For example, a large portion of Japan's limnologic endeavor has been focused on lakes formed by volcanic activity. Also, much of what is known about tropical limnology is based on studies of volcanic lakes in Java, Sumatra, and Bali (Ruttner, 1931) and the Central American lakes investigated by Juday (1916), Deevey (1957), and Armitage (1958).

Volcanoes are common in areas of tectonism. Thus, among the graben lakes of the African Rift valleys, there are many small lakes occupying depressions created by volcanism.

Crater Lakes

The deepest lake in the United States is Crater Lake, Oregon, occupying a *caldera* (Fig. 5-4), a collapsed volcanic crater formed when underlying magma (molten rock) flows out. Crater Lake is an impressive 592 m deep with over 5400 hectares in surface area. Other crater lakes are contained in **maars** (from the German Maare), formed by subterranean explosions that created low rims but left little cinder and lava material. Many of these explosion craters form deep circular lakes with relatively small areas. The Lac d'Issarlès in France is 108.6 m deep with an area of 92 ha. Pulvermaar in Germany holds a lake covering 35 ha and 74 m deep.

In western New Mexico, the Zuni Salt Lake, which Bradbury (1971) studied, is a maar that was formed by volcanic explosions in the late Pleistocene. It is quite different

Figure 5-4 Crater Lake, Oregon, a magnificent caldera lake and the deepest body of water (592 m) in the United States. Photo courtesy of the National Park Service.

from the deep holes occupied by most maar lakes. A subsequent explosion or explosions caused lava and cinders to well up and fill the crater as secondary cinder cones were produced. The main lake occupies an area of 60 ha with a maximum depth of 1 m. From it emerge three secondary cones, one of which is almost perfect, having undergone very little erosion and containing a small circular lake, which seems to be a comparatively unmodified crater rather than a collapsed caldera. Its enclosed pool lies above the level of the main lake and has a mean depth of 3 m. Such a crater lake, present in an unmodified cinder cone, is rare. Most are contained by maars or calderas.

Lakes in Lava Depressions

Usually, lake-holding depressions in lava flows are caused when the upper layer cools, forming a crust over the hotter magma beneath. The magma may drain out, leaving space into which the crust sags. The result is a shallow depression. Collapse following the withdrawal of magma created quite a different type of depression in the magnificent, deep (340 m) Guatemalan Lake Atitlán, studied by Deevey (1957). Cauldron subsidence, the name given to the process whereby this lake was formed, involves the lowering of a great cylindrical block, fractured peripherally, into a chamber created by the outflow of molten rock.

Coulee Lakes

The word *coulee* is commonly used in the American Southwest to refer to dry gulches and arroyos. The geologic word, however, refers to solidified volcanic flow and not to the valley it impounds.

Coulee lakes are formed by lava impoundment. Perhaps the molten material pours across valleys to dam river channels or blocks only gentle depressions. An excellent example of a coulee lake, a newcomer among ancient rift lakes, is Lake Kivu between the African nations of Zaire and Rwanda. A river that flowed north to Lake Rutanzige (Lake Edward) on its way to the Nile was blocked by a late Pleistocene lava flow. The magnificent, deep (480 m) Kivu was the result. It did not breach the lava dam, but it subsequently overflowed southward and now drains to Lake Tanganyika.

A little lake district on the Colorado Plateau near Flagstaff, Arizona, is present because of past volcanic activity. Most of the lakes are nearly extinct, reviving during rainy years. Some are impounded by lava dams or occupy much eroded calderas; others occupy shallow depressions in sheets of lava. Nearly all the methods by which volcanism creates lakes can be demonstrated in this area.

Solution Lakes

Lakes in Carbonate Substrata

Basins created by dissolution and removal of material may later hold water to become lakes or ponds. Usually the carbonates of calcium and to a lesser extent magnesium are the solutes. Although calcium carbonate is only slightly soluble in pure aqueous solutions, it dissolves in acidic water, and karst features may develop in climates where there is ample precipitation. The term *karst* refers to the sum of surface and subsurface aspects that characterize regions where carbonate rocks and particularly the common limestone are exposed. Generally, in karst regions there are underground passages and channels, increasing permeability to such an extent that surface streams are scarce. **Cryptorheism** (hidden drainage) prevails, soils are scanty, and the topography may be rugged and pockmarked by the collapse of surface rocks into subterranean caverns.

Water collecting in surface depressions may seep into joints and cracks in the limestone below, enlarging the lines of weakness and draining the surface. Circulating water removes soluble material and weakens overlying layers, eventually creating depressions called *sink*, **doline** (**dolina** in the singular), or *swallow hole*. These depressions may be conical, shallow and dishlike, or steep walled, resembling wells. At times they connect with underground channels and contain water continuously or irregularly. Soils and debris tend to wash into the sinks, choking them so that they hold water even when the groundwater table drops. Such systems are unstable, however, and there have been exciting instances of plug failure in a sinkhole followed by the abrupt drainage of its lake to subterranean passages.

One of the world's most extensive karst tracts is the level Nullarbor Plain of southern Australia, but at present watering places are scarce in that arid region. The karst terranes of deserts are relicts because corrosion of limestone continues only as long as a supply of acidic water exists. Commonly it is carbonic acid (H_2CO_3) formed by the hydration of CO_2. Karstification proceeds most efficiently, therefore, in humid limestone regions.

The Tertiary limestones capping central Florida have created the best North American example of a large lake district due to solution. Because the topography is low and the climate humid, the water table is high and lakes are common. Many of these are nearly circular, typical of solution-collapse basins. Others lie in basins formed by coalesced doline and are, therefore, less regular in shape, being somewhat like a figure 8 if two were formed together, but much more complex in shape if more sinks were involved. At least one of the Florida lakes is more than 30 m deep, but most of them are probably 5 m or less. An article concerning a detailed study of 55 of these lakes and ponds contains a short summary of the limnologic contributions that have come from the Florida lake district (Shannon and Brezonik, 1972).

Typical also of karst landscapes are large springs situated at widespread intervals, discharging great quantities of water from enormous arterial channels formed from enlargement of solution crevices and fractures in limestone. Silver Springs in Florida, where H. T. Odum (1957) did a well-known study of community energy flow among trophic levels, is one of these.

Most of the Yucatán Peninsula in Mexico has a flat karst topography overlying limestone. Much of the rock is exposed and fluted by solution depressions. There is no surface drainage; therefore, steep-sided solution basins known as *cenotes* serve as water sources. When these circular sinks connect with the water table, they form natural wells. (The ancient Mayans built temples by some of the cenotes that they considered sacred. As a result, these solution lakes have provided rewarding opportunities for modern underwater archaeologic research.) Limnologic research on the cenotes has been rare, but it began with Pearse and associates (1936), who studied 30 of them ranging in depth from 0.5 to 54 m. Because the Yucatán Peninsula is such an extensive karst shelf (300,000 km^2), the underground hydrology is extensive. In addition to thousands of sinkhole cenotes, a series of watery caves are connected to form a subterranean river whose length is over 150 km, only navigated by SCUBA divers in 2013. Kambesis and Coke (2013) wrote an excellent description of this unique landscape and its hydrology.

Parts of the states of Indiana, Tennessee, Kentucky, and Missouri are noted for their karstic topographies and cave districts. Although small water-filled doline are not rare, many of the associated aquatic habitats are underground streams and lakes, and most of the research done on them has been faunistic in its approach. In southwestern Kentucky some dry sinks connect with subterranean waters during the early spring when the water table rises. At that time, water flows out to cover low depressions (which will be cornfields later in the year), and two aquatic communities mingle. The cave fish *Chologaster* emerges from the caverns below to feed on the rich crustacean fauna of the **epigean** temporary pond. Then it returns to the underground habitat as the water recedes, having enjoyed a vernal banquet most cave species never experience.

The Kaibab Plateau in northern Arizona is capped by a remarkably flat, thick layer of limestone. As a result, a little-known lake district composed of sinks occurs north of the Grand Canyon of the Colorado. It must be pointed out, however, that it is composed of small ponds, many of them transitory, that have attracted no biologists except those interested in fairy shrimps, copepods, and larval salamanders. In well-watered areas this "lake" district would receive even less attention.

Perhaps the most remarkable karstic lake is Spain's Lake Banyoles. The lake has six major depressions where contrasting subterranean waters enter. The papers by Casamitjana and Roget (1990, 1993) have references to many publications addressing this multibasin lake.

Lakes in Salt-Collapse Basins

Calcium carbonate and to some extent magnesium carbonate are the common solutes in karst regions. Corrosion, or chemical erosion that requires acidity, has been the main agent. Other substances that are readily dissolved by water without carbonic acid can also play a part in lake formation. For this reason, solution lakes occur at times in noncalcareous regions. Salt karst lakes formed by the dissolution of evaporites such as $NaCl$ and $CaSO_4$ are widespread. They may be more common than hitherto suspected.

In some instances, a lens of salt (e.g., $NaCl$, Na_2SO_4, $CaSO_4$) beneath brittle limestone may be the solute. Water entering the salt pocket may cause expansion, thereby cracking the overlying rocks. If the salt is carried away in solution, the limestone layer may collapse, producing a steep-sided hole.

The origin of the circular Colombian lake called Laguna Guatavita was controversial until Dietz and McHone (1972) asserted that it was not the result of a meteor's impact, but rather that it was formed by groundwater sapping a deep-lying salt pocket. The authors stressed the occurrence of many such sag lakes elsewhere.

The Bottomless Lakes of New Mexico, far from fathomless, are aligned in the Chalk Bluff formation east of Roswell. The lakes are solution basins lying in material composed of very little limestone. They form a small salt karst region in material of two sulfates of calcium, gypsum and anhydrite. These compounds are far more soluble than $CaCO_3$. The Bottomless Lakes are steep-walled, bringing to mind the Yucatán cenotes lying in a limestone region far to the south.

Mound-Spring Lakes

This lake category is based on solution and subsequent precipitation. The name is derived from the Australian mound springs. A great artesian basin occupies 22% of that continent's area. Where water emerges, solutes and colloids precipitate to form mounds that are usually calcareous but are sometimes siliceous or ferruginous. The mound springs, being island-like in distribution, contain many endemic plant and animal species (Ponder, 1986).

Montezuma Well in Arizona (Fig. 5-5) has been described as either a typical solution basin originating from the collapse of a limestone cave or a salt-collapse sag lake owing its origin to the dissolution of a lens of salt beneath a thick layer of limestone. A third explanation, however, appears to be correct, that this is not a typical "sinkhole." Nations and associates (1981) termed the Well a *collapsed travertine mound*. Subterranean water carrying dissolved limestone precipitates a calcareous deposit called *travertine* when it reaches the surface and CO_2 is lost. At first a disc of travertine forms around the spring opening; it grows increasingly convex. As time passes a steep vaulted structure is built with a top central opening through which water passes. If, by chance, a lower lateral opening breaks through and the spring water no longer passes out the mound's top,

Figure 5-5 Montezuma Well, Arizona, a collapsed travertine mound. Photo by Tony the Marine 69-71/Wikimedia.

a cave-like empty dome remains. This is prone to central collapse. This seems to have been the history of Montezuma Well. At present the lateral opening (termed the "swallet" by National Park Service personnel) lies about 24 m below the rim of the collapsed mound. Recent study (Johnson et al., 2012) indicates that the lake is stable, despite increasing demand for water withdrawals from the nearby aquifer. Like an Australian mound spring, Montezuma Well contains endemic plant and animal species.

Piping: False Karst Lakes

The Chuska Mountains lie across northwestern New Mexico and northeastern Arizona. They are topped by a flat layer of sandstone in which there are many unusual water-containing depressions. Some interesting studies in palynology and paleolimnology have come from this lake district (Bent, 1960; Megard, 1964), and the origin of the lakes is especially remarkable. Wright (1964) attributed the depressions to the geologic process of piping, which produces tubular subsurface drainage channels in insoluble elastic rocks (rocks composed of fragments that have been moved individually from their place of origin), simulating what results when solution occurs in calcareous rocks. Cemented sandstone layers collapse into the vacuities produced by the piping of uncemented sand out to the steep escarpments that bound the mountains.

In the case of solution basins formed by collapse, water removes $CaCO_3$ or $CaSO_4$ molecule by molecule. In piping, sand is removed grain by grain by suspension in moving water. The piping could not occur, however, without the proximity of precipitous slopes. The Chuska Mountains are steep-sided mesas. Tunnels developing in their sandstone cap serve to carry loose sand away and permit it to be dumped down the mountain slope, leaving water-filled depressions with such colorful names as Whiskey Lake and Dead Man Lake.

Another lake district that may owe its existence to piping is seen along one part of the Mogollon Rim in Arizona. This steep and scenic escarpment marks the abrupt southern edge of the Colorado Plateau. Along the Mogollon Rim are a series of lakes and lake remnants that can be termed the Potato Lake series after Whiteside's (1965) paleolimnologic study of a small dystrophic lake by that name. Most occupy depressions in the Coconino Sandstone.

Lakes of Aeolian Origin

In arid regions of today or in regions of marked aridity in the past, the erosive force of wind has left its mark; in some instances lake basins have been formed. (The climatic requisite for them is the opposite of that needed for karst formation.) Several categories of lakes occupy concavities fashioned by aeolian forces. Wind-created hollows occur at the top of siliceous dunes. In the moister regions plant growth, organic accumulation, and cemented sand form an impermeable layer and hence a perched water table.

Moses Lake in the cold desert of eastern Washington is an example of a lake that originated by damming of streams by drifting windblown sand. Depressions among dunes may contain water, especially if the wind has scooped away sand to approach or expose the water table. The Sand Hills region in some western counties of Nebraska contains many small, in some instances very saline, waters occupying interdune basins. The perched dune lakes of eastern Australia owe their origin to aeolian forces (see Timms, 1986).

A lake district composed of thousands of shallow depressions in the panhandle of Texas and adjacent eastern New Mexico (the Southern High Plains, Llano Estacado) has been considered an example of wind work (Osterkamp and Wood, 1987; Holliday, 1997). They are sometimes called buffalo wallows, and large hoofed animals certainly use and modify the basins. After a period of precipitation, rain-filled depressions attract herds that trample the soil, thereby destroying stabilizing plant growth and deepening the basin. Later, when the depression is dry, the wind removes dried soil, a process called deflation. In South Africa basins of this type are called pans; Hutchinson and associates (1932) published an important limnologic report on them.

Fluviatile (River) Lakes

Ponding by Deltas

Rivers, which are lotic environments, create standing-water habitats (or **lentic** environments) by various actions. Impoundment of a stream can be effected by wind, ice, lava, landslides, humans, and beaver, but a second river can do the same. A good example in North America is Lake Pepin, a wide place in the Mississippi River where current is slowed and the river somewhat widened. The river lake was formed by the Chippewa River flowing from Wisconsin and depositing material where the current slackened to create a partial dam across the Mississippi River. In the past, Lake Pepin was noted for its pelecypod fauna that supported a nineteenth- and early twentieth-century pearl button industry and for its fish, the most noteworthy being the spoonbill *Polyodon*. The lake is currently noted for its swarms of emerging chironomid dipterans—midge flights that often assume nuisance proportions.

Floodplain Lakes

In the mature reaches of a stream course, where gradient is much reduced and there is a broad floodplain over which the river has wandered in the past, other fluviatile lakes may be found. The Australian word *billabong,* of aboriginal origin, refers to both permanent and temporary waters on riverine flood plains in that continent.

Levee lakes. Levee lakes are shallow, often elongate bodies of water that lie parallel to the streambed and are separated from it by strips of higher land. The barrier is a natural levee of river sediments deposited at times of high water, although such barriers are occasionally breached as water is delivered to the lateral levee lake.

Oxbow lakes. A second type of floodplain lake is the oxbow lake formed from isolated loops of meandering, mature streams (Fig. 5-6). These crescent-shaped basins are usually deeper than the levee lakes because they occupy old segments of the river, which may reach lower than adjacent depressions in the floodplain where levee lakes form. They are the commonest billabong type in Australia according to Hillman (1986).

The North American fluviatile lakes, associated with mature alluvial plains and gentle gradients, are most abundant in the Central Gulf states. Relatively speaking, they assume even more importance there because there are no glacial, tectonic, or volcanic lakes nearby.

Evorsion (Pothole) Lakes

Evorsion is defined as the complex mechanics of pothole erosion, streambed abrasion that occurs under vortices and eddies in torrents or beneath waterfalls. This should not be confused with the small, shallow "potholes" caused by glacial action in an otherwise flat landscape. A category of fluviatile lakes associated with torrential flows and cataracts includes the evorsion lake. Stones, driven by hydraulic pressure and swirling in circular fashion, cut vertically into the stream bedrock, gradually forming a pothole.

Figure 5-6 An oxbow lake. Photo courtesy of the Arkansas Game and Fish Foundation.

In 1956 Eggleton wrote the first detailed limnologic report on Fayetteville Green Lake, New York. Since then several more studies have been made on this unusual lake. Called a "plunge-basin" lake, it was formed beneath a waterfall during a recessional stage of the late Wisconsin ice age. The main portion of the lake occupies a deep cylindrical cavity.

Edmondson (1963) and his students studied another important group of evorsion lakes in eastern Washington. They are the Grand Coulee lakes, lying in linear fashion beneath an enormous escarpment over which the ancestral Columbia River cascaded. The lakes display a sequence of salinity, and studies made on their biota have provided insight into osmotic and ionic tolerances of many aquatic organisms.

In some mountainous areas of southwestern North America occur cylindrical potholes worn in steep rocky washes that are dry much of the year. These are called *tinajas* from the Spanish word for earthen jar. Some of the deeper ones are perennial, but most are temporary aquatic habitats. Kubly (1992) discussed the invertebrate fauna in tinajas of the White Tank Mountains in the Sonoran Desert of Arizona.

Shoreline Lakes

Wave action at shorelines constantly molds and modifies lake basins; this is especially obvious in newly created lakes. There are also good examples of shoreline lakes formed where barriers of sand are thrown up to cut off bays from the sea. These are especially frequent along drowned coastlines. Likewise, sand spits, partitioning a single body of water into two lakes, are not rare. Geologists have been concerned with the origin of such lakes, but no outstanding limnologic contributions have come from shoreline lakes.

Lake Turkana in the desert of northern Kenya is important in anthropology because its shores have yielded consequential primate remains pertaining to human ancestry. Some limnologic reports from this lake emphasize the difference between the main lake and Ferguson's Gulf. The gulf is nearly isolated by a growing spit of sand, molded by wind and wave. Dissolved minerals are less concentrated in the open lake than they are in Ferguson's Gulf, which has its own character even though not yet completely separated as an individual lake by shoreline phenomena.

Beach pools are particularly common examples of shoreline lakes, representing isolated remnants of the main lake. They were included in a most important study by Cowles (1899), who blended geology and botany in a dynamic investigation of sand dunes and vegetation at the south shore of Lake Michigan. (Cowles's paper, published in four installments, contributed much to the development of the concept of community succession.) A minor portion of his work treated the early formation of beach pools and their subsequent fate. These ponds, more or less parallel to the lake shore, result from movement by both wind and wave. Cowles also wrote about the lakes that formed at mouths of rivers along the east shore of Lake Michigan: silt and sand brought by the rivers are dropped in the lake shallows where currents slacken; waves pile the sand up along the beach; and winds pick up the material to form dunes that divert and pond the rivers to produce lakes near their mouths. Thus the birth of these shoreline lakes is due in part to both fluviatile and aeolian forces.

Lake Basins Impounded or Excavated by Organisms

Geologists consider organisms as important geomorphic agents but relatively minor lake builders. Nevertheless, a few kinds of plants and animals have contributed to limnologic habitats, and their contributions should be mentioned.

Dams of *sphagnum* and other bog plants have been reported to impound some Nova Scotian lakes (Hutchinson, 1957). Perhaps many other northern lakes occupy depressions in peaty deposits. Certainly, small pools enclosed by bogmat plants are very common but distinct little ecosystems. At Lake Itasca one can dock a boat and disembark at the margin of floating bogs that have obliterated former bays of the lake. There, in isolated little pools, are peculiar flora and fauna, the latter including the bizarre cladoceran *Bunops*. Members of the biotic community in these little ponds are quite different from their counterparts in the adjacent open lake.

Golubic (1969) detailed the events whereby $CaCO_3$, precipitated from solution in spring brooks, is trapped by algae and mosses. The species that are adapted to being buried and encrusted respond by growing vigorously upward from the cemented layers only to be subjected to further encrustation. There are various results, including the formation of plant-limestone barriers and terraces that create pools. Later Golubic (1973) noted the extreme importance of blue-greens in precipitating calcium carbonate, both in fresh water and seawater.

There are rare Pacific freshwater lakes surrounded by coral reefs, the anthozoans being the primary architects (Hutchinson, 1957). Other invertebrates, including polychaetes and gastropods that build tubes and attached pelecypods such as *Mytilus*, serve to build the walls higher above the coral. Calcareous algae such as *Lithothamnion* and *Porolithon*, especially on the seaward side, raise the rims even higher and serve to cement and protect the invertebrate remains below.

Higher plants also build interesting micro-pools. The biota of tiny pools held by the carnivorous bog pitcher plants and the crustaceans and protozoans in the axils of epiphytic bromeliads in Florida, Jamaica, and other warm humid areas of the New World have been studied (Maguire, 1971).

Some species of mosquito live as larvae in water-filled tree holes. Perhaps seven or eight animal phyla, as well as bacteria, fungi, and algae, are known from such habitats. Maguire (1971) reviewed the literature on this subject and discussed the relationship of such miniscule lakes and their biota to wider ecologic problems of dispersal, immigration, establishment, and extinction, all being problems of colonization.

Mammals are the main lake engineers among the Chordata, but some mention should be made of the pools fashioned by alligators in the southeastern United States. These reptiles excavate only small concavities, but they may be of paramount importance in the survival of alligators and other aquatic animals of the southern swamps during dry periods.

Beaver damming and similar activity by humans are well known. In both instances streams are impounded with subsequent flooding of streamside vegetation and widespread effects on riparian communities. Controversy has accompanied the actions of both *Homo* and *Castor*. The effect that beaver ponds have on trout streams (elevating temperatures of impounded waters and silting in former stretches of moving water) has been attacked as well as defended. Similarly, the ponding of streams by humans has resulted in controversy.

No one knows how many dams have been built and reservoir lakes formed; some date back many years, and the 20th century especially saw a flurry of activity world-wide. In the United States alone, some 75,000 dams together store a volume nearly that of a year's river flow (Graf, 1999). Worldwide, over half (172 out of 292) of the largest rivers are affected by dams, including the eight most biogeographically diverse (Nilsson et al., 2005). The ecological effects are extensive, and include changes in water flow regime, temperature, sedimentation and erosion, meandering, and associated impacts on the biota (Ligon et al., 1995). As a result, natural resource managers have suggested prescribed flow regimes to promote suitable habitat (Richter and Richter, 2000) or, increasingly, dam removal (Hoenke et al., 2014; Van Looy et al., 2014).

In some artificial lakes, the creators have done more than simply impound running water. They have excavated the basin in addition. Bulldozed farm ponds and arid-land stock tanks are typical, although very small and sometimes transitory bodies of water.

Certain artificial excavations contain water, although they were not made for this purpose. Water-filled rock quarries and gravel pits come to mind immediately. The total area of Great Britain's gravel-pit lakes is about five times that of the waters of the English Lake District (Burgis and Morris, 1987). Another interesting example is Hot Lake, Washington, occupying an old epsomite mine (Anderson, 1958b). The mineral was mined intensively during World War I, and later the abandoned diggings filled with water. Similarly, most of Holland's lakes are contained by shallow cavities where peat was harvested. They are examples of resurrected lakes.

Another category is the natural concavity inundated by accident. The Salton Sea of California fills a depression that once held pluvial Lake Cahuilla. Carpelan (1958) has told the story of this unique lake in interesting detail. In the spring of 1905, the inadvertent diversion of the Colorado River's main flow into the dry basin by way of newly dug irrigation canals initiated the creation of a modern lake. Two years later the flow was stopped, but not before a new body of water covering more than 1,000 km^2 had come into existence. Its level has fluctuated since then but now is stable through active management by a government authority. The saline Salton Sea survives to date because the input from irrigation systems approximates the yearly evaporation of about 2 m.

Lakes Formed by Extraterrestrial Objects and Those of Puzzling Origin

The spectacular instantaneous formation of a lake basin by meteoritic impact and explosion may be a more common method of lacustrine genesis than has been suspected. According to Nininger (1972), the high frequency at which stony and metallic objects arrive from outer space had been accepted only recently. In 1967 Barringer listed 31 "identified" meteorite craters and 40 suspected to be of such origin. Recognition of **astroblemes** (loosely translated, star scars) is difficult, because many are badly eroded and cloaked with vegetation. It is significant that when Hutchinson wrote Volume I of *A Treatise on Limnology* (1957), he alluded to *cryptovolcanic depressions,* his designation for some puzzling basins that resemble large calderas but are not in regions of volcanic activity.

The New Quebec Crater Lake is typical of the more easily recognized basins formed by meteoritic impact. Its high rim and circular outline are unique among the irregularly shaped lakes in that barren glaciated landscape more than 61° N latitude near the tip of the Ungava Peninsula in Quebec.

In northern Arizona an astrobleme named Meteor Crater (once believed to be formed by volcanic explosion) is now dry. The floor of the crater has about 30 m of lake sediments, the remnants of a pluvial lake estimated to have existed 11,000 years ago.

The Carolina Bays, distinct water-filled as well as extinct basins numbering up to at least 500,000, are puzzling. They occur in an area of 65,000 km^2 in the southern half of the Atlantic coast of North America. There is an extensive literature on them, and their origin has been attributed to everything from wind action to the nesting and schooling activities of shallow-water fish. One theory holds that they might be the result of our planet's encounter with a shower of stony meteorites when that part of the continent was under shallow seas. Their NW-SE alignment recalls immediately the thaw-basin lakes of Alaska, but there is no reason to believe that the Carolina Bays ever had a permafrost underlay. Their derivation is still unsolved.

Additional Oriented Lakes

The thousands of oriented thaw basins in the Alaskan Arctic Coastal Plain have been known for many years as the prime example of a district of oriented lakes. There are other such districts owing their origins to different phenomena. In the Beni Basin of Bolivia there are 2,000 basins remarkably square to rectangular in outline. Most of them are aligned about N 48° E and at right angles to this, N 41° W. Plafker (1964) explained this lake district on the basis of a system of longitudinal and cross fractures in the crystalline rocks underlying the flat, poorly drained Beni Basin, and he suggested that new lakes are still being created by new faults.

This may not be unique, for there are several hundred basins in the Old Crow Plain, Yukon Territory, Canada, that seem to be this type. Most of them are oriented in a northwest direction, and a smaller number are aligned at right angles.

Important Relict Lakes

An enormous lake formed in North America between retreating Wisconsin ice and the height of land that separates Hudson Bay drainage from the Gulf of Mexico drainage. It covered 540,000 km^2 from eastern Saskatchewan to western Ontario and south to include eastern North Dakota and northwestern Minnesota. This was glacial Lake Agassiz. Its sediments cover a greater area than any lake existing today, although the lake may never have been flooded completely at one time. It lay in a well-drained area termed **exorheic**, an area of open basins whose rivers ultimately reach the sea. Remnants of the lake persist today as separate bodies of water, such as Lake Winnipeg, Lake Manitoba, Lake Nipigon, and Lake-of-the-Woods.

Pluvial lakes, all probably of the Quaternary age, existed south of the continental ice sheet. They were characterized by sensitive responses to climatic changes and by considerable fluctuation in level. In the Great Basin of North America there were at least 120 of these lakes, the two largest being Bonneville and Lahontan. Bonneville is now represented by old shorelines far from water, glistening evaporites of the extensive Bonneville salt flats, and a few saline remnants, of which the Great Salt Lake of Utah is most important. The most striking Lahontan remnant is Pyramid Lake, Nevada, which, along with other bodies of water in the Lahontan Basin, figures in the history of desert limnology (Hutchinson, 1937). Pluvial lakes existing on other continents are similar to our own. For example, the saline Lake Urmia, an Iranian counterpart of Great Salt Lake, occupies an undrained depression of tectonic origin.

The Spanish word *playa* (beach) is applied to a land form—the remarkably flat and barren lower region of arid basins. One estimate holds that there are about 50,000 playas on earth. They occasionally hold water, intermittent flooding that creates a **playa** lake. In many instances, but not always, playas owe their origins to pluvial lakes of the past. But whatever the geologic origin of the depression, lacustrine deposition is the major process in flattening the playa floor. Many of these inland desert basins are bright with salts and are known also as *salinas.* Certain playas are economically noteworthy, yielding valuable salts. Searles Lake, California, is one of these. Any study of these desert basins should start with the 29 important papers on playas and dry lakes assembled by Neal (1975) and lead to Kubly (1982).

Remnants of pluvial lakes occupy closed basins in arid regions characterized by internal drainage. Closed basins can exist only where annual evaporation surpasses the precipitation; in such climatic regions the rivers, if present, never reach the sea. These are termed **endorheic** areas because they are undrained by streams. The realm of desert limnology, by definition, coincides with regions where closed basins are found (Cole, 1968).

Larson (2012) tells the story of a relict lake which has reasserted its ancient basin, just within the time of one limnologist's career. In 1964, Devil's Lake, North Dakota, covered only 80 km^2 with a maximum depth of 3 meters. It has since risen 13 m, with an accompanying surface area increase to 800 km^2, and an astounding 32-fold increase in volume! The resulting flooding has disrupted farms, towns, transportation infrastructure, and has cost hundreds of millions of dollars in losses. Unfortunately, allowing water levels to rise to the elevation of the natural overflow outlet (Tolna Coulee) would create extensive further damages. It is hoped that recent construction of new levees and overflow outlets, combined with active pumping of excess water, will address this serious challenge. A solution to this dilemma has yet to be found.

Case Study: Climate Change in Alpine Lakes

Paleolimnologists are key contributors to the study of climate change, which is considered among the most pressing challenges for society in the new millennium. Lakes are sensitive to climate conditions of temperature and hydrology and are excellent repositories recording those conditions. Studying long-term changes in lakes allows us to place current, human-influenced climate changes in perspective and help predict future climate changes (Cohen, 2003).

Heiri and Lotter (2005) cross-referenced climate and lake sediment records from some 68 Alpine lakes. Figure 5-7 shows the excellent agreement between the air temperature and the biological and chemical conditions in one such lake (Gerzensee). The data set records climate change, both natural and **anthropogenic**, dating to the Bronze Age, 14,000 years BP. The lakes, and montane Europe itself, are warming.

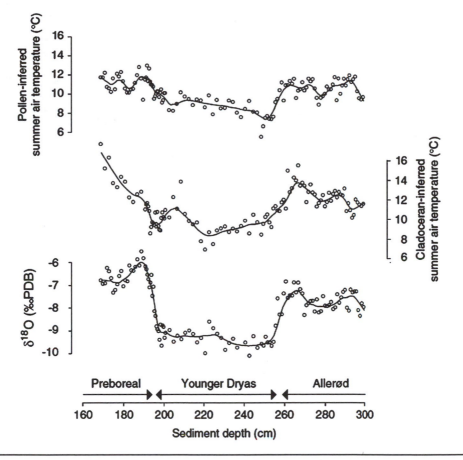

Figure 5-7 Gerzensee (Swiss Alps) area climate as recorded in lake sediments. Deeper sediments (farther right on graph) are older. From Heiri and Lotter (2005). Reprinted with permission.

The same authors analyzed sediments cores from another small lake (Sägistalsee) in the Swiss Alps (Heiri and Lotter, 2003). This study followed Chironomid abundance over 9000 years; their absence indicates human-caused hypoxia in the benthos. The effects of Bronze-Age clearing of nearby land and subsequent agricultural activities are apparent in the sediment records, detailing long-term **anthropogenic** effects.

Climate change effects are sometimes subtle and/or gradual, but not always. A study by Frey and associates (2010) is one of many studies demonstrating a serious hazard associated with warming: sudden, catastrophic lake formation or drainage. These authors used modern remote sensing and geographic information system (GIS) databases to track lakes associated with retreating (melting) glaciers. Sudden failure of the impoundment causes deadly avalanche or mudslides. Again, such occurrences are not confined to the distant past but are expected to present an increasing hazard as the climate warms. Limnologists will help elucidate these processes and their associated hazards.

6

Shapes and Sizes of Lakes

Basin morphometry refers to the measurements of a lake: its size and shape. Most processes in a lake ecosystem are influenced substantially by morphometry, obviously interacting with the location of the basin in the landscape. As noted in Chapter 5, the origin and evolution of a lake is important in determining its shape and size; many factors interact to produce the lake we see today.

This chapter begins with some examples of new technologies useful in morphometry followed by descriptions of the underlying geometry and mapping, together with how they influence the lake environment. We end with a discussion of different definitions of "size," and the implications of size–scale effects.

Lake Morphometry in the Digital Age

Global Positioning System (GPS)

Cars, mobile phones, and portable mapping units are all radio receivers, interpreting time and location signals transmitted by 24 or more satellites situated in orbit 20,000 km above the surface of the Earth. This technology allows precise three-dimensional location to be determined anywhere on Earth, any time of the day.

GPS may be handheld (e.g., used while wading a stream or while on a moving boat) or combined with other equipment. The toxic cyanobacterium *Planktothrix rubescens* was monitored in Lake Zurich by Marie-Ève and colleagues (2013) using a multi-parameter probe on an Autonomous Surface Vessel, navigated in conjunction with GPS. The technology allowed them to successfully track the movements of the bloom in three dimensions in the lake. As another example, an underwater methane bubble "curtain" in Sakinaw Lake, British Columbia, was mapped by Vagle and others (2010) using GPS in conjunction with a remote sensing technology, hydroacoustic sonar. Precise mapping was important to understand the combination of dissolved gas, stratification, and its effect on biota in this deep lake.

Remote Sensing

Many more data can be collected, and at more locations, if remote sensing is employed. Aerial photos from airplanes exist from the 1930s to the present for some locations, supplemented more recently with images from other aerial platforms such as SPOT or Landsat satellite data. Bayley and Prather (2003) used helicopter surveys to determine the nutrient and vegetation status of 148 wetland lakes in boreal Alberta, Canada, classifying them into four categories based on primary production.

The ability to survey dozens of locations, and across a number of years, is a strength of remote sensing. Wynne and colleagues (1996) surveyed 62 northern lakes across eight years to document the relationship between spring ice break-up and latitude and surface area of the lakes. The data were collected by satellite imagery. Satellite imagery over decades were combined with shipboard analyses by Pozdnyakov and associates (2013) to understand the biogeochemistry of Lake Ladoga after cessation of phosphorus enrichment. The authors demonstrated that an anticipated lowering of biological production did not occur, leading to the hypothesis that internal P-cycling drives high biological production, despite the lowered P-inputs.

Geographic Information System (GIS)

An integrated system of computer hardware, specialized software, and data sets allow GIS users not only to make and manipulate maps, but also to query the database to answer questions.

Haltuch and others (2000) utilized data from bathymetry (depth), substrate type, and side-scan sonar in a GIS to map existing distribution and predict future spread of an exotic mollusc, the Zebra mussel (*Dreissena polymorpha*). The animal proved surprisingly adept at colonizing various substrates and has had a considerable impact on ecosystem function.

GIS allows researchers to analyze data in ways that would be difficult or impossible with traditional techniques. Extensive geographic areas with large data sets can be considered and multiple variables studied in concert. In a study of Dissolved Organic Carbon (DOC), Xenopoulos and associates (2003) examined 9 catchment characteristics in 11 geographic regions worldwide and found that wetland presence was the best predictor of carbon inputs.

Although many GIS data "layers" are collected in a digital form, historical data may also be digitized and used in the database to compare changes over time, even the span of centuries. Twining and colleagues (2013) explored the interaction of land use and chemical inputs to New England lakes over 400 years and found that alewives (*Alosa pseudoharengus*) were important, but only one of, the factors driving biogeochemistry in the studied lakes.

The Bathymetric Map and Its Data

Regardless of the source of geospatial data or the tools used to process and analyze these data, they will eventually be represented in a map. Most maps today use the same symbols and rendering developed by cartographers centuries ago, based on the Cartesian *x-y* coordinate system, with one important difference: limnologists are also concerned with a third dimension (z) used to signify depth beneath the horizontal *x-y* surface. Note also that the *x-y* coordinates do not always correspond to the compass directions, so a "compass rose" or other indicator of direction in space is necessary (Fig. 6-1).

An outline of the lakeshore, with submerged contours, drawn to scale is a necessity for calculating morphometric data. Usually such a bathymetric map is easily produced by tracing the shoreline from aerial photographs, thus precluding standard, time-consuming surveying methods. An examination of an aerial photo usually reveals landmarks of various types—perhaps a stretch of straight highway or two promontories demarking a nearly straight line of shore. From these landmarks one can arrive at a scale by driving an auto-

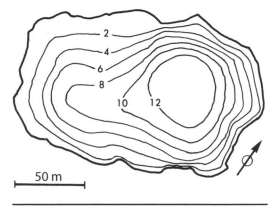

Figure 6-1 Bathymetric map of Lake 230, ELA, Ontario. Adapted from Brunskill and Schindler, 1971.

mobile from point to point and noting the final odometer reading to determine the distance (although the accuracy of this method may be questionable). A calibrated line, or a transit and stadia rod, can be used to measure the distance walked along the shore between conspicuous points. Lind (1985) treated map making in great detail, outlining several ways of arriving at the shoreline configuration and other lake dimensions. Whatever method is used, if all measurements are translated into metric units from the beginning, ensuing calculations will be simplified.

With an outline of the lake (Fig. 6-1) and a proper scale, at least five surface dimensions can be determined about the lake even before subsurface configuration is known. These measurements are explained in the following section; there and in subsequent discussions, a water body (Lake 230) in the Experimental Lakes Area (ELA) is used as a case study to illustrate these morphometric procedures and their interpretation. The ELA is composed of 58 small lakes in a rural area of western Ontario, Canada studied by limnologists since 1967.

The measurements and calculations described here are now routinely performed on computers, with GPS data or digitized images/maps being analyzed by image processing programs such as ImageJ, using rendering by CAD/CAM software, or using GIS. The following describes how these are performed by hand, to illustrate the underlying principles.

Surface Dimensions

Maximum length.   The distance across the water between the two most separated points on the shoreline is, of course, the maximum length (*l*). It can be determined simply from the outline map with a ruler. It may or may not be a significant dimension; the wind-effective length has more limnologic importance. The relation of the long axis of the lake to the direction of prevailing winds (*fetch*) is obviously worth considering, and the greatest distance wind can sweep uninterrupted across the lake surface should be reckoned. The maximum length might be intercepted by islands or promontories, so that the actual fetch of the wind becomes far less. The surrounding topography may be such that the lake is sheltered by hills, resulting in little actual wind action on its surface. The exposure of a lake to wind has a direct effect on water movement, and thus indirect effects on the biota within the lake.

Breadth. The width of a lake (***b***) is determined by measuring, at approximately right angles to the axis of maximum length, a line that connects the greatest distance between two opposite points on the shore. The mean width ($\bar{b}$) can be found if the lake area is known; it is the quotient of area divided by maximum length.

Surface area. Surface area (*A*) is an extremely important dimension, for it is at the surface that solar energy enters the aquatic habitat. Furthermore, many types of data from within the lake are referred to a unit of area, making meaningful comparisons possible among different-sized bodies of water.

With a map of the lake's outline and an appropriate linear scale, there are at least three ways to determine the area. An inexpensive procedure is to draw this outline on graph paper, with the scale of the grid spacing being known (Fig. 6-2). Summing the squares, estimating the fractions thereof, and multiplying by the proper factor would serve to estimate the lake area.

The polar planimeter is an instrument designed for deriving areas of flat surfaces. With it the shoreline is traced in a clockwise direction, and dial numbers are read when completed. These dial readings can be compared with the results of tracing a square or circle of known area. The planimeter is calibrated in square inches or square centimeters, and from the map scale the actual lake area can be calculated. Lind (1985) gives explicit directions for planimeter use and many other morphometric procedures.

Another way to determine area rapidly is to use an accurate balance. If the lake outline is drawn on paper of fairly uniform density, it can be cut out and weighed to yield a mass that can be compared with confidence to the weight of a known area cut from the same type of paper. Ratios derived in this fashion probably have as good a degree of accuracy as those achieved by planimeter.

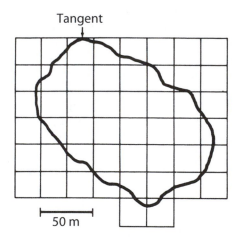

Figure 6-2 Outline of Lake 230, ELA, superimposed on grid. Twenty-six intersections and one tangent shown; grid spacing 25 m. Shoreline = 26.5 × 0.785 × 25 m = 520 m.

The methods of ascertaining area from a map have been discussed in some detail because they can be applied to many other limnologic calculations.

Length of shoreline. The length of the shoreline (***L***) can be determined from the map by stepping off segments with dividers or by sticking pins in place around the shoreline and connecting them with a thread that can be measured later. A map measurer (cartometer) or rotometer may also be used. This instrument is essentially a wheel and dial that indicates the linear distance the wheel rolls.

Olson (1960) outlined another method of finding the dimension *L*, which leads us back to the map sketched on the graph paper grid (Fig. 6-2). It is preferable that the long axis of the lake be diagonal to the grid lines rather than parallel to the grid lines. One counts the times that the shoreline crosses grid lines, summing all horizontal and vertical intersections and tallying any point of tangency as one-half of an intersection. The sum of intersections multiplied by $\pi/4$, or 0.785, and then by the actual distance represented by the grid spacing yields *L*. Olson discusses many concepts of morphometry and reveals methods of achieving all lake dimensions from a map imposed on a grid without specialized equipment.

Shoreline development index. The development of the shoreline (D_L) is a comparative figure relating the shoreline length to the circumference of a circle that has the same area as the lake. The smallest possible index would be 1.0. The formula from which the index is derived is

$$D_L = \frac{L}{2\sqrt{\pi A}}$$

Both *L* and *A* must be in consistent units for this computation—meters and square meters, or kilometers and square kilometers.

Solution basins, volcanic lakes in various craters, some deflation basins, those bodies of water in collapsed travertine mounds, and most of the rare depressions owing their origin to meteoric impact and explosion are all nearly circular, with indices approaching unity. Good examples are Arizona's Montezuma Well (see Fig. 6-10), D_L = 1.04; Crater Lake, Oregon, a volcanic caldera, 1.27 (closer to 1.10 if an island that adds to shoreline and decreases area is ignored); the Javanese maar Ranu Pakis (Ruttner, 1931), 1.02; the south African deflation basin Avenue Pan (Hutchinson, 1957), 1.01; and the modern rim of the dry Meteor Crater, Arizona, with a D_L of 1.02.

Some glacial lakes, such as those in kettles, are nearly circular, but others have irregular shorelines characterized by higher shoreline development indices. The elongate Finger Lakes of New York, formed by glacial ice scouring and deepening valleys, have high shore development values. For example, Cayuga Lake has a D_L value of 3.3. Similarly the elongate, somewhat rectangular lakes found in grabens have high D_L values; for example, that of Lake Baikal is 3.4.

River impoundments, whether natural or artificially created, often back water up tributaries, creating irregular dendritic outlines with much shoreline in relation to surface area. By contrast, despite marked shoreline regularity, a narrow, elongate lake can have a high index (Fig. 6-3 on the following page).

Although a newly formed lake basin derives its outline from the events that gave rise to it, the shoreline will change with aging. Currents and waves within the lake tend to erode the promontories, reducing shoreline irregularities. Protected bays are the first lake parts to fill in and become portions of the terrestrial environment. As the bays are eliminated and irregularities are removed by events occurring within the lake itself, D_L values are reduced. Conversely, the arrival of sediment via an incoming stream might result in a delta projecting into the lake itself. If part of the delta eventually emerges as water level lowers, it contributes more irregularity to the shoreline and, perhaps, increases the D_L index.

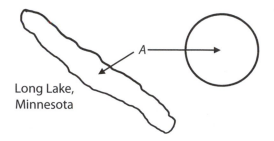

Long Lake,
Minnesota

Figure 6-3 Elongate lake showing comparable circle with the same area, *A*. The shoreline-development index of the lake is greater than the D_L of the circle, 1.0.

The development of shoreline may play some role in determining the trophic nature of the lake because shallow water is the most productive. Most photosynthesis occurs in the upper, illuminated stratum of a lake. Furthermore, in this upper zone near shore, there is a proximity to the decomposition products from the bottom sediments that cannot be equaled in offshore regions. Moreover, the arrival of terrestrial nutrients to the lake is, to a great extent, a shoreline function. It is tempting to theorize that, of two lakes alike in most features except shoreline development, the one with the highest D_L would be more productive. We must know other facts about lakes, however, before placing much significance on a D_L value. Certainly a low index reveals an approach to circular outline. But when one learns that the shallow, productive Canadian Lake Winnipeg has the same shoreline development index (3.4) as the far different, tectonic Lake Baikal, it becomes obvious that caution should govern interpretation of the index.

Subsurface Dimensions

Contour mapping.   To this point, morphometric data have been derived entirely from an outline map drawn to scale. Subsurface contours must be established and plotted before the bathymetric map is complete and before one can proceed with other morphometric work.

There are several methods of finding depths and their positions within a lake. With echo-sounding equipment readily available, it is possible to move by boat from one landmark to another, keeping the speed constant and timing the entire journey. Time intervals are assumed to be proportional to linear spacing for the purpose of plotting depths along the transect. Also, the position of the sounding station on the lake surface can be estimated by triangulation or GPS. Echo-sounding devices permit the rapid accumulation of many depth data.

On northern lakes during winter, holes can be bored or chopped in the ice at measured intervals along transects between known points on opposite shores. If echo-sounding equipment is unavailable, sounding with a weight and calibrated line can prove effective.

A simple method employed in surveying small lakes is to row a boat on a straight course between points on opposing shores, sounding the bottom at intervals, such as at every 10 strokes. One should transect the lake many times, keeping in mind that a shallow stratum contains greater volumes of water than a deeper layer of the same thickness; therefore, shallow strata should not be neglected during sounding operations.

Whatever method is used, a matrix of points will eventually appear on the map. Contour lines can be plotted by interpolation or, more rapidly, by computerized programs. In small lakes a contour interval of 1 or 2 m may be appropriate. In large lakes the contour lines may be shown at 5 or 10 m intervals.

With contours plotted at regular intervals, the area bounded by each line (or **isobath**) is determined by one of the methods suggested earlier for obtaining area bounded by z_0, the total surface area, A. With these data assembled, it is possible to learn more particulars of a lake's morphology.

Maximum depth.   Depth is indicated by the symbol z. Thus z_0 is the surface, a depth of zero; its contour is the shoreline. The first subsurface contour would be z_1, followed down the basin slope by z_2 and so forth. While sounding the lake, a measurement of popular interest, the maximum depth, z_m, should be found. This is the datum most people search for and inquire after in studies of lake dimension.

It is interesting to note that people are prone to accept fantastic tales of enormous depths. The so-called Bottomless Lakes, occupying a string of solution basins in gypsum rocks, are said to have acquired the name from New Mexican cowboys, who sounded them with lariats tied to inadequate weights and could not sense when bottom contact was made. Henry David Thoreau wrote of Concord farmers fathoming Walden Pond with rope, measuring their gullibility rather than the lake depth z_m; as their test line coiled on the sediments below, they continued to measure it out from above. Thoreau, we must believe, did a better job; about 100 years later, Deevey (1942a) found Walden's z_m to be 31.3 as compared with Thoreau's datum, 31.1 m.

It is apparent that as a lake ages and accumulates sediments, the maximum depth lessens and the volume of deep water diminishes even more strikingly (Deevey, 1955).

Relative depth.   Rarely in the literature on lake dimensions have authors presented the relative depth (z_r). This is the ratio of the maximum depth in meters to the average diameter of the lake surface. The following equation has the computation for percentage built into it, for the ratio is expressed in that manner.

$$z_r = \frac{88.6 \times z_m}{\sqrt{A}}$$

Among the larger lakes of the world the relative depths are less than 1%. In calderas, maars, fjord lakes, and solution basins there are some high values for z_r. The caldera Crater Lake has a notably great relative depth, about 7.5%; this is very similar to that of the circular, deep meteoritic Pingualuit (New Quebec) Crater, 7.6%. Montezuma Well has a surprising z_r of 17%, but this is outmatched by Devil's Ink Well, one of the Bottomless Lakes; this small circular solution basin's z_r is 23.2%. Small maar lakes, such as Pulvermaar in Germany and d'Issarlès in France, have high relative depths. The dimensions given by Hutchinson (1957) permit the calculation of 11% and 10% for the two explosion craters, respectively. Lake Kanyangeye (A, 0.5 ha), one of 16 crater lakes in western Uganda studied by Melack (1978), occupies one of the vents of two merged cones and is noteworthy for its z_r, 71.5%. The record, however, is still held by an Hawaiian volcanic crater lake called Kauhako; its dimensions, $A = 0.35$ ha, $z_m = 250$ m, permit calculation of a relative depth of 374%. This lake, then, can be likened to a posthole with a diameter of 15 cm and a depth of about 56 cm!

Cryptodepressions. Frequently the maximum depth of a lake is below sea level; the portion of the basin beneath sea level is called a cryptodepression (z_c). For example, the surface of Cayuga Lake lies 116 m above mean sea level; its maximum depth is 133 m; and, by subtraction, it has a z_c of 17 m. The surface of the saline Dead Sea is 399 m below sea level; therefore, its entire basin is a **cryptodepression**.

Volume. The volume of a lake (V) can be calculated when the area circumscribed by each isobath is known. One method involves equations such as the following:

$$V_{z_1-z_0} = \tfrac{1}{3}\left(A_{z_0} + A_{z_1} + \sqrt{A_{z_0} \times A_{z_1}}\right)(z_1 - z_0)$$

From this, the volume of water between the shoreline contour (z_0) and the first subsurface contour (z_1) is found. A_{z_0} is the total area of the lake, and A_{z_1} is the area limited by the z_1 line. If the values are at 1 m intervals, the value of $z_1 - z_0$ would be 1 m, the vertical contour interval. Volumes of succeeding strata are now determined one by one and summed. The next would be $V_{z_1-z_2}$, the volume of water occupying depths bordered by the first two subsurface contours. This method is based on adding layer after layer of water, each a truncated cone or one frustum of a large imaginary cone that is the entire lake. Though a tedious, cumbersome procedure, it eventually yields the total number of cubic meters of water within the lake.

Hypsographic curves. An easier method of computing V is to construct a curve on graph paper (Fig. 6-4). The area of this curve, or of parts of it, can be determined by any one of the methods outlined before. In this instance, however, area beneath the curve designates volume, for the area bounded by each contour line has been plotted against the depth of that contour. Thus, at zero depth, z_0, A is plotted; at z_1 the area limited by that contour, A_{z_1}, is designated, and so forth. The points are connected, ending finally with a curve at z_m where area is zero. The total area beneath the curve, essentially the product (cubic meters) of square meters × meters, is the important lake dimension of volume, V.

This hypsographic curve provides many other data as well. For example, in Fig. 6-4 it can be seen that vertical lines from all depths intersect the curve; from this any stratum in the lake can be delineated so that its volume can be computed. Furthermore, a horizontal line from any point on the curve intersects the

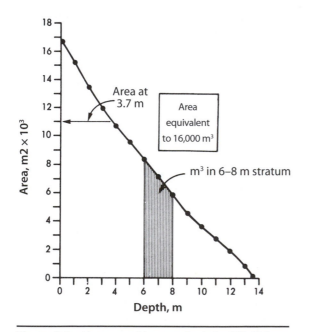

Figure 6-4 Hypsographic curve for Lake 230, ELA. Area beneath curve represents the volume of the lake in cubic meters. Further data that can be acquired from the plot are indicated.

vertical axis to show the area inside a contour drawn at that depth. Therefore, if the area of a plane 6.2 m below the surface is needed, a vertical line drawn from 6.2 m on the horizontal axis would intersect the curve above at a point marking that area.

Further useful data that could be obtained and tabulated with the depth-area curve of the lake include: the area at each depth, its percentage of the total area (A_{z_0} or $A_0 = 100\%$), the volume of each stratum, and the percentage of the volume contained in each stratum (Tables 6-1 and 6-2).

Table 6-1 Morphometric data from Lake 230, Experimental Lakes Area, Ontario

Depth (m)	Area (ha)	Percent	Product (Depth × Area)
0	1.67	100.0	0
1	1.52	91.0	1.52
2	1.35	80.8	2.70
3	1.19	71.2	3.57
4	1.07	64.1	4.28
5	0.957	57.3	4.78
6	0.837	50.1	5.02
7	0.713	42.7	4.99
8	0.588	35.2	4.70
9	0.460	27.5	4.14
10	0.359	21.5	3.59
11	0.274	16.4	3.01
12	0.192	11.5	2.30
13	0.094	5.6	1.22
13.6	0	0	0

Data from Brunskill and Schindler, 1971.

Table 6-2 Morphometric data from Lake 230, Experimental Lakes Area, Ontario

Stratum (m)	Volume ($m^3 \times 10^4$)	Percent	Cumulative percent
0–1	1.60	15.38	15.38
1–2	1.43	13.75	29.13
2–3	1.27	12.21	41.34
3–4	1.13	10.86	52.20
4–5	1.01	9.71	61.91
5–6	0.90	8.65	70.56
6–7	0.77	7.40	77.96
7–8	0.65	6.25	84.21
8–9	0.52	5.00	89.21
9–10	0.41	3.94	93.15
10–11	0.32	3.08	96.23
11–12	0.23	2.21	98.44
12–13	0.14	1.35	99.79
13–13.6	0.02	0.19	99.98

Another direct hypsographic curve is drawn by plotting on one axis the volume above each subsurface contour and on the other, the contour depths. Percentage hypsographic curves are constructed by plotting the percentage of lake surface area bounded by every contour line against the respective contour depths, or in the case of the percentage volume curve, plotting against depth the percentage of total lake volume lying above each contour (Fig. 6-5).

An additional dimension, the so-called center of gravity (z_g), can be found from the percentage volume curve. This is done by marking the depth opposite the point where a perpendicular line from the 50% mark on one axis intercepts the curve (Fig. 6-5). The result is a depth at which a horizontal plane would divide the lake volume into two equal parts. If the density of the water is homogenous, the same plane would separate the mass of the lake into two equal portions. The center of volume, center of mass, and center of gravity are *points*. The techniques shown in Figs. 6-5 and 6-6 for establishing this so-called center of gravity, z_g, actually only reveal a center of volume depth above

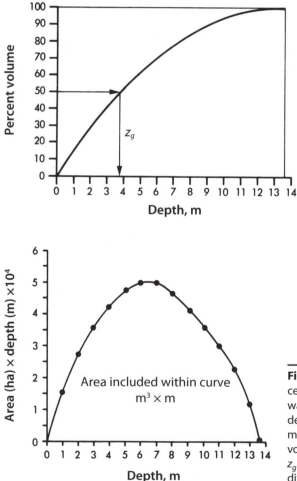

Figure 6-5 Percent volume hypsographic curve for Lake 230, ELA. The depth where a horizontal plane would separate the lake into two equal volumes is shown where the 50% volume on the y-axis intercepts the curve. This has been termed loosely the center of gravity (z_g) when the water is uniform in density throughout.

Figure 6-6 Curve for deriving the so-called center of gravity for Lake 230, ELA, when the water is of uniform density. The plot of area × depth against depth, on the x-axis, yields cubic meters × meters. When this is divided by the volume of the lake in cubic meters, the depth z_g in meters is revealed. It is the depth that divides the lake volume into two equal parts.

which the volume equals the volume below. If the density of the water is uniform, we can say something about the depth z_g and the level that separates equal amounts of mass; only in a uniform gravitational field does the center of mass equal the center of gravity. Despite the fact that the moments of all the particles in the body of water are neglected, and therefore true centers of mass, volume, and gravity are not found, the parameter z_g has had use in certain calculations pertaining to physical limnology.

Mean depth. The lake volume divided by its surface area obviously will yield the mean depth ($\bar{z}$) if the units employed are the same. Thus, the volume in cubic meters must be divided by area in square meters, or two other comparable units should be used in the formula $V/A = \bar{z}$.

Mean depth has been considered an important dimension since Thienemann (1927) proposed that in German lakes a boundary between what he defined as oligotrophy and eutrophy lay at about 18 m. Lakes with z greater than 18 m showed features he had assigned to oligotrophy; shallower lakes were more productive and belonged to the eutrophic series. Similar data were documented by Rawson (1953, 1955) and Hayes (1957). In a series of diagrams, Rawson showed the dry weight of plankton, bottom fauna, and fish crops from many lakes plotted against the mean depths of the lakes. Rawson's plots form L-shaped curves, similar to hyperbolas (Fig. 6-7). There are two arms to the curves; their junction is not far from 18 m.

Earlier, Deevey (1941) had presented a diagram of bottom-fauna weights plotted against mean depths from 116 lakes in Europe and North America. That graph also suggests a group of shallow lakes characterized by variable but high biomass and a group of lakes deeper than 20 m with a relatively small mass of bottom fauna. Deevey came to the conclusion, however, that the supposed hyperbolic curve was the "result of noncorrelation within several very diverse bodies of data"—that correlation between bottom fauna and mean depth alone was poor, since the situation was much more complex and based on additional interrelated factors. With the data he had available to analyze from a series of Connecticut and New York lakes, Deevey found that, in addition to mean depth, the amount of phytoplankton beneath a unit area of lake surface and the average oxygen content of the hypolimnion were closely correlated with the biomass of bottom fauna.

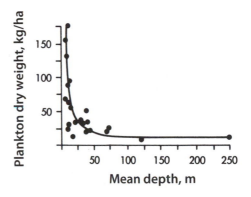

Figure 6-7 Standing crops of plankton in kilograms per hectare plotted against mean depths in various lakes. Data from Rawson (1955).

Gorham (1958a) further discussed mean depth and biologic productivity, pointing out that Rawson's data fit a curve of photosynthetic rate per surface area plotted against mean depth. He suggested that $\bar{z}$ is an important dimension in determining what proportion of a lake's volume is well lighted and within the photosynthetic zone.

With photosynthesis as the basis of most productivity within the typical autotrophic lake, Gorham's conclusions seem reasonable.

Volume development. A further index to basin shape is the volume development, D_v. This compares the shape of the basin to an inverted cone with a height equal to z_m and a base equal to the lake's surface area. In a lake with a volume equal to this hypothetical cone, $D_v = 1.0$; a lake with a relatively greater volume would have an index greater than 1.0; and a basin with a smaller volume than such a cone would have an index below 1.0 (Fig. 6-8).

To arrive at D_v keep in mind that the volume of a cone is one-third the product of basal area and height; A is substituted for the area and z_m for the height. Since the actual volume of the lake is $A(\bar{z})$, the ratio of actual volume to the volume of the theoretical cone:

$$\frac{A(\bar{z})}{\frac{1}{3}A(z_m)}$$

equals volume development. By cancellation,

$$D_v = 3\frac{\bar{z}}{z_m}$$

From this it is apparent that the relation of mean to maximum depths is the case of an ideal cone, with D_v of 1.0, is 0.333.

Hayes (1957) examined morphometric data from 500 lakes and found the average D_v to be about 1.27. Gorham (1958a) assembled figures from many lakes of northern Britain and elsewhere that revealed volume developments greater than 1.0 as typical. Neumann (1959) used dimensions, given by Hutchinson (1957) for 107 lakes, and found that the mean value of $\bar{z}/z_m$ was 0.467; this implies a D_v of 1.40. Koshinsky (1970) studied the morphology of 68 glacial lakes sharing a common history and geologic environment on the Precambrian Shield in Saskatchewan. Their average volume development was 1.23.

These averages show us that most lakes for which there are good morphometric data occupy depressions that are U-shaped in cross section. Deevey (1955) pointed out that vertical profiles of at least deeper parts of lakes are more like segments of spheres than parts of cones. The flat part of the spherical slice is an imaginary subsur-

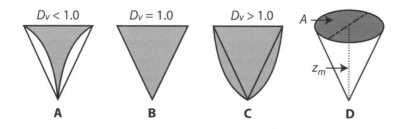

Figure 6-8 Volume developments (D_v). $D_v = 1.0$ (**B**), where volume of lake is the same as that of an imaginary cone (**D**), which has the same area, A, and maximum depth, z_m, as the lake.

face plane parallel to the lake surface; the rounded part represents the bottom contour, and the resultant geometric figure has $D_v > 1.0$.

Neumann's analyses convinced him that the ideal lake is close to an elliptic sinusoid. This is a geometric body, its base (the lake area) an ellipse, and its surface (the bottom contour) a sinusoid, rounded like a sine curve. This model has a D_v of 1.39.

The precision of the frustum-of-cone formula to calculate volume is reduced because of error caused by convexity of the lake sides. If the lake is divided into many thin, horizontal slices, each considered a frustum of the conical model, the error caused by convexity will be lessened, but the construction and analysis of a depth-area curve (Fig. 6-4) is probably the best method for determining lake volume.

Can we look at a volume development index and visualize the type of lake from which it came? Probably not. In Minnesota, some rich lakes that lie in basins of calcareous glacial drift often have D_v of less than 1.0, while some lakes in the northeastern part of the state occupying granitic ice-scour basins have indices greater than 1.0. It is easy to generalize that high D_v values are equated with elongate, steep-sided, deep basins, and with oligotrophy. When we consider the elongate Finger Lakes of upper New York, our conclusions are reinforced—Cayuga Lake, for example, has a D_v of 1.23, which is not remarkable but substantially above unity and not much lower than Lake Baikal's D_v of 1.29. Then we examine Lake Tahoe, within a deep, elongate graben and extremely oligotrophic; it exhibits a D_v of 1.87. Oligotrophic Crater Lake, although not elongate, is the deepest body of water in the United States; occupying a steep-sided circular caldera, it is characterized by a high index, 1.65.

The generalization, however, proves untenable when we note that shallow lakes with relatively great areas have the highest indices—these are saucer-shaped lakes rather than bath tubs or deep, rounded pits. Lake Winnipeg and other remnants of glacial Lake Agassiz occupy shallow depressions and have D_v values even greater than those of deep and trough-like basins; Winnipeg's is 2.07 and shallow Lake Nakuru's is 2.46 compared to Tanganyika's D_v of 1.17.

But what about low volume-development indices? Do they reveal precise information about basin configurations? Low D_v values are the result of different morphologic phenomena. Extensive broad shoal areas surrounding deep central waters yield D_v indices <1. In many hard-water glacial lakes in Michigan and Minnesota, for example, thick beds of chalky material called **marl** have been laid down in the shallows. These littoral marl shelves serve to form basins characterized by peripheral shallow areas that slope precipitously to deeper water on their inner borders.

A small, originally funnel-shaped depression, its side slumped and now convex toward the water, will show a low D_v. Demming Lake, Minnesota, has a D_v value of 0.65 (Hooper, 1951). This is the result of a very small volume of deep water and also, to a great extent, of the shape of the original ice mass that formed the kettle that Demming occupies. Perhaps there was subsequent slumping of the shallower sides to accent the relatively small deep region. Also the Florida solution basin, Little Lake Barton, may owe its relatively low D_v of 0.77 to slumping and marginal shoaling as well (Scott and Osborne, 1981).

A lake basin characterized by a more or less uniform bottom with the exception of an anomalous deep hole somewhere will have a low D_v. For example, Mountain Lake, Virginia, a landslide lake, has a D_v of 0.93 (Roth and Neff, 1964). This is because its deepest water lies in a small depression near the rocky debris that dams the lake. Theoretically, an impounded lake sloping uniformly to the deepest part at the dam would have a volume development greater than 1.0.

In large lakes early morphometric data may have been inaccurate, and we see some indices change with time. Lake Michigan serves as an example: Hutchinson's data (1957) gave a D_v of 1.12; Ragotzkie (1974) showed U.S. Army Corps of Engineers data from which a D_v of 0.91 can be inferred; and Ristic and Ristic (1980) gave it a D_v of only 0.87. Less than 1% of the lake's volume lies below 250 m, and that 1% represents a small, deep pit.

A similar phenomenon may explain why the larger Saskatchewan lakes on the Precambrian Shield have lower indices than the smaller lakes. For example, the mean D_v for the four lakes with areas surpassing 1,000 km^2 is only 0.57. Koshinsky (1970) considered the possibility that greater wave action and shore erosion associated with ice movements could have destroyed the U-shaped basin of large lakes. His final conclusion, however, was that the progressive lessening of D_v, which was associated with increasing area, may simply reflect the greater chance of larger basins encompassing aberrant, localized depressions.

Two very different pieces of laboratory glassware, a Petri dish and a 1,000-ml graduated cylinder, have identical D_v indices, 3.0. Thus, volume development conveys useful information only if we know other facts about the lake. It is a comparative value that can be instructive, but caution must be observed in its interpretation. By contrast, A, V, z_m, z, and D_L indices near 1.0 are data that tell something real about the lakes from which they come.

Basin slope.   The degree of slope from contour to contour can be found if the lengths of all isobaths are known. The methods for determining the lengths of these contours are the same as those outlined for finding the shoreline. The slope, or tangent (Fig. 6-9), between any two contours is

$$\frac{L_{z_1} + L_{z_2}}{2} \times \frac{z_2 - z_1}{A_{z_1} - A_{z_2}}$$

In this formula L_{z_1} is the upper contour's length, and L_{z_2} is that of the deeper contour; $A_{z_1} - A_{z_2}$ is the area enclosed by the two; and $z_2 - z_1$ is the height between the two isobaths, or the contour interval. The resultant calculations will be in decimal form and must be multiplied by 100 to be expressed as percentage.

The average slope of the entire lake bottom is sometimes determined, although it may have less ecologic import than the mean slope from shore to a given depth or the slope circumscribed by any two subsurface contour lines. Furthermore, there may be significant contrasts between the average slope and individual parts of the basin (Fig. 6-10).

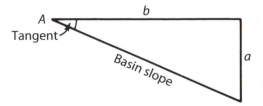

Figure 6-9   Slope between contours in a lake shown as the tangent, A. In this diagram b is the distance from the shore to the spot above depth, a.

The mean slope of the entire lake is found by the following formula, n being the total number of contours:

$$\frac{\left(\tfrac{1}{2}L_{z_0} + L_{z_1} + L_{z_2} + L_{z_3} \ldots + L_{z_{n-1}} + \tfrac{1}{2}L_{z_n}\right)z_m}{nA}$$

In simplest terms, mean slope is the average change in depth per unit horizontal distance. It is usually expressed as a percentage, but there are at least two other ways to designate it.

A lake bottom, for example, that has plunged to a depth of 30 m when it is at an average of 30 m from shore shows 100% slope within that distance; the angle of slope

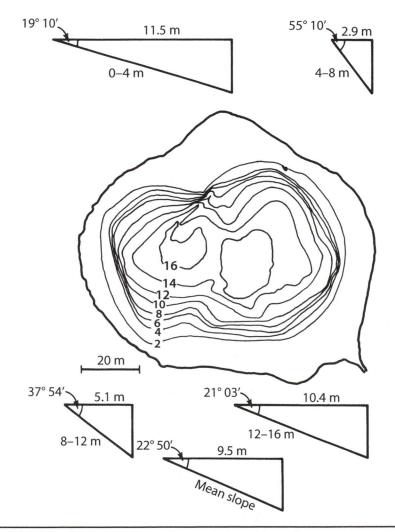

Figure 6-10 Bathymetric map of Montezuma Well, Arizona, showing 2 m contour intervals. The differences in slope from 0 to 4 m, 4 to 8 m, 8 to 12 m, 12 to 16 m, and the mean slope of the entire basin are shown. Tangent and b values are indicated; all a values are 4 m.

is, therefore, 45°. Henson and associates (1961) reported that the mean slope of the entire Cayuga Lake is 5.2%; this is an angle of 2° 58′. Expressed in another manner, the average slope of the lake is 52 m/km or 5.2 cm/m.

A simpler approach to approximating the mean slope of a basin is founded on the assumption that the lake surface is circular, and, therefore, $\pi r^2 = A$. The following formula is derived from this assumption:

$$\text{Percent mean slope} = 100\,\frac{z_m}{\sqrt{A/\pi}}$$

The steps used to arrive at this equation are

$$\pi r^2 = A$$

$$r^2 = \frac{A}{\pi}$$

$$r = \sqrt{\frac{A}{\pi}}$$

Therefore

$$\frac{z_m}{r} = \frac{z_m}{\sqrt{A/\pi}}$$

The trigonometric function for a right-angled triangle is (see Fig. 6-9):

$$\text{Tangent } A = \frac{a}{b}$$

The depth, z_m, is a; the radius, r or $\sqrt{A/\pi}$, becomes b; and the angle of A is assumed to be the average for the entire lake.

Uses of Morphometric Data

If a particular body of water is to be studied in detail, the results of all morphometric procedure should be tallied and kept on record for future use. This applies also to the hypsographic curve, the depth-area curve discussed earlier (see Fig. 6-4). These data will be useful in faunal, floral, and productivity studies. Some specific examples follow.

The accepted manner of expressing amounts of bottom fauna harvested by dredge is number per hectare or per square meter; or, on the basis of mass, expressions such as kilogram per hectare or gram per square meter are used. It is also becoming common for these and comparable data to be expressed in terms of energy: calories per unit area appears regularly in limnologic literature. Whatever the unit of measure, lake morphometry is useful.

Assume that the profundal zone, where the dredging is being done, ranges from a depth of 22 to 40 m and that 18 one-meter strata from 22 to 23 m and 39 to 40 m have been sampled. Each dredge haul encompasses a fraction of a square meter (most Ekman dredges commonly used to sample soft, fine-grained sediments grab an area of about 225 cm^2), so that counting and weighing the catch allows one to express numbers and masses in square meters, hectares, or square kilometers. It would be valid to state, therefore, that in the sediments lying at a representative depth of 32 to 33 m there are, on the basis of sampling procedure, so many animals per square meter of bottom. The average number collected per dredge would have been used to calculate numbers within that stratum. It is not valid, however, to find the average per dredge throughout the entire profundal zone and simply express this without correcting for unequal areas at different depths. First, the numbers of animals (or their mass, if required) per unit area of any stratum must be multiplied by the total area of that stratum. If all profundal strata are treated thus and summed, the total number of animals in the profundal zone is calculated. Dividing that sum by total area of the 22 to 40 m region yields a permissible average figure; it can be stated that there are so many animals per hectare in the profundal zone because the areas of the various depths within it have been properly weighted.

Similarly, knowing the lake's area and the volumes of the various strata below it permits proper expression of other limnologic features. Assume, for example, that photosynthesis occurs from the surface to a depth of about 5 m, below which the process cannot be demonstrated. If we know the milligrams of O_2 per liter (grams per cubic meter) produced per hour within each stratum, we can find the total produced within the stratum by multiplying its volume by oxygen value. With summation, the total amount of oxygen produced within the lake during a certain period of time is revealed, each stratum having been weighted properly. If the sum is divided by the surface area of the lake, a figure useful for comparative purposes is derived, photosynthetic rate expressed as grams of O_2 per square meter per hour.

This method can be applied to many other data, to mass or numbers of plankton organisms, to iron or sulfate, or to whatever is being measured and studied. The main point is that morphologic details of the lake must first be known.

Schindler (1971b) used morphometric procedure to explain differences in trophic status among a group of neighboring glacial lakes in Ontario. He reasoned that in most instances the nutrient input to the lakes was only via runoff from the drainage area (A_d) and precipitation on the lake surface (A_o). The total catchment area $(A_d + A_o)$ would be proportional to the yearly supply, while lake volume would serve to dilute it. Thus, the ratio

$$\frac{A_d + A_o}{V}$$

should be proportional, first, to levels of nutrient and, second, to the biologic productivity dependent on nutrient supply. The correlations were direct and excellent when the chemical nature of the sediments and water column, primary production, biomass, and some other indices of productivity were compared with the ratio. Such correlations lend support to the concept that change in morphology, as a lake fills, leads sometimes to an increase in nutrient per unit volume and to a resultant rise in productivity.

The Morphoedaphic Index (MEI)

A blending of morphologic and edaphic factors to indicate potential lake productivity was proposed by Ryder (1965) and reviewed in detail by Ryder and others (1974). It combined the principle of mean depth as an index (Fig. 6-7) with the amount of dissolved material in the water to serve as "a rough indicator of edaphic conditions" (Rawson, 1951). The morphoedaphic index, or **MEI**, is the quotient or ratio of total dissolved solids (**TDS**) in milligrams per liter and mean depth in meters.

The original proposal applied especially to commercial fish yield from northern temperate lakes, and there is good agreement between the MEI and fish harvests in Canadian and Finnish lakes. In Canada the most productive lakes have indices lying between 10 and 30, with the theoretically most favorable index being about 40 in that climate (Ryder et al., 1974). There are, however, morphologic and salinity restraints to be considered. An index of 40 could be found in a lake with a TDS of 4,000 mg/liter and a mean depth of 100 m, or in some other lake with a mean depth of only 10 m and a TDS of 400 mg/liter.

Lakes in north temperate regions, with mean depths of more than 18 m, may have their productivity controlled by excessive depth, as suggested by Thienemann (1927) and Rawson (1953, 1955). This probably applies to Great Slave Lake, Lake Baikal, and the deeper Laurentian Great Lakes, even though their edaphic components are somewhat less than productive Lake Erie's (Fig. 6-11). At the other extreme, many northern lakes with mean depths of less than 5 m may be subject to thick ice and snow cover with the threat of fish kill each winter and a lowered harvest as a consequence.

Geography and climate now enter the picture, for in the tropics mean depth may not be so important. Kilham and Kilham (1990) pointed out that the mixing depth does not affect annual primary productivity in some African freshwater lakes. Fig. 6-11 illustrates some of their unusually high fish productivity despite comparatively great mean depths. The East African Lake Victoria lies on a TDS vs z_{mean} plot (Fig. 6-11) somewhere between Great Slave Lake and Lake Superior, yet it produces 6 and 26 times their annual fish crops, respectively. Similarly, the more saline Lake Turkana, Kenya, has a mean depth of much more than 18 m, but its annual fish harvest is 30 times greater than high-yielding Lake Erie. At the other end of the spectrum, another tropical lake, the extremely shallow ($z = 2$ m) Ugandan Lake George (Lake Dweru) has no threat from ice cover, and its production is about 20 times that of the northern temperate Erie.

Morphologic lake dimensions play little part when stresses from salinity are great. Hypersaline waters present so many ionic and osmotic problems for freshwater organisms that none but the hardiest of euryhaline forms survive and fishes are absent. Great Salt Lake in Utah, the Dead Sea in Israel, and other hypersaline arid-land waters typify this condition (Fig. 6-11). Extremely low concentrations of minerals also hinder productivity. Waldo Lake, Oregon, has an average depth greater than 18 m, but the impossibility of a fish harvest is owed to its dilute waters (TDS = 1.7 mg/liter).

Despite favorable climatic-geographic locale, morphologic dimensions, and salinity strengths there are other factors that preclude fish survival. Two lakes are plotted in Fig. 6-11 to illustrate this. Montezuma Well is fishless because of high CO_2 tensions in the water, and Anguish Lake, Ontario, lost its fish fauna as its waters recently became progressively more acid, finally reaching a pH of 4.0.

The Shape of Lakes at z_0

Various reasons for lake surfaces appearing circular with shoreline development indices approaching 1.0 have been discussed. Similarly, outlines of elliptic lakes are owed to natural events. Dendritic patterns are the result of river waters backing up tributaries, and fluviatile oxbow lakes are typically crescent shaped.

A lake configuration rarely noted is ringlike, and it is the product of several phenomena, there being no underlying common mechanism. One type of ring lake is

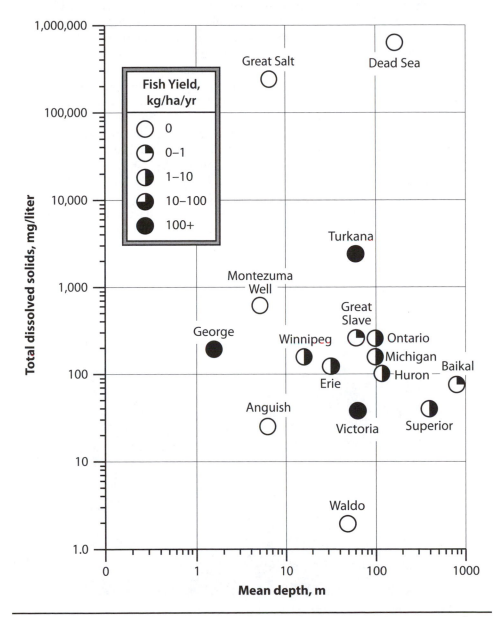

Figure 6-11 Plot of total dissolved solids in milligrams per liter vs. mean depth in meters for various lakes shows morphoedaphic index (MEI). Annual fish yield in kilograms per hectare per year are shown by graduated quarter-filled circles.

found encircling volcanic structures. Perhaps the outflow of magma leaving a shallow moatlike depression around the volcano accounts for these lakes; the mountain's great weight would depress a much broader area. One of the best-known volcanoes on earth, the beautifully symmetric Fujiyama of Japan, is ringed by five separate lakes. Similarly, some small lakes nearly surround a volcanic-ash cone in the southern Libyan desert; were the water level a bit higher, the ring would be complete (Beadle, 1981). In northern Sonora, Mexico, in the spectacular Pinacate volcanic field, a cinder cone called Cerro Colorado is ringed by a depression that holds water after extensive rains.

Meteoritic impact often creates a concavity, circular in outline (New Quebec Crater Lake, for example), but not always. Some hard-to-recognize scars on the earth's surface also are categorized as astroblemes. One type has a raised central area. You might see the same effect if you tossed a stone into mud of such consistency that the splash remained partially elevated—"frozen." A puzzling structure 720 km northeast of Montreal, Quebec, once was believed to be an old volcano encircled by a ringlike lake. Now this is thought to be an astrobleme, the elevated central region rising above the surrounding water. This central, high land is a "splash," not an eroded cone. Fig. 6-12 shows the ring lake around this Canadian astrobleme known as Manicouagan.

Another type of ring lake is formed when an island of some kind is raised above the center of a more or less circular lake. Thus a secondary crater in the Japanese caldera lake, Tôyako, is comparatively large and makes the lake ringlike (see Hutchinson, 1957, Fig. 92).

Finally, a small pingo rising in and above what seems to be a small thaw-basin lake forms a ring lake. Alternatively, there may have been a depression and melting around the base of a pingo to create such a configuration.

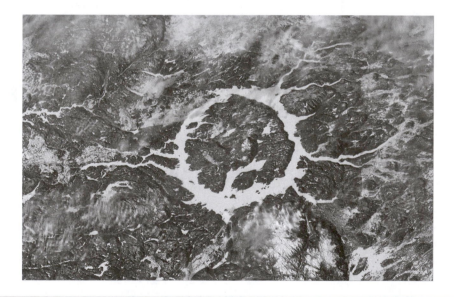

Figure 6-12 Ring-shaped lake in the Manicouagan astrobleme, Quebec, Canada. Lat. 51 ° 30′ N, long. 69° 00′ W. The photo was taken from by astronauts on the Space Shuttle mission STS-9. NASA Photo ID: STS009-048-3139.

Lake "Size"

When people ask the dimensions of a lake, they wish to know the area, maximum length, and especially the maximum depth. Volume is rarely questioned, and the common way of ranking the great lakes of the earth does not consider that dimension. We speak of "bodies of water," yet rate the lake's size on the basis of surface area. This is something that, in retrospect, may make little sense. On the other hand, there is only so much surface to the planet Earth, and the amount occupied by lakes is significant to us. Moreover, solar energy arriving at the lake surface marks the beginning of lacustrine energy exchanges from one trophic level to the next. As a result, we express productivity in terms of surface area: for example, annual photosynthesis as carbon fixed per square meter or yearly fish growth as kilogram per hectare. Even the total heat absorbed by the lake during the warming period of the year is expressed as calories per square centimeter of surface.

Some past controversy about the relative importance of surface area in determining the trophic status of a lake led to work that has bearing on lake size. It had been said that the enriching effect of run-off, entering at the shoreline, is diluted in lakes of large area, and indeed, Thomas (1969) pointed out that central European lakes of small area are more susceptible to pollution than larger ones. However, the relationships between area and other dimensions must be considered here. Hayes (1957) attempted to clarify the effects of area and other lake dimensions on such indices of productivity as fish yield and bottom-fauna mass.

His analyses of morphometric data from 500 lakes led to the presentation of a formula and a plot showing a direct relationship between surface area and mean depth. This shows that large lakes are usually deeper than smaller ones, and we can generalize that lakes of great expanse are apt to have great volumes also. The relationship, Hayes found, did not apply well in small lakes; those with an area of 0.3 km^2 or less often had mean depths (and hence volumes) greater than expected. Also, the two great lakes of Nicaragua, Lake Managua and Lake Nicaragua, are anomalous. They are too shallow for their surface areas to fit on Hayes' plot.

Hutchinson (1957) presented, in tabular form, morphometric data derived from the greatest of the world's lakes. From these facts one can devise a list of the 10 largest lakes, with respect to area alone. Eight of these 10 would also appear on a list of the 10 largest, ranked according to volume.

Thus, lakes as three-dimensional bodies of water can be arranged with some success on the basis of their two-dimensional surfaces. There are exceptions such as the Nicaraguan lakes, but the most exceptional may be the African equatorial Lake Chad. This tremendous lake covers more than 16,500 km^2, but its mean depth is only about 1.5 m. This reveals that its volume is only about 25 km^3.

Lake Baikal has a volume of 23,000 km^3, surpassing by 11,000 km^3 the volume of Lake Superior. Superior, however, has an area of 83,300 km^2, as compared with only 31,500 km^2 for Baikal. Which is the "larger" lake?

7

Wetlands

Wetlands Are Difficult to Define

Wetlands represent zones of transition (**ecotones**) between typical terrestrial ecosystems and aquatic habitats such as lakes and seas. As neither land nor water, they have proven difficult to live or work in, or to classify and categorize, and therefore slip between the established ecological disciplines. Nevertheless, these systems have fascinated researchers for many years and are now increasingly recognized as rare and important. In fact, wetlands are the only category of ecosystem protected by the US government; in 1971, an international treaty (the Ramsar Convention) was established to protect important wetlands across the globe.

Ramsar defines wetlands as "areas of marsh, fen, peatland or water, whether natural or artificial, permanent or temporary, with water that is static or flowing, fresh, brackish or salt, including areas of marine water the depth of which at low tide does not exceed six meters." This definition describes many systems, both limnological and marine, across many geographic and climatic conditions. An obvious shortcoming of this definition is that it neglects to specify *how* to recognize these systems.

The US Fish and Wildlife Service urged protection for wetlands with the "Circular 39" publication of Shaw and Fredine (1956); wetlands were defined as lands covered shallowly, and sometimes only intermittently, with water. In 1987 the US Corps of Engineers took the concept further and identified wetlands as areas inundated (flooded), or with soil saturated by water sufficiently to support characteristic plants. The fact that a wetland may, in fact, never be covered by water causes understandable confusion. Regardless, if a plant's roots and the surrounding soils are frequently and sufficiently wet, it is a wetland---standing water or no.

This legal definition is necessary because Section 404 of the Federal Clean Water Act directed the Corps to regulate activities causing significant alteration to aquatic systems ("dredge and fill"). Subsequent court rulings extended such protection to wetlands. The 1987 Corps of Engineers wetland delineation manual defines wetlands and describes in detail how **delineation**—drawing a line around the wetland—is accomplished. The procedure specifies how *field indicators* of hydrology, vegetation, and soils characterize an area as wetland.

Hydrology

Certainly, the single most important determinant of wetland function is hydrology, the dynamics of water in the wetland. The depth of standing water or the depth to the groundwater below the soil surface affects soil chemistry, plant germination and growth, animal behavior, and every other ecological process. It is no wonder that a plot of water-level surface elevation, known as a **hydrograph**, is considered the "signature of the wetland" (Mitsch and Gosselink, 2007). Different wetland types will have characteristic patterns of hydrology, but each site will be unique and variable over time (Fig. 7-1). The vertical axis of a hydrograph indicates height of the water surface relative to the substrate; negative values indicate the water surface is the water table depth in the soil.

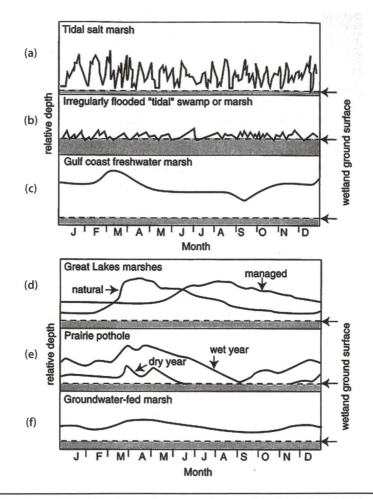

Figure 7-1 A comparison of hydrographs in several types of wetlands. (a) tidal salt marsh, (b) "tidal" swamp, (c) coastal freshwater marsh, (d) Great Lakes marshes, (e) marsh from Prairie Pothole region with surface inputs, (f) prairie pothole marsh with groundwater inputs. From Mitsch and Gosselink (2007). Reprinted with permission.

Usually this is recorded at the deepest part of the basin, using a float rising and falling within a pipe situated perpendicular to the soil surface. When considering the water depth on the hydrograph, remember that a small change in water depth can mean a large change in area of standing water, if the basin has sides with a shallow slope.

Water comes to the wetland from the sky (precipitation), runoff from surrounding ground, a nearby lake or stream, or seepage from the ground itself. Determining the movement of water can be difficult, as underground features such as *lenses* of sand or gravel can act like pipes, causing groundwater to seep into the system. The wetland vegetation can move significant volumes of water through living tissues (transpiration) and even dead, undecomposed plants (peat) can wick up water, bringing it to a bog's surface; a "perched" bog mat can be wet, even though it sits higher that surrounding ground or surface water. A complete hydrologic budget will consider not just water depth, but all inputs, outputs, and storages of water. This explains the movement of nutrients or other chemicals through the system, the role of biota in water dynamics, and the role of wetlands in the landscape. For example, a wetland might be a site of groundwater seeping out to the surface (discharge wetland) or surface water entering the ground (recharge wetland), or might vary between these conditions at different times or circumstances.

Hydrology exists within an environmental context: the wetland experiences climate patterns and perhaps flooding from adjacent water bodies (lakes or streams), which will interact with any groundwater inflow or outflow. As expected, wetlands in coastal areas can be profoundly affected by tides; remarkably, these can extend to wetlands well away from the coast itself. The Kakadu wetland complex in the Northern Territory of Australia experiences tidal influences up to 100 km (more than 60 miles) inland, including a massive tree die-off from saltwater intrusion (Woodroffe et al., 1986). Usually inland wetlands influenced by tides are more typically **oligohaline**, so that water levels rise and fall with tides while salinity remains low.

Flooding presents both opportunity and danger to the inhabitants of wetlands. A flood pulse brings with it not just water, but often silts and sediments, nutrients (a *subsidy*), eggs or juvenile animals, seeds or other plant propagules, among much else. Wave energy can be destructive, uprooting even large trees and carrying them away. But just the extreme variability in water availability, from long periods of dry conditions to deep and prolonged flooding, challenges wetland organisms in ways unknown in typical terrestrial or aquatic habitats.

Wetland Organisms

The wetland soil (described in the biogeochemistry section) differs from the adjacent upland soil, and only a few of the world's plants (**hydrophytes**) can colonize it and survive. Thus, biodiversity in wetlands is generally lower than in other ecosystems, often containing rare species found nowhere else. Paradoxically however, the few species that *are* found can be highly productive, in part because of the input of nutrients and water from flood pulses combined with removal of wastes and toxins from the system (Batzer and Sharitz, 2006). Among the highest average primary production values of any ecosystem were recorded in wetlands: 1050 g $C/m^2/year$ for freshwater marshes (Rocha and Goulden, 2009).

Typically the wetland soil is anaerobic, with only a shallow oxidized layer overlying it. As a result, the vegetation is threatened by the danger of root anoxia. Air spaces in roots and stems have evolved and serve to conduct oxygen; hydrophytes often have a stem functioning as a kind of "snorkel" to bring air from above the water down to the lower parts of the plant. Some plants have also developed adventitious roots that are important in acquiring oxygen rather than anchoring the plant. It is believed that the so-called "knees" of bald cypress trees possibly play the same role. These are conical or rounded processes rising from the roots and extending above the water into the air.

Animals must also adapt to the challenges of intermittent flooding and drying. Animals in many cases can complete a life cycle quickly, burrow to find available water (typical of some crayfish or amphibians) or simply leave the system as needed (many fish fry develop in marshes adjacent to large lakes where the adults are found). Insects like mosquitoes or midges may include a short juvenile stage in water, followed by emergence into the air. Another strategy is found among many waterfowl: most ducks and geese, for example, nest in wetland areas even if they swim in larger water bodies at other times. It may be useful to think of organisms as migrants vs. residents, or for sessile organisms, obligate vs. facultative wetland species.

A characteristic of all life is to actively change the environment; the biota of wetlands are no exception. Activities such as trapping sediments, accumulating organic matter in the soil (peat), damming streams, and transpiring water will all modify the system. In the long term, such activity may direct and accelerate the aging of the ecosystem (*hydrarch succession*), perhaps filling it and forming a terrestrial system over time. In the short term, nearly every chemical processes occurring in the system are subject to the actions of the biota.

Biogeochemistry

Typical terrestrial (upland) soils are complex and very much alive, filled with spaces (voids) literally crawling with insects, worms, fungi, bacteria, and anchored by plant roots. The chemical processes such as mineral weathering, nutrient transport, and pH shifts are critical to ecosystem function. As anyone who has killed a plant by overwatering knows, waterlogged soil can be deadly, suffocating roots, killing beneficial symbionts, and accumulating toxins. Flooding soil for as little as a few days will drive out the air, and the resulting anaerobic conditions cause unique chemical reactions. The replacement of the normal microbiota by a distinctive wetland microbial community further alters the chemical processes in water and soil (Richardson and Vepraskas 2001).

The normal chemical processes in soil include absorption of water and dissolved substances by plant roots; decomposition of dead organic material by bacteria and fungi; cellular respiration by plant roots, soil animals and microbes; and nitrogen fixation. The lack of oxygen in flooded soils limits much cell activity and generally slows these biogeochemical transformations.

Prolonged lack of oxygen available to the biota (**anaerobiosis**) leads to a shift in **redox potential** (E_h, measured in electrical units of volts or millivolts; see Chapter 15 for a discussion of electrochemistry). As the oxygen is consumed, the soil conditions become progressively more chemically reduced, and the microbes successively use alternative electron acceptors: nitrogen, then manganese, iron, sulfur, and eventually carbon. A wetland giving off bubbles of reduced carbon "swamp gas" (methane, CH_4) has low redox potential indeed, less than -0.2 V (Mitsch and Gosselink 2007).

The slowed decomposition processes associated with prolonged submersion results in a buildup of undecomposed organic matter (**peat**). Some peat-building wetlands can store the carbon-rich deposits for millennia. In particular, bogs contain floating mats of water-saturated peat built up by *Sphagnum* moss, sometimes several meters thick (Rosa et al., 2008). The actively growing plants are perched on a mat of porous, sponge-like dead moss.

Sphagnum is unusual in producing organic acids and tannins as it grows. The low pH and these naturally preservative chemicals further the inhibitory effect of anaerobiosis on decomposition, aiding in the accumulation of peat. Interestingly, these conditions also preserve the organic matter of the human body and clothing. At numerous sites in northern Europe, peat harvesting has uncovered so-called "bog people," some perfectly preserved for thousands of years (Fig. 7-2). The tannins preserve the tissues and artifacts exceedingly well, in a manner similar to tanning leather.

Worldwide, carbon storage in peatlands is large: boreal and subarctic peatlands alone store about 15–30% of the world's soil carbon (C) as peat (Limpens et al., 2008). An area of active research concerns the fate of this carbon, since possible changes in climate, especially warming and drying of those peatland-rich regions, could create a "positive feedback"—causing the peat to release carbon (CO_2 or CH_4) to the atmosphere, thereby accelerating climate change, and even more carbon release, in an increasing spiral.

Many organisms are engaged in processing nitrogen, a subject discussed in Chapter 15. Some are solely anaerobic; others are facultative anaerobes. In general, the anaerobic environment in wetland soils limits nitrogen in two directions: less nitrogen input (relatively lower rates of fixation than elsewhere), and accelerated loss through **denitrification**, the use of oxidized nitrogen (nitrate, NO_3^-) as a terminal electron acceptor by microbes when oxygen itself is unavailable. Denitrification allows wet-

Figure 7-2 The unique chemistry of a peat bog in Denmark perfectly preserved the remains of "Tollund Man," who died approximately 2,400 years ago. Photo by Sven Rosborn/Wikimedia Commons.

land microbes to use organic matter as a carbohydrate energy source with nitrate as an oxidizer. Since many of our surface and groundwaters are polluted by nitrates from fertilizer application, denitrification is an important means for pollution removal.

The characteristic environmental conditions of wetlands lead to the development of distinctive **hydric soils**. These soils are often dark-colored (gray, blue, or black from reduced iron), sometimes with unusual colored "inclusions" such as red/orange stripes of pore linings, where oxygen leaks out of roots and rusts the iron in the adjacent soil. Even mineral soils will often have considerable organic matter content (Richardson and Vepraskas, 2001).

Hydrology and chemistry combine in the consideration of physical processes in wetland sediments. Since wetlands collect river or lake overflow and overland runoff, water entering the wetland can bring considerable silt and sediment. Combined with primary productivity by algae and plants in the wetland, considerable buildup of soil substrate, a process called *accretion*, occurs. Accretion can be directly measured by using horizon markers, a layer applied to the substrate (perhaps in several replicated 1-m^2 plots). Subsequent samples in the plots will clearly indicate accretion deposited onto the marker layer. Markers can be laid down using white clay (feldspar), plaster, or shiny glitter.

It is important to realize that some of the "water purification" in wetlands is simple physical processes, not chemical transformations. An important example is phosphorus: the element may be retained by the wetland acting as a "sink," but most often this probably is simply precipitation from the water column or adherence (sorption) to soil particles. Such processes may involve little direct biological storage, and phosphorus removal is determined by simple physical limitations. When the element saturates the available storage capacity, it simply flushes out of the system in outflow (Mitsch and Gosselink, 2007), and the wetland becomes a "source" instead of a "sink."

Wetland Classification and Diversity

Among the challenges of studying wetlands is terminology. Many people are only lately even using the term "wetland," and names for the various types vary in their use. For example, in North America, a "swamp" is nearly always a wetland whose dominant plants are woody (trees or shrubs). But in other parts of the world, wetlands dominated by grasses like reeds (*Phragmites*) are commonly known as swamps. A discussion of this confusion is described in the classic wetlands text by Mitsch and Gosselink (2007).

Cowardin and associates (1979) of the US Fish and Wildlife Service established a comprehensive, hierarchical classification system to divide wetlands into categories of coastal or inland, each then grouped by factors such as proximity to a river or lake, tidal influence, and other factors. This scheme became the basis for the National Wetlands Inventory (NWI), a systematic cataloging and mapping of the wetlands of the United States. The terminology, which is standardized throughout the US federal government and widely throughout the country, is used here.

Marine or estuarine marshes are located near enough to the ocean to be influenced by coastal processes—in particular, tides. *Freshwater tidal marshes* are inland from the coast, so salinity is low or absent, yet a hydrologic connection causes water

Figure 7-3 The Pa-Hay-Okee nature trail in Everglades National Park straddles two wetland types: the wooded area on the left of the photo is NWI code PSS1/3C, a Palustrine Scrub-Shrub with Broad-Leaved Deciduous (1) or Broad-Leaved Evergreen (3) vegetation, and seasonally flooded; on the right is PEM1F, a Palustrine Emergent, Persistent, Semipermanently Flooded system (sawgrass prairie). Photo courtesy of the National Park Service.

levels to rise and fall with tidal cycles. *Salt marshes* are located on the coast itself, and have high salinity—some are even *hypersaline*, more salty than seawater.

Depending on the particular salinity gradient, coastal wetlands are dominated by graminoid (grass-like) plants, including true grasses (Cordgrass, *Spartina*; Reed, *Phragmites*; Saltgrass *Distichilis*) or Rushes (*Juncus*); and Cattail (*Typha*) or Glasswort (*Salicornia*). The proximity to the ocean allows a variety of marine animals to inhabit or visit, including fish, molluscs, wading birds, and waterfowl.

Mangrove wetlands occur along the coasts in tropical and subtropical regions of the world. There are about 70 species of mangroves, from unrelated plant families; they are grouped by a shared growth form and ecology. Mangrove trees form tangled masses of intertwined trunks and extensive prop roots (Fig. 7-4 on the next page) which together brace the tree from wave action. Mangrove trees are sensitive to extreme tidal currents and wave action; many mangrove swamps develop in coastal areas protected by offshore coral reefs. Their ability to catch shifting sediments and anchor the substrate makes them important in the stabilization of barrier islands. Mangroves house a variety of snails, crabs, and fish. These forests provide ideal nesting habitat for many birds, including the wood stork (*Mycteria americana*), bald eagle (*Haliaeetus leucocephalus)*, osprey (*Pandion haliaetus*), and brown pelican (*Pelecanus occidentalis*).

Freshwater marshes are an extremely diverse group of habitats. The category includes all non-forested inland wetlands except for those classified as peatlands. When permanently flooded, they may be referred to as shallow ponds; yet some sys-

Figure 7-4 A mangrove swamp. Photo by 9comeback/Shutterstock.

tems may only very rarely have standing water. Freshwater marshes contain neither the extreme dissolved mineral concentrations nor the pH levels typical of many peatlands. They are categorized by water source: the outward fringe of a lake's littoral area (lacustrine), in a floodplain near a stream (riverine), or isolated from other surface waters (palustrine).

As with coastal marshes, the graminoid plants dominate marshes: grasses, sedges, rushes, and cattails; other important "emergent" plants include smartweeds (*Polygonum*) arrowhead (*Sagittaria*), or pickerelweed (*Pontederia*). Areas with deeper water include rooted floating-leaved plants (spatterdock, *Nuphar*, waterlily, *Nymphaea*) or submersed plants (pondweed, *Potamogeton*; coontail, *Ceratophyllum*). Marshes are significant breeding areas for migratory waterfowl, including the many familiar ducks, geese, and swans. These habitats also produce many wading birds, sparrows, and blackbirds. Obviously, these sites are important for mammals such as muskrat (*Ondatra zibethicus*) or beaver (*Castor canadensis*) and many amphibians, reptiles, snails, fish, and arthropods.

Peatlands or *mires* include any peat-forming wetland (including bogs and fens). They are found in the world's boreal zones that once were glaciated, and in a few wet areas in mountains. The topography is such that there are no inlet or outlet streams. Despite this, the water table is high. Most bogs are **ombrotrophic**; they receive their water and minerals via precipitation. Certain peatlands are affected by minerals they receive from the surrounding terrain and groundwater; these are **minerotrophic** bogs or mires and are called fens.

Mosses and other acidophilic plants grow in bogs, contributing to their acidity and to the peat soil that accumulates. Mosses of the genus *Sphagnum* are especially important and they bring the pH levels down below 4.0 in some bogs. Fens are less acid and generally richer in nutrients; they accumulate peat from non-moss plants such as sedges. The scarcity of nitrogen in bogs is overcome by specially adapted plants with an unusual trait: carnivorousness! Sundews (*Drosera*), pitcher plants (*Darlingtonia, Nepenthes, Sarracenia*), and Venus flytrap (*Dionaea*) all trap animals (usually insects). Even though the plants are photosynthetic, the animal corpses provide organic nitrogen.

Swamps are wooded wetlands, and like marshes are of several types. *Southern deep water swamps* are characterized by rather large trees that do not tolerate saline water. Typical are species of cypress trees (*Taxodium*; Figure 7-6 on the following page) and tupelo (*Nyssa*) trees. Germination of these trees ("regeneration" of the forest) is generally restricted to exposed mud or debris; after germination, the tree may spend most of its life with roots under water. In northern regions, coniferous swamps of Arborvitae (*Thuja*), Spruce (*Picea*), or Larch (*Larix*) are locally common. Buttonbush is a cosmopolitan genus of shrubs growing in wetlands across the world; the North American species (*Cephalanthus occidentalis*) is widespread in wetlands across the continent.

Figure 7-5 A peat-building wetland system. Sphagnum bog, Dolly Sods, West Virginia. Photo from forestwander.com/.

Figure 7-6   Southern deepwater swamp. Note the flared, buttressed roots at the base of the trunks, and the snorkel-like "knees" protruding above the water. Photo by Steve Bower/Shutterstock.

Riparian wetlands are found along rivers, and in arid regions they may support the only trees to be found over great expanses. As a result of their location bordering streams, they are elongate in outline and have a high water table. They receive their nutrients and minerals from upstream sources. In some European countries peat deposits occur in riparian zones. Some of the trees are huge. In the southwestern deserts of North America, sycamore trees of the genus *Platanus* are typical. (There are other unrelated trees called sycamore in other parts of the world.) Because the treed section of the floodplain may be some distance from the channel itself, bottomland forests are classified as *palustrine forested* (PFO) in the NWI system.

Gleasonian (Cyclical) Succession

The concept of succession, so important to ecology, is applied in a unique way to wetlands. Here a form of Gleasonian (individualistic) succession known as **cyclic succession** is an organizing principle, allowing us both to understand patterns of ecological development and to better restore or manage ecosystems (van der Valk, 1981). Gleasonian succession involves species groups forming and changing over time, based on individual tolerances and changing circumstances. The process does not form the typical stable climax community, however: the system instead degenerates or is profoundly modified by biota (such as muskrats) and remains an open-water environment of limited emergent plant cover and richness, until a dry period results in a drawdown with exposed mud (Fig. 7-7). The drawdown allows germination from the mud (where a so-called "seed bank" persists) and regeneration of the plant community. This concept was developed by workers studying inland freshwater marshes

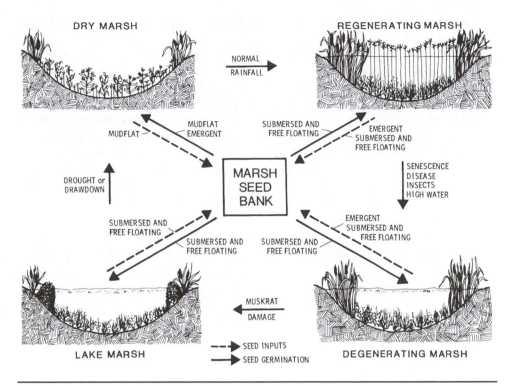

Figure 7-7 The Gleasonian, cyclic succession of a wetland plant community. From van der Valk and Davis. Reprinted with permission.

(van der Valk and Davis. 1978) but was expanded to wetlands of other types and regions. More importantly, it has led to the concept of the environmental sieve, in which plants are recruited from a dormant pool of seeds or other propagules as environmental conditions allow. We can then predict and perhaps influence the community of a new or established wetland by manipulating the hydrology and other conditions based on a knowledge of assembly rules, to encourage a desired vegetation (Keddy, 1992; Casanova and Brock, 2000).

Wetlands and Human Society

A feature shared by the many types of wetland ecosystems is the threat from human activity. In the United States alone, probably 50% of the wetlands that were present when Europeans arrived in North America have been destroyed. Requirements for agriculture, building, and highway construction have led to the draining and filling in of countless watery areas.

Although not widely recognized, wetlands provide valuable services to society. As described above, the biota of wetlands include furbearers, waterfowl, fish, and molluscs, which can all be valuable. The biogeochemistry of wetlands make them effective "sinks" to remove eroded silts and sediments, nutrients, and other pollutants from water. Zedler and Kercher (2005) describe another important benefit: absorbing excess river flow to lessen flooding of downstream areas. Such flood abatement in seldom recognized but substantial.

As mentioned previously, wetlands are protected by law in the United States. Congress passed the Clean Water Act in 1972, and subsequent court rulings have included wetlands in the protection of "Waters of the United States." Section 404 of the Act directs the Corps of Engineers to regulate activities (i.e., dredge and fill) that could significantly impact wetlands. The Corps published a wetlands delineation manual in 1987, which remains the guide for recognizing a wetland and delineating a boundary around it. Each of the three criteria (hydrology, vegetation, and soils) has characteristic qualities in wetlands, and "field indicators" are used to determine the condition of each.

- *Hydrology* is indicated by watermarks (bleaching or staining of tree trunks), matting of leaves or sediment deposition, cracked mud, and deposited flotsam in drift lines.

- *Vegetation* in a wetland consists mainly of hydrophytes, plants specially adapted to flooding. These are listed in a government publication (Reed, 1988, and updates) according to the probability of occurrence in a wetland, from obligate (99% or more in wetlands) to facultative to upland (less than 1% presence in wetlands).

- *Soils* become hydric when they are regularly submersed for extended periods; they are identified by high organic content, color, texture, and smell (particularly sulfidic, "rotten egg" odor).

Given the realities of human society and development, it is impossible to preserve every wetland entirely, so regulators of wetland policy use a process of *mitigation* to lessen the environmental impact of a wetland loss. This often includes replacement of a loss from construction or other human activity, with a newly constructed or restored wetland elsewhere. Building a functional ecosystem from scratch is a daunting task, but it has been done quite successfully (Mitsch et al., 1998). Ensuring appropriate hydrology at the site is key: soils will develop and biota will colonize, but only if water is present in appropriate amounts and patterns.

Case Study: Murray-Darling Wetlands

One approach to addressing these challenges is provided in the Murray-Darling basin in Australia (Reid and Brooks, 2000). This vast drainage basin covers nearly 14% of the continent's land, and water flows have been regulated for over a century. Since Australia is the driest inhabited continent and has both a growing population and intensive agriculture, water resource demands are problematic. The indigenous Aboriginal peoples have both historical-cultural and practical rights to water use. Through careful study, planning, and communication by all interested parties, the governmental authority has established hydrologic regulation that includes "environmental water allocations" designed to protect the numerous and diverse wetlands of the basin. It is hoped that wise management will successfully balance the competing demands.

How do we measure success in creation and restoration of wetlands? The hydrogeomorphic (**HGM**) approach appears to be the means to answer the question. The first step is to develop a list of well-characterized reference wetlands of the various types in each region. Human-built wetlands are measured against these *reference systems* to assess success in ecological function (Smith et al., 1995). Development of this system is a decades-long undertaking, unfortunately complicated by the extensive human impacts on many remaining "natural" reference wetlands.

Wetlands around the world face pressures from a growing human population with an extensive environmental footprint. Drainage and filling to provide agricultural land and building sites, harvesting timber, and other direct impacts continue, despite a growing realization of the benefits wetlands provide. Indirect impacts of altered hydrology by water resource utilization or changes from climate change are also substantial (Junk et al., 2013).

8

Streams
The Lotic Ecosystem

Every year, the world's rivers (excluding Antarctica) deliver about 37,000 km^3 of water to the oceans, about 35% of the precipitation falling on the continents (Dolman, 2008). Along their lengths, these streams provide various habitats for biota, cut valleys and canyons on the Earth's surface, transport many tons of soil and sediments, and provide water and waste removal for much of the human population. These **lotic** eco-systems of flowing water are known as streams by ecologists; depending on size and other characteristics, they may commonly be known by a wide variety of terms such as river, creek, brook, run, drain, arroyo, or others.

Bridge (2003) provides an excellent background to the geology of rivers and their floodplains. A river begins in the headwaters at higher elevations in the interior of the continent, spills out an alluvial fan where the stream meets the plains, snakes back and forth in meanders through broad valleys, and finally meets the sea (or large inland lake) in either an estuary or a fan-shaped delta. Over millennia, a flat landscape (such as a glacial till plain) is dissected and drained by incising rivers that cut channels and drain the landscape with ever-greater efficiency.

The area drained by a particular stream is its watershed, also known as the drain-age basin. Small streams (first-order) may start from intermittent sources (e.g., washes in arid regions), groundwater seeps, or outflows from lakes (Fig. 8-1 on the following page). Two such first-order streams join at a confluence to form a second-order stream; two second-order streams join to form a third-order stream, and so on. Note that adding a smaller-order tributary to an existing stream does not change its order. A diagram of a river with numerous branch-like tributaries resembles a tree, and that drainage is called *dendritic* after the Greek word for tree (*dendros*).

155

Case Study: A Bridge over the Amazon River

The great rivers drain vast areas of land and annually discharge large volumes of water. The mightiest of all is the Amazon, draining much of South America; its flow is nearly 15% of all the world's river flow—greater than the combined flow of the next five rivers in Table 8-1! Remarkably enough, the first bridge to cross the Amazon River (the Manaus Iranduba Bridge) only opened in late 2011. Although officials from Brazil insist the Amazon River is the world's longest (beating the Nile for the title), definitive length measurements—or, incredibly, even the ultimate source of the river—are only now being revealed (Contos and Tripcevich, 2014).

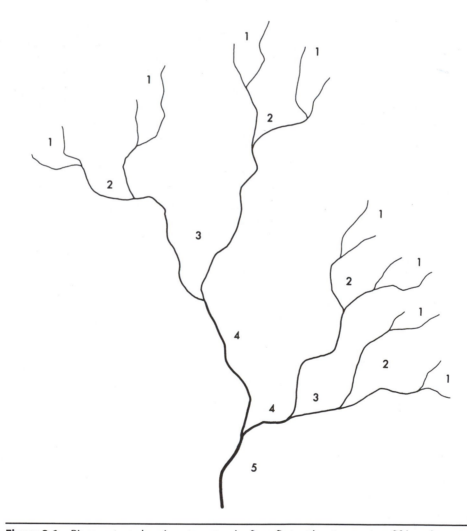

Figure 8-1 River system showing stream order from first-order streams to a fifth-order stream.

Table 8-1   The 15 world's greatest river systems, as measured by flow rate

Name	Country(ies) of Location	Mean discharge (km³/yr)	Length (10³ km)	Drainage area (10⁶ km²)
Amazon	Brazil	5444	6.44	5.78
Congo	Congo / Angola	1270	4.70	4.01
Orinoco	Venezuela / Colombia	996	3.46	0.88
Changjiang (Yangtze)	China	907	5.98	1.94
Brahmaputra	India / Bangladesh	643	2.90	1.62
Yenisey	Russian Federation	588	5.54	2.59
Mississippi	United States	552	6.02	3.22
Lena	Russian Federation	532	4.40	2.42
Paraná	Brazil-/ Paraguay / Argentina	517	3.94	2.31
Ob	Russian Federation	402	5.41	2.48
Ganges	India / Bangladesh	371	2.53	1.08
Tocantins	Brazil	347	2.64	0.80
Mekong	China-/ Burma (Myanmar) / Laos / Thailand / Cambodia / Vietnam	312	4.00	0.80
Amur	Russian Federation	307	4.44	1.84
Mackenzie River	Canada	286	1.74	1.81

Data from Dai et al., 2009.

Types of Flow

Anyone would be curious about the rate of flow in a stream; measurement is discussed in the next section. Less obvious is the fundamental manner in which water flows within the channel. There are two principal types of flow, *laminar* and *turbulent*. Of the two, **laminar flow** is by far less common. It is characteristic of very viscous fluids and in water occurs only when it is moving very slowly. Then, all liquid units progress in parallel lines with respect to their neighbors and at the same speed. It resembles motionless water in that whatever mixing occurs is brought about by molecular action. As velocity increases, **turbulent flow** arises and it is characterized by irregularity. Neighboring liquid units may move in different directions and at different velocities from the average of the flow. It is an erratic and mixing progression of the fluid, and most of a stream's flow pattern is of this kind. **Turbulence** explains why streams do not achieve greater velocity as they slide down channels marked by an even gradient. Accelerating forces are checked by the turbulence. Another factor is the channel roughness, for this induces turbulence, the force that resists and retards acceleration.

The **Reynolds number** is a dimensionless figure that marks the boundary between the smoothness of laminar flow and the eddying sinuosity of turbulent flow. In the stream channel it is the value derived from dividing the product of the average diameter of the flow by the *kinematic viscosity* of the fluid (the quotient obtained when dynamic viscosity is divided by density, and corrected for temperature). Reynolds numbers less than 2,000 describe the rare laminar flow; numbers greater than 2,800 are much more prevalent and indicate turbulence or low viscosity.

Although laminar flow is an infinitesimal part of a stream's course, small plants and animals can experience it. A riverbed stone has a thin laminar film at its upper surface, although its upstream and especially downstream more-or-less vertical surfaces may be washed by turbulent jets. At the channel bed velocity is zero; somewhere above it there is a sharp break that marks the line between laminar and turbulent motion. Below this critical depth is the world where small organisms escape much of the turbulence. Most mechanical velocity or current meters are too bulky to detect microcurrents over and at the surfaces of submerged objects, but McConnell and Sigler (1959) overcame the problem. They established the rates at which standard sodium chloride tablets lose weight while immersed one minute in various flows.

The **Austausch coefficient** is a measure of turbulence, the mixing in excess of what would be caused by molecular diffusion alone. It is an attempt to quantify all the aimless movements that are counter to the main direction of flow. We discuss this later in a section concerned with turbulent mixing in stratified lakes.

Estimating Flow

The rate at which water moves downstream, **discharge** (Q), refers to the total volume of water flowing in a given segment (a particular "reach"). Discharge is calculated by multiplying the cross-sectional area of the stream A_{xs} by the speed of the water's current in that stretch. So, a hypothetical ditch-like stretch of stream with a rectangular channel 2 m wide and 0.25 m deep has an A_{xs} of 0.5 m^2, and if current is 2 m/s, the flow is 1 m^3/s.

The complexities of stream flow are far from obvious to the untrained observer. In addition to gradient steepening, an increase in volume or depth will result in greater velocity. Surprisingly, as shown by Hynes (1970), a river 200 m wide and 4 m deep with a slope of only 0.5 m/km would have a velocity almost 1.2 times that of a small stream 20 m wide and 0.5 m deep, flowing down a gradient 10 times steeper. Flow proceeds at a faster rate on smooth channels than it does on rough-bottom channels. A sudden narrowing of a stream channel raises the velocity abruptly as the discharge holds constant. In higher **stream orders** the channel widens, discharge is augmented, and velocity remains about the same or increases. Thus, the current of a river far from its source does not move at a reduced rate as is often assumed. Also, we may foster the concept that water flows more rapidly than it actually does. Stream flows rarely exceed 6 m/sec, and velocities are usually less than 3 m/sec (Hynes, 1970).

The velocity of a stream should be determined for most physical, chemical, and biologic studies. For establishing the mean current velocity, it is not adequate to time the progress of a floating object. The current at about six-tenths of the stream depth is considered a close approximation of the average velocity. A better result is gained from averaging the velocities at two-tenths and eight-tenths of the stream's depth. This is an approach to accounting for boundary friction, turbulent eddies, and the slowing effect of surface tension at the very top of the water column. An orange or lemon sinks to such a level that its progress downstream reflects the mean velocity in shallow brooks.

Using many thousands of field measurements over more than a century, hydrologists empirically quantified relationships between environmental factors and predicted flow rates in a stream. The resulting tables of numerical constants are consulted to provide values for the equations of the Rational Method (see recent improvements by Grimaldi and Petroselli, 2014), and peak discharge using the SCS-Curve Number method (modified by Mishra and Singh, 1999).

Such empirical models predict the input of water to a stream and therefore the likely discharge. The Rational Method uses the equation $Q = CiA$, where discharge is determined as the product of the runoff coefficient C (found on a published table, or entered into a computer program) times the rainfall intensity (i) and the area (A) of the river's catchment basin. Obviously, the coefficient is indexed to the units used for intensity and area.

Similarly, peak flow can be determined in more situations by the SCS-CN approach; however, the equation is more complex, incorporating soil factors (including area of soil covered by impervious surface). Q is obtained by consulting discharge curves on published charts; they are incorporated into the convenient computerized discharge calculators. Peak flow is increased when land in the watershed is cleared and urbanized; hydrologists have developed models for this phenomenon too, as discussed in the later section on human impacts to streams.

Transportation of Materials

Materials are transported by running water in three principal states: as *dissolved* matter, as *suspended* solids, and as the *bed load*.

Tremendous amounts are carried in solution. From Livingstone's (1963b) estimate of the mean chemical composition of the world's rivers, one can say that about 56 metric tons of the common inorganic ions are dumped into the seas each second by just the largest rivers.

Smaller alluvial items are carried not in solution, but in suspension; they are often visible to us as what we call **turbidity**. Most of the other substances carried by the running water can be classified as either organic detritus or **alluvium**, which is rocky debris. Clay, silt, sand, and small pebbles are moved by the current according to the stream's **competence**. This word describes the grain size that a current can barely move. The Hjülstrom curve (Angelier, 2003) mathematically describes competence; it was developed in Europe but has applicability in streams worldwide. Hjülstrom studied the transport of different-sized sediments (from tiny clay particles through sand up to boulders) and noted the ability of water moving at particular current velocities to move each sized particle (Fig. 8-2 on the next page). So, for example, a sand with a grain size of 0.3 mm will be picked up and transported if velocity exceeds about 20 cm/s but will drop out of the water and deposit if flow falls below 3 cm/s. Between those speeds, water transports the sand.

The role played by turbulence is important here. It projects alluvial fragments from the bed up into the current that sweeps them downslope. In addition to velocity and turbulence, the size, shape, and specific gravity of a suspended bit determine how

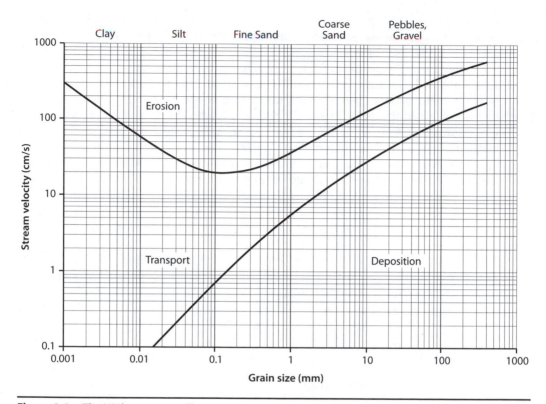

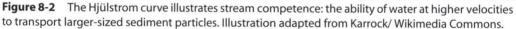

Figure 8-2   The Hjülstrom curve illustrates stream competence: the ability of water at higher velocities to transport larger-sized sediment particles. Illustration adapted from Karrock/ Wikimedia Commons.

long and how far it will be carried. The relationship between velocity and alluvial transport is exponential; the size of a rocky fragment that can be moved varies as the sixth power of the velocity of the water. Increasing the velocity two times makes it possible to roll a stone 64 times as large as one that could be moved at the beginning. There is an exception to this. The finest of clay and silt particles (diameters, 3 to 30 μm) are no more easily dislodged than are normal sand grains 100 times larger.

The coarser-sized alluvial fragments may rarely move, and then only by rolling and sliding, scarcely off the channel's bed except for brief jumps. They make up **bed load**, the fraction of the load consisting of alluvium too large to be carried via suspension.

The Longitudinal Profile

Many streams gradually flatten from headwaters to mouth. The mouth may widen into a delta where the deposition of alluvium fans out, having been carried by erosion from the upper reaches. Local differences in gradient, as both harder and less resistant rocks are crossed and as pools alternate with riffles, interrupt the smoothness of the profile and prohibit us from making generalizations about the "typical river."

The principle of grade should be brought up here. (This word should not be confused with gradient.) A stream depositing its excess load, that load greater than its

capacity, is said to be *aggrading* its channel. Similarly, if a particle's size exceeds the stream's *competence*, it is deposited. An underloaded stream (in terms of its theoretical capacity), however, might tend to erode its channel, picking up alluvium from the bed and thus *degrading* its channel. A **graded stream** section, or a stream at grade, is one in dynamic equilibrium between erosional and depositional forces. No river is graded throughout its profile, although all rivers are trending toward the condition where they move over a bed composed of alluvium brought and deposited by their own flows.

Also accompanying the longitudinal profile are changes in the channel cross section. Even if reduced to the same scale, the cross sections from places upstream and downstream in the vertical longitudinal profile could be distinguished.

The size of alluvial grains transported decreases rapidly during the first few kilometers; in the headwater reaches the decrease may be exponential. In general, the deposition of material downstream reduces the roughness of the channel bed and its sides. This smooth scheme is blemished by local irregularities, for within a segment of stream there can be found alternating riffles and pools; the latter are regions of sedimentation far more than are the former. In a 1-km section of a stream 20 m wide it would not be unusual to find six or seven pool-like spots separated by riffles. The pools have fine-grained bottom deposits, whereas the faster shallower riffles are paved by stones and gravel from which silt is washed away. These adjacent geomorphic features also are ecologically different. Theoretically, a lentic fauna is found in the pools, and quite different invertebrates and fishes are found in the riffles. In reality, there is much overlap of species that can be collected in both parts of the stream. In North American streams the ictalurid madtoms of the genus *Noturus* are more of a nocturnal riffle form, spending the daylight hours in quieter regions. Many of the percid fishes known as darters, and especially those belonging to *Etheostoma*, are riffle species. A few species of the holarctic sculpin family (Cottidae) also are typical of shallow, fast current. In the riffle-inhabiting species of the last two examples there has been an adaptive loss of the gas bladder. Furthermore, the well-aerated riffles serve as nurseries for the eggs and embryos of several fishes, whereas the pools are shunned. The pools, however, are important refugia for a variety of aquatic species at time of very low water.

Meanwhile, the bed and bank substrata, the flow velocity, and the qualitative and quantitative attributes of the load determine the shape of the channel. In straight reaches the bottom is fairly flat and level; where bends occur, there is less symmetry and the bottom tips downward toward the outside of the curve.

The Patterns of Channels

The river follows various courses as it progresses seaward. If the distance traversed is less than 1.5 times the linear span between an upstream and a downstream point, the reach is considered virtually straight despite any gentle sinuosities. Even when the river, confined by its bank, is fairly straight, the line of maximum depth may be remarkably winding. This deep channel goes by the old German name **thalweg** (valley path). Here at the very bottom of the river bed, currents are reduced and it is believed that salmonid fishes moving upstream find the thalweg to be the easiest route. Mossop and Bradford (2006) found evidence that juvenile salmonid fish, at least, utilize the thalweg for migration.

It is more common for streams to follow paths greater than 1.5 times the shortest distance. Then the sinuosities are called **meanders**. The meander loops or bends are complex, creating more habitats for the river biota than one might expect. In a single bend there is a convex bank bounding a blunt point and opposite it a concave bank, presenting a concave surface riverward and marking the outer side of the curve. The current velocity is greatest at the concave bank, and the water surface may be slightly elevated as the turning stream piles against it. There the bank may be undercut and obviously eroding; in some instances bedrock is exposed here. Strong crosscurrents near the bed convey alluvium toward the convex bank, and typically the water is shallower there. Meanders demonstrate that stream erosion is not all vertical, although that type of corrasion would dominate where gradients are very steep. Lateral corrasion occurs as rivers swing back and forth within their channel walls, and it surpasses the concurrent erosion at the bed.

Meanders commonly are cut within the river floodplain. This is the broad flat area bordering the channel and composed largely of alluvium deposited by the river itself during those times it has flooded its banks and wandered over the adjacent lowland. There are, however, meandering streams cut deep in rocky substrata.

When water flows through channels composed of extremely fine-grained sediments, there is a lessening of sinuosities; the stream's competence may be insufficient to move such material. The clay and silt particles that make up the Mississippi's banks below New Orleans could account for the unusually straight channel the river follows there.

Another pattern described by rivers seems to be a function of the alluvial load. This is expressed as the **braided stream**. There are many small channels rather than just one large one; they divide and rejoin, separating and anastamosing to form a complex multiple channel. Parallel elongate bars and islands separate the channels. Braided streams carry great loads of suspended alluvium. The Platte River, Nebraska, has extensive braided reaches and carries enormous loads. Some abandoned channels of the Mississippi system, now almost obscured by forests, exist in the southern United States. They are braided channels, quite in contrast to the nearby modern meanders. The old braided network carried tremendous amounts of debris as the Pleistocene glaciation came to a close.

At the mouth of most large rivers there is a delta, built up of alluvium. Exceptions are seen in two large North American rivers that have no deltas for good reasons. The Columbia River dumps directly into the Pacific and its alluvium is dispersed by waves and currents; the Saint Lawrence is relatively short and, therefore, does not carry enough alluvium to build a good delta.

In those rivers that have deposited a substantial delta we see a further channel pattern, one that resembles tributary patterns in reverse. The main river breaks into branching lesser streams called **distributaries**. These fan out to distribute the river's discharge to the sea at many points. The distributaries of the Mississippi River delta are bordered by natural levees and extend out into the Gulf of Mexico. These narrow extensions of land reach beyond the delta's border, forming a *bird's foot delta*.

The River's Mouth and the Estuary

Here at the end of the river's journey is a region of contrasting density currents. The river water flows out over the dense sea water, and countercurrents of saline water slide upward toward the river's mouth or farther, depending on the channel morphology and bottom roughness. The two water types are mixed by various irregularities to form a zone of brackish water in the true sense of the adjective. Here in fluctuating salinity gradients are found the euryhaline plants and animals typical of estuaries. Some of the highest biotic productivities are found in well-illuminated estuaries, where nutrients from the land are mixed with seawater minerals. The annual commercial harvest of some estuarine species (the blue crab, *Callinectes*, for example) is enormous. With tides rising and falling, a daily rejuvenating process occurs in estuaries. The tidal currents and oscillations transport plant nutrients and animal food; they circulate waste products, and in many instances they expose mud flats where intense photosynthesis by epipelic algae occurs. The estuary thus differs from the river and the sea that contribute to its character. A diversity of niches is provided for the biota, and an enrichment of its faunal and floral elements is owed to both marine and river sources. Moreover, the so-called freshwater fauna is enriched by brackish water representatives from several phyla that move into coastal rivers and marshes (Pennak, 1989).

Upsetting the estuarine dynamics by impeding river discharge can have far-reaching effects. A large sea bass, *Totoaba macdonaldi*, endemic to the Gulf of California, spawns in the dilute waters of the vernal runoff at the mouth of the Colorado River. Now in many years the multi-impounded Colorado does not reach the Gulf, and there is no reproduction for the specialized and endangered *totuava*, as it is known to Mexican fishers, during those years. The multiple damming of the Colorado River has changed the characteristics of the upper Gulf; it is no longer the estuarine world in which the *totuava* evolved.

The River Continuum

The concept of the river continuum (Vannote et al., 1980) relates changes in lotic communities to the downstream gradient of abiotic factors from the headwaters to the mouth (Fig. 8-3 on the following page). Successive, intergrading geomorphic, physical, and chemical factors in a lotic system are accompanied by communities adjusting and adapting to each particular state of hydraulic and morphologic elements. Streams can be grouped roughly into headwaters (stream-order positions 1 to 3), medium-sized streams (4 to 6), and large rivers (greater than 6). Broad generalizations have been made about the communities that form in these stream-system zones. As a general theory of trophic dynamics and biogeochemistry, the theory has proven useful for decades, in a variety of rivers worldwide (Minshall et al., 1983; Grubaugh et al., 1997; Tomanova et al., 2007).

Streamside vegetation has marked effects on headwater brooks by shading the water and contributing leafy and woody detritus. As a result, allochthony surpasses autochthonous production of organic matter, and the stream is essentially heterotrophic. The ratio of gross photosynthesis to community respiration is low, P/R <1.0. But this is too simplistic, for there are temporal and spatial contrasts within headwater systems, and although the majority of stream studies have been carried out in mesic regions where headwaters are almost hidden beneath the canopy of riparian trees, the relationships differ in other climates.

In Hope Creek, North Carolina, the arrival of leaves from and the shading by the riparian forest are seasonal (Hall, 1972). The mean annual P/R quotient is 0.6, but

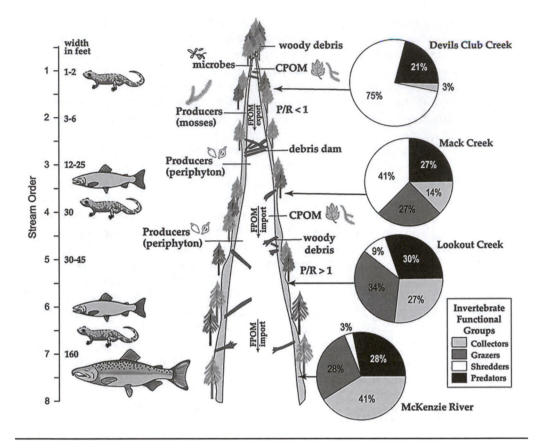

Figure 8-3 The river continuum illustrated: changes in food webs and corresponding organic materials along a river course (increasing order). Adapted from an image created by Gregory Stanley and Sherri Johnson after Vannote et al., 1980.

the stream is not always heterotrophic. In early spring, algal photosynthesis peaks sharply beneath the bare trees and the community is autotrophic, but by May new leaves are beginning to shade the stream and it becomes heterotrophic. In October heterotrophy reaches its greatest height when falling leaves serve as a new source of potential energy. Later the winter sun strikes the water through denuded tree branches and some lotic photosynthesis is again possible.

In arid regions the impact of riparian vegetation may be far less significant. Busch and Fisher (1981) investigated the community metabolism of Sycamore Creek in the Sonoran Desert of Arizona, finding the summertime P/R equal to 1.7. Daily gross primary production reached 8.5 g O_2/m^2, equivalent to the fixing of about 2.7 g C/m^2. This is high productivity for a nonthermal desert stream; Tecopa Bore, a thermal stream in the Mojave Desert of California, averages 3.0 g C/m^2 per day (Naiman, 1976). The lack of shade, the intense sunlight, and a high biomass of algal periphyton brought about the remarkable photosynthetic rates in Sycamore Creek; the low inputs of organic litter made contrasts between P and R even greater. There is no allochtonous organic detritus in Tecopa Bore, its trophic structure being based on a benthic cyanobacterium, *Oscillatoria*. In these two desert streams the downslope flow of organic matter surpassed what had been derived from upstream or from direct detritus input; they are both autotrophic systems. In Sycamore Creek there was striking spatial as well as temporal heterogeneity in the P/R quotients; the riffles were predominantly autotrophic; the pools were usually heterotrophic.

Turning now from the special case of desert streams, where headwaters are unshaded in contrast to the upper reaches of their counterparts in forested regions, we find the situation changes downstream in the latter. Stream widening accompanies an increase in stream-order rank, and much of the water surface becomes exposed to the sun even in woodland settings. Here the relative importance of organic matter from the immediate terrestrial watershed decreases and the detritus from upstream is more important. Meanwhile, where the canopy is open and the water is still clear and shallow, hydrophytes appear along the river margins, other submerged angiosperms take root in the coarse rubble of the stream bed, and there is a shift from heterotrophy toward **autotrophy**. The water plants often find the conditions downstream less favorable, for deeper water, fine-grained and less stable bottom material, and increasing turbidity preclude their optimal growth. Ideally, the situation in the middle reaches, where the hydrophytes thrive, would be reflected by a rise in the P/R quotient. Sometimes, however, local conditions prevent more solar radiation from striking the stream surface as the river broadens, and gross primary production is hindered by shade. The deeply incised canyons of Chevelon Creek, Arizona, cast shadows on the stream, reducing periphyton photosynthesis abruptly in some downstream reaches of the creek (Blinn et al., 1981).

By the time water accumulates and flows down to river channels ranked above a stream-order number of 6, it has "aged." This movement through time and space is the essence of lotic systems in the eyes of Rzóska (1978). Lentic conditions are approached by the large river system, and at last the plankton community can develop in its waters. Now the effects of phytoplankton primary production come to light. The P/R rises, although the littoral hydrophytes' effect may be less than in the smaller reaches upstream. The fish fauna now includes some planktivorous forms for the first time in the continuum. In the headwaters the sparse fish fauna consisted largely of insectivores, and in the medium-sized stretches some piscivorous species made their appearance.

Dissolved organic material (DOM) arriving in the big river reaches from upstream is more homogenous than the DOM in the headwaters. It is composed of a greater percentage of hard-to-digest material, more resistant to bacterial and fungal enzymes. Meanwhile, in the particulate category there is a lessening in the relative amount of coarse detritus—the quotient of the coarse particulate organic matter to the fine particulate organic matter (CPOM/FPOM) has decreased on the journey through time and space to the big river. The idealized continuum is complicated if flooding tributaries introduce "young" water with a diversified array of detritus into the large stream.

Another point to consider is the turbidity of large rivers. Sometimes it is greater than upstream, and it reduces light penetration so much that the P/R quotient falls below 1.0. The organic matter arriving from upstream and nearby tributaries may also contribute to the respiratory burden of the river, acting toward the lowering of P/R values. Thus, there may be a return to heterotrophy in turbid, large rivers. A remarkable example of a tributary's effect on the main stream is seen when the Little Colorado River is in flood. This very turbid tributary enters the Colorado River in Arizona, raising the turbidity to such an extent that vertical light transmittance is reduced to 0.001%, and the opacity persists for 350 km to the Colorado's mouth in the upper portion of the impounded Lake Mead.

It is perhaps easy to fall into the trap of considering lotic P/R values as indices to the relative ratio of autochthony to allochthony. Perhaps this is because allochthonous detritus is sometimes conspicuous. Minshall (1978) presents evidence and argues for the idea that the autotrophic contribution to stream metabolism has been grossly underrated and that much more of the stream ecosystem is founded on autotrophy than is evident at first. Periods when photosynthetic production is very high may be followed by heterotrophic times, although the stream metabolism remains based essentially on the preceding autotrophy. Detritus from autochthonous plant tissues can be used by heterotrophic organisms long after the material was produced. This is fairly obvious in the fragments of macrophytes that accumulate here and there in streambed depressions. It is less apparent when the substance being respired is autochthonous DOM, released both at the time the green autotrophs are active and after their deaths. When the P/R is below 1.0, one cannot assume the material being respired is allochthonous. Also, Minshall (1978) pointed out that equating biomass with production had proved a pitfall to stream biologists (including himself), and that this had led to underemphasizing the role of autotrophy in stream ecosystem dynamics. Because the standing crop of periphyton algae may be insignificant when compared with allochthonous organic matter, it might follow that it would be incapable of supporting the stream fauna. The rapid turnover rate of the tiny algae, however, can support far greater biomasses of animals than seems obvious at first.

The Flood Pulse and Lateral Aquatic Habitats

Another general principle of riverine landscape ecology, perhaps an extension of the river continuum, is the flood pulse concept (Junk et al., 1989). This theory states that beyond the channel, the river connects with the floodplain, forming a unique interface between river and land, having some characteristics of a littoral lentic habitat. This dynamic zone has high productivity, rapid chemical cycling, and variable hydrology. The pulse of seeds, floating animals, deposited sediments and associated nutrients involves a "batch process" rather than the flow-through stream channel in the river continuum. Developed originally in large tropical river systems, the flood pulse is also known from other regions (Tockner et al., 2000) and for floodplain wetlands (Middleton, 2002). The concept is useful for management of river flow and habitats (Bayley, 1995).

Recent work attempts to expand both the river continuum and flood pulse concepts into a unified picture of how rivers and their associated floodplains change in time and space (Amoros and Bornette, 2002). It will be necessary to consider the river in a landscape context: a stream is part of a floodplain and a larger river basin. Current research aims to understand the dynamics of the river channel and associated floodplain water bodies, the lateral habitats (Reese and Batzer, 2007). Species richness and population sizes increase with connectivity between lateral systems and the stream (Gallardo et al., 2008); in a study modeling lateral habitats, Gallardo and others (2009) found distance to the river channel and flood duration were the most important determinants of macroinvertebrate composition, whereas flood magnitude and water level variability best accounted for the variance in zooplankton and phytoplankton compositions. The specifics of this topic must still be investigated, but it is clear that lateral habitats are important for biodiversity in the river itself and for the surrounding basin (Karaus et al., 2013).

Adaptations to the Lotic Environment

Plants and animals that live in the unidirectional flow of the lotic aquatic environment show some remarkable adaptations for such existence. These adaptations are usually behavioral or morphologic. Some species clearly show combinations of the two; others serve as excellent examples of either one of the adaptation types.

Behavioral Adaptations

In the behavioral category, the commonest adaptation is simply avoidance of swift currents. A common way to do this is to seek shelter in masses of debris or beneath stones or simply to burrow in the stream substratum or tunnel into an attached aquatic plant. Many burrowers inhabit stream pools, where the sediments are soft and rarely scoured by flood waters. There are other locations in a stream where flow is reduced and where animals can escape severe water velocities. For this reason we find contagious (clumped) rather than random or uniform distribution of organisms, when we sample an arbitrary cross section of a stream. Therefore, the number of organisms beneath unit area (for example, m²) of stream surface must be used with caution and explained. For example, in an Ontario stream less than 3% of the *Gammarus* population was found farther than 3 m from the banks. The surface flow varied from 0 to 25 cm/sec within 1 m of the shore, where 75% of the amphipods were found, compared with 50 to 80 cm/sec in the middle of the stream, about 6.75 m from the bank (Marchant and Hynes, 1981). At times of flooding it appeared that the amphipods sought shelter even closer to the bank, where the current was slowest. Similarly, salmon on spawning runs upstream are usually found swimming along the bottom in the thalweg of the stream. Here, protected by the deepest water, they contend with a flow far less than that at the surface. Other behavioral patterns thought to have adaptive value are upstream flight by ovigerous female insects of species whose immature stages are aquatic, upstream movement by some small insect larvae and other invertebrates, and drifting down slope with the current. These are discussed in the section on organic drift.

The behavioral adaptations discussed above show only one side of the coin. Many animals seek current and are stream dwellers because of the benefits derived from current. The rheophiles derive a constant oxygen supply, have their waste products swept away, and receive food as a function of water movement. Moreover, the contagious distribution of organisms that typifies the cross section of a stream is owed to more than current avoidance—leaf packets, periphyton patchiness, and substrate variation are obvious factors.

Morphologic Adaptations

A striking feature of the biota of fast-flowing water is morphologic adaptation. An early paper by Nielsen (1950) dealt with this phenomenon in the invertebrates of torrential streams, but much of the following comes from Hynes (1970) and Resh and Solem (1978). Hynes listed 11 categories under this heading and the last two authors tallied 12 that could be considered thus, although there is difficulty in distinguishing some from behavior and/or physiology. Furthermore, many of the so-called adaptations for lotic conditions are found also in species from standing waters; there has been more interest in looking for beneficial structural modifications in stream animals.

Reduction of body size permits animals to escape the current by creeping or hugging close to the substratum, where water velocity is greatly reduced. For example, in brooks with stony rubble, a boundary layer at the exposed surface of each stone is a region of slow current. Small size alone seems to permit tiny species—protozoans, nematodes, and rotifers—to deal with current by occupying the relatively quiet boundary zone or by escaping it in rocky cracks and fissures; they show no special morphologic adaptations that set them apart from their lentic relatives. Hydrophilid and dystiscid beetles have small representatives in lotic habitats when compared with the large lentic types, and most fast-water coleopterans are small (Resh and Solem, 1978). Furthermore, the water mites of streams, where many spend their immature stages on insects, are said to be smaller and less hairy than the swimming Hydracarina of ponds and lakes. Size reduction does not apply to all fishes living in fast water. Adaptive changes in body shape are often accompanied by a relative increase in size when compared with their slow-water relatives.

Some mayflies, planarians, and leeches are remarkably adapted to withstand current through flattening. Perhaps, the best example, however, is the immature member of the beetle family Psephenidae. This clinging, flattened larva has given its family the common name, water penny beetle (Fig. 8-7, *F*). Its flattened body makes possible not only inhabiting the boundary layer of stones and other debris but enhances its ability to crawl beneath the stones or into crevices that would be denied to more corpulent individuals. Actually, the probability is greater that a water penny larva will be found in a crevice or beneath a stone than on the surface exposed to the current.

There are some flattened forms that have little to do with current and make for exceptions. The planarian flatworms, for example, live in quiet spring waters as well as running waters; their thin bodies make gas exchange possible. Among the microscopic plants, however, flattened algal thalli, closely fixed to the stony substrate, are typical of the periphyton of swift water. The aquatic isopod crustaceans are characterized by dorsoventral flattening, even more pronounced than in their terrestrial relatives. Members of the genera *Asellus*, *Caecidotea*, and *Lirceus* occur in lakes, temporary ponds, streams, and subterranean waters. Their flattening makes it possible for them to crawl beneath rocks or into leaf packs; it does not seem to be a unique adaptation for life in running water, although it does help them escape currents. Another group, the stenasellid isopods, are narrow as well as flattened. They are restricted, however, to quiet subterranean waters or springs where they often crawl through interstitial spaces.

Streamlining has been adopted by a few invertebrates and many fishes. The salmon and trout that swim against or maintain themselves in strong currents are often cited as examples of this adaptation; their bodies are fusiform and round in cross section, offering reduced resistance to flow. The larvae of black flies (Simuliidae), most of which inhabit swiftly flowing water, are streamlined (Fig. 8-9, *B*). Species of *Baetis* are among the most streamlined of the Ephemeroptera, although some species live in quiet water; they are active swimmers, which may account in part for the adaptation. Perhaps the slimmest and most tapered of all the aquatic immature insects belong to the dipteran family Ceratopogonidae (Heleidae), the biting midges (Fig. 2-11, *E*). Many of their larvae, however, are burrowers in the sediments of lentic environments; some are found in unusual habitats—tree holes and at the margins of saline lakes and hot springs.

In torrential streams some fishes show morphologic configurations that permit them to hug the bottom in the face of current. One of the most striking examples in North America is the rare humpback chub, *Gila cypha*, from the Colorado River of Arizona (Miller, 1946). Its body is grotesquely humped just posterior to the head, an arching that is thought to act as a hydrofoil holding it down as it faces upstream (Fig. 8-4, *D*). In addition, the body of *G. cypha* is streamlined and round in cross section; its fins are large and falcate; its caudal peduncle is long and narrow, and it has lost nearly all its scales, a friction-decreasing modification. A close (and also rare) relative, the bonytail chub, *G. elegans* (Fig. 8-4, *C*), is another swift-water form, showing further adaptation than *G. cypha* in some respects. Its caudal peduncle is more slender and relatively longer, and its dorsal fin contains more rays (Minckley, 1973). An intermediate stage from slow-water forms to the two *Cypha* species is depicted by another endangered species from a different fish family, the razorback sucker, *Xyrauchen texanus*, of the Catostomidae (Fig. 8-4, *B*). It has disappeared from much of its former range, but persists in the lower reaches of the Colorado River. Its caudal peduncle is strong and thick, its dorsal hump is more keel-like, and scales are often absent from the anterior hump margin.

Some fishes living in torrential streams of Asia have gone further in adapting to current stress than any North American species. The hill-stream loaches (family

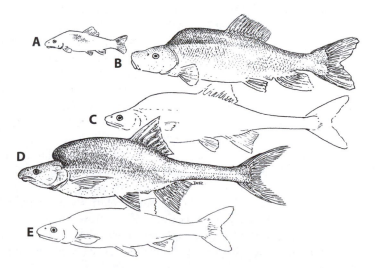

Figure 8-4 Fishes of fast-flowing waters of the Colorado River drainage. **A**, Mountain sucker, *Pantosteus or Catostomus clarki* (Catostomidae); **B**, razorback sucker, *Xyrauchen texanus* (Catostomidae); **C**, bonytail chub, *Gila elegans* (Cyprinidae); **D**, humpback chub, *Gila cypha*; **E**, Colorado River roundtail chub, *Gila robusta*.

Homalopteridae) have flat ventral surfaces that, combined with expansion of pectoral and pelvic fins along the lateral margins, form a large, complex sucker. With this the fish maintains itself in rapid flow while it feeds on epilithic algae. Members of another Asiatic family use the mouth as a hold-fast organ, necessitating rerouting the course taken by respiratory water, which usually enters orally and leaves laterally through the gill opening beneath the operculum. In these fishes, the gyrinocheilids, the lateral aperture has been partitioned into an upper incurrent opening, substituting for the mouth, and a lower excurrent opening, the water passing over the gill capillaries while flowing between the two orifices.

A rare type of structural modification for invertebrate life in swift current is shown by the flattened larvae of the net-winged midges (Blephariceridae). Most species are found in the upper reaches of mountain streams, where they cling to rocky or woody surfaces from which they scrape algae. The remarkable thing about them is the fusion of head, thoracic segments, and first abdominal segment to form a body region much like each abdominal segment posterior to it; thus there are six major body regions, each of which bears a median ventral sucker. These discoid structures function, on smooth surfaces at least, to hold the larva as it feeds in the current.

Reminiscent of the blepharicerids are the mountain midges (Deuterophlebiidae) whose larvae have seven pairs of lateral prolegs, the blunt ends of which serve as crude sucking discs (Fig. 8-7, *H*). The prolegs are ringed distally with concentric rows of minute hooks.

A morphologic feature somewhat different from the blepharicerid suctorial disc is seen in the naiads of some ephemerellid and heptageniid mayflies, families noted for clinging, flattened naiads. A ventral roughened patch or pad that serves as a friction organ effectively holds the insect in place as it makes contact with surfaces. This is typified by the ephemerellid species *Drunella doddsii*. In some heptageniids, *Rhithrogena* species for example, the gills are modified in such a fashion that they form a ventral adhesive pad. This is a widespread adaptation, for Bayly and Williams (1973) describe a genus of rheophilous mayflies from New Zealand that use their seven pairs of gills to form the ventral friction pad. On the other hand, similar structures that serve as friction pads are known in insects that do not inhabit torrential streams.

Various types of hooks that serve as holdfasts are found among the insects of lotic waters. The riffle beetle larvae (Elmidae) bear a pair of hooks concealed in a terminal segment, which along with filamentous gills are revealed when a posteroventral operculum opens. The tiny adult elmids have well developed tarsal claws, but this can be said for many beetles that do not live in flowing water. Most larval trichopterans are equipped with claws on their anal prolegs—claws that help them maintain position within their cases. The members of the caddisfly family Rhyacophilidae, however, have free-living, predaceous larvae that build no cases and are marked by mobile anal prolegs bearing hook-shaped claws. Streamlined larval black flies spin silken meshes that stick to rocky substrates (perhaps a physiologic rather than morphologic adaptation). They then hold their position by grappling into the silk with a circlet of posterior hooklets. Thus, anchored in the relatively quiet boundary layer, they reach up to catch suspended particles drifting in the flow above.

All stages of the spectacular dobsonflies (Megaloptera, Corydalidae) are terrestrial except for the large predaceous larvae (length 6.5 cm or more). All genera of the family are characterized by paired terminal claws on each anal proleg, but *Corydalus* is especially well equipped. Although *Corydalus* can swim in search of prey, its posterior hooks serve to hold it as it climbs about and clings to coarse detritus and rubble on the stream bottom.

Most of the angiosperms that have become aquatic grow rooted in soft bottom deposits. Their existence is precluded from fast currents, where the substratum is being eroded and fine particles cannot settle. Only two groups of primarily stream-dwelling angiosperms have evolved: Podostemaceae and Hydrostychaceae. The river-weed family (Podostemaceae) is essentially tropical, but at least one species of *Podostemum* is found in streams of eastern North America. Superficially it resembles a macroscopic alga, thriving in swift water, attached to streambed stones by fleshy discoid holdfasts rather than by roots.

Ballasting with stony cases in some caddisflies or forming massive shells in the case of bivalved molluscs serves to maintain position. Some caddisflies build cases of tiny, rocky fragments with two or more ballast stones on each side according to species. *Helicopsyche* (Fig. 8-7, *E*) secretes sticky material with which it glues its stony, snail-shaped case to the substrate; it would not be accurate to say that the density of the tiny case would hold this caddisfly in place by itself. The protection against predators is an important function of such ballasting structures, but the attention of most workers has been focused on adaptation to current stress.

The loss of projecting structures is often used as an example of adaptation to running water, but this is certainly not true of the corydalids and some mayfly species from lotic habitats. Burrowing mayflies, the species *Hexagenia,* for example, are unusually hairy, perhaps as a response to conditions in the sediments rather than to current.

Birds and insects from oceanic islands are noted for their flightless species. Perhaps reduction in flying ability serves to keep them from being blown out to sea and lost; it would have a significant selective value. It has been theorized that the not-uncommon decrease in flight powers seen in stream insects is owed to a similar phenomenon. This may be applicable to some cases in lentic environments that may be likened to islands; an endemic species of flightless grebe in Lake Atitlán, or the neo-tenic, wingless stonefly from Lake Tahoe (Jewett, 1963). Perhaps the reason for flight reduction in lotic habitats is that they are not so discrete as lakes and ponds; they are elongate, continuous waters. The need for overland dispersal ability might be far less than in the lake species. On the other hand, the immature stages of many flying forms live in certain sections of a stream, and there would be an advantage in the ovigerous adult being able to fly to that region in case it were displaced. This brings us back to the possibility that dispersal away from a particular section of stream would be disadvantageous; and flightlessness would be selected for, as it is on ocean isles. The loss of hindwings in the adult elmid beetles, the reduction of wing size or complete loss of them in some stoneflies, and the flightless females in a few trichopterans are associated with the upper, erosional reaches of streams. Bayly and Williams (1973) state that the only apterous stonefly in Australia spends its immature stages in small streams near the summit of a Victorian mountain. The danger of being blown away would always be a threat to flying adults. Flightlessness in small ponds at high altitudes also has adaptive value, but we tend to become especially interested in lotic systems and overlook similarities in the morphology of lentic and lotic representatives.

Organic Drift

Many years ago it was observed that a net, when properly placed, would capture material drifting along with the current. The most remarkable part of the drift phenomenon was that living things were stopped by the net in addition to various types of debris and that daily and seasonal patterns seemed to prevail. Drift, as it is now called, is a normal feature of running-water environments, and a fairly diverse group of animals participates. The aquatic young of the insect orders Ephemeroptera (mayflies), Trichoptera (caddisflies), Diptera (true flies), Plecoptera (stoneflies), and Coleoptera (beetles) are commonly represented in drift collections. Crustaceans referable to a few genera of the Amphipoda also are conspicuous drifters and have been studied in Asia, Europe, and North America. Other less closely related forms include snails, oligochaete worms, various microcrustaceans, and even fish fry. Usually, however, drift organisms are members of the stream benthos.

Drift proceeds throughout the day but peaks in numbers and biomass occur during the dark hours (Fig. 8-5). Furthermore, drift is more pronounced during the warm summer months and this is related, in some aquatic insects at least, to the later life-history stages. High temperatures increase drift amplitude in a few aquatic insects and abrupt temperature changes have been known to induce catastrophic losses (see Waters's review, 1972; Ward and Stanford, 1982a). Some forms are caught in nets only when water velocity is great, suggesting that drifting is not part of their normal life patterns; in other species there is no such correlation, although total downstream movement of organisms is usually greater at times of high discharge. In general, however, drift is not a simple random event.

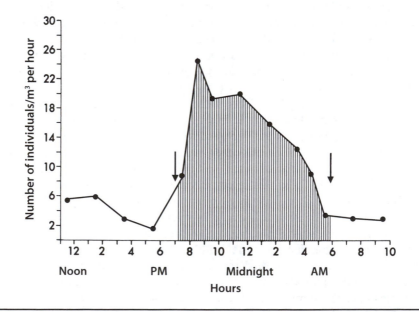

Figure 8-5 Invertebrate drift in an Australian stream. Sunset and sunrise are indicated by arrows; dark hours are shaded. Data from Morrissy (1967), Table 4.3, in Bayly and Williams (1973), Table 8.6, p. 165.

Regardless of the cause, even a slight loss of animals downstream to drift would, over time, tend to depopulate the upstream populations; this "drift paradox" vexed researchers for decades. It was predicted that during the life cycle, some upstream migration, possibly of adults, would compensate for drift. Such behavior was termed *colonization cycle* many years ago, when Müller (1954) described female insects flying upstream to oviposit in certain Swedish streams. Finally, Hershey and others (1993) used stable isotope tracers in the Kuparuk River of Alaska to trace the movement of *Baetis* mayflies, finding that 1/3 to 1/2 of adults flew 1.6–1.9 km upstream after hatching, enough to compensate for drift in the aquatic nymphs. Anholt (1995) then explained the evolutionary advantage of upstream migration, combined with density-dependent dispersion. Overall, upstream dispersion of adults appears to be an effective strategy from an evolutionary standpoint for adult aquatic insects (Kopp et al., 2001).

Rader (1997) attempted to organize drift organisms into a classification scheme based on functional ecology. He grouped 95 invertebrate taxa into *Baetis* plus three mixed taxa guilds, and he classified organisms based on behavior alone (intentional drift) and three behavioral/morphological traits. He related these classifications to ecology of the stream, in particular exposure to salmonid predation. This is a good start to organizing our understanding of invertebrate drift ecology, but we require further research to place this in the context of other river ecology theories (Power and Dietrich, 2002).

Functional Classification of Lotic Animals (Trophic Roles)

Aquatic animals have been placed in functional, rather than phylogenetic, categories according to their roles in food webs. Although most categories apply also to lentic organisms, the classification stemmed from stream studies where vertebrates and invertebrates contribute in various ways to the complex "processing" of organic matter as it flows from reach to reach. The functional categories of Cummins (1978) and Cummins and Klug (1979) are followed here. Included in the groups are herbivores, detritivores, and predators—and, of course, there are parasites and commensals benefiting from association with all three. Because an animal may be omnivorous, in many instances it is impossible to assign it to a single class. Furthermore, there is still so much unknown about the food habits of some stream animals that accurately ranking them in categories is impossible.

Carnivores

The carnivores, preying on all the rest, can be placed in two groups, the *piercers* and the *engulfers*. The former seize their prey and suck fluids from tissues and cells. Those true bugs, order Hemiptera, that occur in aquatic habitats are the best examples (Fig. 8-6, *E*). Their piercing, beaklike mouthparts and the raptorial front legs (of many species) make successful attacks possible, even on small fish by some of the larger forms—belostomatid bugs, for example. The person who is careless while handling aquatic hemipterans soon learns respect for them; their elongated rostra not only draw out fluids but also inject painful digestive enzymes and anticoagulants.

Another group of piercers are quite different structurally; these are the predaceous diving beetles, Dytiscidae (Fig. 8-6, *C*). They have long, sharp, curved mandibles best described as falcate, or sickle-shaped. Each mandible is grooved and capable of penetrating prey, and through the grooves flow digestive enzymes. As is true for large belostomatid bugs, some dytiscid beetles attack small fish in addition to their usual invertebrate victims.

True specialists among the piercing predators are the spongillaflies (Fig. 8-6, *E*) belonging to the family Sisyridae of the order Neuroptera (lacewings). They clamber about on colonies of freshwater sponges (one genus of which is *Spongilla*) and with elongated, styletlike mandibles they stab the sponge cells and suck their fluids. Another type of piercer is represented by those herbivores that suck the juices of vascular plants or of filamentous algae. Many corixid bugs and some so-called microcaddisflies typify this group.

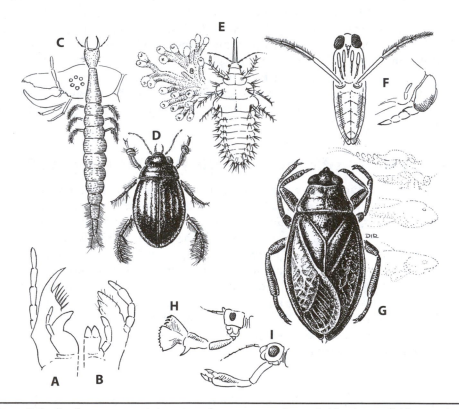

Figure 8-6　Predaceous aquatic insects and some comparisons. **A**, Mouthparts (labium and maxilla) of engulfing-predator stonefly naiad, *Isoperla*. **B**, Same mouthparts for shredder-scraper stonefly naiad, *Pteronarcys*. **C**, Larva of predaceous diving beetle *Dytiscus* (Dytiscidae); background, lateral view of the head of another dytiscid larva, *Hydroporus*. **D**, Adult *Dytiscus*. **E**, Larva of spongillafly *Sisyra*, piercing predator, and its freshwater sponge prey. **F**, Ventral view of piercing predator *Buenoa* (Hemiptera, Notonectidae) and lateral view of head showing the sharp rostrum. **G**, Adult piercing predator *Belostoma* (Hemiptera, Belostomatidae). **H**, Head of dragonfly naiad *Macromia* (Odonata, Anisoptera), lateral aspect showing the grasping labium. **I**, Head of dragonfly naiad *Aeschna*, lateral aspect showing a narrower grasping labium. In the background of **G**, some prey items are faintly shown—mayfly, stonefly, salamander, and fish.

Other predators bite off chunks of their prey or devour their captives completely; these are the *engulfers*. Fishes, turtles, and salamanders are the main vertebrate carnivores in this category. Others in the stream ecosystem are typically dragonfly nymphs, the dobsonfly larvae, many of the stonefly nymphs, some caddisfly and beetle larvae, and even a few mayfly naiads. Among the Diptera, the dance flies (Empididae) and the biting midges (Ceratopogonidae) have predominantly predaceous larvae; a few species of the crane flies (Tipulidae) and some chironomid larvae occur in lotic habitats, where they play the role of engulfers.

Large Grazers and Detritivores

Scrapers are herbivores that graze on algae attached to stony and organic surfaces. These periphyton feeders include gastropods and, in rapid water especially, the snails called limpets (Fig. 8-7, *G* on the next page). Many of the aquatic insect orders include herbivores that are primarily scrapers: the mayflies; a few stoneflies; many trichopterans and some microcaddisflies, including the rheophilous *Helicopsyche* (Fig. 8-7, *E*) that builds a snail-shaped case of fine stony fragments as well as the glossosomatids (Fig. 8-7, *D*) that use tiny stones to construct portable cases resembling turtle or tortoise shells. Among the beetles in the scraper category are the water penny larva, both larval and adult riffle beetles, dryopids and hydroscaphids; in the Diptera, the rheophilous larvae of the mountain midges (Deuterophlebiidae, Fig. 8-7, *H*), the net-winged midges (Blephariceridae), and the solitary midges (Thaumaleidae) are predominantly aufwuchs scrapers. A few of the familiar chironomid midge larvae also are scrapers. Labeling the scrapers as herbivores is probably too narrow an approach, for the aufwuchs (or periphyton) community is diverse: in addition to the green algae, chrysophytes, and cyanobacteria there are protozoans, tiny metazoans, fungi, bacteria, and FPOM. The scraper probably derives nourishment from them all.

Shredders include a group of herbivores and detritivores that chew large organic particles (Fig. 8-8 on p. 177). The herbivorous shredders bite the leaves, petioles, and stems of aquatic plants; some tunnel their way completely into the living tissue to feed. These occur in ponds, in the littoral regions of lakes, and in the lower reaches of rivers. The second group of shredders are typical inhabitants of small, shaded streams where fallen leaves and woody material provide nutrition. They munch on allochthonous CPOM and the associated fungal and bacterial populations. These detritivores include amphipod species referable to the genus *Gammarus* (Fig. 8-8, *G*). In North America, two species have been studied especially: *G. pseudolimnaeus* and *G. minus*. In Europe, where there is a far richer amphipod fauna community, several species are known to be shredders, but most attention has been directed toward *G. pulex*. Two of the common lotic insect orders, the Plecoptera and the Trichoptera, include many coarse-detritus shredders. The ptilodactyl beetles in lotic settings and members of the elmid genus *Lara* burrow into woody detritus. Some crane flies (Tipulidae) are lotic shredders, and as might be expected, the versatile chironomids include some leaf shredders.

Collectors

The *collectors* feed on decomposing FPOM and the associated bacteria. One group, the *filterers,* are especially represented by the Hydropsychidae of the Trichoptera (Figs. 2-10, *A*, *B*; 8-9, *A*). Species of the genus *Hydropsyche* spin silken nets that catch tiny particles swept down from above; subsequently these are devoured by the *Hydropsyche* larva waiting in an associated lair or retreat. Three other caddisfly families include net makers, and some chironomid larvae are net spinners also, feeding on

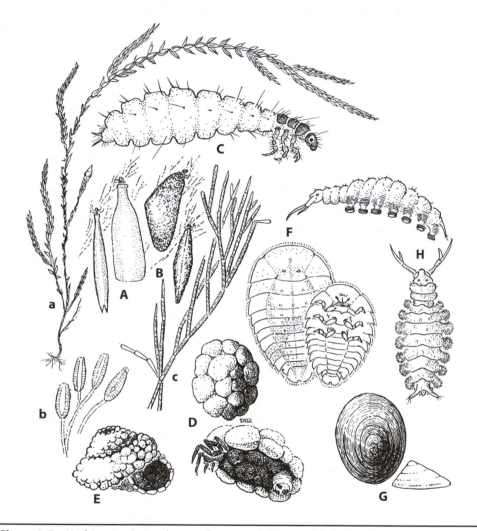

Figure 8-7 Herbivorous invertebrates of streams. **A**, Microcaddisfly *Oxyethira* larva (Trichoptera), dorsal and lateral aspects; feeding on algal mat. **B**, Microcaddisfly *Hydroptila* larva, dorsal and lateral aspects; feeding on algal mat. **C**, Microcaddisfly *Hydroptila* larva removed from case. **D**, Trichopteran scraper *Glossoma*, dorsal and ventral aspects of the case, showing the larva in the latter. **E**, Empty case of *Helicopsyche* larva, trichopteran scraper. **F**, Water penny larva *Psephenus* (Coleoptera), dorsal and ventral aspects; lotic scraper. **G**, Limpet *Ferrisia* (Gastropoda) dorsal and lateral aspects of shell; scraper. **H**, Mountain midge *Deuterophlebia* (Diptera), lateral and dorsal aspects; lotic scraper. Plants illustrated—**a**, *Fontinalis,* moss. **b**, *Gomphonema,* diatom. **c**, *Cladophora,* green alga.

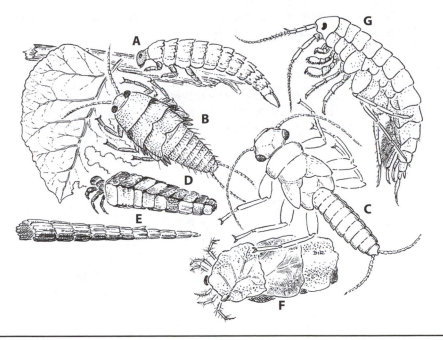

Figure 8-8 Invertebrate shredders. **A**, Larva of elmid beetle on coarse woody detritus. **B**, *Peltoperla,* plecopteran naiad shredding leaf of cottonwood (*Populus*). **C**, Plecopteran naiad, *Capnia.* **D**, Trichopteran larva *Lepidostoma;* case composed of bark and woody fragments. **E**, Case of trichopteran larva *Triaenodes;* slender pieces of green plant material arranged spirally. **F**, Trichopteran larva, *Phylloicus;* case composed of leafy material. **G**, *Gammarus* (Crustacea, Amphipoda).

small particles brought downstream to them in suspension. The appropriately named *Rheotanytarsus* is one of these dipterans.

The black flies (Simuliidae) characteristic of fast-flowing waters constitute 13 genera in North America. The larvae of almost all collect FPOM with a pair of cephalic fans that arise dorsolaterally from the head capsule (Fig. 8-9, *B* on the following page). The bivalved molluscs are filtering collectors, with some of their densest and most diverse populations occurring in streams. In North America the unionid clams are the predominant native group, but the alien *Corbicula*, the type genus of another family, is now widespread. It lives in lentic and lotic environments, feeding on a wide range of particle sizes.

The filtering collectors leave plenty of sestonic particles for the second type of collector, the *gatherer* or *deposit feeder.* In many streams the filterers may reduce the quality of seston by selectively taking animal particles, but quantitatively the seston may be little diminished by their activities. The fine particles that escape them settle to the bottom deposits to be gathered by such collectors as some oligochaete worms and certain caenid and ephemerid mayfly naiads (Fig. 8-9, *D*).

Much finely grained organic material may be trapped on the water surface, where it is gathered by the tiny wingless insects called springtails (order Collembola). Although the collembolans avoid swift water, some forms do skate about on the surface film near the stream bank collecting FPOM, and perhaps scavenging the bodies of other invertebrates trapped on the film.

Processing of Detritus

Rivers are the "gutters down which flow the ruins of continents," or so wrote Leopold and associates (1964). The world's rivers' annual dumping of 2.25×10^{10} metric tons of dissolved and particulate matter into the seas, having altered the land

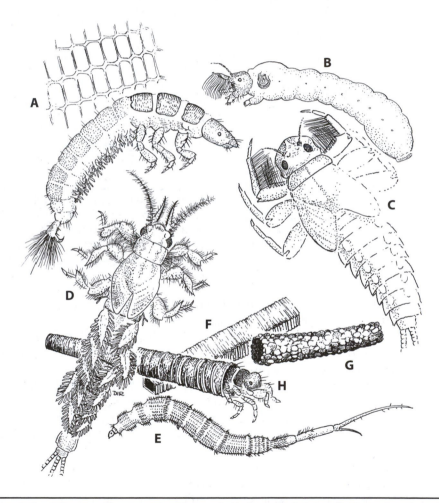

Figure 8-9 Collectors, filterers and gatherers. **A**, *Hydropsyche* larva and enlarged section of its silken net (Trichoptera); net filterer. **B**, *Simulium* larva (Diptera), filtering with cephalic-fans. **C**, *Lachlania* naiad (Ephemeroptera); filtering with specialized forelimbs. **D**, *Hexagenia* naiad (Ephemeroptera), a burrowing gatherer. **E**, Larva of *Bittacomorpha*, phantom cranefly (Diptera, Ptychopteridae), a burrowing gatherer. **F**, Case of *Brachycentrus* larva (Trichoptera), filterer. **G**, Stony case of *Setodes* larva (Trichoptera), gatherer. **H**, Larva of *Amiocentrus* (Trichoptera) in its silken case, gatherer.

by erosion to make this possible, seems to bear this out. The stream is more, however, than just a transporter of materials. It should be thought of also as a processor of materials, for the biota take up, convert, use, and set free the materials that come to them (Fig. 8-3). That which arrives at the river's mouth is far different, quantitatively and qualitatively, from what was present in the waters of the upper reaches. The organic material that moves seaward is both dissolved and particulate (DOM and POM). When it commences its journey, it is remarkably heterogeneous; at the river's mouth the diminished material is more homogenous.

There are dangers inherent in generalizing about downstream drift of organic material from watershed to watershed, and even within sections of the same stream. The drift of POM varies from season to season, from one station to the next and, in the case of invertebrates, from hour to hour. In addition, the type of material carried by the current may be very important in its movement. Some examples of this come from the work of Dance and associates (1979), who monitored two Ontario streams for 13 consecutive months. In north temperate regions, perhaps 90% of the annual drift occurs during the spring flooding, and there are definite peaks in the seasonal aspects of algal and green-detritus drifts. Some allochthonous materials do not travel far from the place of input; cedar (*Thuja*) detritus, for example, does not move far from its place of origin, and this is true for larger chunks in general. By contrast, the leaves of deciduous trees such as beech (*Fagus*) and maple (*Acer*) move nicely over smooth bottoms but are held up by rough, stony substrata, where leaf packets may form and remain, in some instances relatively far from their place of input. Usually a much greater fraction of the annual, solid POM occurs in the form of FPOM rather than as CPOM. The amount of DOM, however, surpasses those combined. High proportions of DOM vary from over 97% of the total organic material in the Ontario streams to about 69% in the thermal desert stream Tecopa Bore (Naiman, 1976).

Coarse Particulate Organic Matter

In some climates much of the allochthonous material is woody. The Oregon streams studied by Naiman and Sedell (1979) lie in forested regions. As a result, more than 90% of the benthic organic detritus is ligneous debris. In an Australian mountain stream *Eucalyptus* bark and branch litter amounted to 60% of the total annual mass of forest debris; this amounted to about 62% of the annual energy contributed by the forest (Blackburn and Petr, 1979). Size categories are altered when ecologists rank woody particles. The litter is considered fine if its dimensions lie between one and 100 mm; chunks larger than 10 cm are classified as coarse. Many of the coarse fragments are "stored" upstream, becoming lodged and water soaked. The wood is persistent material, breaking down slowly over the span of one or two centuries, meanwhile serving as a source of FPOM for downstream communities. Organic dams structured by coarse woody debris dissipate much of the stream's erosional potential, forming pools that alter the flow of DOM and POM while providing semilentic habitat for fishes and invertebrates (Bilby and Likens, 1981).

Woody and leafy particles arriving in the headwaters are colonized by bacteria and aquatic fungi (mostly hyphomycetes) and attacked by their enzymes. Certain insects tunnel their way into the coarse woody substance, and other invertebrates begin to shred the leafy material. The fungi are more important than the bacteria in settling on the newly arrived coarse detritus in the upper reaches of the stream. Many so-called shredders actually derive most of their nutrition from the fungal mycelia on the autumn leaves, twigs, and branches. Amphipod crustaceans belonging to the genus *Gammarus* congregate in bundles of leaves, prospering more on fungus-enriched material than on sterile leaves (Bärlocher and Kendrick, 1973; Kostalos and Seymour, 1976). Furthermore, the fungi possess varied enzymatic powers, which explains why certain species are preferred over others by the amphipods.

Other shredders, however, such as some larger crane fly maggots, bite and chew the leaves into smaller bits, many of which are lost to these somewhat inefficient feeders. The surface area is much increased on these finer particles, and now the bacteria become increasingly important. They settle on the leafy bits and begin extracellular digestion, breaking down the easily hydrolyzed substances, leaving the more refractory material with its coating of nutritious bacterial cells to be devoured by other detritivores.

Dissolved Organic Matter

DOM has several sources; it is introduced directly by groundwater and surface runoff, released from woody-leafy detritus, excreted by invertebrate shredders and fungus scrapers, and leaked from algal and higher-plant cells. This DOM pool serves as a culture medium for other bacteria. In this way DOM becomes FPOM as the material is incorporated into bacterial cells. The DOM that continues seaward is more resistant to bacterial enzyme activity than that transformed to finely particulate bacteria. In general, in both POM and DOM the carbon/nitrogen ratio increases and lignin percentages mount, the longer the woody debris remains in the water.

Autumnal leaves sinking below the stream surface rapidly lose weight as soluble organic material leaks out, even as they are being colonized by fungi (Hynes, 1969). A fraction of the DOM becomes particulate when bacteria incorporate it, when it aggregates in agitated waters to form particles greater than 0.45 µm, and by later accretion to tiny clumps of FPOM 10 to 100 times larger.

Leaching of DOM from the leaves occurs rapidly and varies among tree species. Also, preferences of the shredder-processors are species related. Alder (*Alnus*) leaves, high in nitrogen content, are attacked preferentially by invertebrates and are rapidly processed even in water temperatures near freezing (Short et al., 1980). By contrast, leaching from conifer needles is slower, and they are chosen less readily by shredders. In addition, leaves from deciduous trees, and especially those of *Alnus*, are assimilated with greater efficiency by the invertebrates.

Much of the leaf leachate is taken up rapidly or converted in the upper reaches; small brooks are efficient processors (Fisher and Likens, 1973; Fisher, 1977).

Fine Particulate Organic Matter

Moving downstream, then, are fine autochthonous particles that have come from shredding activities of the stream fauna, the animals' finely divided feces, bacterial cells, organic aggregates derived from DOM, and DOM more refractory to enzymatic hydrolysis than that found upstream. The small detrital particles, ranging from ultrafine to fine (UFPOM–FPOM), are coated with microbes that belong to a world where detritus and detritivore are hardly distinguishable. Awaiting are other processors, the collectors that feed on finely divided matter.

In the river-continuum model the relative abundance of the collectors increases with stream order. Outnumbered two to one in the headwaters, the collectors are at least equal to shredders in medium-sized rivers. The FPOM is not always gathered by collectors waiting downstream. At times they can be found in leaf packets close to the shredders that slough off the fine leaf particles and produce the fecal pellets that support them (Short et al., 1980).

One must not think of the various collector-filterers as gathering identical particles, even though we define FPOM simply—particles smaller than 1 mm. In the Tallulah River, Georgia–North Carolina, there are many filter feeders including net-spinning trichopterans. Wallace and others (1977) published excellent pictures, based on scanning electron microscopy (SEM), of some of the different nets. The mesh opening sizes ranged from only 1 × 6 μm for one species up to 403 × 534 μm for another; the latter lived in much swifter water. There were many intermediate mesh sizes depending on species and instar, each designed to function best in a specific current velocity and to intercept certain sizes of particles. By contrast, *Corbicula*, the introduced Asiatic clam which is very abundant in the river, collects a wide range of particle sizes. On the other hand, some stream entomologists question the precise selectivity of different-sized trichopteran nets, and most filterers may collect a variety of particles even though the interstices of their collecting devices differ.

Collectors are suspension feeders (filterers) or deposit feeders, those that pick up and ingest tiny bits that have settled from the water. They and the shredders are dealing with hard-to-digest polymers, except for the associated fungal and bacterial biomass. Many of them harbor microbial symbionts in their guts. The stream invertebrates can take in material that seems inadequate for their nutrition—matter with a high carbon/nitrogen quotient; the gut microbiota with greater enzymatic capability attack the substance, and ultimately the host animals absorb low C/N food stuff with the essential amino acids.

Similarly, the feces are colonized by microbes, both bacteria and hyphomycetes. They break down many of the chemical bonds that the feces-producing animals failed to break down. Some break up into finely divided bits that can be taken up by filter-feeding collectors. They, in turn, may produce feces that can be described as compacted—larger particles result from the intake and consolidation of FPOM. The fecal material is nutritionally valuable. In certain Rocky Mountain streams, the lotic insects produce feces with protein comprising 12% of the ash-free, dry-weight protein, compared with 8% for leaf litter and 13% for other detritus (Shephard and Minshall, 1981). Before that, Rossi and Vitagliano-Tadini (1978) had demonstrated an indirect role played by feces in the nutrition of young isopods, *Asellus aquaticus*, in Europe. The hyphomycetes in the feces of the adult isopods continue to survive in plant detritus on which the young feed. Without the fungi the detritus is an inadequate food source for them.

Autotrophic Organisms and Predators

Despite the obvious allochthony in the headwaters, autotrophic organisms begin to appear. If the stream is fed by springs and is not heavily shaded, it is not uncommon to find dense beds of watercress (*Nasturtium*) at the source. Usually, however, the bryophytes and epilithic algae (periphyton) constitute the bulk of the phototrophs. Periphyton communities demonstrate the diversity of lotic environments, a heterogeneity that is obscured, perhaps, by the river continuum concept. Smoothly changing homogeneity is not typical within a short stream segment. The activities of limnephilid caddisflies underscore the spatial and nutritional patchiness of benthic algal producers. These grazers move from patch to patch, quickly abandoning those where previous feeding has lowered food value and lingering where there has been no recent grazing and food levels are high (Hart, 1981).

The plant life that develops along a river course is attacked by invertebrates assigned to the scraper and piercer functional categories. Bits of plant material are sloughed off inadvertently, and they add to the FPOM pool. DOM leaks from the plants directly and is exuded and excreted by the herbivores.

Flocculation and bacterial uptake of DOM contribute to the particulate pool over and over again, and as other bacteria attach and initiate polymer breakdown with their enzymes, the detritus is ingested again by other consumers.

Meanwhile, all along the stream bed as it progresses toward the sea, the predators have been taking their toll. They are sustained by high-quality proteinaceous food, the resources offered by the lotic detritivores and herbivores. They, in turn, contribute feces to the particulate stores and excretions and exudations to the DOM pool.

Stream Metabolism and Efficiency Indices

Unlike the situation in standing waters, the cycling and recycling of nutrients in a stream does not occur in the same or approximate place. As an atom of P or N, for example, goes sequentially from the soluble to particulate phase, to being taken up by an invertebrate consumer and to being subsequently released, it is displaced downstream. The downslope vector added to nutrient cycling creates the picture of an imaginary spiral path. **Nutrient spiraling** is the downstream transformation of the aqueous, particulate, and consumer compartments of a nutrient cycle. A method of quantifying the efficiency of a stream in using its nutrient resources, **spiraling length**, is the sum of the mean distance the nutrient atom travels in each of the three compartments of the cycle: in the watery, sestonic, and consumer forms (Newbold et al., 1981). This index of recycling is linear rather than being based on time as in lentic assays. The shorter the spiral length (in m or km down the stream profile), the more efficiently the nutrient is used and recycled within a given reach. One expects tighter coils in the spiral as efficiency increases.

Fisher (1977), viewing the stream as an active biologic system that metabolizes organic matter, proposed a method to compare different stream systems or stream segments. The method has direct bearing on downstream processing. Comparisons are made on the basis of the stream metabolism index (SMI), the same as his loading maintenance efficiency (LME). The indices are founded on the concept of stream loading: if the organic matter leaving a stream segment is no greater than that which arrived via the main stem, loading is zero; if there is more OM leaving the reach than entered by way of the main stem, loading has occurred. To prevent loading, the storage of OM in the segment and the respiratory activity of the ecosystem must be adequate to eliminate the sum of OM that entered by way of tributaries, groundwater, terrestrial litter, wind and precipitation, and primary productivity. The ratio of observed respiration (plus or minus storage) to the respiration required to bring about zero loading is the SMI or LME. If all OM received in the stream section is oxidized so that the outflowing water contains no more than was present at the upstream site, there has been no loading and the SMI or LME is 1.0. Fisher compares Hubbard Brook with Fort River, Massachusetts, referring their efficiencies to 1 km of flow. The Hubbard Brook segment does not load, but Fort River does, its efficiency being only about 0.65 rather than 1.0. The relation between SMI and LME and the concept of spiraling length is obvious.

Were it not for all the complex processing phenomena occurring throughout a stream's course, the loading at its mouth would be far greater than it is. Thanks to fairly high SMIs and LMEs, tremendous amounts of detritus are removed from the 2.25×10^{10} tons of dissolved and particulate material that is poured into the seas each year.

Human Impacts

Streams and rivers worldwide have been modified by humans for more than 4,000 years and in dozens of ways (Gregory, 2006). Common modifications discussed here include changes to the structure of the channel itself, usually straightening or channeling one or more segments; altering amount or timing of water flow, either inputs and withdrawals; and impoundment by dams. Changes to the water quality (in particular, water chemistry) are discussed in other chapters.

Structural Alterations of Channel

Streams can be dynamic and unpredictable, characteristics inconvenient for human society. We frequently strive to drain water off of land quickly after a storm and keep the streamflow in a well-defined, strictly confined, channel. As a result, channels are often straightened and reinforced, and streambeds smoothed or paved. The ecological effects have been known for years; an excellent catalog was compiled by Brooker (1985). Nearly every ecological process and biological taxon mentioned in this chapter are affected by the alteration of the channel when straightened, deepened, or reinforced. It is difficult to know how widespread these modifications are, but for example, if 53–75% of floodplain vegetation in the continental United States were removed, it is likely that the majority of stream channels would be likewise impacted (Swift, 1984).

Next to enclosure in a pipe, the most severe stream modification is into a concrete ditch. A famous example is Los Angeles River in southern California; most have seen it featured in a Hollywood movie (often for a chase scene). Over 80 km have been channelized in response to damaging floods in the early 20th century (Gumprecht, 1999). Nevertheless, even this river has its friends, and concerned citizens are working to maximize the ecological functioning of the stream.

Flow Alterations

The hydroperiod of the stream (discharge volume over time) is modified by humans either directly or indirectly. Direct changes occur because flow is augmented artificially, such as through a factory or wastewater treatment plant effluent. Direct reduction in flow is usually the result of diversion of water from tributaries.

Humans indirectly alter flow by changing surrounding land use and drainage. Land development includes rapid and efficient removal of water from the surface of the land. The use of drain tile, ditches, and sewers sends overland and shallow groundwater quickly to streams and sends more of the available water to the stream (it is less likely therefore to evaporate or infiltrate to deeper groundwater). The result is predictable: streams become "flashy," with larger flood pulses and more frequent floods. The effect is quite well documented, and in fact hydrologists have excellent empirical models for the phenomenon (Fig. 8-10).

Impoundment

Over centuries, we have dammed streams to control floods, impound water in reservoirs for recreation or water storage, or for power. Yet the pace increased dramatically in modern times: More than 45,000 large dams were built in the 20th century (World Commission on Dams, 2000), and approximately 60% of the world's large river basins are now impounded by dams (Nilsson et al., 2005). Rivers no longer run freely, and flow is "homogenized" on dammed rivers (Poff et al., 2007). Remarkably,

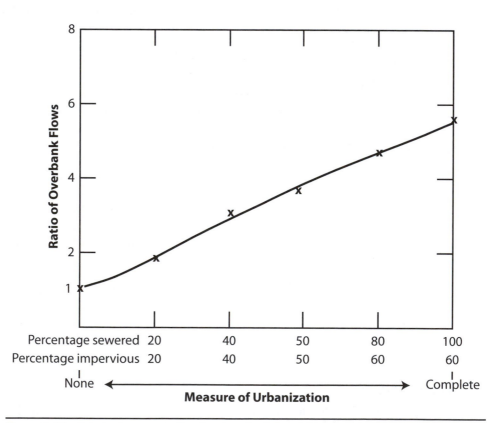

Figure 8-10   Relative increase in flooding (overbank flows) due to urbanization. An increase in either the area of land drained using storm sewers, or land covered with surfaces such as pavement, of 40% will roughly triple the flood potential. Adapted from Leopold (1968).

the scale of this has increased as demand for hydropower increases; the largest dam ever built (the Three Gorges Dam in China, with a reservoir length in excess of 600 km) was completed in 2003. The project has pronounced environmental impacts (Yang and Lu, 2013): more than 500 earthquakes (associated with the gravitational force on the Earth's crust from the weight of water). and more than a billion tons of trapped sediments. The ecological effects include habitat alteration and impacts on a biodiverse aquatic system (Wu et al., 2004).

Case Study: The Disappearing Aral Sea

In addition to the Colorado River mentioned earlier, a famous case in human flow alteration is the Aral Sea in central Asia. Beginning in the 1960s, the Soviet Union diverted water from nearby rivers to irrigate cropland in this very arid region. The lake was once the fourth largest on Earth, with a surface area of 68,000 km², but it is now 89% smaller (Löw et al., 2013). The disappearance of the Aral Sea has created a saline desert, the Aralkum.

Stream Assessment Methods

As described in Chapter 2, a common worldwide approach to stream assessment is to use indices of biological integrity (IBI). Muhar and Jungwirth (1998) present a series of general criteria for stream bioassessment internationally; they stress the importance of "integrative parameters" that incorporate several factors at once and connect the river to its watershed. In the United States, a majority of the states utilize a version of IBI, some in use for decades. Rankin (1995) presents a history of the approach and compares the results of different indices. A particular stream should be compared with "reference systems" from the same geographic area, since habitats and associated biota will differ from region to region. The U.S. Geologic Survey has employed standardized protocols, the National Water-Quality Assessment (NAWQA) Program (Fitzpatrick et al., 1998). The habitat is described by including the morphometry of the stream segment (reach), surrounding/overhanging vegetation, bank features, surrounding land use, and flow characterization.

IBI were originally developed using fish taxa, but Karr (1991) called for the development of more comprehensive tools with regional versions incorporating the ecology of local taxa and their response to habitat degradation. The specific taxa vary but often include a mix of fish, mollusc, algal, macrophyte and other organisms, but above all, macroinvertebrates. As described previously in this chapter, macroinvertebrates are key to understanding stream function due to their numbers and variety, their sensitivity to pollution (in particular, low oxygen levels and siltation), their short life cycles, and their multiple trophic roles.

As we learn more and more about the ecology of stream habitats and their organisms, it is hoped that our use of bioindicators and stream monitoring will become increasingly effective.

9

Light and the Aquatic Ecosystem

The arrival of solar radiation at the lake surface marks the beginning of the photo-synthetic process below. In some lakes and streams organic detritus from outside sur-passes the production of organic material within the water. Even in such instances, however, sunlight falling on a terrestrial surface somewhere was the ultimate source of energy in the carbon-containing compounds delivered to the lake. We indeed live on a solar-powered planet.

Except for well-known uranium fission and hydrogen fusion devices, one is hard pressed to list energy sources that do not depend on the sun. Perhaps the earth's internal heat and moon-pulled tides are the only other nonsolar sources of energy we can list.

It interests the limnologist to know first, how much radiation falls on the lake sur-face within a span of time; second, how far it penetrates; and third, how it can be used or how it affects aquatic organisms.

Solar Constant and Nature of Light

To quantify the energy arriving from the sun, a value called the *solar constant* has been established on the basis of many careful assays. It has varied somewhat but is presently 1.94 cal/cm^2 per minute. It is a tremendous quantity of energy, which, if available to the phototrophic plants and used with 100% efficiency, would produce about 325 tons of plant biomass on a square kilometer of the earth's surface each hour. Much of this energy, however, does not reach phototrophic organisms, for the solar constant applies to the sun's rays arriving outside the atmosphere at the earth's mean distance from the sun. Energy is lost as the rays enter the atmosphere: sunlight is scattered, reflected, and absorbed by molecules of carbon dioxide, ozone, and water. More than half the energy is lost, although some absorbed radiation appears again as long waves (greater than 5,000 nanometers), playing no direct part in photosynthesis.

Light arrives as a pulsating field of electromagnetic force composed of an endless series of waves speeding from the sun at more than 186,000 miles per second (299,792 km per second is more nearly exact). The light contains quanta of energy that are directly proportional to the frequency, or number of waves produced per second. There are myriad individual rays within the solar radiation, each with its own wavelength and characteristic frequency. A broad electromagnetic spectrum is formed by the entire range of waves radiated from the sun, but only a small fraction reaches Earth. About half these waves constitute visible light, although the *pyrheliometer,* the instrument employed to measure radiation, is equally sensitive to all arriving wavelengths.

Photosynthetic plants capture some of this electromagnetic force and convert it to chemical energy rather than let it escape as heat or fluorescence. They are not especially efficient at this conversion; perhaps almost 99% is lost.

The length of each wave in the mixture that is sunlight can be expressed in various ways. The Angstrom unit (Å) has been much used. It is 10^{-10} m. The nanometer (nm) is employed more often now; it is 10^{-9} m. At times the micrometer (μm), or 10^{-6} m, is the unit used to measure wavelengths. Thus, a certain blue-green wavelength, containing the highest energy content of waves within the visible spectrum, is about 4900 Å, or 490 nm, or 0.49 μm. Representative colors selected from the continuum of the spectrum are violet, 400 nm; blue, 460 nm; green, 520 nm; yellow, 580 nm; orange, 620 nm; and red, 700 nm. The longest ultraviolet waves and the shortest of the infrared rays are about 350 and 750 nm, respectively. Within this range, or perhaps from about 380 to 780 nm, the waves that can be detected by the human eye are defined as "light." Rays shorter or longer than those within visible range are designated as radiation, not as light, even though other animals can detect them. This anthropocentric approach is commonly used.

Energy in Light

The lengths and frequencies of the waves within light vary. As a result, their energy content differs. The size of a quantum of electromagnetic-radiation energy is directly related to its frequency. The short waves contain larger quanta than the longer, because they have a greater frequency. During the brief span of time that light travels 299,792 km, enough waves must be present to fill this enormous distance. Red light near the limits of our visual capabilities is about 750 nm; about 400 trillion of these waves span the distance light travels each second. Extreme violet light comes in waves about half the length of red light. Therefore, twice as many waves will be needed to fit into the great distance that is a light-second. About 789 trillion violet waves of 380 nm are required. Thus, the frequency of visual light ranges from almost 400 trillion at the red end of the spectrum to nearly twice that at the violet extreme.

Planck's quantum theory states that the relationship of wavelength to energy is inverse, so the direct relationship of frequency to energy is given by the equation:

$$E = hv$$

E is the energy in a single quantum of radiation, a *photon; h* is Planck's constant, 6.6255×10^{-34} joule seconds; and *v* is the wave frequency, in cycles per second, known as hertz (Hz). Frequency of a light wave is the quotient of the speed of light, *c,* and the wavelength λ:

$$v = \frac{c}{\lambda}$$

The speed of light is 3×10^8 m per second, and is a constant in the numerator. The denominator, of course, should have the same linear unit. Red light with a wavelength of 750 nm has a frequency, therefore, of

$$v = \frac{3 \times 10^8}{7.5 \times 10^{-7}} = 4.00 \times 10^{14}$$

Similarly, violet light, with a wavelength of 380 nm, is characterized by a frequency of 7.88×10^{14} cycles per second. Substituting the two frequencies in the equation $E = hv$, we get 26.500×10^{-20} and 52.205×10^{-20} joules, respectively. From this it appears that the energy in a quantum of the shorter wave is 1.97 times the energy in the longer wave.

Another constant, serving to quantify the energy in light, is based on the fact that an electron is ejected from a plant pigment following the absorption of a suitable quantum. Photosynthesis is initiated when the energy of a photon excites a single electron from a single pigment molecule; the collision of the photon and electron results in the electron's gaining the quantum of energy lost by the photon. A mole of a pigment compound, containing Avogadro's number (N) of molecules, would absorb 6.024×10^{23} photons of energy to initiate a photosynthetic reaction. The total energy absorbed by a mole of pigment is an einstein; it is Nhv, or NE. The red and violet light waves discussed earlier contain $6.024 \times 10^{23} \times 26.50 \times 10^{-20} = 15.96 \times 10^4$ joules, the energy in 1 einstein of 750 nm light, and $6.024 \times 10^{23} \times 52.205 \times 10^{-20} = 31.45 \times 10^4$ joules, the energy in 1 einstein of 380 nm light. In caloric terms, 1 kcal is the equivalent of 4.186×10^3 joules. This reveals the energy in a so-called mole of red light at 750 nm is 38.13 kcal, and similarly, a mole of violet light at 380 nm contains energy equaling 75.13 kcal. The energy relationship between the two wavelengths remains simply

$$750 \text{ nm} / 380 \text{ nm} = 1.97$$

There is almost twice as much energy in the short-wave light; its wavelength is a little more than half the length of the longer wave.

Light at the Earth's Surface

There are two sources of radiation, direct and diffuse. Direct radiation is from the sun; diffuse is the light from the clouds and sky.

Unfiltered pyrheliometers can measure incoming waves up to 5,000 nm, but most radiant energy from skylight and sun ranges from 290 to 3,000 nm, and there is little evidence that cyanobacteria and eukaryotes can use light outside the band of 400 to 700 nm; this is the photosynthetically active range, or **PAR**. Only absorbed light can be used for powering chemical reactions. Chlorophyll, which is the key to photosynthesis in the cyanobacteria and eukaryotic plants, efficiently absorbs red and violet waves, reflecting the intermediate greens and yellows. This is, of course, why plants appear green to the eye, and why photosynthetic efficiency is highest in violet-blue and orange-red light.

Ecologists have described solar energy in terms of calories or kcal (i.e., 10^3 cal); the flow of calories from one trophic to the next, in terms of unit area of lake surface, has been a common denominator in community studies. Formerly the langley (ly) was used, but it is outmoded now. One langley equals 1 cal/cm^2 or 10 kcal/m^2. Thus, the solar constant presently is 1.94 ly per minute. Using the International System (SI) of units, the radiant flux would be expressed in terms of energy and power as joules/m^2

per second and watts/m². Similarly, the lumen and the lux can be referred to the SI unit of luminous intensity, the candela. If focusing on aquatic photosynthesis, we measure irradiance as quanta, usually in µmol/m² per second, measured with a submersible quanta meter.

Light at the Lake Surface

A portion of the energy reaching the lake surface does not enter the water, but is immediately reflected. This amount depends first upon the sun's angle of incidence, varying with the hour, the season, and latitude. From studies of an Oklahoma impoundment, Lake Hefner, Anderson (1952) developed the following formula to quantify reflectance (R):

$$R = aS_A^{-b} = 1.18 \, S_A^{-0.77}$$

In the above, S_A is the sun's altitude, or angular height in degrees, while a and b are constants, 1.18 and 0.77, respectively. From Anderson's formula one can calculate almost no reflectance when the angle of incidence is zero (the sun is at its zenith, directly overhead) to a loss of 20% by reflectance when S_A is 10° (i.e., the angle of incidence with the water surface is 80°). Reflectance expressed as a percentage is $R \times 100$.

The calculation of reflectance is complicated by factors such as wind sweeping across the water surface. Gentle ripples allow more light to enter than do surfaces whipped to whitecaps by strong winds. Moreover, the sun's angle becomes relatively less important with increasing cloud cover. The ratio of diffuse radiation from the sky to direct solar import increases with cloud cover.

In Phoenix, Arizona, the mean annual loss by reflectance is about 5% in summer and 12% in winter. This includes the back scattering from particles immediately below the surface, taken into account in Anderson's formula.

Light Below the Water Surface

Although water is a transparent liquid, light passing through it is weakened and eventually extinguished. A discovery made in the first half of the eighteenth century, now termed Bouguer's law, applies here. The principle is often attributed to Lambert, who rediscovered it soon afterward, in 1760. The law explains how light passing through various thicknesses of an absorbing medium is diminished. When a parallel beam of monochromatic light enters chemically pure water, for example, it is absorbed exponentially; the absorption varies directly with the logarithm of the thickness of water through which the light passes. No exceptions are known to this rule. Light entering a homogeneous absorbing medium is decreased at a constant rate as it proceeds through each infinitesimally thin layer. These ideal conditions are not to be found in nature, where polychromatic light strikes a pond surface from many angles, is refracted sharply downward, and penetrates heterogeneous layers.

Beer's law expresses the relationship between absorptive capacity and concentration of a uniform solution through which light passes. The absorbing capacity is directly proportional to the number of absorbing entities. This principle lies at the heart of the familiar spectroscopic method used in some chemical analyses. The light-path distance is held constant as concentrations of solute are varied. A beam of mono-chromatic light passing through a given distance of a solution is absorbed exponen-tially according to the concentration of solute. This is shown when standard curves of different strengths of a solution are prepared. The transmittance of light through the solutions, plotted on a semilog graph, reveals a straight line, an exponential decrease from weak to strong solutions. The technique works best with an array of colored solutions and approximately monochromatic light. This method is weakened by enti-ties that scatter rather than absorb light. Obviously, nature seldom matches with the conditions of Beer's law. Light composed of many wavelengths falls on the lake sur-face from many angles and is deflected as well as absorbed by materials that change concentration in the downward light path.

Despite imperfections, a combination of the two laws, called the Beer-Bouguer law, is used in limnology and oceanography. A *vertical absorption coefficient* quantifies the disappearance of light as it passes from the water surface to some light-sensitive device arranged horizontally to receive the light from above. Here another factor com-plicates strict application of Bouguer's law, for the length of the light path is not unequivocally known. The average distance traveled by rays arriving at a depth of 1.0 is at least 1.2 m, much of the light having followed oblique paths: the angle of the sun, the diffuse radiation from the sky, and the scattering of light by particles within the water contribute to increasing mean distances traversed to a given depth. Vertical absorption coefficients are expressed, then, no matter what attenuates the light, as

$$k = \frac{\ln I_0 - \ln I_z}{z}$$

This formula is derived from the equation

$$I_z = I_0 e^{-kz}$$

which shows the remaining intensity of light, I_z, having passed through a water of thickness in meters, z. The original intensity at zero depth was I_0.

Some confusion exists in nomenclature; the coefficient of absorption is often termed the *coefficient of extinction*. These terms, when used interchangeably, usually refer to calcu-lations using natural logarithms. However, the coefficient of extinction accurately applies when a logarithm to the base 10 is used in the formula rather than a natural logarithm. The absorption coefficient, therefore, is 2.3 times the coefficient of extinction, since

$$\ln x = 2.303 \log X \quad \text{— or —} \quad e^x = 10^{x/2.303}$$

In some lake research, data on optical density have been presented. This term, properly used, applies to procedure in photometry where light path, z, is expressed in centimeters.

$$OD = \log \frac{I_0}{I_z}$$

The sum of three coefficients contributes to the *total coefficient of absorption* in natu-ral waters, so that the intensity of light at depth z is

$$I_z = I_0 e^{-k_w} + I_0 e^{-k_p} + I_0 e^{-k_c}$$

Water can be analyzed in the laboratory, reducing the light path to 1.0 m. The coefficient of absorption due to pure water (k_w) can be approximated from data based on past experimentation. The remainder is caused by suspended particulate matter (k_p) and dissolved substances (k_c). Filtration or centrifugation eliminates the former and permits estimation of k_w and k_c by subtraction. The total coefficient of absorption is the sum of these three and is usually what is implied when limnologists and ocean-ographers write of extinction coefficients, absorption coefficients, and attenuation coefficients. Whatever the terminology, the log base used in calculations should be stated clearly. Although the term *absorption* is in widespread use, especially for equations and coefficients, light is deflected too. A more accurate term for decrease in illumination through the water column, *attenuation*, is also in common use and may be more descriptive of the process.

Rather than thinking only of light that is diminished in its journey through lake water, one can consider the part transmitted. Transmittance, T, is the percentage of incident light passing through 1 m. It is $100\,e^{-k}$, where k is the vertical absorption coefficient. The transmittance of light with a given wavelength (λ) differs from that of other wavelengths. Thus, when considering red light of 680 nm, k for $\lambda = 680$ in distilled water is known to be 0.455; from this, $T\,[\lambda_{680}] = 100e^{-k} = 100 \times 2.718^{-0.455} = 63.44$.

The following took place in an afternoon survey of Saguaro Lake, Arizona. A limnology class found that the light of mixed wavelengths remaining at a depth of 10 m was but 1.3% of the surface light. From this, the mean vertical coefficient k, owing to absorption, back scattering, and other diminution factors, is 0.434 per m and transmittance, expressed as a percentage of radiant energy passing through each meter to that same depth, is $T = 100\,e^{-0.434} = 100 \times 2.718^{-0.434} = 64.79$.

As the mixture of wavelengths making up the incident solar radiation penetrates lake water, it is extinguished exponentially, but wavelengths are absorbed differentially so that the spectral nature of light changes with depth. The ambient light may contain a spectrum of waves from below 300 nm to well over 800 nm. If it were to pass to a depth of 100 m in some improbable lake of distilled water, where the coefficient of absorption is due to water only, there would remain a range from about 350 to 570 nm. Within this, the greatest intensity would be in blue waves approximately 473 nm. The dim light at the depth would be mostly blue with a little violet and green (Fig. 9-1).

The abundant long rays in the incident light are absorbed rapidly. About 65% of the visible red rays are gone by the time the light commences the second meter of its vertical path in the water. The infrared waves, best described as heat, are absorbed in the first meter to an even greater extent. As a result it can be said that, during the course of a summer, sunlight could warm hardly more than the upper meter of a body of water.

At the other end of the spectrum there are some waves shorter than 350 nm. These are the ultraviolet waves that were not filtered out by the atmosphere. Their relative intensity is not great, about 7%, and these waves become a tiny fraction of the subsurface spectrum.

Natural lake and sea waters obviously contain substances not found in pure distilled water, and these substances cause additional light attenuation. Dissolved yellow substances (mixtures of plant breakdown products) in some near-shore marine waters and many lakes cause loss of 70% to 80% of the total photons absorbed by the water.

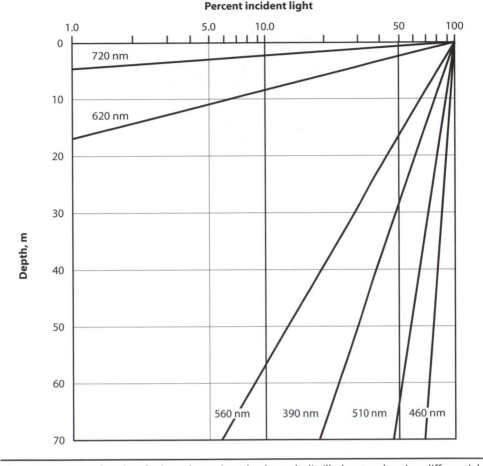

Figure 9-1 Vertical paths of selected wavelengths through distilled water, showing differential penetration. Adapted from Clarke, 1939.

Not only is the penetration of light reduced, but the alteration of spectral composition differs from that seen when light passes through pure water. Dissolved and suspended materials have selective actions on light that vary from lake to lake (Fig. 9-2 on the next page and Fig. 9-3 on p. 195). Of the three components of the total attenuation coefficient, only k_w remains the same. A generalization can be made that the higher the vertical attenuation coefficient, the greater the transmittance of longer wavelengths. Red, for example, penetrates relatively farther than would be expected. The light reaching the depths of what seem to be very clear lakes is often a green-yellow mixture, while in heavily stained waters orange may penetrate farthest. Vollenweider (1961) showed that, within limits, knowing the mean vertical extinction coefficient of the light allows for estimation of its spectral distribution. Fig. 9-4 (on p. 195) portrays this generality and reveals the relative increased transmittance of longer waves as the attenuation coefficient rises. Confirmation for this idea comes from later studies on an extremely pro-

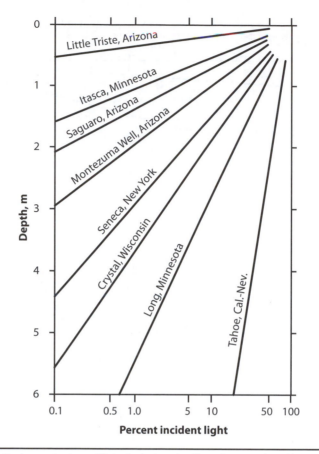

Figure 9-2 Vertical penetration of light in various bodies of water, showing percentage of incident light remaining at different depths.

ductive Scottish lake (Bindloss, 1976). The turbid waters had total coefficients of "absorption" ranging roughly from 0.7 to 4.0; blue light had the highest attenuation value, while the minimum was usually in the yellow-orange spectral region of 590 nm.

Vertical Visibility, The Secchi Disk, and the Euphotic Zone

During the 1860s a report was published that detailed some experiments an Italian oceanographer had made with white disks. The investigator was named Secchi, and devices similar to those he worked with are now called **Secchi disks**. A disk, with the flat surface horizontal, is lowered into water on a calibrated line, and the exact depth at which it disappears is noted. This is the Secchi disk transparency, expressed as a depth in meters, Z_{SD}. Obviously, it is half the distance light travels to the disk and back up to the observer's eye.

One of Secchi's disks was more than 2 m in diameter; since then, people have experimented with various sizes. Usually in limnologic work a platter 20 cm in diam-

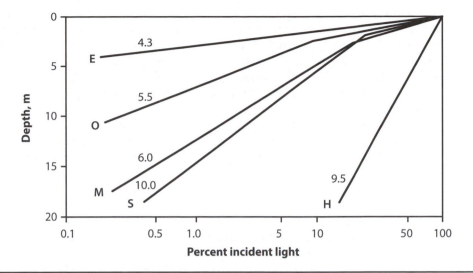

Figure 9-3   Vertical penetration of blue light (400 to 490 nm) in the Laurentian Great Lakes: **E**, Lake Erie; **O**, Lake Ontario; **M**, Lake Michigan; **S**, Lake Superior; and **H**, Lake Huron. Figures on lines refer to typical Secchi disk transparencies in meters. Orange light penetrates farthest in **E** and **O**; green light in **M** and **S**; and blue light penetrates deepest in **H**. Adapted from Beeton, 1962.

Figure 9-4
Transmission curves as functions of absorption coefficients (indicated by figures on each curve). Dotted line shows wavelength with greatest transmittance, illustrating a shift toward increased transmittance of long waves as absorption coefficient increases. Adapted from Vollenweider, 1961.

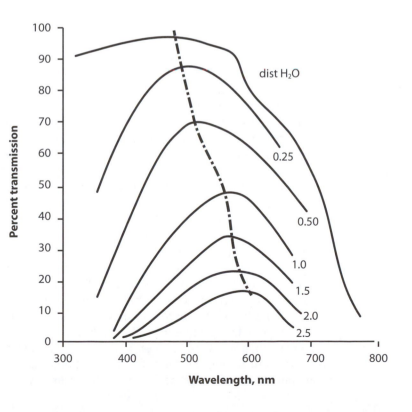

eter is used. In lakes, a version with alternating black-and-white quadrants is frequently used. To make comparisons possible, the transparency should be determined between 10 AM and 2 PM and observed off the shady side of the boat. Secchi disk transparency depends on several factors: the eyesight of the viewer, the contrast between the disk and the surrounding water, the reflectance of the disk, and to a lesser extent its diameter. Still it is a valuable tool, and every lake investigation should include Z_{SD} data.

Secchi disk transparencies are extreme in certain lakes. The oligotrophic lakes Pingualuit (New Quebec Crater), Waldo and Crater (both in Oregon), and Tahoe are noted for their transparencies: the Secchi disk is visible for 40 m at certain times. For many years, aptly named Crystal Lake, Wisconsin, was one of the clearest lakes known, with recorded Z_{SD} of 15 m; polluted Crystal Lake, Minnesota (near Minneapolis), belies its name with Secchi disk visibility only about 1 m. These data show how lake transparency varies.

The minimum intensity of subsurface light that permits photosynthesis has been set at about 1.0% of incident surface light. Thus, the region from surface to the depth at which 99% of the surface light has disappeared is called the **euphotic zone**, or simply the photic zone. The energy flux in this level is well within the photosynthetically active spectral band of 400 to 700 nm, and the irradiance is about 1.4 watts/ m². The depth of the euphotic zone changes throughout the day and it is different greatly from lake to lake, depending on the clarity of the waters. Below this somewhat arbitrary depth, primary productivity is usually considered nil. However, shade-adapted algal communities carry on photosynthesis beneath the snow and ice of Alaskan and Swedish lakes in light intensities well below 1.0% of ambient light. Hobbie (1964) showed that about one-half of the annual production of Lake Schrader, Alaska, came from winter photosynthesis at very low light intensities.

The euphotic zone can be ascertained between 10 AM, and 2 PM with a subsurface photometer connected to a microammeter within the boat. The response of the photometer (in micromol) at the water surface is considered as 100% of incident light, and subsequent readings at increasing depths are referred to this value. From the light percentages plotted on a semilog graph, it is easy to establish the depth intercepted by the line where 1% of surface illumination remains (see Figs. 9-1 to 9-3).

Two British oceanographers, Poole and Atkins (1929), presented important data on Secchi disk transparency and the vertical absorption coefficient. They empirically determined a constant, 1.7, which, when divided by Z_{SD} in meters, gives the vertical coefficient of light absorption. Although this 1.7 constant has wide application, it does not always hold true, because it was derived from the average illumination at the Secchi disk depth in British Channel waters: specifically, 16% of the surface light. This percentage is not invariable for all waters, but it yields a 1% level of 2.7 times Z_{SD}, very close to the rule-of-thumb factor of 3.0 used by limnologists. From 3.0 can be inferred a Secchi disk transparency to a level where 1% of the surface light remains.

The factors 3.0 and 2.7 discussed above represent the ratio of the photic zone depth to the Secchi disk transparency (Z_{SD}). This ratio varies with stained and colored water and may not apply at all in waters where thick, turbid layers occur *below* the Secchi disk level. Illustrating this latter condition, Edmondson (1956) reminded us of some data published in the first half of the 20th century by the Wisconsin team, Birge and Juday (1929). In Crystal Lake, some Secchi disk depths from different seasons were identical, but the photic depths differed. At one time, they determined a ratio as low as 1.07. In that instance, a large population of the pigmented chrysophyte *Dinobryon* was found lying at and below the Z_{SD}, causing a rapid attenuation of light below that level.

At Montezuma Well, the mean lower photic zone depth throughout the seasons is 3.1 times the Secchi disk transparency, with little variance (Cole and Barry, 1973). Padial and Thomaz (2008) recorded over 2,000 Z_{SD} readings during four years in South America, finding a value of 2.26, which is close and yet proved significantly higher than the accepted ratio. Verduin (1956) presented data to show that 5.0 was a good factor to translate Secchi disk level to 1.0% surface light in Lake Erie, but this is unusually high for lakes. In some very turbid waters the euphotic zone/Z_{SD} ratio is substantially above 3.0. In such cases, the white disk rapidly disappears from sight, but diffuse light, scattered by particulate matter, penetrates deeper than one would expect. Thus William T. Barry, in some unpublished research in Arizona, found that a shallow, argillotrophic impoundment called Little Triste Pond had a mean Secchi disk visibility of only 6.0 cm. The average euphotic depth was 4.6 times deeper.

Many times the Secchi disk has been used to estimate chlorophyll content of the phytoplankton community through which the observer peers as the apparatus is lowered. It seems to work fairly well because, borrowing some ideas from Edmondson (1980), the size and chlorophyll content of cells are neither infinitely variable nor independent. This may be why Talling and others (1973) could generalize about 70 mg of chlorophyll per m^3 being near the level that limits a community's photosynthetic ability, most light being absorbed within a 1-m stratum. If, however, we imagine two science-fiction phytoplankton populations with identical chlorophyll content beneath each square meter of lake surface, and both populations are composed of chlorophyll-packed, spherical cells, we can find some discrepancies. Let the average cell radius in one population be 0.5 mm; the other lake has cells with an average 5-mm radius. It takes 1,000 of the small cells to match the volume of one larger cell. Because Secchi disk visibility is sensitive to the number of particles that absorb and scatter light, the first population would produce a lesser Z_{SD}. Edmondson (1980) warned of such problems and wrote that he occasionally found deep Secchi disk values in Lake Washington when the chlorophyll content was high; the phytoplankton was dominated by large, bundled colonies of *Anabaena* and *Aphanizomenon* rather than by many scattered, tiny cells.

Effect of Ice and Snow Cover

Although an ice cover terminates circulation and isolates the lake from the atmosphere, it does not put an end to light penetration. Actually, clear ice transmits light better than the water beneath it. When a surface layer freezes, particulate and especially dissolved electrolytic materials are reduced in the ice. This may be demonstrated by weighing the total residue following evaporation of a melted ice sample and comparing it with the greater amount of material after drying an equivalent volume of water taken from just beneath the ice.

A snowfall, however, alters conditions drastically, as Curl and associates (1972) demonstrated in a study of the absorption of solar radiation in alpine snowfields. The intensity of light present at a given vertical distance within new-fallen, powdery snow is less than in some types of older and denser snow. In the former, 99% of the incident solar radiation may be lost within 18 cm, but transmittance is increased sixfold in older, crystalline snow. In both instances, however, high reflectance or *albedo* from the white surface is an important factor. A series of snowstorms at close intervals especially serve to darken the winter lake. The result in a rich shallow lake may be drastic oxygen depletion and a faunal catastrophe (see Chapter 12).

Opaque Layers and the Horizontal Transmission of Light

Ordinarily the vertical transmittance of light can be plotted on a semilog graph (see Figs. 9-1 to 9-3) with little or no obvious inflection in the resultant line, especially beneath the upper meter where there is greater relative absorption of the fractions of total light. This is explained by the rapid loss of long rays, with perhaps 50% of the total illumination disappearing in the upper 100 cm. Below this there is usually relatively little change in the percentage absorbed from meter to meter. Lake waters, however, do not fulfill the requirements of Bouguer's and Beer's laws in that they are not homogenous and do not present a uniformly attenuating medium. In some instances this is obvious when vertical light path is measured; a denser layer of particles may lie below the surface and absorb light much more effectively than overlying strata (Fig. 9-5).

With an instrument now called the transmissometer the vertical stratification of light-reducing entities can be shown. In the 1930s the physicist L. V. Whitney (1938) made some underwater measurements in Wisconsin lakes using a device that was the forerunner of this modern instrument. Briefly, it consists of a light source projecting a horizontal beam to a photocell receiver. Usually a distance of 1 m separates the two, but in waters of less clarity the distance can be lessened. A battery power source and a microammeter, receiving impulses from the photocell, are both in the boat. Transmittance just above or below the lake surface, or sometimes through a neutral filter, can be considered 100%; and as the instrument is lowered stage by stage, readings from the microammeter are recorded on this basis. Some workers calibrate the microammeter exponentially and record the results as optical density, the initial reading being zero. Plotting OD data yields a more meaningful curve because layers where marked light-extinguishing factors occur are shown as positive peaks (Fig. 9-6).

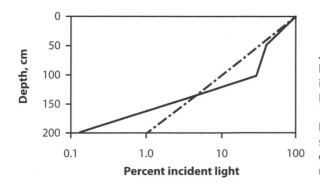

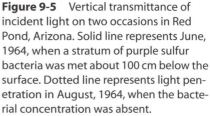

Figure 9-5 Vertical transmittance of incident light on two occasions in Red Pond, Arizona. Solid line represents June, 1964, when a stratum of purple sulfur bacteria was met about 100 cm below the surface. Dotted line represents light penetration in August, 1964, when the bacterial concentration was absent.

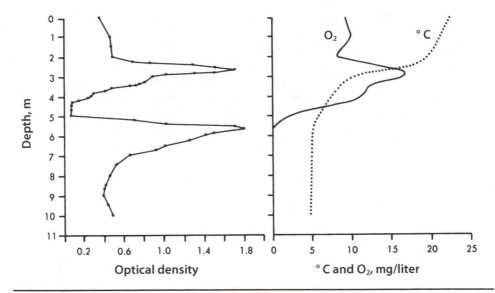

Figure 9-6 Data from Arco Lake, Minnesota, June 27, 1970: optical density from use of transmissometer at left; temperature and oxygen profiles at right. The peaks in OD were caused by a concentration of *Oscillatoria* at the upper level and a dense assemblage of sulfur bacteria below.

The transmissometer often reveals that the subsurface environment is indeed far from uniform; somewhat opaque layers may lie between layers of marked clarity. The phycologist Allan J. Brook and his students were among the first to study algal communities with a transmissometer, calling it a turbidimeter; their results were published in a series of articles (Baker and Brook, 1971; Brook, Baker, and Klemer, 1971). Samples taken where peaks of subsurface OD occurred contained larger concentrations of phytoplankton, especially when from the lower metalimnion. Cyanobacteria were responsible for many peaks, but eukaryotic algae and sulfur bacteria accounted for others.

Color

Light coming from the lake surface yields an *apparent color* that is the result of many factors. Of these, the materials in solution and the particulate matter, both living and nonliving, are most important. The particulate segment can be termed sestonic. Reflections of subsurface objects in shallow ponds may also impart apparent color—light-colored sandy bottoms and dense, dark vegetation, for example.

Some reddish ponds are the result of suspended inorganic particles eroded from the surrounding soil. Organisms account for other red waters. A shallow saline pool in Arizona (Cole et al., 1967) is known as Red Pond, presumably because of the purple sulfur bacteria not far below the surface (see Fig. 9-5); at times schools of the brine shrimp *Artemia* are so thick that they intensify the reddish hue of the pond.

Oscillatoria rubescens, the cyanobacterium that heralds the onset of extreme eutrophication, sometimes gives a lake a reddish tint. "Blood lakes," although ordinarily small ponds, owe their name to surface scums of the flagellate *Euglena sanguinea*. Similarly, some ponds are bright green because of a superficial layer of some other species of *Euglena*. Phytoplankton populations consisting of diatoms or of the common dino-flagellate *Ceratium* give a dark yellow tint, and heavy blooms of cyanobacteria impart a green color. An exception to the last statement, along with the case of *Oscillatoria rubescens*, was reported from Uganda by Kilham and Melack (1972). Lake Mahega is orange colored because of the cyanobacterium *Synechococcus*, which blooms about a meter below the surface. *Synechococcus is* a unicellular type that is remarkable at times for its blue color. The unusual African strain is reddish orange!

Filtration removes the solids contributing to apparent color, leaving the so-called *true color* of the water. This ranges from clear blue to dark brown, or even to a blackish hue in hypolimnic waters containing ferrous sulfide.

The bluest waters are transparent and ultraoligotrophic, low in both dissolved humic material and seston. The caldera Crater Lake, Oregon, serves as an example. Much of the color of such deep blue water comes from molecular scattering; light waves with the highest frequencies, especially, are scattered upward to view.

Many people who have rafted down the Colorado River in the gorge of Arizona's Grand Canyon have seen the lovely blue waters of the Little Colorado River entering the main stream. About 20 km above the confluence, water from Blue Spring pours into the Little Colorado. The spring water, high in carbonic acid, carries dissolved ancient limestone, but when aerated, CO_2 is lost and tiny particles of calcium carbonate appear (see Chapters 13 and 14). These scatter the solar light rays that enter the water and the short, high-frequency waves come to view. Like the sky above, the Little Colorado River below Blue Spring appears bright blue although no pigments are involved. This phenomenon is not obvious when the river is flooded; then you can expect only unattractive brown turbulence.

Waters with dissolved organic materials—humic substances leached in from soil, peat, or lake sediments—represent a cline of concentrations and color from green oligotrophic lakes, through the yellow of eutrophic waters, to brown, tea-colored bog pools.

Color standards by which natural waters can be compared and categorized have been developed. The Forel-Ule scale of colors is most often used. It consists of mixtures of different proportions of three solutions, ranging from the rare blue color I, a pure solution of cuprammonium sulfate, to the extreme brown color XXII. Brown XXII and other browns and yellows are made from mixtures of cobalt ammonium sulfate, potassium chromate, and the above copper compound. An array of green shades, serving as standards, is made from various proportions of the yellow potassium chromate and the blue cuprammonium sulfate.

Absorption of Light by Plant and Bacterial Pigments

Chlorophyll *a* is the master pigment in blue-green and eukaryote photosynthesis. In living cells it absorbs light in two peaks, one between 670 and 680 nm and the other at about 435 nm. It is clear that long rays, making up the former peak, would be present in abundance in shallow water and that the short waves would penetrate deeper, permitting photosynthesis to occur at many levels. Since plants contain assortments of accessory pigments, other light waves that travel vertically in the lake also play a part in photosynthesis.

Accessory pigment molecules, called photosensitizers, absorb energy quanta from light waves, become "excited" and energy-rich, and pass their excitation energy on sequentially.

This energy eventually reaches chlorophyll a, the photoreductant in cyanobacteria and eukaryotes, although some has been lost by fluorescence at each transfer. Pigments absorbing short rays pass some of their energy to those absorbing longer waves, but not vice versa. The yellow carotenoids absorb blue light and pass it to the reddish phycoerythrins that absorb the longer wavelengths composing green light; they in turn pass energy to the phycocyanins, absorbing in the yellow-orange part of the spectrum. Chlorophyll a, found in all photosynthesizing plants and cyanobacteria, is the final energy recipient. In addition, the molecules of the photosensitizers, chlorophylls b, c, and d, have absorption peaks near but different from those of chlorophyll a. The result of all these organic pigments is that energy from light waves ranging from near 400 nm to at least 700 nm can be used in primary production by the higher plants. Because the photosensitizer chlorobium chlorophyll absorbs strongly near 755 nm and the bacterial photoreductant bacterio-chlorophyll a absorbs in the red bands of 800, 850, and 890 nm, the spectrum from below 400 to 900 nm is available as an energy source to phototrophic organisms.

The response of aquatic photosynthesizers to light is more sophisticated than previously thought. McKew et al. (2013) compared phytochemistry of a microalga grown under high- and low-light conditions, and found that both the mix of pigments and the special proteins of a light-harvesting "antenna" system showed adaptation. The algae achieved high photosynthetic yield under low light and protection from cell damage under intense light. A similar protective response to high light was shown by four cyanobacterial taxa using carotenoid pigments in a study by Schagerl and Müller (2006).

Responses of algae to constantly changing light conditions are both long and short term. Talarico and Maranzana (2000) listed responses from whole-body to molecular levels, most importantly phycobilisomes (proteins associated with internal membranes and excellent antenna), transferring energy to chlorophyll a of photosystem II. This is an excellent adaptation to the light spectra at depth in the water column. However, a dozen distinct responses were listed in the red algae alone.

Extremely rapid responses (i.e., less than a minute) to intense light are known from the diatoms. Lavaud (2007) describes switching between pigment types, formation and decoupling of pigment antennae, free-radical scavenging, and others. Overall, it is clear that algae have evolved a suite of effective responses to both high and low light intensity, and spectral quality as well.

Over evolutionary time, photosynthesizers have adopted niches of light use in the water. A recent study by Schwaderer and colleagues (2011) that compared growth of 56 species of freshwater phytoplankters from 527 lakes across the United States found that the organisms of different taxa (Kingdoms, Phyla) had distinctive light-use strategies. The growth-irradiance curve showed efficient growth under low light in cyanobacteria but an intolerance for intense illumination; Chlorophytes tolerated full sun the best of all six Phyla.

Light Penetration and Aquatic Plant Zonation

Relationships exist among Secchi disk transparencies, the euphotic zone, and the depths to which pondweeds and other macrophytes extend. Shading, which inhibits these plants, is the basis of a technique used to control them in fish-rearing ponds. Nutrients are added to increase the phytoplankton crop to such an extent that it acts as a light barrier.

The remarkable light penetration in Lake Tahoe led to the establishment of an unusually deep bed of macrophytes (Frantz and Cordone, 1967). Unfortunately, early signs of eutrophication appeared since Charles R. Goldman began intensive studies of the lake in 1968. After decades of decreasing transparency, the Secchi disk depth measurements have stabilized at an average of about 22 m, much less than the 30 m recorded in the late 1960s. Runoff from the much-disturbed Tahoe watershed and the arrival of nutrients have led to the reduced transparency.

Schindler and Comita (1972) reported some abrupt changes in Severson Lake, Minnesota, that further illustrate the interactions of light penetration and plant growth. During a decade of summer observations, Secchi disk visibility always had been from 0.5 to 1.0 m, and submersed macrophytes were confined to depths no greater than 1.0 m. In 1965, following the first winter during which the lake became anoxic beneath the ice, *Daphnia pulex* made its initial appearance. Members of the growing population of this cladoceran grazed the phytoplankton to such an extent that by June a Secchi disk could be seen 4.5 m below the surface and 5.5 m below in July. The rooted plants responded rapidly to the change in light environment and colonized the bottom to a depth of 3 m. Likewise, Stuckey and Moore (1995) found a dramatic increase in Secchi depth (over 3.5 times deeper: 0.8 m to over 3 m) due to a combination of pollution control and invasion by the exotic zebra mussel into Lake Erie. The increased light through the water column allowed 14 species to prosper and 9 others to reappear after a 25-year absence.

Table 9-1 shows various data on Secchi disk transparency (Z_{SD}) and the greatest depth inhabited by the associated macroscopic hydrophytes. Some of the most remarkable deepwater communities known are included.

Table 9-1   Relation between Secchi disk transparency and depths to which aquatic macrophytes grow in selected lakes

Lake	Secchi disk (m)	Deepest plant growth (m)
Crystal Lake, Minnesota	0.32–1.5	1.75
Sweeney Lake, Wisconsin	0.6–1.0	2.25
Lake Itasca, Minnesota	1.8	3.5
Montezuma Well, Arizona	3.1	7.5
Walden Pond, Massachusetts	6.0+	16
Long Lake, Minnesota	8	11
Weber Lake, Wisconsin	8	13.5
Lake Ontario, USA-Canada (1912–1914)	12 (est.)	46
Crystal Lake, Wisconsin	14	20
Waldo Lake, Oregon	28	127
Crater Lake, Oregon	38	120
Lake Tahoe, California-Nevada	33–41	136

Most macroscopic vegetation disappears before subsurface illumination has dwindled to 1% of the surface value, but other factors must be considered. Some green angiosperms function more efficiently in dim light than do other closely allied species, but for tracheophytes in general the deepest limit down the basin slope is surpassed by lower plants. The charophytes, bryophytes, and benthic algae extend to the greatest depths; no plant records greater than 8 m in Table 9-1 are from ferns or angiosperms. Hydrostatic pressure rather than light seems to be the limiting factor (Hutchinson, 1975). There is, however, a report of sparse stands of *Elodea* at 12 m and *Potamogeton robinsii* at 10 m below the surface of Lake George, New York, where the maximum Secchi disk transparency has been 13.5 m (Sheldon and Boylen, 1977). Earlier, Hutchinson (1975) cited an 11 m record in Lake Titicaca as the greatest depth achieved by an angiosperm.

Light and Aquatic Animals

Light is important to aquatic animals as well as to plants, and an extensive literature exists on these relationships. Animals can detect far lower intensities than those to which plants respond. Thus the larvae of *Chaoborus*, buried in the sediments well below the euphotic zone, rise to the ooze-water interface at sunset to reconnoiter the fading light. In its absence, or at some low threshold of light, a nocturnal migration to the epilimnion may occur, but too much light inhibits such a journey (LaRow, 1969).

An early paper by Smith and Baylor (1953) reported experiments with light and Cladocera. They showed that diminishing intensities provoked upward swimming, while brightening caused downward movements. Moreover, they showed responses to blue, yellow, and red wavelengths that have ecologic significance. For example, blue light evokes a downward migration (this response is blocked in cold water, such as that of the thermocline layer) until sunset dimming causes upward movement again. The reasons for some of these responses were reviewed by McNaught (1971), who pointed out that most cladocerans have visual pigments that absorb strongly at peaks of 430, 560, and 670 nm. In addition, they can respond to ultraviolet light because a fourth pigment absorbs maximally at 370 nm.

Photoperiodicity is a part of aquatic life cycles. Some less obvious examples are the initiation and cessation of diapause in *Daphnia* embryos (presented in a series of articles by Stross in 1971) and the termination or inducement of similar resting stages in freshwater cyclopoid copepods.

The reddish color of some microcrustaceans is owed to various carotenoids, which serve to protect the animals from photooxidation. B-carotene is the source of the colored carotenoids of which astaxanthin is the reddest. The bright red hue of some temporary-pond copepods, *Diaptomus stagnalis* and *D. sanguineus*, for example, may lessen the dangers of sunlight in shallow water. Similarly, the red pigmentation of *D. shoshone* in permanent but shallow lakes at high altitudes may protect it from photooxidation. Hairston (1979a, b) discussed the carotenoid pigmentation of two *Diaptomus* species in saline lakes in eastern Washington. He showed that the ability to synthesize carotenoids from dietary carotene can be selected against as well as for; a large species with intense coloration is an easy target for vertebrate planktivores, and this selective force may outweigh the benefits of being capable of efficiently using plant carotenes to produce the protective pigmentation. Both species were intensely red in late winter and early spring when reproduction occurred. The early naupliar stages did not feed but nevertheless were protected from the sunlight in surface waters where they swarmed. Their body carotenoids were received from their mothers via egg material.

A remarkable example of protection from photooxidation was reported from a Danish lake by Whiteside and Lindegaard (1982). The chydorid *Alonopsis elongata* reached its peak abundance at the lake margins in less than 1-m water depth, where no other cladocerans were collected. This dark-colored species does not shed its carapace at each moult. It appears that the multilayered shell and the coloration protect against ultraviolet radiation.

From the above it appears that *Chaoborus* may respond negatively to light because it lacks protective pigmentation. The equally transparent cladoceran *Leptodora* also lacks hemoglobin and carotenoids; it is near the lake surface only at night.

Neuston-filtering species, such as the cladoceran *Scapholeberis* and the ostracod *Notodromas*, creep along beneath the surface film with their pigmented ventral sides up (in contrast, their dorsal surfaces are pale). Similarly, marine organisms floating at or near the surface show an almost universal presence of blue carotenoproteins or dorsal white sheens. These might be adaptations for lessening the danger of solar radiation and especially ultraviolet rays.

10

Density, Layering, and Lake Mixing

Temperature Stratification, Lake Regions, and Water Density

As solar radiation passes downward from the surface of a lake, it disappears exponentially (see Figs. 9-1 to 9-3), and the heating wavelengths are usually absorbed very rapidly. At the end of the yearly heating period, one might expect the vertical temperature curve to resemble the light curve. This is not the case; in fact, a moderately deep lake sheltered from extreme winds would have a temperature profile quite different from the light curve (Fig. 10-1 on the next page). The difference is readily explained by the wind mixing the upper layers of water and distributing downward the heat that had been absorbed in these surface strata. The result is a curve such as that shown in Fig. 10-2, also on the next page. It portrays **direct stratification**, with dense cold water lying beneath lighter warm layers, and shows the lake divided into three regions. The upper warm region, mixed thoroughly by wind to a more-or-less uniform temperature, is the **epilimnion**. At the bottom lies a colder region of heavier water little affected by wind action and, therefore, traditionally considered stagnant. This is the **hypolimnion**. Separating the two lake regions of more constant temperature is an intermediate zone, where temperature drops rapidly with increasing depth. Various names have been applied to this middle zone. Birge (1897) called this zone the **thermocline**, an appropriate term signifying a temperature gradient. The name **metalimnion** (changing lake), a contribution from Brönsted and Wesenberg-Lund (1911) and effectively accepted by Hutchinson (1957), is often used and has merit because it is not as rigid a definition as thermocline. Birge's term limited the middle zone to a region where temperature drops at least $1°$ C with each 1 m increase in depth. Regardless of terminology, with a typical direct stratification there are, in a sense, three lakes in one: an upper lake, the epilimnion; a middle lake, the metalimnion (perhaps in the present sense more appropriately termed the **mesolimnion**); and a lower lake, the hypolimnion.

Hutchinson (1957) preserved Birge's word *thermocline*, but no longer to designate a broad region. Instead, he followed Brönsted and Wesenberg-Lund (1911), considering the thermocline to be an imaginary plane within the lake. It is located at a level intermediate between the two depths where the rate of temperature decrease is greatest. This planar thermocline lies within the metalimnion, or Birge's classic thermocline (Fig. 10-2).

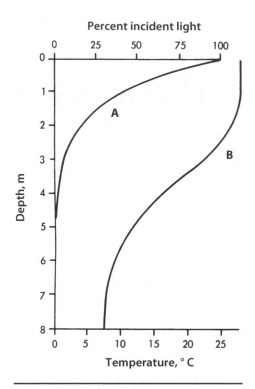

Figure 10-1 Comparison of the vertical day-time light curve, **A**, and the temperature curve, **B**. Tom Wallace Lake, Kentucky, June 25, 1952.

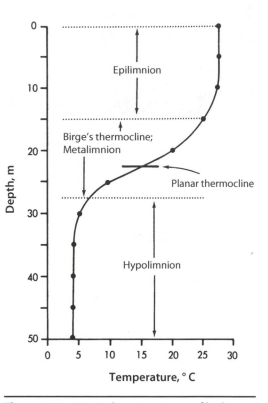

Figure 10-2 Vertical temperature profile showing direct stratification and the lake regions defined by it.

We imagine a plane that separates an irradiated, wind-stirred upper lake (where primary productivity prevails) from a darker, less turbulent lower lake. Organic material is synthesized in the trophogenic epilimnion; much of it sinks to the tropholytic hypolimnion to be mineralized by bacteria. An imaginary plane, the area of which can be calculated and through which organic particles theoretically drop, is useful in techniques of estimating lake productivity.

Because the word thermocline implies a gradient, perhaps it is not the best choice for such a plane, but there are also unusual situations where the modern term metalimnion is awkward. For example, a body of water much protected from air currents would have a summer temperature curve somewhat similar to the exponential light-absorption profile. In certain lakes that have come under study this is the case: a pronounced summertime temperature gradient starts at the surface; the epilimnion is poorly defined, shallow, or practically absent; and the metalimnion is the uppermost stratum. Here the word thermocline is more useful to describe the zone of decreasing temperature starting at the surface.

In addition to their effects on animal and plant metabolism, temperature differences are significant because of the density contrasts that accompany them (Table 10-1). In general, we can say that under normal pressure, water is heaviest at 4° C when

1 mL has a mass of 1 g, the standard unit of density. Because of water's unusual molecular constitution, it becomes lighter as it cools below 4° C, forming floating rather than sinking ice. Also, as temperature rises above 4° C, density once more decreases. Direct stratification (Figs. 10-1, *B,* to 10-3) shows that warm water is

Table 10-1 Some temperature-density relationships of pure water at one atmosphere pressure and with density 1.0 at 4° C

Temperature (°C)	Density of second temperature (g/cm^3 × 10^{-7})	Density change (g/cm^3 × 10^{-7})	Ratio*
0–1	9999267	+588	7.25
1–2	9999679	+412	5.08
2–3	9999922	+243	3.00
3–4	10000000	+78	0.96
4–5	9999919	−81	1.00
5–6	9999681	−238	2.93
6–7	9999295	−388	4.79
7–8	9998762	−533	6.58
8–9	9998088	−674	8.32
9–10	9997277	−811	10.01
10–11	9996328	−949	11.71
11–12	9995247	−1081	13.34
12–13	9994040	−1207	14.90
13–14	9992712	−1328	16.39
14–15	9991265	−1447	17.86
15–16	9989701	−1564	19.30
16–17	9988022	−1679	20.73
17–18	9986232	−1790	22.00
18–19	9984331	−1901	23.46
19–20	9982323	−2008	24.79
20–21	9980210	−2113	26.08
21–22	9977993	−2217	27.37
22–23	9975674	−2319	28.62
23–24	9973256	−2418	29.85
24–25	9970739	−2517	31.07
25–26	9968128	−2611	32.23
26–27	9965421	−2707	33.41
27–28	9962623	−2798	34.54
28–29	9959735	−2888	35.65
29–30	9956756	−2979	36.77

*The ratio above is (ignoring signs):

density change between adjacent temperatures
      density change between 4 and 5 C

Thus, between 24° and 25° C the density change is 2517 × 10^{-7}; between 4° and 5° C it is 81 × 10^{-7}. The ratio is 2517/81 = 31.07. This means that the change between 24° and 25° C is more than 31 times the density change between 4° and 5° C. By summation, the density change between 12° and 16° C is 5546 × 10^{-7}.

buoyed up by cold, but a natural **inverse stratification** can be found beneath an ice cover. The ice at 0° C floats on water that warms progressively with depth to 4° somewhere above the bottom.

The adjectives **isothermal** or **homoiothermal** describe a body of water of uniform temperature. The temperature curve would be a straight vertical line, and theoretically the density would be the same throughout.

Factors Modifying Density of Water and Temperature Gradients

Temperature

Our knowledge of thermal phenomena in lakes and ponds has been derived mostly from studies of relatively dilute waters. In density computations the water is usually considered pure, and temperature is the major factor affecting density. Table 10-1 shows that under these conditions the density change between 24° and 25° C is more than 31 times the change between 4° and 5° C. This point is emphasized further by Fig. 10-3, in which the density gradient caused by pronounced temperature stratification in a cold-water lake is matched by the gentle inflection in a warmer lake.

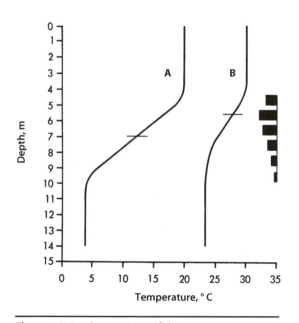

Figure 10-3 Comparison of the temperature profiles of a cold lake, **A**, and a warmer lake, **B**, having identical density gradients caused by temperature. Density changes shown by histograms at right. Horizontal lines in curves represent planar thermoclines based on maximum temperature change.

Pressure

Pressure lowers the temperature of maximum density so that in a very deep lake the lowest stratum may be composed of water substantially below 4.0° C and even below 3.98° C, a more precise estimate of the densest pure water at one atmosphere. At the surface (at sea level) an atmosphere approximates a column of water 1,000 cm × 1 cm², 1 kg/cm². Actually, it is 1.032 kg, and it appears that every 10 m increase in depth equals roughly another atmosphere of pressure. Conversely, decreases in elevation are equivalent to the pressures of shorter water columns. In Lake Tahoe, lying at an elevation of 1,890 m, the pressure at a depth of 10 m equals that found at roughly 8 m in a lake at sea level. Vascular plants might be expected to extend another 2 m below the surface in Lake Tahoe than they would in a transparent lake along the sea coast.

On average the temperature of maximum density is lowered about 0.1° C per 100 m. More specifically, the temperature of maximum density below 500 m of water is 3.39° C, and below 1,000 m it is 2.91° C.

Solutes

Rawson and Moore (1944) studied some saline lakes in Saskatchewan and reported summer temperature profiles that on cursory examination appear normal: epilimnion, thermocline, and hypolimnion are delineated clearly. But unlike the typical situation, the temperature of greatest density is depressed by high salinity, and hypolimnion temperatures fall below 0° C. For example, in the deeps of Manito Lake, the freezing point has been lowered to –1.1° C, and the maximum density is at –0.3° C. These apparent summertime anomalies are simply explained by the winter ice cap that melts at an unusually low temperature to expose a cold, salty body of water. Stirred by vernal winds, the lake becomes isothermal below the normal freezing point of water and does not gain much heat throughout before the surface strata warm enough to shape a stratification that traps cold water in lower levels.

Materials in solution, if abundant enough, change density dramatically, enough to cause stratification. Specific gravities are increased by solutes. In lakes with marked salinity with layering, the usual temperature-induced density relations can be reversed: warm water may persist below colder strata if it contains enough extra salt. The outcome is an inverse stratification far from the 0° to 4° C range considered normal. Fig. 10-4 illustrates such an improbable condition in a shallow brine pool in Mexico; here an inverse temperature curve is explained by salt layering. A shallow pool of salty water concentrated by evaporation became overlain by a dilute layer originating from rain and runoff. The high temperature in the deeper part is readily explained by direct solar radiation penetrating the upper stratum. The saline bottom water accumulated heat, while the dilute upper layer reduced outgoing radiation and prevented evaporation from it. The contrasting densities prohibited anything but very slow mixing of the two layers.

Abrupt temperature gradients often accompany stratified saline waters, although decreases amounting to 0.1° C per centimeter are rare in the metalimnia of freshwater lakes. Small inland salt ponds often display gradients amounting to perhaps 0.5° C per centimeter but differ from the typical metalimnion because the temperature rises with depth.

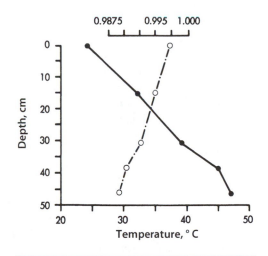

Figure 10-4 Inverse temperature stratification in a saline desert pool. Solid line indicates temperature, °C; broken line shows density of distilled water at the same temperatures. Adapted from Cole and Minckley (1968).

In addition to salt, lakes may contain other dissolved solutes causing stratification. Von Rohden and colleagues (2010) studied two small lakes (pits) in eastern Germany that were formed by coal mining. Fed by inputs of groundwater high in iron, the lower layer of the lake was more dense, leading to a persistent stratification across seasons. The researchers were able to measure slow internal circulation within the hypolimnion, with dissolved iron forming an oxidized precipitate falling to a low-oxygen benthos and redissolving into solution. A similar situation occurs in Lake La Cruz (Spain), with dissolved bicarbonate and calcium from the karst rock of the basin adding to a stable hypolimnion based on chemical stratification (Rodrigo et al., 2001).

Suspended Particles

Materials need not be dissolved to increase water density; suspended particles also are effective. A common example is that of a turbid, muddy stream entering a lake or joining a clearer stream. The muddy flow maintains a degree of integrity and can be readily seen. Thus, a **density current** due to particles in temporary suspension can enter a lake, cut through colder water, and continue its downward path until reaching a level where its density is matched or surpassed. Unusual temperature stratification often accompanies such an event but is not lasting.

Anderson and Pritchard (1951) discussed the annual cycle of the turbid Colorado River as it flowed through the Grand Canyon to enter Lake Mead. During summer a down-lake flow along the bottom was observed and attributed to sediment-laden water. As the water cooled in autumn and winter, the flow was even more conspicuous, its density being augmented by lower temperatures. Salinity also played a role, and the spring runoff water flowing down the Colorado revealed its dilute nature by spreading out over the lake rather than sinking. Since the 1951 report, there have been alterations that may have changed this annual pattern. Because of the impoundment of the stream to form Lake Powell (which serves as a sediment trap), the Colorado River no longer flows through the gorge to Lake Mead as a remarkably turbid stream, except when swollen, silt-laden tributaries (e.g., the Little Colorado River) enter.

Another example of a density current due to turbidity comes from Lake Mendota. A bathymetric map constructed from soundings made many years ago by Birge and Juday had some bizarre contour configurations. A later survey using sonar equipment for establishing depths showed that the soundings of the early workers had revealed a real feature, a subsurface gully. This had been formed by density currents from an inflowing stream eroding the bottom slope. One summer, following an unusual spell of heavy rainfall, the swollen stream, densely laden with silt and mud, flowed down the gully to unusual depths, introducing some oxygenated water to the hypolimnion (Bryson and Suomi, 1951).

The classic example of a density current entering a lake is seen in Lake Geneva (Forel's La Léman) where the entering Rhone has cut an elongate gully in the bottom of the lake for more than 6 km.

Particulate matter, like solutes, can bring about sharp temperature profiles in addition to imparting density. The gradient usually is the reverse, however, from what has been described from briny stratified pools. Very abrupt thermoclines have been reported from small turbid ponds in hot sunny areas. The particles in the upper layers absorb heat rays so effectively that pronounced vertical contrasts occur.

Stability of Stratification

Direct stratification persists through much of the summer because the different densities caused by either temperature or dissolved substances can be mixed to homogeneity only by a tremendous amount of work. The summer wind is not powerful enough to do this, and we say that the lake exhibits **stability of stratification**.

The concept of stability (S) is owed to Schmidt (1915, 1928). Very simply, S is the amount of work that would be required to mix an entire lake to uniform density without adding or subtracting heat in the process. If density is uniform from top to bottom, stability is zero; no work must be performed to promote homogeneity. Somewhere in the lake at any given time there is a depth below which the water mass equals the mass above; it is a depth where an imaginary plane would separate the mass of the lake water into 50% portions. This has been termed the center of gravity (z_g), although this is not technically accurate, since the exact mass balance in the lake is not determined. In any case, stability begins to increase as z_g lowers (i.e., is found deeper in the lake), and this is the result of vertical contrasts in density. Warming of the upper layers, introduction of dissolved material to the lower layers, or the accumulation of suspended particles in the lower strata are common phenomena that make for density stratification and increased stability.

Theoretically, S equals the amount of work necessary to lift the entire lake from the actual z_g up to the level of z_g that would exist at uniform density (Figs. 6-5 and 6-6). The work, implying a force and distance through which it moves a mass, is reduced to the absolute unit of work and reported per unit area of lake surface: J/m^2 (or kJ/m^2).

Idso (1973) discussed stability based on Schmidt's original formula, pointing out that, although the formula gives correct integrated values for the lake as a whole, there are inherent errors in it that lead to mistakes in some subsequent calculations. Idso proposes the following formula:

$$S = \frac{1}{A_0} \int_{z_0}^{z_m} \left(\rho_z - \overline{\rho} \right) \left(A_z \right) \left(z - z_{\overline{\rho}} \right) dz$$

A_0 = A, the surface area
A_z = the area at some depth z (considered positive)
$\overline{\rho}$ = the final or mean density that would result from stirring the lake to uniformity
ρ_z = the depth where the final or mean density ($\overline{\rho}$) exists prior to mixing
z_m = maximum depth
z_0 = surface, or zero depth

Stability measures were determined by O'Reilly and colleagues (2003) to understand the effects of climate change on the ecosystem in Lake Tanganyika, and by Jankowski and others (2006) to describe the effects of a heat wave on dynamics in two Swiss lakes (Zurich and Greifensee).

Figure 10-5 on the following page shows how stability changes due to weather conditions. Holzner and associates (2009) found that the Alpine Lake, Lugano, had a stable stratification for over 40 years until two cold, windy winters (2004–2005 and 2005–2006) destabilized the layering which already had a small Schmidt value. As S fell to nearly zero (and then, completely to zero), winds were sufficient to circulate water top-to-bottom. Isotopic analyses and tracer movement provided evidence that water from the hypolimnion could be found throughout the lake.

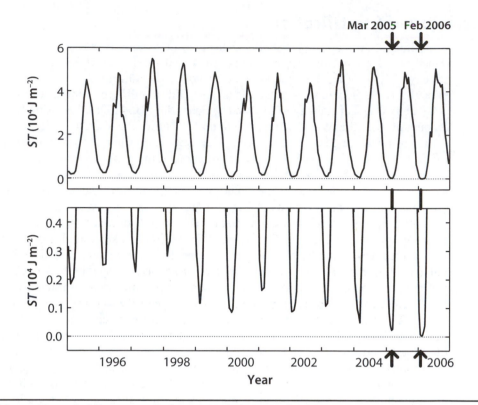

Figure 10-5   Stability curve of Lake Lugano, Italy/Switzerland, between 1995 and 2006. In 2005 and 2006, the stability value (*ST*) was so small that the lake experienced complete overturn, which had not happened in 40 years. From Holzner et al. (2009). Reprinted with permission.

A related measure of "local stratification" (Denman and Gargett, 1983) uses the difference in density between the epi- and hypolimnion to calculate the volume of water from one layer to the other. This *buoyancy frequency* (*N*) is discussed further in the next chapter.

Classification of Annual Circulation Patterns in Lakes

The times when circulation occurs are, of course, periods of mixing. The noun *mixis* and its adjective *mictic* are used in compound words to categorize lakes according to annual circulation patterns. Such a classification is superimposed on older schemes; as a result some confusion exists. Moreover, it is not easy to classify many lakes because their behavior varies from year to year.

Amixis

Some bodies of water never circulate. These are the *amictic* lakes, permanently ice covered and immune throughout to the stirring effects of wind. Hutchinson and Löffler (1956) diagrammed a theoretical scheme showing that amixis is improbable at sea level until latitudes of slightly more than 80° are reached, and at the equator below 6,000 m elevation.

There is very little land above 80° in the Northern Hemisphere, but still there may be as many as 10 amictic lakes in northern Greenland and adjacent Ellesmere Island. One of these, Greenland's Anguissaq Lake, is 590 m above mean sea level at a latitude of slightly more than 77° N (Barnes, 1960). The ice thickness varies from 1.8 to 4 m, capping water that attains a maximum depth of 187 m. Temperatures are very low and constant (0.1° to 0.7° C from surface to bottom), implying a lake quite isolated from outside influences.

Lake Hazen (81° 50' N) on Ellesmere Island is a large lake at least 280 m deep. Although usually amictic, during some summers it is free of ice. In 1957, for example, it was open for 54 days (McLaren, 1964).

In Antarctica, perennially frozen lakes have been known since the time of the 1907 to 1909 British Antarctic Expedition. At least five of them were named on Ross Island's Cape Royds, almost 79° S lat, where the explorers wintered. A few protozoans, rotifers, tardigrades, and nematodes were observed, but no intensive work was done on Antarctic amictic lakes until much later, when some of the Ross Island coastal ponds were revisited as part of a general limnologic reconnaissance of the McMurdo Sound region (Armitage and House, 1962). Goldman's review of the history of Antarctic limnology (1970) has revealed that two lakes, Bonney and especially Vanda, have received most attention. A little more than 77° S lat, their basins lie in a strange dry valley system in Victoria Land. More recently, Laybourn-Parry and Pearce (2007) described the ecology and evolution of over a dozen lakes across the continent.

Layering beneath the ice covers in these lakes is by no means simple. Saline stratification and unusual temperature profiles imply increments of heat from the bottom, from an inflow of saline thermal waters, or from the sun. Solar radiation is theorized to account for the temperature profiles of Lake Bonney; but in Lake Vanda, where temperatures attain 25.5° C more than 67 m below the ice, geothermal effects are important (Angino et al., 1965).

Other amictic lakes are found at high elevations and relatively low latitudes. We owe much of our knowledge of these lakes to the Austrian limnologist Heinz Löffler (1964). His summary of tropical high-mountain limnology lists amictic lakes in the Peruvian Andes and at least one on Mount Kenya, Africa. The last, Curling Pond, is interesting because it is Africa's highest known lake (about 4,800 m elevation) and is located approximately on the equator. It lies below the theoretical 6,000 m limit proposed by Hutchinson and Löffler (1956) for equatorial amixis, and there is evidence that occasionally its ice cover partially melts.

It is obvious that permanent amixis can exist only in those rare cold climates where ice produced by unusual duration of subzero temperatures is not lost at another season. *Ablation*, the loss of ice through evaporation and melting, is largely a function of air temperature. Cold must prevail through the year to promote amixis.

Holomixis

Holomixis is a typical lake phenomenon whereby wind-driven circulation mixes the entire lake. Overturn periods involve the total water mass.

Oligomictic lakes. Originally Hutchinson and Löffler (1956) proposed the term **oligomixis** to describe conditions in certain equatorial lakes at low elevations. These lakes have warm water at all depths and are subject to very little seasonal change. Rather effective stability is brought about by the warming of surface strata, thereby creating great density differences between them and the colder layers below (see Fig. 10-3). Occasional cooling of surface strata destroys density differences, reducing stability as a prelude to mixing. Circulation periods in these warm oligomictic lakes are unusual, irregular, and short in duration.

A new term, **atelomixis**, was coined by Lewis (1973) following a 14-month study of the thermal regimen in tropical Lake Lanao (Philippines). At irregular intervals, quiet warming periods caused the formation of high-lying secondary thermoclines in what had been the epilimnion of this 112-m deep lake. Thermoclines above 20 m were easily formed and destroyed, a light wind bringing about their destruction. Thermoclines forming from 20 to 30 m were destroyed by brief squalls, but only rarely were storms severe enough to obliterate a more persistent thermocline at a depth of 40 to 60 m. When either one or two secondary thermoclines lasted for a few weeks, the upper region of the lake became chemically stratified also. When mixing occurred, layers of water with different chemical properties were homogenized without disturbing the hypolimnion stratum. This irregular, incomplete mixing may be typical of tropical lakes and could occur in lakes of temperate regions occasionally. Lewis's term, *atelomixis*, is based on the Greek word *ateles*, meaning imperfect.

The geography of oligomictic lakes must be broadened when the behavior of some of the large subalpine European lakes are taken into account. For example, Maggiore, a large Italian lake formed by glacial corrasion, its surface 193 m above sea level and at 46° N lat, circulates for a few weeks during February and March. This occurs, however, only once every 5 to 7 years during unusually cold and windy winters (Vollenweider, 1964). Nearby Lake Como (Italy) and, perhaps, Lake Geneva (France/Switzerland) behave similarly. Also, Hadzisce (1966), who studied Lake Ohrid for 24 years, found that this ancient Balkan lake circulates completely about once every 7 years, depending on climatic conditions.

Lake Tahoe (California/Nevada), lying 1,890 m above mean sea level at 39° 09', can probably be categorized with the large European lakes as oligomictic. Because of its tremendous depth (501 m) and volume (156 km^3) it only mixes every 3–4 years, although it is predicted to mix less frequently due to climate change (Sahoo et al., 2012).

This second type of oligomictic lake is characterized by comparatively cold water. Lake Tahoe, for example, circulates at about 4.5° C. It and the other members of the category are deep, voluminous bodies of water requiring tremendous work for holomixis; surface cooling and the resultant convection currents bring about incomplete circulation in the absence of adequate wind. The climatic environment of these lakes is such that they experience seasonal contrasts. Thus, their periods of circulation are fixed but occur only with unusually low temperatures and high winds, unlike the equatorial or warm oligomictic lakes that experience a diel climate and circulate at irregular times throughout the year.

Monomictic lakes. Lakes that experience one regular period of circulation occurring sometime within the year undergo a process called **monomixis**. There are two main types, separated here on the basis of whether the overturn occurs during summer or in the cold season. Monomictic lakes usually exist in climates where seasonal changes are pronounced.

COLD MONOMIXIS. Cold monomictic lakes are frozen over during the winter months when the water is shielded from the mixing effect of wind. They are characterized by stagnation only during winter, when they are temporarily amictic. Ideally, these lakes exhibit inverse temperature stratification, with water ranging from 0° C below the ice to about 4° C in the deeps. With the melting of ice in spring, stability is extremely low and circulation commences. Throughout the summer they are essentially isothermal, warming but never stratifying to any extent.

Originally, cold monomixis applied only to those lakes where surface waters do not rise above 4° C during the warm season. These were the so-called **polar lakes** in a classification devised by Forel. Though not common, there are some of these that exist at high elevations in lower latitudes far from the poles. Berg (1963) discussed the problem in detail, de-emphasizing the 4° C threshold and broadening the definition of cold monomixis. He cited the case of windswept Oneida Lake, New York, which is at most poorly stratified during the ice-free season, sometimes being isothermal as high as 24° C.

Even the small, shallow Imikpuk at Point Barrow (71° 17′ N lat) on the Arctic coastal plain cannot be classified as a polar lake. It is open only about 60 days of the year, but during that short circulation period it warms 4 to 8 degrees above 4° C, depending on summer meteorologic conditions (Brewer, 1958). Perhaps as much as 75% of the incoming vernal solar radiation serves to melt the ice blanket; after that, only a small amount is needed to raise the water temperature above 4° C in such a shallow lake.

The cold monomixis of deep Pingualuit (New Quebec Crater) Lake, although farther south (61° 17′), fits the definition of a polar lake. The thermal requirement to warm its great water mass above 4° C during the summer overturn period is much too great. Martin (1955) recorded summer surface temperatures of only 3° and 4° C in this lake.

Lake Hazen was free of ice during 54 days in August and September of 1957, and its open water did not exceed 3° C (McLaren, 1964). That year it was a cold monomictic lake of the polar type rather than amictic.

Some shallow volcanic lakes on Anderson Mesa near Flagstaff, Arizona, freeze over during winter months and circulate much of the summer, warming to at least 22° C. Successive calm days result in warm upper strata and far more stability than would be found in Imikpuk, for example; but strong winds are the rule, and they soon spring up to destroy the transitory stratification. With some reservations, these lakes are classified as cold monomictic.

Also, according to Eddy (1963), Minnesota lakes such as Mille Lacs, Lake Winnibigoshish, and Leech Lake (all of which range from 15 to 32 km across) show little summertime temperature stratification. They are relatively shallow and exposed to the wind. Their winter stagnation, due to ice cover, and their typical mixing throughout the remainder of the year leave little choice but to designate them cold monomictic despite their departure from the 4-degree threshold of older classifications.

WARM MONOMIXIS. The term to describe monomictic lakes that lack an ice cover and that circulate in winter is **warm monomictic**. Direct stratification, arising during early summer, puts an end to complete circulation until the thermocline is destroyed sometime in autumn.

There are borderline regions where the classification breaks down at times. Cayuga Lake, New York, stratifies during the summer months but remains open and isothermal during the average winter when an effective wind plays on its surface. Tremendous quantities of heat must be lost to bring the surface near freezing, and it is only during exceptionally cold winters, perhaps marked by periods of unusual calm, that a sheet of ice forms to stop circulation.

Warm monomixis is a usual occurrence south of about 40° latitude in North America where lakes are deep enough and sheltered enough to stratify during summer. The winters are mild, and the lakes circulate at comparatively high temperatures. Elephant Butte Reservoir, an impoundment on the Rio Grande in New Mexico, circulates at about 8° C. Farther south and at lower altitudes, impoundments on the Salt River in central Arizona cool to about 11° or 12° C during the winter mixing period. Central American lakes such as Atitlán (Guatemala) and Gúija (El Salvador) dip to about 20° C during their winter circulation periods. Obviously, the summer hypolimnia of these lakes can be no colder than the lowest temperature of the winter circulation and are far above 4° C (Deevey, 1957).

Other warm monomictic lakes, however, are not so warm. Cayuga Lake circulates at temperatures below 4° C. Henson and coworkers (1961) reported isothermy at 2.8° C in March and cited earlier records of freely circulating water at only 1.3 ° C. The nearby Seneca Lake behaves similarly. Also, three of the Laurentian Great Lakes—Michigan, Superior, and Huron—are largely open during winter months although icebound marginally, and they exhibit temperature stratification during summer. Their winter water temperatures commonly fall below 4° C.

The huge, relatively shallow Lake Victoria, although lying across the equator, behaves like a monomictic lake of temperate regions. One would expect it to exhibit almost continuous circulation, but through an unusual mechanism it stratifies during the rainy season when its inflows introduce cool, dense water to the lake. Its direct stratification is brought about by cooling of bottom layers rather than by the familiar method of warming the surface. This phenomenon may be widespread in equatorial lakes.

Dimictic lakes. Dimictic lakes have two mixing periods, the vernal and autumnal overturns, each year. The idealized temperate-zone dimictic lake with an annual temperature cycle is commonly portrayed as a typical lake. However, there are important limnologic regions where dimictic lakes are not found. Australian limnologists, for example, have no such bodies of water to study (Bayly and Williams, 1973).

The typical dimictic lake stratifies directly during the warm months; if it were shallower or exposed to forceful wind or both, it might exhibit cold monomixis. The onset of cold weather starts the cooling of surface water that eventually destroys the stratification and initiates complete circulation. During the fall overturn, chilling of the entire water mass occurs until, ideally, the whole lake is uniform at about 4° C. On some cold, calm night the surface gives off enough heat to the atmosphere to cool to the freezing point, and a film of ice is formed. This inaugurates the winter stagnation period when wind-induced circulation is impossible. With the warming days of spring the ice melts, and a cold lake, with stability near zero, lies exposed to wind action. The result is the vernal overturn, a circulation period that continues until ended by the direct temperature stratification of summer. This is the dimictic cycle (Figs. 10-6 to 10-11).

During the autumnal circulation period, loss of heat at the air-water interface creates dense water that streams downward, aiding the wind in the mixing process. After the entire water mass is 4° C, however, further cooling creates lighter water at the surface, and convection currents no longer move downward. In most instances, freezing soon follows. The stability of this stratification is so slight that wind can drive the cold stratum downward before freezing occurs. The result is a lake chilled well below 4° C before an ice cover forms. Berg (1963) discussed this and pointed out that in some large, exposed dimictic lakes there is prolonged autumnal cooling resulting in isothermy at 1° or 2° C prior to winter stagnation.

Just as wind prolongs autumnal circulation, so it lengthens the vernal mixing period in warm monomictic and dimictic lakes. In years with stormy spring weather the circulation continues farther into the warming period; the whole water mass gains heat, and its temperature rises until it approaches ranges where slight differences in temperature are marked by significantly different densities. Then a calm day or two may initiate layering and summer stagnation. In years when calm weather prevails during spring, the heat gained in upper strata is not driven downward, and the density differences between surface and lower waters may end circulation while bottom waters are still cold from conditions of the previous winter. Hypolimnion temperatures during early summer, therefore, reveal information about vernal weather conditions.

An excellent example of the vernal wind's effect on summer hypolimnion temperatures is seen by comparing the small Tom Wallace Lake, Kentucky, with Lake Itasca,

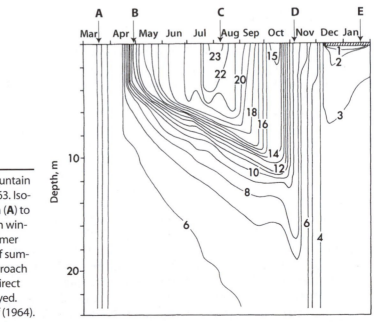

Figure 10-6 Dimixis in Mountain Lake, Virginia, during 1962–63. Isotherms from spring overturn (**A**) to inverse stratification beneath winter ice cover (**E**). **B**, Early summer stratification; **C**, the height of summer stratification; **D**, the approach of the autumn overturn as direct stratification is being destroyed. Adapted from Roth and Neff (1964).

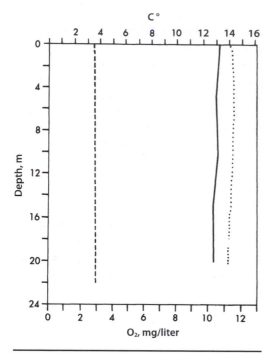

Figure 10-7 Spring overturn, Mountain Lake, Virginia (**A** in Fig. 10-6). Temperature, dashed line; oxygen, solid line; 100% saturation oxygen, dotted line.

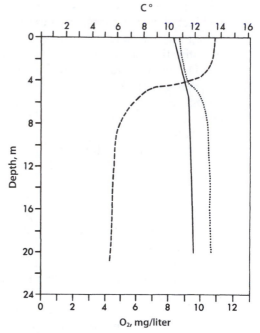

Figure 10-8 Early summer stratification, Mountain Lake, Virginia (**B** in Fig. 10-6). Temperature, dashed line; oxygen, solid line; 100% oxygen saturation, dotted line.

Minnesota, more than 9° farther north. Tom Wallace is only 8.7 m deep and during mild winters does not freeze. Lake Itasca, with a maximum depth of 14 m, lies in a northern region where prolonged ice cover is the rule. The Kentucky lake has a small surface, a maximum length of 0.33 km, and is protected by steep, wooded hills that rise abruptly from its shores. Itasca, by contrast, is 4 km long and exposed to the wind. The differences in topography, morphology, and wind factors result in marked contrast in the temperatures of these two lakes. The June hypolimnion in the southern lake is no more than 7° or 8° C, whereas Lake Itasca's hypolimnion is 11° to 15° C (Fig. 10-12 on p. 221).

Polymictic lakes. Lakes that have many mixing periods or continuous circulation throughout the year undergo **polymixis**. Polymictic lakes are influenced more by the changing diel fluctuations in temperature than by seasonal changes. The original examples described by Hutchinson and Löffler (1956) are close to the equator. In some small Andean ponds at fairly high elevations the upper waters gain heat to stratify diurnally. Cold nights cool the upper layers to such an extent that down-welling convection currents destroy the stratification, and nocturnal circulation takes place until terminated by the following day's solar input.

A similar diel cycle was described by Foster (1973) in quite a different region. In small Arizona desert ponds averaging about 1 m in depth, intense solar radiation in

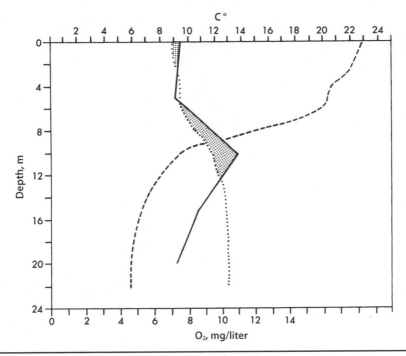

Figure 10-9 Late summer stratification, Mountain Lake, Virginia (**C** in Fig. 10-6). Temperature, dashed line; oxygen, solid line; 100% oxygen saturation, dotted line; shaded area represents oversaturation of oxygen.

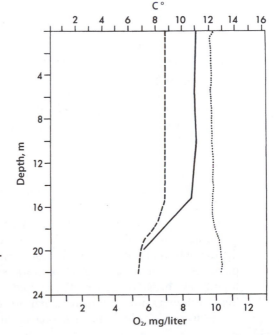

Figure 10-10 The approach of the autumnal overturn and the destruction of summer stratification nearly complete, Mountain Lake, Virginia (**D** in Fig. 10-6). Temperature, dashed line; oxygen, solid line; 100% oxygen saturation, dotted line.

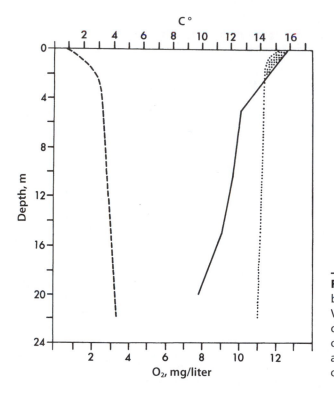

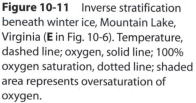

Figure 10-11 Inverse stratification beneath winter ice, Mountain Lake, Virginia (**E** in Fig. 10-6). Temperature, dashed line; oxygen, solid line; 100% oxygen saturation, dotted line; shaded area represents oversaturation of oxygen.

all seasons brings about pronounced diurnal density differences. At night the cooling of superficial layers sets up convection patterns that obliterate the stratification. In that arid region the nocturnal cooling is, to a great extent, a function of evaporation. Evaporation occurs, of course, throughout the 24-hour period and results in an annual loss of about 2.2 m; however, during daylight its cooling effects are surpassed by solar heating.

The heat of vaporization for water at 30° C is about 580 cal/g. Thus, the cooling effects derived from the evaporation of a film 1 mm thick beneath 1 cm² of pond surface would be 58 cal if it were distilled water. More heat exchange is involved in the evaporation of saline water and at lower temperatures (Table 10-2). Because the latent heat of vaporization is supplied from the uppermost layer of the water, it results in a pronounced cooling effect in some climates. Where the vapor pressure gradient at the air-water boundary is marked by strong differences, the evaporation rate serves to cool water bodies far beyond what occurs in humid climates. A shallow Arizona pond monitored one August day by Denton Belk (personal communication, 1978) lost by evaporation 41% of the calories that arrived via radiation at the lake surface; during the daylight hours the air temperature rose 29° C while the mean temperature of the pond rose no more than 7.5° C. By contrast James E. Schindler and his students in the humid southeastern part of the United States find their pond temperatures rather high, following ambient air temperatures closely; this, in part, is a result of less efficient "sweating" (Schindler, personal communication, 1981).

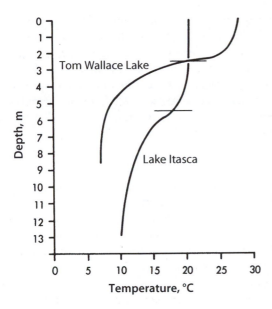

Figure 10-12 Comparison of summer temperature profiles from Tom Wallace Lake, 26 June 1954, and Lake Itasca, 26 June 1970. Horizontal lines indicate planar thermoclines. Although Lake Itasca is 9° farther north, it has a warmer hypolimnion and a higher mean temperature because of a prolonged vernal circulation period.

Although few data are available, it seems that large, shallow Lake Nicaragua represents another polymictic type. At night and during the morning hours some stratification occurs, although the lake has a mean depth of only 12.4 m. Temperature differences from 28° to 24° C, the range from surface to bottom at these times, imply density contrasts as great as those to be found, for example, in a range from 4° to 16° C. Each afternoon strong winds arise, reducing the morning stability to zero.

There are undoubtedly many shallow, unprotected bodies of water that circulate almost continuously in mild climates where ice blankets never exist. Unusual conditions account for lack of stratification in others. Pearse (1936) described some Yucatán cenotes that were unstratified, presumably because of water continuously entering and leaving somewhere below the surface. Montezuma Well, Arizona, resembles the cenotes in that it is usually isothermal although 17 m deep, is remarkably protected from the wind, and has a relative depth of 17% (Cole and Barry, 1973). Water amounting to about 10% of the well's volume enters from deep subsurface crevices and leaves each day at a surface outlet, precluding all but the most transitory layering. Subject to a climate quite different from the Yucatán, Montezuma Well would freeze over during winter months but for the fact that its entering water supply averages about 24° C.

Polymixis seems to describe another condition, although quite different from the original use of the word. In regions near the boundary of warm monomixis (in the

Table 10-2 Latent heat of vaporization for pure water at different temperatures

°C	Calories per gram	Joules per gram
0	597.3	2500.8
10	591.7	2477.3
20	586.0	2453.5
30	580.4	2430.0
40	574.7	2406.2

sense of Bayly and Williams, 1973) and dimixis or cold monomixis, there may be unusual winters when ice lies across lakes that are usually open throughout the year. As a result, lake categories change. On very rare occasions, for example, Cayuga Lake becomes ice covered and, therefore, falls into the dimictic class for that year. When there are multiple ice covers that come and go during a winter, the phenomenon must be termed a type of polymixis. One winter there were three separate freezings, punctuated by thawing and open water, at Tom Wallace Lake (Krumholz and Cole, 1959). There were five distinct periods of circulation that year.

Meromixis

Meromictic lakes circulate at times, but incompletely. In contrast to holomixis, the entire water mass does not participate in the mixing. A dense stratum of bottom water remains stagnant and characteristically anaerobic. Findenegg (1935) coined the useful adjective meromictic to apply to some Austrian lakes he had investigated.

A unique terminology has evolved to describe the regions of a meromictic lake. First, there is a permanently stagnant layer containing a markedly greater concentration of dissolved substances than that found in the overlying water; this is the **monimolimnion**, a term we owe to Findenegg. The upper layer, being much more dilute, is mixed by the wind and shows seasonal changes; this is the **mixolimnion**. In a sense it is comparable to an entire holomictic water mass. Between the mixolimnion and monimolimnion there is a zone where salinity increases rapidly with depth, the **chemocline**.

The vertical density gradient in the chemocline is analogous to the thermocline in the sense of Birge, although the great stability of a meromictic lake is a function of materials in solution rather than of temperature. Many authors use the generic term **pycnocline** to describe a density gradient, no matter what its origin. Any phenomenon that brings about loss of soluble material from the dense monimolimnion, or contributes significantly to the salinity of the mixolimnion, decreases the stability and may eventually terminate meromixis.

The words mixolimnion and chemocline came from Hutchinson (1937). They were derived from his study of Big Soda Lake, Nevada, a meromictic member of a group of lakes he investigated in the Lahontan Basin. His published report, cited above, contributed much to both arid-land limnology and our knowledge of meromixis.

Biogenic meromixis.　There is a classification of meromictic lakes based on the manner by which solutes accumulate in the monimolimnion. The first type to be observed in North America was the biogenic meromictic lake. In this kind there is an accumulation of substances derived from bacterial decay, diffusion from the sediment and from photosynthetic precipitation of carbonate. There is no striking accumulation of salts such as NaCl. The waters are typically the calcium bicarbonate type, and the dissolved materials in the monimolimnion are much like those found in the hypolimnion of a eutrophic lake toward the end of summer stagnation.

Eggleton (1931) described unusual summer temperature profiles and deep anoxic water, foul with H_2S and devoid of metazoan life, in Fayetteville Green Lake, New York. Eggleton did not expand on the unusual features he had discovered until later (Eggleton, 1956). Now Fayetteville Green Lake ranks as one of the most intensively studied meromictic lakes in the world.

The plunge-basin Green Lake and its neighbor Round Lake, of similar origin, illustrate some of the unusual requisites for biogenic meromixis. Morphologic, topographic, and meteorologic conditions that hinder overturn are necessary, either singly or working together, for the accumulation of the substances of biogenic origin that impart density to the bottom waters. In the case of Eggleton's lake, steep, wooded slopes rise abruptly from the shores for 120 m, serving to protect the surface from air currents in three directions. Moreover, the depth (59 m) of Fayetteville Green Lake is great in relation to its area, about 27 ha. Its relative depth, z_r, is a surprising 10%; this is typical when meromixis is due to biogenesis.

An instructive comparison can be made between Sodon Lake, Michigan (Newcombe and Slater, 1950), and Tom Wallace Lake (Cole, 1954). Sodon Lake, a pit lake in thick glacial drift, is meromictic; Tom Wallace was formed by artificial impoundment and is holomictic. Both are surrounded and protected by wooded hills and have nearly identical areas, about 2.31 and 2.34 ha, respectively. Sodon Lake is more nearly circular and has a maximum length of 192 m, compared with the Kentucky lake's wind-effective length of 328 m. The most striking difference between the two, however, is their maximum depths: Sodon Lake is 17.7 m deep; Tom Wallace is 8.7 m deep. Their respective relative depths are 10.3% and 5.0%. Sheltered Tom Wallace Lake shows several effects of insulation from wind, but it lacks the morphologic extremes that established meromixis in Sodon Lake. Table 10-3 further emphasizes the high relative depths found in most biogenic meromictic lakes.

One of the most celebrated of the biogenic meromictic lakes is the Austrian Längsee, one of the Carinthian lakes that Findenegg studied in his early investigations of meromixis. It is a fairly shallow lake with an unusually low (for meromixis) relative depth of only 2.13% and is shielded to some extent by hills. The key factor, however, in its meromixis is meteorologic. In that part of Austria calm weather with low wind velocities characterizes the autumn when the fall overturn is expected. The American Frey (1955) studied the sediments, pollen grains, and animal relics in a 9.1 m core taken near the center of the lake and arrived at a theory to account for the start of meromixis in Längsee. The uppermost 1.5 m of sediment is **sapropel**, a shiny black material formed under intense anaerobic conditions. This chemically reduced ooze is confined to the monimolimnion and represents an estimated 2,000 years of continuous meromixis.

A transition zone below the sapropel contains bands of clay, pollen from cultivated plants and agricultural weeds, and a smaller number of beech pollen grains than are found at lower levels in the core. Below this lies **gyttja**, the typical organic sediment found in the profundal zone of a eutrophic, holomictic lake. This structured material, only partially oxidized and semireduced, may be a forerunner of sapropel if subjected to strongly reducing conditions. Frey concluded that clearing land by cutting the beech forest and breaking the soil for agricultural purposes led to increased erosion and importation of clay to the Längsee hypolimnion. The slowly settling particles, lingering in the bottom water, increased its density to such an extent that, with accumulation of the normal summer metabolic products from the sediments and the epilimnion above and a comparatively windless autumn, no overturn occurred. Thus, meromixis was triggered by human activity two millennia ago. The annual contribution from lake organisms more or less mineralized by bacteria has served to increase stability, and today the lake is considered a model of biogenic meromixis.

Small particles derived from erosion had different effects in Hall Lake, a small kettle lake that appears as an example of biogenic meromixis in Table 10-3. Turbid winter inflows in the 1960s ended meromixis by introducing density currents that destroyed stratification (Culver, 1977). Moreover, the arrival of oxygen in the monimolimnion decreased its density further by oxidizing dissolved ferrous iron compounds that subsequently were sedimented as ferric compounds.

The *termination* of meromixis effected by density currents in Hall Lake fits nicely with Löffler's (1975) interpretation of sediment cores from Carinthian lakes. Löffler theorized that the *onset* of meromixis in some of these Austrian lakes (including Längsee) coincided with the cessation of substantial inflows. Furthermore, it might have accompanied a change from cold monomixis to **dimixis**, a shift that would have increased the total annual period of deep-water stagnation and, perhaps, accumulation of density-augmenting materials there.

It has been established that some protected lakes in the Minnesota forests are good examples of partially meromictic lakes (Eddy, 1963; Baker and Brook, 1971). One of these is Deming Lake (Fig. 10-13), where the meromixis is biogenic and generally persists for years; occasionally there is a complete overturn. Findenegg (1937) discussed at length this intermediate type of lake, lying somewhere between holomixis and permanent meromixis. Difficulty arises in classifying partially meromictic bodies of water because oligomictic is an adjective that also describes them.

Certain deep African Rift Valley lakes, lying near the equator and experiencing little seasonal temperature change, are extremely stable and meromictic. Lake Tanganyika (about 5° to 7° S lat) is the most striking of these, with nine-tenths of the lake permanently stratified and deoxygenated from 200 m to the bottom at 1,430 m (Beauchamp, 1964). Lake Malawi (11° S lat) is shallower, having a maximum depth of 760 m. It is said to mix completely every decade or two, although usually water below 300 m is anaerobic for years.

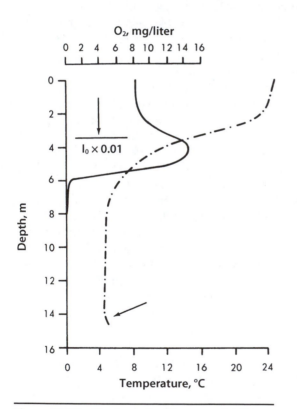

Figure 10-13 Deming Lake, Minnesota, 3 July 1963. Arrow points to slight dichothermic temperature curve. A positive heterograde oxygen curve is present with a metalimnetic peak, and the lower level of the euphotic zone ($I_o \times 0.01$) is indicated.

Ectogenic meromixis. Effective contrasts in density can be brought about by delivery of water to a lake from outside sources. This leads to what is termed **ectogenic meromixis**. The water can be delivered in two forms:

first, as a dilute superficial layer that comes to lie above a preexisting saline body of water; or second, as saline water that finds its way to the bottom of a freshwater lake. Both phenomena maintain meromixis in the coastal Lake Suigetsu, Japan; river water supplies the light upper layer, and sea water creates the dense monimolimnion (Mori, 1976). The topographic, morphologic, and meteorologic requirements for biogenic meromixis are needed to a far lesser degree for ectogenic meromixis (Table 10-3).

The stability of ectogenic meromixis is a function of salinity differences. Ectogenic meromictic lakes are much more saline, containing relative proportions of ions that differ from the ratios typical of biogenic meromictic lakes. These lakes are usually found in arid regions where water is concentrated by evaporation, or in coastal areas where sea water contributes to density extremes.

Table 10-3   Relative depths (z_r) of types of meromictic lakes in North America

Type of meromictic lake	Relative depth (percent)
Ectogenic and crenogenic lakes	
Big Soda, Nevada	1.17
Blue, Washington	2.90
Hot, Washington	2.70
Long (northern tip), Alaska	2.00 (?)
Lower Goose, Washington	3.76
Nitnat, British Columbia	3.37
Ogac, Baffin Island	4.40
Pingo, Alaska	5.50
Powell, British Columbia	3.60
Sakinaw, British Columbia	4.38
Soap, Grant Co., Washington	1.16
Soap, Okanogan Co., Washington	1.29
Tessiarsuk, Labrador	2.05
Vee, Alaska	0.70
Wannacutt, Washington	2.77
Partial meromixis, ectogenic	
Cinder Cone Pool in Zuni Salt Maar, New Mexico	4.49
Green Pond, Arizona	6.54
Red Pond, Arizona	4.27
Biogenic meromictic lakes	
Arco, Minnesota	7.67
Budd, Minnesota	6.77
Canyon, Michigan	9.89
Deming, Minnesota	7.92
Fayetteville Green, New York	10.03
Hall, Washington	8.46
Josephine, Minnesota	5.27
Mary, Wisconsin	20.36
Sodon, Michigan	10.03
Squaw (protected bay only), Minnesota	22.80
Stewart's Dark, Wisconsin	9.03

An artificial and relatively recent condition is causing meromixis in unexpected geographic locations. An early report (Judd, 1970) showed some effects of the runoff of winter salt from city streets. The salt used to deice winter roads is flowing to the bottom of freshwater lakes in the northern United States, creating a temporary layer of dense, saline water. Incipient meromixis is the result.

The small Hot Lake, Washington, lies in an artificial basin rich in $MgSO_4$ that accounts for dense saline water. Runoff from snow melt each spring adds a dilute mixolimnion and serves to maintain perennial stability (Anderson, 1958b).

Ectogenic meromixis tends to diminish through time when there is no rejuvenating event to compensate for the slow mixing of saline and fresh water. In Hot Lake the rejuvenating event is the arrival of melt water each spring; in Judd's lakes it is the winter salt from nearby highways that restores a fleeting meromixis. Two other examples come to mind. The pool in the cinder cone rising from the Zuni Salt Maar, New Mexico, contains very saline water derived from springs within the maar. At irregular intervals rainfall added directly to the pool creates meromixis that lasts for months. Without the addition of more dilute water, the pool eventually becomes isohaline and holomictic (Bradbury, 1971). Twenty-four kilometers to the west, the twin salt ponds at the Long-H Ranch, Arizona, show a temporary meromixis that seems to have a seasonal pattern (Cole et al., 1967). The monimolimnion in each lake is formed by ectogenic freshwater, concentrated by evaporation in a closed basin. During cool seasons the same dilute water reaches the basins and adds a mixolimnion. The meromixis lasts only a few months; during the intense summer heat, evaporation interrupts the arrival of freshwater from springs in the banks above, increases the salinity of the mixolimnion, and reduces pond volume. At the same time monimolimnion salts are diffusing upward into the chemocline until the pond becomes isohaline-isothermal in late summer. The regularity of this cycle suggests a type of monomixis.

Some fjord lakes, of which at least five have been described from North America, are meromictic because of occasional invasion by sea water that sinks to the deepest part of the basin to establish a monimolimnion. In some instances, as in Tessiarsuk in Labrador, the introduction of marine water occurs almost daily (Carter, 1965). Williams and associates (1961) believe that seawater was introduced in Powell Lake, British Columbia, about 13,000 years ago. Similar ancient monimolimnia derived from seawater exist in Norwegian fjord lakes that experienced isostatic uplift of perhaps 9,000 BP (Barland, 1991).

Crenogenic meromixis. Subsurface flows of saline, dense water into a basin bring about the crenogenesis of meromixis. The continuous import from one or more deep springs contrasts with a surface freshwater inflow. If equilibrium between the two types of water is established, the water becomes persistently stable; an outflowing stream makes this possible in some instances.

The African Lake Kivu is similar to some of its rift neighbors: it is meromictic and contains anoxic water below 60 m. It differs from Malawi and Tanganyika by having about 4.5 times more salinity in its deep water than in its mixolimnion. It is believed that the saline water is introduced by deep sublacustrine springs.

Dussart (1966) lamented the loss of two excellent examples of crenogenic meromictic lakes to further scientific study because of modification and utilization for hydroelectric power. These are Lake Girrotte (France) and the Swiss lake, Ritom. The saline subsurface inflows became visible in each lake when it was partially drained in preparation for construction of the power facilities.

Likens (1965) described an unusual pond in Alaska that may be the world's shallowest meromictic lake. Vee Pond has a maximum depth of only 1.6 m and a relative depth of 0.7%. Its monimolimnion, with an upper boundary between 0.5 and 1.0 m, is maintained by saline springs marked by light bottom areas here and there in the pond's dense carpet of benthic vegetation.

Distinctions are not always clear cut when the origin of meromixis is sought. Hot Lake, Washington, for example, may be considered ectogenic because dilute water comes from external sources to float above heavy magnesium sulfate water in the basin. The edaphic factors account for the presence of $MgSO_4$ because the lake occupies old epsomite ($MgSO_4 \cdot 7H_2O$) diggings. If springlike seepage of ground water into the concavity is a phenomenon at Hot Lake, the meromixis is due to crenogenic as well as ectogenic agents.

Other phenomena establishing meromixis. Goldman and coworkers (1967) studied Antarctic lakes and proposed cryogenesis as a method of delivering salts to a monimolimnion and reducing the salinity of upper waters. This is a simple freezing-out process that may be very effective at times. Cole and others (1967) concluded that temporary meromixis during winter in two Arizona saline ponds was, in part, based on such a phenomenon. Freezing occurred each night, followed by daytime melting. Melted ice from the ponds contained a mean of 11.6 g/liter of filterable residue, whereas immediately beneath the ice there were 41 g/liter. The diurnal ice melt left a layer of dilute water over the lower saline strata.

A similar phenomenon in saline waters during winter is the precipitation of Na_2SO_4 even without freezing. This compound's solubility is decreased with lowered temperature, and many examples are known where large crystals of mirabilite ($Na_2SO_4 \cdot 10 \, H_2O$) are precipitated during cold weather. The surface waters become diluted as material is sent to the bottom, where it may go into solution again at higher temperatures. Increased stability and meromixis thus may be brought about through low temperatures at the lake surface.

A lengthy discussion and modified typology presented by Walker and Likens (1975) is recommended for additional information about meromixis and lake circulation patterns. Their classification of meromixis rests on its various origins. Very often, however, meromixis is owed to more than one factor, and distinctions between categories are blurred.

Unusual Temperature Profiles

In earlier pages isothermy and direct and inverse stratification were explained. Also, an unusual inverse stratification brought about by salt layering was presented. A specific vocabulary has grown up around some bizarre curves found in meromictic lakes. We are indebted to the Japanese limnologist Yoshimura (1936, 1937) for the terms pertaining to these curves.

Dichothermy

Eggleton (1931) graphed unusual summer temperature profiles for Fayetteville Green Lake, making little comment about them. A year later Juday and Birge (1932) called attention to the unusual character of a similar temperature curve in Lake Mary. Their words describe what we know now as a dichothermic curve, "The coldest stratum . . . was found at an intermediate depth and not at the bottom in summer."

Dichothermy is illustrated by a vertical temperature curve with an inflection within it; there is a low point bounded above and below by warmer water (Figs. 10-13 and 10-14, *D*).

Dichothermy is a characteristic of meromictic lakes so that, if a slight temperature elevation is detected in the bottom waters in summer, at least partial meromixis is implied. (Such a rise near the bottom in winter might signify nothing more than the slow cooling of sediments that had been warmed the preceding summer.)

The origin of heat in the monimolimnion is complex, but in many meromictic lakes it is biogenic, bacterial metabolism as the major source. Zobell and coworkers (1953) estimated that in holomictic Lake Mead the heat produced by microorganisms amounted to 30×10^{-12} cal per cell per hour. Pamatmat and Bhagwat (1973) estimated that 1 g of sediment from the deeps of Lake Washington released from 0.003 to 0.017 cal per hour, and much of this was through anaerobic activity. With the monimolimnion not participating in circulation, any heat accumulating there can be lost only by slow conduction.

Geothermal sources in relatively deep basins, such as Lake Vanda, and direct solar radiation in shallow, saline meromictic lakes account for the rise at the lower extreme in dichothermic temperature curves.

Mesothermy

In waters with unusual salt stratification it sometimes happens that a vertical temperature profile shows a high point within the curve. The temperature increases with depth, momentarily suggesting inverse stratification, but decreases before the bottom is reached. The result is a mesothermal contour indicating a warm stratum sandwiched between upper and lower colder layers (Fig. 10-14, *B*).

In some instances mesothermy is transitory. For example, in the twin salt ponds at the Long-H Ranch, Arizona (Cole et al., 1967), an annual cycle includes a time in early winter when dilute water is extruded onto an isothermal, dense, saline pool. Solar radiation penetrates the clear layer and is absorbed rapidly by the upper part of the saline stratum, effecting a mesothermal profile. This may be a prelude to the establishment of inverse stratification brought about by the slow diffusion of heat downward. Before a complete inverse profile is achieved, however, vernal warming of the surface layers of these ponds proceeds to such an extent that dichothermy is produced.

The amictic Antarctic Lake Bonney shows a mesothermal curve that may persist through the year. Interestingly, it increases from 0° to a maximum of 7.9° C and then decreases to –2.8° C, the effect of high salinity depressing the temperature of maximum density. Vertical water motions in the lake are very small, and the ice is so transparent that direct solar heating is responsible for the warm inflection in the curve 14 m beneath the ice (Ragotzkie and Likens, 1964).

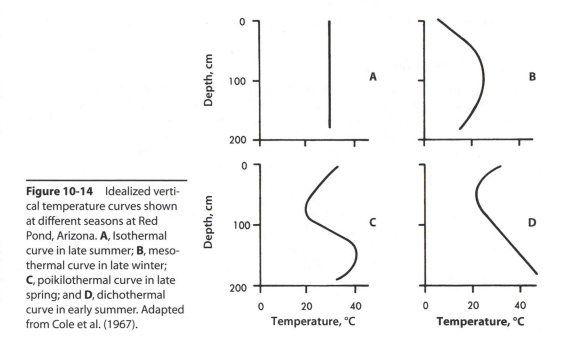

Figure 10-14 Idealized vertical temperature curves shown at different seasons at Red Pond, Arizona. **A**, Isothermal curve in late summer; **B**, mesothermal curve in late winter; **C**, poikilothermal curve in late spring; and **D**, dichothermal curve in early summer. Adapted from Cole et al. (1967).

Perhaps one of the highest temperatures ever reported in a stratified saline pond came from a mesothermal situation. Solar Lake (Sinai) owes its origin to the isolation of a lagoon that was once connected to the Gulf of Elat; most of the year it is meromictic, heat accumulating in subsurface saline strata. Cohen and colleagues (1977) reported 56.6° C at the high point of its mesothermal temperature profile.

There seem to be good examples of organisms playing roles in the genesis of mesothermy. Kilham and Melack (1972) believe that the mesothermal peak of 40° C in African Lake Mahega is caused by solar heating of a dense subsurface population of the Cyanobacterium *Synechococcus.*

Poikilothermy

Complex temperature curves, marked by the inclusion of at least one low point and one high point, have been described from saline-layered waters (Fig. 10-14, *C*). In the Long-H ponds, poikilothermic curves represent the transition from late-winter mesothermy to dichothermy. Before heat has diffused downward from the warm mid-pond layer to raise the temperature of the lowest level, the days of early summer heighten surface temperatures. The result is, from top to bottom, the following sequence: a warmed upper layer, perhaps mixed downward to some extent; a layer where temperature diminishes, representing some of winter's cold; a region where temperature rises to a high point, signifying the accumulation of solar heat during winter when mesothermy ruled; and a cooling trend to a bottom sheet not yet warmed by conduction of heat from above.

Data from Hot Lake (Anderson, 1958b) indicate dichothermy, mesothermy, poikilothermy, and a hot inverse stratification beneath the winter ice. Such an array of seasonal curves probably is displayed only by saline meromictic lakes and small briny pools.

The Thermal Bar

In large, cold lakes where temperatures fall below 4° C during winter circulation, an interesting event occurs in the spring. As the shallow water near the shore warms, the deeper water continues to circulate isothermally as it had been doing throughout the winter. The two lake regions are separated by a narrow zone of down-welling water—the most dense in the lake. This zone is marked by a vertical 4° C **isotherm**, the **thermal bar**, which serves as a barrier between the colder, less dense offshore water and the warmer (much lighter) water of the shallows. Nutrients derived from watershed runoff are trapped by the thermal bar, and there, in the warm water near the shore, vernal phytoplankton blooms occur, quite in contrast to the situation farther out. Phytoplankton abundance, species composition, and chemical and optical properties differ between the two lake regions until the bar moves offshore and the entire lake becomes stratified (Rodgers, 1965; Mortimer, 1974; Stoermer, 1978; Hohman et al., 1997; Rao et al., 2004).

11

Heat Energy and Water Movements

In the previous chapter we considered how water temperature relates to ice cover, layering and mixing, and other important physical properties of lake water. Temperature also influences hydrology, especially evaporation from lakes (Gianniou and Antonopoulos, 2007). These factors also effect the chemical reactions in the water; most students have observed changes in a solution when heated. This extends to the biota, since metabolism, digestion, reproduction, growth, and other life processes are likewise affected by water temperature. For example, the common planktonic grazer *Diaptomus pallidus* develops more than 4.5 times faster at 25° C than at 10° C (Geiling and Campbell, 1972).

Heat versus Temperature

Heat is both a noun and a verb. The noun refers to a form of energy, while the verb refers to a transfer of that energy (a flux). Heat content, like other energy, is expressed in Joules or calories; flux incorporates a rate at which that energy transfers, e.g. J/s (= Watt, W). Heat flux can be calculated for an entire lake, but for many studies we refer to a unit measure and compare the standard units between water bodies. For example, Lofrgen and Zhu (1999) compared heat flux (W/m^2) standardized across the surfaces of all five North American Great Lakes using data collected over three years.

The thermodynamics of heat and flux are unusual, because water is a remarkable substance. It takes a great amount of energy, 1 cal/g (4.18 J/g) at 14.5° C, to raise water one degree. This so-called heat capacity (also called thermal capacity) is four times that of air, and 10 times that of typical metals. A practical consequence is that lakes resist temperature change, keeping nearby land warmer in the autumn and cooler into the spring.

Heat energy always "flows downhill"—in other words, from objects or fluid masses with a higher heat content towards those with a lower heat content. The flux along this gradient is spontaneous and obviously governed by the laws of thermodynamics. *Temperature* is a measure of the heat content of an object or mass; heat transfers away from bodies with higher temperature towards bodies with lower temperature. We read temperature via its effects on matter, for example, a change in the volume of liquid inside a thermometer or electrical current conducted through the wire in a probe.

Mean Temperature and Heat Content

Temperature is an important parameter, but to understand lake warming and cooling overall, we must profile both temperature and depth simultaneously. As we saw in the previous chapter, temperature profiles are quite variable, so heat flux must take into account the heat content of the various layers based on their respective volumes (see basin morphometry, Chapter 6).

Heat content is the energy in a lake stratum or a unit volume. We assume in many calculations that density is constant (1 mL = 1 g), and so is thermal capacity (1 cal/g). We realize that natural waters have solutes affecting heat capacity, but the effects are of minor importance in most situations.

Table 11-1 shows data taken in August 1951 from Tom Wallace Lake. Ten temperatures were measured from top to bottom in the deepest part of the lake. The mean of these is 17.0° C, but this is no more than the average temperature of a sampled vertical column in the lake. It is not the lake's mean temperature. Obviously, there is a much smaller volume of water in the deeps at 8.8° C than there is in the upper 1-m stratum between 28.9° and 27.2° C. The temperatures are not weighted properly for determining the mean lake temperature, and the average of the 10 profile readings does not apply.

Fortunately, morphologic details of Tom Wallace Lake permit further treatment of the data. The volume is 90,645 m^3, and the area is 2.34 ha or 23,400 m^2. From these data we get the mean depth, volume divided by area, or 3.87 m.

In Table 11-1 the volume of every stratum of Tom Wallace is shown and summed in column IV. In addition, the volume of each stratum relative to the total is given as a decimal fraction in column V. In column VI, the temperature from each 1-m stratum is shown; this was found by averaging the upper and lower temperatures of each layer.

Multiplying the items in column IV by those in column VI approximates the heat content, the product of volume and temperature. The sum of these products divided by lake volume yields the average or mean temperature, 21.9° C, almost 5° C higher than the results derived from averaging the vertical profile without taking volume into account.

Had we simply multiplied column V by the temperatures in column VI and summed the products as shown in column VIII, the lake's mean temperature would have been revealed immediately. By the same method, the mean temperature on June 26, 1954 can be calculated easily from data in Table 10-2. It was 19.5° C.

The products of water volumes and temperatures shown in Table 11-1, column VII, are heat. If the lake volumes had been converted to cubic centimeters, the total heat content in calories would have been found as the sum of column VII. Remembering that 1 m^3 equals 10^6 cm^3, we can state that the total heat content of Tom Wallace Lake on August 3, 1951 was 1,988,140 × 10^6 cal. This is of some interest but of little value for comparative purposes. It would be possible to find a larger lake with, let us say, 5° C water throughout that would have a heat content identical to that of summertime Tom Wallace.

For comparative purposes, heat content is referred to unit surface area. Tom Wallace Lake's surface is 2.34 ha, or 2.34 × 10^8 cm^2. The calories beneath the average square centimeter of lake surface, then, amount to the quotient of total heat and area:

$$\frac{1,988,140 \times 10^6 \, \text{cal}}{234 \times 106 \, \text{cm}^2} = 8,496 \, \text{cal/cm}^2, \text{ or } 3.56 \times 10^5 \, \text{kJ/m}^2$$

Table 11-1 Some thermal data from Tom Wallace Lake, Kentucky, August 3, 1951

I Depth (m)	II Temperature (°C)	III Stratum (m)	IV Volume (m³)	V Relative volume	VI Mean temperature of stratum (°C)	VII* Heat content (IV × VI)	VIII (V × VI)
0	28.9						
		0–1	21,460	0.237	28.1	603,026	6.66
1	27.2						
		1–2	17,990	0.198	26.9	483,931	5.33
2	26.7						
		2–3	14,980	0.165	25.6	383,488	4.22
3	24.4						
		3–4	12,280	0.135	20.3	249,284	2.74
4	16.1						
		4–5	9,775	0.108	13.3	130,007	1.44
5	10.6						
		5–6	7,195	0.079	10.3	74,108	0.81
6	10.0						
		6–7	4,645	0.051	9.4	43,663	0.48
7	8.9						
		7–8	2,090	0.023	8.9	18,601	0.20
8	8.8						
		8–8.75	230	0.003	8.8	2,041	0.03
8.75	8.8						
TOTALS	170.4		90,645	1.000		1,988,140	21.91

*This assumes thermal capacity and density are both at unity; otherwise these values should be multiplied by $(C \times p)$.

Another way of approximating the heat beneath the average square centimeter of lake surface is by multiplying mean depth by mean temperature. In Tom Wallace, $z =$ 387 cm, meaning there are 387 cm³ on the average beneath each square centimeter of lake surface. This could be expressed as 387 cm³/cm², or 387 g/cm². When multiplied by mean temperature, the results are 387 g/cm² × 21.9° C = 8,475 cal/cm², a discrepancy of less than 1%.

Heat Distribution: Work of the Wind

The August temperature curve from Tom Wallace Lake was translated to vertical distribution of heat when volumetric data were taken into account. The dispersion of calories contrasts sharply with the condition during the spring overturn when uniformity prevailed and also with the pattern of heat supplied by solar radiation during the warming period, as reflected by a typical vertical light curve (see Fig. 10-1). Work, in addition to energy supplied by the sun, must have been invoked to explain this unusual pattern of heat. After the vernal overturn when isothermy and uniform density prevailed, solar radiation heated the surface; work must then have been required to push the lighter layers of water downward as they were formed. This process is analogous to forcing down sheets of floating cork.

Birge (1916) developed the concept of the wind work (buoyancy, B) necessary to accomplish this downward distribution of heated water (through water of uniform density) to a particular level of stratification. In a sense, this wind work is the opposite of stability, for it is an expression of the force necessary to create rather than destroy stratification. For our calculations, Birge's work of the wind is expressed as gram-centimeters per square centimeter; this is logical since we define energy in calories as heat in a cm^3 of water. However, the values are later converted to standard SI units. The buoyancy (wind work) is calculated as:

$$B = \frac{1}{A_0} \int_{z_0}^{z_m} z \left(\rho_i - \rho_z \right) A_z dz$$

Where

A_0 = A, the surface in square centimeters
A_z = the area (square centimeters) at some depth z
z = depth, considered positive and in centimeters
ρ_i = the initial density, constant at all depths
ρ_z = observed density at depth z, created by the work of the wind

Table 11-2 includes some Tom Wallace Lake data from which the work of the wind can be calculated and serves for the ordination of points used in Fig. 11-1, a so-called direct work curve, representing the work spent on each layer by the wind in pushing buoyant water down to that depth.

Table 11-2   Data and calculations for establishing work of the wind, Tom Wallace Lake, Kentucky, June 26, 1954.

I	II		IV	V	VI	VII
z	T_z	III	ρ_z	$\rho_i - \rho_z$	$A_z/A_0 \times z$	(V × VI)
cm	C°	A_z/A_0	g/cm³	g/cm³	cm	g/cm³
50	27.7	0.9188	0.99634	0.00366	45.9	0.16799
150	26.7	0.7692	0.99662	0.00338	115.4	0.38998
250	20.9	0.6410	0.99804	0.00196	160.2	0.31409
350	14.0	0.5278	0.99927	0.00073	184.7	0.13485
450	10.3	0.4188	0.99970	0.00030	188.5	0.05654
550	8.1	0.3098	0.99987	0.00013	170.4	0.02215
650	7.3	0.2009	0.99992	0.00008	130.6	0.01045
750	7.0	0.0940	0.99993	0.00007	70.5	0.00493
850	6.9	0.0043	0.99993	0.00007	3.7	0.00026
TOTAL						1.10355

$$B = \sum_{z_0}^{z_m} z \left(\rho_i - \rho_z \right) \frac{A_z}{A_0} \Delta z = 1.10355 \frac{g}{cm^2} \times 100 \ cm = 110.4 \ g - cm/cm^2$$

The dyne (g-cm) is converted to standard SI units of force (Newtons, N) and then to work (N-m or Joules) per m^2 of lake surface. The conversion is 1 dyn = 10^{-5} N, so 110.4 g-cm/cm^2 = 11.04 Newton/square meter.

The distribution of the summer heat is not necessarily or entirely a result of the wind's work. Factors other than solar heating and its distribution by air currents at the lake surface account for the summer heat content. As surface temperatures heighten, back radiation and evaporation dissipate heat. Also, direct solar warming of shallow, littoral sediments creates unstable inverse temperature gradients resulting in upward transfer of heat to the water. This could be relatively important in small lakes. Ricker (1937) put forth as possible a further factor: the daily vertical migration of enormous numbers of zooplankters. Having warmed in the epilimnion during night hours, they transfer heat, amounting to 0.07 cal/cm² per day, while moving downward to the hypolimnion toward dawn. The plankton predator *Chaoborus* may also be responsible for such heat exchanges.

The summer heat income, then, as evidenced by the difference between the highest mean temperature and the temperature at the spring overturn, is accomplished and modified by Birge's work of the wind as well as other factors.

The direct stability curve can be combined with the direct wind-work curve (Fig. 11-1) to yield a direct total work curve (Fig. 11-2). This has been symbolized *G*. Expressed in gram-centimeters, the temperature stratification of Tom Wallace Lake on June 26, 1954, implies that work in the amount of 105.75 g-cm for every square centimeter of the lake surface had been performed to change the density stratification

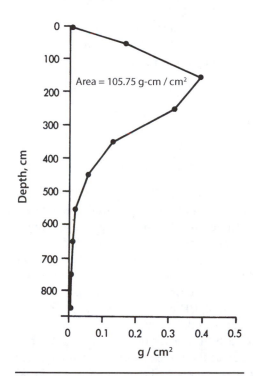

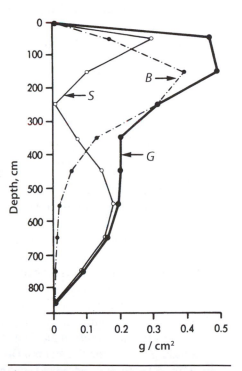

Figure 11-1 Plot of Birge's "work of the wind." Direct work curve for Tom Wallace Lake, Kentucky, June 26, 1954. Based on data in Table 11-2.

Figure 11-2 Total direct work curve, *G*, for Tom Wallace Lake, Kentucky, June 26, 1954. *S*, Stability; *B*, work of the wind, from data in Table 11-2 and Fig. 11-1; *G*, the sum of *S* and *B*, 201.8 g-cm/cm².

since the time of uniformity at 4° C. Similarly, the stability shows that 96.05 g-cm/cm^2 are needed to mix the water to a new uniform temperature and density, about 21° C and 0.99799 g/cm^3. The total work involved in the stratification is, therefore, 201.8 g-cm/cm^2 (Fig. 11-2). This is the total work that would have been needed to distribute heat from the time of vernal overturn in order to keep the lake unstratified and isothermal at the mean temperature observed on June 26.

Heat Budgets

Calculations of Various Budgets

The review of lake heat budgets by Ragotzkie (1978) is an old, but good summary of the subject to be discussed in this section. **Heat budgets** refer to the heat absorbed by a body of water during some period of time. Of these, the *annual heat budget* is the major category and is the sum of other types. It is the total amount of heat that enters a lake from its lowest mean temperature of the year to its highest. Certain symbols have become useful in designating the heat budget. Θ refers to the total heat content in calories; *a*, in this case, refers to the annual gain, and *b*, according to the standard set by Birge, refers heat to unit surface area. Thus Θ_{ba} is the symbol for the annual heat budget: total heat gained during the year expressed as calories per square centimeter.

If the peak summer heat content in Tom Wallace Lake were represented by the August 3, 1951, datum of 8,490 cal/cm^2, and if that year the coldest mean winter temperature had been 4° C, as is sometimes the case, Θ_{ba} could be computed. The lowest heat content of the year would be

$$\frac{V, \text{cm}^3 \times 4.0°\,\text{C}}{A, \text{cm}^2} = \frac{90,645 \times 10^6 \times 4.0°\,\text{C}}{234 \times 10^6 \text{ cm}^2} = 1,549 \text{ cal/cm}^2$$

The difference between highest and lowest heat contents is

$$\Theta_{ba} = 8,490 - 1,549 = 6,941 \text{ cal/cm}^2 = 2.91 \times 10^5 \text{ kJ/m}^2$$

Hutchinson (1957) suggested a method of calculating Θ_{ba} by planimetry or any other way of determining area beneath a curve. Using observations over several years at Tom Wallace Lake, this is demonstrated in Table 11-3. It shows the greatest temperature differences ever recorded at each meter depth, and these data are used in plotting Figure 11-3. It represents a theoretical extreme, not a mean annual heat budget for Tom Wallace, because the temperature difference at each level was based on more than one year's observation. The area beneath the curve (Fig. 11-3) roughly equals 1.971×10^9 cal. Dividing by *A*, in this case 234×10^6 cm^2, gives 8,423 cal/cm^2 as some sort of extreme yearly budget. In SI units this is 3.53×10^5 kJ/m^2.

In some lakes the annual heat budget is clearly composed of two portions. With *w* standing for winter, the symbol Θ_{bw} is translated as the birgean *winter heat income*. It is defined as the difference between the lowest heat content and that found at the vernal overturn when the lake is isothermal at 4° C. In dimictic and cold monomictic bodies of water much of the winter heat income is used to melt the ice. In those water bodies that fit the definition of a polar lake, never warming above 4° C, the annual heat budget is composed of only the winter income; it is fairly small because the annual range

Table 11-3   Areas and temperature extremes observed at various depths in Tom Wallace Lake, Kentucky, used for calculating heat budget, Θ_{ba}*

| Depth (m) | cm² × 10⁻⁶ | Temperatures (°C) | | | Range × area × 10⁻⁴ |
		Lowest	Highest	Range	
0	234	0	30.7	30.7	7,183.8
1	196	3.6	29.2	25.6	5,017.6
2	164	3.6	28.0	24.4	4,001.6
3	136	3.6	23.3	19.7	2,679.2
4	111	3.8	22.2	18.4	2,042.4
5	85	4.0	19.4	15.4	1,309.0
6	60	4.0	17.8	13.8	828.0
7	34	4.0	10.5	6.5	221.0
8	10	4.0	10.0	6.0	60.0
8.5	1	4.0	10.0	6.0	6.0

*Also see Fig. 11-3.

in temperature is not great. Recalling the heat exchange when pure water freezes or when ice thaws to the aqueous phase, about 80 cal/g, we see that a mantle of ice 1.0 m thick would absorb about 8,000 cal/cm² in the final conversion to water.

A monomictic lake such as Cayuga, although not often freezing, cools below 4° C during winter circulation; its Θ_{bw} is considerable, being about 24% of Θ_{ba}, yet involving no ice melt.

When oligomictic Lake Tahoe circulates, its temperature is very near 4° C. It has, therefore, a winter income of zero. Problems arise in quantifying the winter heat income of warm monomictic lakes circulating well above 4° C. Hutchinson (1957) resolved this by suggesting a negative winter income based on the circulating temperature above 4° C. On the Rio Grande the monomictic impoundment called Elephant Butte mixes at 8° C; from this value and its mean depth of 18 m, a negative winter heat budget is calculated to be −7,200 cal/cm². Farther south, the Guatemalan Atitlán, with a $\bar{z}$ of 183 m, circulates near 20° C during the winter; its negative Θ_{bw} is −288,300 cal/cm².

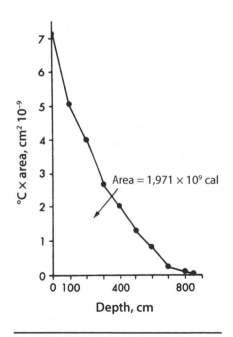

Area = 1,971 × 10⁹ cal

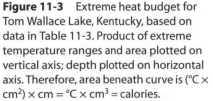

Figure 11-3 Extreme heat budget for Tom Wallace Lake, Kentucky, based on data in Table 11-3. Product of extreme temperature ranges and area plotted on vertical axis; depth plotted on horizontal axis. Therefore, area beneath curve is (°C × cm²) × cm = °C × cm³ = calories.

The summer heat budget (Θ_{bs}) is that part of the annual income extending from the coldest isothermal condition to the maximum summer heat content. In a warm monomictic lake circulating at 4° C or above, Θ_{bs} equals Θ_{ba}. In dimictic and cold monomictic lakes the summer heat budget is one segment of the annual heat budget. The summer income is mostly wind-distributed heat, although other factors are operative, as pointed out earlier.

The preceding discussion refers to apparent heat budgets that are calculated from changes in water-mass temperature alone, known as *sensible heat* changes. Any careful, analytic study considers all avenues by which heat is gained or lost from the lake to arrive at an annual balance sheet. Heat is gained from the internal energy content of influent waters, from freezing, and from the sediments. Evaporation, amounting to considerable loss of heat (see Table 10-2), back-radiation to the atmosphere, and loss through outflow, melting ice, and conduction to the sediments are some of the ways by which heat, gained and distributed by the wind, is dissipated. Any changes in the physical state of water, such as melting or freezing, are separate from those measured by temperature, and are called *latent heat* fluxes; these are important in accurate energy budgets.

In a small lake apparent annual heat budgets are founded on underestimation of calories taken up by the lake, since there is considerable heat exchange between water and sediments. Krumholz and Cole (1959) found that during winter there was a decreasing temperature gradient from the profundal ooze to the overlying water in Tom Wallace Lake. They concluded that the sediments were a source of heat that had been gained during summer and the fall overturn. The classic picture of such a phenomenon was presented by Birge and associates (1928), who inserted temperature probes into the bottom deposits of Lake Mendota. Seasonal temperature changes 5 m deep in the bottom deposits were marked by undulations with an amplitude of 1.6° C. These were compared with a variation of 21.5° C from summer to January at the lake-bottom interface, which was 8 m below the lake surface. In addition, the seasons were out of phase deep in the deposits; the high point, for example, was reached in January when the heat gained in summer finally reached that depth. Beneath 23.5 m of water, the annual variation 5 m deep in the bottom deposits was only 0.7° C. The rate at which heat was conducted through the sediments was very similar to what might occur in still water unstirred by currents.

Heat budgets have become more important in recent years, as scientists attempt to understand the role of lakes in global climate change. As the Earth's atmosphere warms, we expect individual lakes to likewise warm, although the changes will be complex. Tropical and sub-tropical lakes will likely have increased water loss through evaporation, as demonstrated by Lake Vegoritis in Greece, which has greater energy loss through latent heat flux (evaporation) than through warming (Gianniou and Antonopoulos, 2007). Lake Tanganyika had considerable warming over the last century, affecting water density and therefore stratification and mixing; the epilimnion is less productive as a result (Verburg and Hecky, 2009).

Regional Heat Budgets

Lakes in tropical and polar regions have low heat budgets because annual temperature contrasts are not great. Low budgets are especially obvious in equatorial zones.

Deevey (1957) found a remarkable exception: Lake Atitlán (14°40′ N lat) has a budget as high as those of some lakes at latitudes much farther north. The strong daily winds on Atitlán mix a tremendous volume of water, distributing a quantity of heat comparable, for example, to Θ_{bs} of Cayuga Lake in New York, 28° lat to the north-29,480 cal/cm^2.

Gorham (1964) studied heat budgets in 71 temperate lakes for their possible relationship to the effects of morphology. He found the expected good positive correlation between volume and budget. As lake volume increases, there is a rise in heat uptake. The greater the water column, the more heat taken up each year.

The largest annual budgets calculated to date are for Lake Baikal and Lake Michigan; they amount to 65,500 and 52,400 cal/cm^2, or 2.74×10^6 and 2.19×10^6 kJ/m^2, respectively.

Meromixis tends to lower heat budgets because part of the lake does not participate in overturns when heat is gained or lost. The monimolimnion also stores significant quantities of heat so that mean winter temperatures remain unusually high. Anderson's (1958a) comparison of meromictic Soap Lake, Washington, 27 m deep with Θ_{ba} 14,902 cal/cm^2, with its shallower, holomictic neighbor, Lake Lenore, 11 m deep with Θ_{ba} 16,100 cal/cm^2, illustrates this tendency. The partially meromictic Long-H ponds in Arizona have apparent annual heat budgets of about 1,000 cal/cm^2 (Cole et al., 1967). Heat stored in the temporary monimolimnia works against a summer-winter differential; in early February their mean temperatures are over 20° C, although they lie in a region where the mean ambient temperature is only about 2.8° C for that time. A hypothetical dilute-water holomictic pond, having the same dimensions as the Long-H ponds (z_m, 2.0 to 2.5 m) and ranging from 4° to 25° C annually, would have an apparent annual income of 2,500 cal/cm^2.

In small ponds there is a relatively significant sediment heat budget, and in arid regions, especially, evaporation is a factor to be taken into account. Three examples point this out. The sediment budget of shallow ($\bar{z} = 2.4$ m) Dunham Pond, Connecticut, was quantified by Rich (1978), illustrating its importance in small bodies of water. Below a depth of 2.5 m, 78% of the solar input during the heating period was taken up by the bottom deposits. In saline ($\bar{z} = 3.5$ m) Lake Werowrap, Victoria, Australia, the apparent annual heat budget is increased from 1,712 to 2,930 cal/cm^2 when the sediment budget is considered (Walker, 1973). The heat taken up by the mud of Larue Swamp, Illinois, equaled 32% of the summer income in the overlying meter of water, found to be 3,204 cal/cm^2 by Parsons (1975). In addition, the marked effects of salinity on evaporation should be considered. The lowering of the surface levels of the Long-H ponds by some 30 cm during the summer required about 300 cal/cm^2 per day; in a dilute freshwater pond, 250 cal/cm^2 would have sufficed.

Analytic Energy Budgets

Complete analytic energy budgets are relatively rare, perhaps because limnology is dominated by researchers trained in biology rather than physics. Nevertheless, a few published accounts from Lake Mead (Anderson and Pritchard, 1951), some lakes of Israel (Neumann, 1953), Lake Ontario (Rodgers and Anderson, 1961), and Lake Werowrap (Walker, 1973) are evidence of some researchers' interest in physical problems of lakes. The English translation of a book by the Russian Pivovarov (1973) contains technical details of heat balances in lakes and rivers, as well as most aspects of heating, cooling, ice formation, and ice melting.

The data from research at Lake Mead and the lakes of Israel demonstrate an important effect of evaporation in desert waters. Mean temperatures begin to decline before the heating season has come to a close in Lakes Mead and Kinneret (formerly Lake Tiberias or the Sea of Galilee). The cooling effect of intense evaporation is greater than the effect of solar warming. This may be much less pronounced in Lake Mead today than it was at the time of Anderson and Pritchard's investigation because Lake Mead receives colder water from the deeps of Lake Powell.

Analytic heat budgets take into consideration every factor contributing to gain or loss of heat, expressing them uniformly as calories per square centimeter or joules per square meter. The goal is to balance the various positive and negative agents of heat transfer. The following serves as a general equation:

$$Q_T - Q_R - Q_E - Q_S - Q_V = 0$$

Q_T symbolizes the gain or loss of energy revealed by increasing or decreasing water temperatures; it can be expressed as a rate by introducing a time factor. Thus, Q_T could be the change in heat content per unit area per day. Q_R is the net radiation and is the result of direct solar or diffuse radiation from the sky, less reflectance at the water surface and back-scattering from beneath it; Q_E is the energy in evaporation and condensation, usually the former and therefore a loss; Q_S is the conduction of sensible heat from the water to the atmosphere or vice versa and the net transfer of long-wave energy between water and the air, depending on their temperature differences; and Q_V is the net advected energy, such as that entering via influents as balanced against loss in effluents. Symbols could be used for each factor, the equation thereby becoming more complex.

In Lakes Mead and Kinneret, ice would not be taken into account in annual heat budgets. In more northerly lakes, however, ice formation and its subsequent melting are a regular part of analytic and annual heat budgets. Ice is considered "negative heat" storage and is quantified in terms of the heat required to melt it, about 80 cal or 334.7 J per gram.

Arhonditsis and associates (2004) calculated a complete heat budget for Lake Washington using data from the most recent 35 years and found a warming trend consistent with that associated with global climate change, 0.9° C/yr. The heat energy changed most dramatically during the growing season. Recent work in weather and climate models and forecasting have shown that lake heat budgets may have global implications; deep lakes or lake districts with many water bodies especially need consideration of individual lakes, including volume (Venäläinen et al., 1999).

Langmuir Circulation

In 1938 the American physical chemist Langmuir wrote of observations he had made at sea and the subsequent experiments they inspired in a New York lake. His work is an excellent example of the close relationships that exist between oceanography and limnology. During an Atlantic crossing, he noticed parallel, elongate lines of seaweed where the water seemed to be converging. Upon his return, he carried out experiments in Lake George to explain the streaks and brought to light an important mechanism whereby wind distributes surface heat and materials downward.

Elongate, spinning spirals of water, oriented with the wind, are now called **Langmuir cells**. They rotate about horizontal axes (parallel to the water surface and the wind's direction). These cylindrical helices of water alternate in direction of roll from clockwise to counterclockwise where they adjoin (Fig. 11-4). Surface streaks mark parallel lines of convergence and downwelling. Upwelling areas lie between the streaks but are less conspicuous because water diverges there, and floating materials do not accumulate. In many instances oily material gathers in the converging, downwelling lines, accounting for the name "surface slicks." Langmuir spread oil on the surface of oligotrophic Lake George and observed its subsequent convergence and accumulation as elongate slicks, moving downwind with a velocity greater than the adjacent interstreak water. The streaks are sometimes marked by floating debris—autumn leaves in Lake George and windrows of *Sargassum* in the Atlantic.

The streaks mark downwelling lines of great vertical velocity, up to more than 9 cm per second in Lake Ontario, according to Harris and Lott (1973). Langmuir observed that autumn leaves converged in the streaks, and the less buoyant leaves moved downward at rates far in excess of natural sinking, the vertical velocity decreasing with depth. The mechanisms of streak formation are complex, but the downwelling velocities correlate well with wind speed.

Langmuir cells are less obvious in small bodies of water but permit the generalization that action of wind on lake surface is not uniform and has a nonrandom pattern. Langmuir found that helices descended 10 to 15 m in Lake George, mixing the summer epilimnion to those depths. At the autumnal overturn the helices were diffuse but reached the bottom.

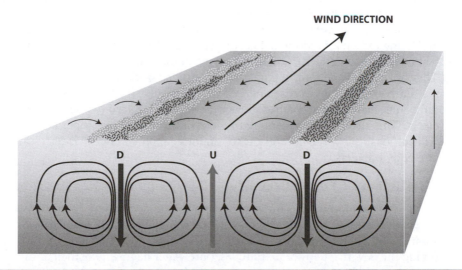

Figure 11-4. Langmuir cells spiral just beneath the surface of the water, driven by wind (oriented in a direction moving away from the reader—see the arrows in the figure). Where Langmuir spirals merge, a downwelling (**D**) is formed, and a slick appears on the surface. The surface above the upwelling (**U**) is clear.

Wind forces, then, account in part for patchiness in the distribution of plankton and surface nutrients. So, care must be taken in sampling. For example, samples from upwelling areas might contain considerably more phosphate than the average upper layers. In sampling near-surface organisms, more representative information will be obtained from plankton-net tows across the wind (crossing upwelling and down-welling, as well as intermediate waters) than from tows aligned with wind direction.

The active role played by the animals in their spotty distribution was demon-strated by Stavn (1971), who induced miniature Langmuir circulations in small tanks containing *Daphnia* and studied the reactions of these cladocerans. During daylight they clumped in low-velocity downwellings, swimming upward against the current. At higher velocities the *Daphnia* assembled in the upwelling side of the Langmuir spi-ral, swimming downward. They clumped at the bottom of the spirals, swimming hor-izontally against currents of intermediate velocities.

In the sea, the accumulation of *Sargassum* and an associated, specialized marine fauna is well known, but congregations of thousands of the poisonous sea snake *Pelamis platurus* in surface streaks of the Pacific are more spectacular (Dunson and Ehlert, 1971).

Austausch Coefficients

The symbol *A* comes from the German *Austausch*, meaning exchange. It pertains to various types of turbulent transport by eddy systems, including heat conductivity, diffu-sivity of dissolved materials, and mass-momentum transfer in various directions. A clar-ification of two important concepts is necessary in any discussion of Austausch coefficients. First is the understanding of turbulence, which is irregular, unsteady motion, in contrast to the smooth, regular laminar flow of fluids or gases. Turbulence is induced when the velocity of laminar flow is increased to a point where friction assumes prominence as fluids rub solid substrates or against each other (see Chapter 8). It is the result of shearing stress. The velocity of turbulence at any point fluctuates in direction and magnitude in a random and chaotic manner. Second, the concept of an **eddy** must be understood; it applies to currents moving in directions different from the main flow. They are circular and swirling motions that may be vertical as well as horizontal.

Early descriptions of turbulent eddy flows emphasized their analogy to molecular diffusions, and molecular diffusivities were simply replaced by the greater eddy coeffi-cients. This kind of comparison with molecules is useful but imperfect. One example will suffice to show the contrast: a continuous supply of energy is necessary to main-tain eddy turbulence.

Coefficient of Eddy Viscosity

The mass-momentum transport refers to the coefficient of eddy viscosity, the excess over normal molecular viscosity, the degree to which water resists flow when some force, including friction, is applied. The viscosity of a fluid is usually tested by timing the flow through standard tubes, pure water having a coefficient of 0.01 (dyne-sec/cm^2) at 20° C.[1] In the desert waters of Mono Lake, viscosity is 20% greater than

---

[1] The *poise* (η) is the standard viscosity unit, equaling a dyne-sec/cm^2. The absolute viscosity of pure water at 20° C is 1.002 centipoises. The viscosity of water at 35° C is less than half the viscosity at 5° C and is 72% of the absolute viscosity.

distilled water and must be taken into account in the study of physical limnology of that lake (Mason, 1967). But normally, viscosity can be ignored in limnologic studies.

Turbulence acts to give the effect of greater internal friction by slowing the descent of sestonic particles and speeding the transfer of momentum. The effect of speeding the transfer of momentum is seen in some data from physical oceanography (von Arx, 2005). If only molecular viscosity of water were taken into account, it would take centuries for a wind velocity of 20 m per second to generate a current of 1 m per second to a depth of 100 m. In the oceans such a current can be developed within a few days, because eddy viscosity can surpass the molecular viscosity of sea water (about 0.0107 at 20° C) by factors ranging from 100 to 10^{10}.

Coefficient of Eddy Conductivity

More effort in limnology has been devoted to the study of coefficients of eddy conductivity—those gentle turbulent currents involved in heat transfer that are mostly inferred from temperature changes—than to coefficients of eddy viscosity.

Limnology owes an important theoretical consideration and treatment of heat transport to the oceanographer McEwen (1929), whose article stimulated further work by Hutchinson (1941). Since then, a pertinent literature on the subject has slowly grown.

The data for calculating coefficients of eddy conductivity are provided by a series of vertical temperature measurements, preferably made at close intervals of time during the heating season. The total temperature change, θ, at a depth, z, divided by the number of days (or some other unit of time) since the start of observations, gives a value $d\theta/dt$, which can be plotted on semilog paper against depth (the depth on the arithmetic axis). From the ordination of points for every depth, a curve may result in which there is a straight segment, showing that the rate of heating falls exponentially with depth (Fig. 11-5 on the next page). Hutchinson (1941) called the layer of the lake represented by this straight line the **clinolimnion**. In the epilimnion above, the points show effects of direct solar heating, wind-induced mixing, and other losses and gains by atmospheric exchange. The results lead to irregularity in the plot, although the points often show somewhat similar heating rates at different depths.

Below the clinolimnion is the deepest part of the lake, which Hutchinson termed the **bathylimnion**. The $d\theta/dt$ plot in this region is usually puzzling because it suggests a turbulent region, the reasons for which are obscure. The rates of heating for bathylimnion depths are fairly uniform. Since the bathylimnion is far removed from direct wind action, several other factors—density currents of chemical origin, turbulence induced by internal waves, and heat gained from the sediments—must influence this divergence from the clinolimnion curve.

The main feature of interest is the clinolimnion, however. It resembles a completely undisturbed stratum of water through which heat is moving; the heat flux would be due to molecular motion alone. For pure water, this molecular thermal conductivity is usually taken to be 0.12×10^{-2} g/cm per second. Over a span of time, then, temperature changes in quiet water would suggest heat progressing at that rate due to molecular thermal conductivity. The difference between this rate of flux and a higher one would be caused by eddy currents and is symbolized A, the *coefficient of eddy conductivity*. In a sense, this conductivity is analogous to the conductivity of some metal alloy:

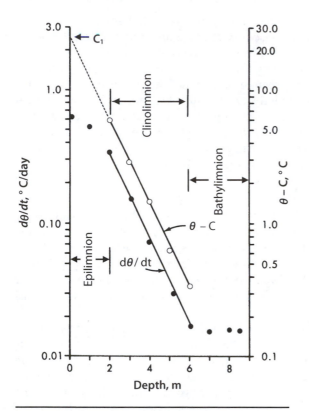

Figure 11-5 Semilog plot of heating rates ($d\theta/dt$), solid circles, and temperature increments ($\theta - C$), open circles, plotted against depth. Heating season of 1951, Tom Wallace Lake, Kentucky. Epilimnion from 0 to 2 m; clinolimnion from 2 to 6 m; bathylimnion from 6 to 8.75 m. C_1 at 25° C indicated by arrow at top left.

downwelling water distributes heat exponentially in the same manner as a solid conductor. The value A (in excess of 0.12×10^{-2} g/cm per second) is the effective thermal conductivity of the eddy current. It is also a *coefficient of turbulence*, carrying gases and other substances downward in excess of the normal diffusion rate. Therefore, *coefficient of diffusivity* is another name for this type of exponentially decreasing eddy. The method Hutchinson used for solving A is permissible only in the clinolimnion. Furthermore, the constancy of A is suspect if a semilog plot of $(\theta - C)$ versus z is not parallel to the $d\theta/dt$-depth line in the clinolimnion (Fig. 11-4). C was the temperature at the start of the experiment, preferably the vernal overturn, and $(\theta - C)$ is, therefore, the temperature increment.

Usually a great many observations are needed to accumulate enough data to calculate Austausch coefficients. Despite this, Idso and Cole (1973) found that very few data were necessary to achieve good results in the sheltered Tom Wallace Lake, and concluded that the mean eddy conductivity there (0.94×10^{-2} g/cm per second) may very well be a real heat transport mechanism. Such may also be true for other small lakes not exposed to strong winds. In any event, the strange eddy currents operating below the epilimnion are probably ultimately wind induced, owing much to the energy of standing waves.

Theory and methodology of eddy conductivity were discussed in detail by Hutchinson (1957), and some subsequent literature on computing coefficients of eddy conductivity was cited by Idso and Cole (1973). The value for A can be solved if within the clinolimnion, the points showing rate of heating ($d\theta/dt$) and those designating the temperature gain since spring overturn ($\theta - C$) lie on straight and parallel lines, decreasing exponentially with depth. This indicates that A is practically constant throughout the clinolimnion and validates the use of the following:

$$\frac{d\theta_z}{dt} = Aa^2 \left(\theta_z - C\right)$$

C, the temperature observed at spring overturn, is a constant, as is a^2, derived from

$$\frac{1}{z}\ln\frac{\theta_z - C}{C_1} = -a$$

In the above, z is depth in centimeters for each point, and C_1 is a constant found by extending the $(\theta - C)$ versus z line to zero depth where it intersects the log axis. This intersection gives what Hutchinson called the *virtual surface temperature* (Fig. 11-5).

Modern techniques and equipment have made it possible to resolve and record vertical temperature profiles at intervals of approximately 1 cm. Even with coarser resolution much can be learned from the "stair steps" and inversions seen in the microstructure of vertical profiles—profiles that are not the smooth plots usually presented. Dillon and associates (1975) reviewed the methodology whereby careful study of small-scale irregularities in temperature-depth profiles can lead to estimates of eddy diffusivity values. Their work on Lake Tahoe showed good agreement between calculations from profile microstructure and the classic method founded on observations made at close intervals throughout the heating season.

Because clinolimnetic turbulence is inferred from temperature changes, direct solar heating could invalidate some A values. However, few authors have made corrections for this possibility. For example, the eddy conductivity coefficient 85.6×10^{-2} g/cm per second for Cultus Lake, British Columbia, is high (Ricker, 1937; Hutchinson, 1957). The relatively large surface area of this lake may be responsible for this high value, since clinolimnetic eddies may ultimately be a function of wind and area. On the other hand, Cultus is a very transparent lake, and direct solar heating could be a factor at times.

Other ways of calculating eddy coefficients are more direct than the various temperature methods, no matter how refined these temperature methods may be. With use of oceanographic techniques the horizontal as well as the vertical eddy diffusivity can be appraised by using a natural radioisotope, the gas radon 222. This tracer, with a half-life of 3.82 days, emanates from radium 226 in the sediments, there being practically no ^{226}Ra in the water. (Moreover, there is essentially no ^{222}Rn to be gained from the atmosphere.) Distance from the sediment source can be related to time through analysis of radiodecay: places where the Rn concentration has decreased by a factor of 2.0 are markers. The molecular diffusion of ^{222}Rn is about 10^{-5} cm^2 per second at 10° C. If the distance at which Rn content becomes halved surpasses its rate of molecular diffusion, the coefficient of eddy diffusivity can be estimated by subtraction. Imboden and Emerson (1978) used the ^{222}Rn tracer technique to quantify Austausch values in the Swiss lake Greifensee. This may have been the first such effort in limnology, although Schindler and associates (1972) used radon to study gas flux at the air-water interface. Eddy coefficients have become important to those limnologists interested in rates at which nutrients are returned to upper waters from profundal sediments and the hypolimnion.

Currents during Stagnation

The word *stagnation* is used to describe conditions beneath an ice cover or thermocline. It does not refer to pollution, anaerobiosis, or saprobic situations; it applies to the absence—or the supposed absence—of water movements.

Bryson, his students, and other coworkers applied meteorologic methods to lacustrine situations, summarizing some results that show that seemingly steady currents exist in the quiet hypolimnion (Lathbury et al., 1960). Currents averaging 7.7 cm per second move above the profluidal floor during summer stagnation in Lake Mendota. The cause of these currents is varied and not well understood. Bicarbonate and other ions and compounds diffusing from sediments could impart density to adjacent water, which then streams down the basin slope. Internal standing waves and turbulence set up by Langmuir cells may also transmit some energy necessary for currents, but they seem to be an insufficient explanation for all of the observed motions in the deeps of Lake Mendota.

Beneath the ice, where the wind is ineffective, there are, surprisingly enough, movements of water and isotherms that refute the concept of a uniform inverse stratification. Samples taken through a single hole in the ice over the deepest part of a lake have yielded most of our knowledge about winter limnology. At Tom Wallace Lake, Krumholz and Cole (1959) drilled hundreds of holes along various transects to study events beneath the ice one cold winter. The results of a series of measurements made during a period of 49 days showed that most water was warmer than 4° C, and bizarre temperature profiles existed that could not be explained by temperature-density relations. The arrangement of water masses beneath the ice must have been unstable, however, for temperature data indicated water movement and especially horizontal currents. Heat gains from the sediments in deeper parts of the basin and from solar radiation through the ice were inferred from abrupt temperature gradients. The evidence was indirect, but the concept of winter stagnation was contradicted by the changing temperature conditions from day to day.

Likens and Hasler (1962), going a step further, placed ^{24}Na, a radioisotope of sodium, beneath the ice of a Wisconsin lake and followed its movements. Within the first day there were rapid lateral movements from 15 to 20 m, and by three days radioactivity was demonstrable 30 m from the source. This was a direct manifestation of water motion. Stewart (1972), after studying temperature beneath ice-covered lakes in New York and Wisconsin, expressed the opinion that there is no absolute stagnation with respect to water motion at any time of year.

There are several ways by which currents are induced beneath the ice. Chemical density currents can form so that water masses stream down the basin slope, reaching equal or slightly greater density layers, where they are deflected horizontally. Small volumes of water beneath littoral ice may become anaerobic or at least rich in carbonic acid. This acid dissolves sedimentary marl, which flows toward deeper levels as calcium bicarbonate in solution. In Tom Wallace Lake a yellowish red mass of ferric iron, precipitated and lying on the bottom down-slope from a subsurface seep, marks the course of a density current carrying invisible, soluble ferrous iron from anaerobic subterranean stores.

Warming of water beneath the ice due to the greenhouse effect increases its density and initiates a downward flow. This is accented in the shallows where solar heat can be absorbed and stored by the sediments, thus warming overlying water even more.

In addition, water just below the ice may cool enough to freeze. As the ice cover thickens and the newly frozen stratum at the bottom gives up many electrolytes, the freezing-out effect increases density of the top water. Subsequent solar warming of this layer would have more consequence than raising the temperature of dilute water.

Seiches

The External Seiche

The word *seiche* (pronounced sāsh) is of uncertain origin but is almost certainly from the French; it was used by lakeshore dwellers at Le Léman (the narrows of Lake Geneva) since at least the sixteenth century and was brought to prominence by Forel (1895).

If a shallow area that was covered by water a few hours earlier is suddenly dry, or at least laid bare and exposed, and sometime later floods again, a seiche has been observed. Usually where the shallows have an almost imperceptible slope, a slight fall in water level exposes conspicuous areas of littoral bottom; yet one of the most obvious seiches occurs in Le Léman and has long been known. There, near the city of Geneva, a funneling effect accents ebb and flow.

In simplest terms the **seiche** is a free oscillation of water, reestablishing equilibrium after having been displaced. The least complicated seiche would be a standing wave oscillating on a single node. Such an oscillation could have been energized by several mechanisms. Actually, any force that can pile water up at one side of the lake and cause an instability may lead to creation of a seiche. The high water eventually streams back to restore equilibrium. It overshoots the mark and returns, the surface rocking with an ever-decreasing amplitude until no free energy remains.

Forel explained the elementary seiche as the result of two long waves traveling simultaneously in opposite directions through each other. In a uninodal seiche, each wave has a distance between crests equal to twice the length of the lake, or at least twice the distance be-

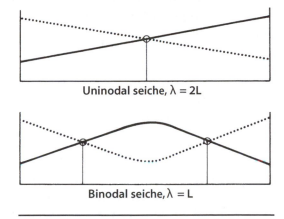

Figure 11-6   Diagram of uninodal and binodal seiches.

tween shores located at right angles to the wave (Fig. 11-6). Sometimes harmonic binodal, trinodal, or polynodal waves make up a seiche, and it may be complicated further by the coexistence of multiple seiches.

In all such instances the wave length (λ) would show the following relation to lake length (L):

- Uninodal $\lambda = 2\,L$
- Binodal $\lambda = L$
- Trinodal $\lambda = 2/3\,L$
- *n*-nodal $\lambda = 2/n\,L$

The periodicity of a seiche is a function of lake-basin morphology. In a simple rectangular pan or aquarium the period could be demonstrated by generating a wave at one end of the tank. It arrives at the other end in time equal to $L/(gh)^{1/2}$, where L is the length, g is the acceleration due to gravity, and h is the uniform depth. The velocity is $(gh)^{1/2}$. Reflection sends the wave back to the original point after the same period of time. The total periodicity (T), therefore, is $2L/(gh)^{1/2}$, twice the length divided by velocity. Using this formula, we can learn something about a wave generated by tipping an ordinary rectangular cake pan 28 cm long and holding 3.5 cm of water. If we assume that the acceleration caused by gravity is 980 cm/sec per second, the velocity of the wave is 58.566 cm per second and the periodicity is 0.956 second:

$$ T = \frac{2L}{\sqrt{gh}} = \frac{56 \text{ cm}}{\sqrt{980 \text{ cm/sec}^2 \times 3.5 \text{ cm}}} = \frac{56 \text{ cm}}{58.566 \text{ cm/sec}} = 0.956 \text{ sec} $$

This example demonstrates the largest possible free oscillation, a wave rocking on one node. The periodicity of such a system with n nodes would be $1/n$ times that of a uninodal wave. It is far more complex in natural bodies of water because they are not contained in vertical-sided pans of uniform depth. In water contained by a bowl parabolic in dimensions, the periodicity of a binodal seiche would be somewhat greater than half the uninodal time. The data from most lakes indicate that the concavities they occupy are more like parabolas than rectangular depressions of uniform depth.

Energy for the Seiche

A common agent supplying the energy for a seiche and working at the lake surface is the wind—a lapsing wind. Strong air currents pile up water at leeward shores and then cease; the pent-up water surges back toward the windward shore as a returning current. A restoring force is acting to reestablish equilibrium following displacement.

A seiche is strengthened by the passage of small barometric fluctuations over the lake, fluctuations of a period matching that of the seiche. Closely associated with this prevalent phenomenon would be the rapid changes in air pressure as a squall blows across the water or as gusts of wind varying in velocity and pressure pound at the surface.

An unusual event brought about by strengthening of a seiche occurred at about 9:30 AM on June 26, 1954, along the southwestern shore of Lake Michigan at Chicago. The water rose rapidly, attaining heights 3 m above normal in some places. Several people lost their lives as a result. A seiche, reflected from the shore near the southeastern border of the lake, had sped at velocities in excess of 100 km per hour, traveling to the scene of the Chicago tragedy in 80 minutes. An explanation for the spectacular wave was pieced together from meteorologic reports by Ewing and associates (1954). A squall line from the northwest, moving with the same speed and in the same direction as the seiche, had energized it. A pressure jump (best described as a high-pressure zone at the squall front) accompanying the high winds had transferred energy to the wave. If the seiche velocity had been less, the squall would have passed over it without such a remarkable effect.

Sudden heavy and localized precipitation—hail, snow, or rain—on a portion of a body of water and the flooding in of a swollen stream at one end of the lake represent other forces that create the gradients and inequalities that power a seiche.

The most spectacular of seiche energizers, although rare, is the earthquake. Seismic tilting in bayous and rivers along the northern coast of the Gulf of Mexico followed the far-distant Alaskan earthquake of March 27, 1964.

One might consider adding the pull of the moon to the list of agents supplying energy to the seiche, but even in the largest of lakes the greatest tidal amplitudes observed are usually no more than 10 mm, and there is evidence that they often are damped out by seiches.

Seiches of noteworthy amplitude might occur if many powering agents are working in harmony, but amplitudes are unusual if they attain 2.0 m.

Internal or Temperature Seiches (Internal Waves)

Mortimer (1952) published results of a study of vertical temperature profiles in Windermere, including some interesting data from a day in late May. At 11:20 AM the 10° C isotherm was about 4 m below the surface; by 4:30 PM it was 4 m deeper, or about 8 m below the surface, and warmer water lay at the 4-m depth. This could mean either marked absorption of heat in 5 hours or vertical eddy currents distributing warm water downward at a remarkable rate. Neither explanation was adequate. The temperature-time data had revealed tilting of the thermocline. A particular type of movement, the internal seiche, had been observed. Pioneer studies on internal rocking in lakes had been detailed 45 years earlier by Wedderburn (1907), who concentrated especially on phenomena in Loch Ness. Because there is no lake bottom exposed, the term internal seiche should probably be replaced by internal wave.

In thermally stratified lakes oscillations of different layers occur. The standing wave generated by wind may not have a great amplitude (measured by surface deviations from equilibrium), but its effect on the stratum of dense water below, marked by the top of the thermocline, is great. The thermocline, or metalimnion, is depressed by piled-up water and is released to bound back when surface waters surge toward the opposite shore. Thus, the rocking thermocline starts out of phase with the standing wave above as it is forced downward by the wave crest over it. The resulting amplitudes and periods are much greater than the oscillations of an external seiche although not readily observable except by measurements of temperatures or, in some instances, by the rise and fall of plankton aggregations. The explanation for the magnified amplitude lies in the very small density differences between waters at the boundary of the epilimnion and metalimnion, compared with the great contrast at the air-water interface. This is demonstrated by a generalized equation for the energy (E) in the amplitude of an oscillating two-layered system:

$$E = \tfrac{1}{2} \, (\rho - \rho') \, g \, (\text{amplitude})$$

In the above equation, ρ is the density of the lower layer and ρ' is that of the upper layer; g is the acceleration due to gravity. If the temperature of a metalimnion is 10° C and that of an epilimnion is 20° C, $(\rho - \rho') = 0.0015$, and the difference between the densities of the epilimnion and of the overlying air is about 0.9970. This shows that for identical amplitudes, the surface wave has about 665 times more energy than the internal wave.

The depression of the lower interface caused by the amplitude of the surface wave is

$$\text{surface amplitude} \times \frac{\rho}{\rho - \rho'}$$

This would be 666.4 for the two temperatures of 10° and 20° C. Thus, a surface seiche of 10 mm could depress the thermocline 6.66 m. A most remarkable internal wave, observed in Lake Baikal, had a period of 38 days and an amplitude of 75 m. Theoretically, a surface seiche amplitude of about 10 cm could have energized this wave.

The above discussion of internal waves has pertained to a two-layered system, but a stratified body of water is roughly three layered. In some studies there has been evidence of two internal oscillations in addition to an external seiche. The lower interface of the metalimnion was displaced so that a second internal wave was generated.

The seiche with all its complexities (harmonic and otherwise) is important to limnology. It is believed that turbulence generated by the free oscillations of the seiche and resultant internal waves create eddies and currents of fundamental importance in mixing and transporting heat, dissolved gases, and nutrients.

Surface Waves

All standing waves, like simple seiches, are characterized by a trough and a crest fixed in space but alternating in time. The water particles move along different paths, although in phase. Vertical movements are pronounced at the crest and trough. At the node only horizontal oscillations prevail. At a time halfway in the period of such a wave all water particles are briefly at rest.

Another type is the progressive wave, crest following crest in such a manner that the water mass within each crest seems to be moving in a given direction with the wind. Actually, all the water particles in such a wave are moved up and down and back and forth so they follow identical paths during a wave period. The water movements describe orbits that are usually ellipses. Such waves have been of great interest to oceanographers, who have developed much of the theory pertaining to them and other forms of water motion. They have not attracted the attention of limnologists and are cursorily dealt with here.

Surface waves start with gentle air currents that form the familiar ripples called capillary waves. Ideally, these tiny swells have a length no more than 1.72 cm between crests, with heights of about 0.022 cm from trough to crest; their maximal amplitude, therefore, is 0.011 cm. Once established, they travel forward and persist until dampened by surface tension. Uneven water surface adds to the effectiveness of the wind. ·

As wind velocity increases, ordinary gravity waves replace capillary ripples, and the relative importance of surface tension is reduced as gravity assumes importance in restoring the wind-induced waves. As winds mount, the waves may reach critical steepness, theoretically when the wave length is seven times the height. In oceans and large lakes where winds are strong, wave length increases faster than height, and eventually this brings about the long swells that escape the influence of the wind that raised them.

The preceding is probably of less limnologic import than what occurs in the shallows where vertical movements in the wave are restricted and the predominant movements are backward and forward. When a wave enters shallow water, it slows down and its leading slope steepens. Its crest suddenly heightens as it overtakes the trough in front of it and then breaks. The breaking of the crest occurs at a place where water depth beneath the trough is about 77% of the height. Theoretically, a wave traveling at

6 m per second would have a height of 55 cm and would, as a result, break when its trough is 42 cm above the bottom. The surf zone formed where the crests break adds a forward component to the waves, which crash against the shore with erosive effectiveness. If the waves are large, a considerable amount of water is thrown up on the beach and flows back to the lake as a return current.

It is in the littoral region that surface waves have the most effect in lakes. Return currents carry eroded shore material lakeward to create subsurface banks that have abrupt drop-offs. Swells approaching the shore obliquely tend to form longshore currents parallel to the shoreline. In this manner, spits are formed that may isolate bays to form shoreline pools. Debris is swept from the bottom, and in windswept shore regions there may be a paucity of aquatic macrophytes compared with the stands in protected bays. Drift lines of debris may be concentrated to provide habitats for various aquatic invertebrates. These are just a small sample of the wave effects in the shallows and at the lake bank.

12

Oxygen and Other Dissolved Gases

Introduction to Chemical Limnology

The remaining four chapters concern chemicals in water: dissolved gases in this chapter, then various solutes thereafter. We'll consider the most important and ecologically significant chemical species among many thousands of potential importance. Any particular lake or stream, for example, could contain pollutants from human activity; recent improvements in analytical techniques allow us to detect legal and illicit drugs (Bartelt-Hunt et al., 2009), endocrine disruptors (Writer et al., 2010), insect repellant, caffeine, and dozens of others (Kolpin et al., 2002).

Regardless of which chemicals or their origin, any study must use recognized methods and proper quality assurance and control (QA/QC)—otherwise, the numbers are meaningless. Important considerations include use of "analytically clean" glassware for sampling and testing, selection of proper time/place/means of sample collection, sample preservation until analysis, and reporting of units of measurement and chemical species with each number. The limnologist may need to use a protocol developed by a particular expert in a field, a government agency, or refer to a published collection of procedures. The most widespread compilation of analyses, *Standard Methods for the Examination of Water and Wastewater*, is published jointly by three water analysis associations with regular updates.

Unlike most dissolved chemicals, dissolved gases are not stable in storage. A bottle of water taken from a pond will almost certainly have lost dissolved gas by the time it arrives at a lab for analysis. For this reason, field tests are common, although some protocols allow "fixing" a gas by chemically combining it with a reagent, stabilizing for later assay.

Atmospheric Source of Gases

Although we may be inclined to associate gases with the atmosphere rather than with water, it must be noted that there are at least five or six important gases dissolved in lakes, streams, and the seas. They all have biologic and physicochemical functions, but they differ from one another in behavior and origin.

Surface waters in contact with the mixture of gases and water vapor called "air," absorb some of its components. Nitrogen, oxygen, and carbon dioxide are especially

Table 12-1 Gaseous composition of atmosphere in percent by volume

Gas	Percent
Nitrogen (N_2)	78.084
Oxygen (O_2)	20.946
Argon (Ar)	0.934
Carbon dioxide (CO_2)	0.040

Table 12-2 Solubility of the common atmospheric gases in pure water at 10° C, mL/L

Gas	At theoretical 1 atmosphere pressure	At normal partial pressure
Nitrogen (N_2)	18.61	14.53
Oxygen (O_2)	37.78	7.90
Argon (Ar)	41.82	0.39
Carbon dioxide (CO_2)	1,194.00	0.39

important because of their essential biologic roles. Nitrogen and oxygen, in addition, are the most abundant constituents of the atmosphere, about 78% and 21%, respectively, at sea level (Table 12-1). Carbon dioxide has about 1/28 the atmospheric abundance of argon, for example, but it is at least 15 times more soluble in water (Table 12-2). For that reason, carbon dioxide, dissolved as a gas by absorption at the lake surface, occupies about the same percentage of volume in water as argon despite the greater presence of argon in the atmosphere.

Among the atmospheric gases found in trace quantities are molecular hydrogen, carbon monoxide, nitrous oxide, ozone, methane, ammonia, sulfur dioxide, and such inert gases as krypton and neon. Water vapor is present in varying amounts, ranging up to 3% by volume.

Henry's Law and Gas Solubility

To understand the physics of gases in natural waters, certain fundamentals must be kept in mind. The first of these is the notion of gas solubility in a liquid. This is a function of the characteristics of the individual gas itself, modified by pressure, temperature, and salinity. Here we must recall Henry's law, which states that at a constant temperature the amount of gas absorbed by a given volume of liquid is proportional to the pressure in atmospheres that the gas exerts. The following equation shows this and will be useful in later pages:

$$c = K \times p$$

Here c is the concentration of gas that is absorbed; it may be expressed in milligrams per kilogram, μ moles, millimoles, milligrams, or milliliters per liter. The partial pressure that the gas exerts is p, while K is a solubility factor, differing from gas to gas.

Most gases obey Henry's law fairly well, and one can predict the amount of an atmospheric component to be found dissolved in lake water. Carbon dioxide, however, may combine with various cations upon entering natural waters to become more abundant than Henry's law would dictate. It is found both free and in combined states.

Effect of Elevation

With an increase in elevation and a thinner atmosphere, the value p in the formula decreases. Therefore, solubility, expressed as the amount of gas dissolved at equilibrium with the air, decreases. Table 12-3 shows that with each 100-m rise above sea level the mean atmospheric pressure decreases by 8 to 9 mm Hg, the amount of gas dissolved at saturation levels decreasing roughly about 1.4% with each 100-m ascent. Thus, solubility of a gas is clearly a function of its partial pressure, p, in the solubility formula.

The pressure reduction with increasing elevation is not as simple as stated above. Actually, from sea level to the first 600 m, the reduction is about 4% for every 300 m; from 600 to 1,500 m the rate is less, about 3% for each 300 m; and from that elevation to 3,050 m the decrease is 2.5% on the average for each 300 m. Moreover, it should be pointed out again that the international standard unit of pressure is the pascal (Pa), not shown in Table 12-3. This is 9.87 atmospheres.

Table 12-3 Factors for correcting partial pressure (p) and relative saturation of gases at different elevations

Elevation (Feet)	(Meters)	Pressure (mm Hg)	(Partial (p) factor)	Solubility factor
0	0	760	1.000	1.00
330	100	751	0.988	1.01
655	200	742	0.976	1.02
980	300	733	0.965	1.04
1310	400	725	0.953	1.05
1640	500	716	0.942	1.06
1970	600	707	0.931	1.07
2300	700	699	0.920	1.09
2630	800	691	0.909	1.10
2950	900	682	0.898	1.11
3280	1000	674	0.887	1.13
3610	1100	666	0.876	1.14
3940	1200	658	0.866	1.16
4270	1300	650	0.855	1.17
4600	1400	642	0.845	1.18
4930	1500	634	0.835	1.20
5250	1600	626	0.824	1.21
5580	1700	619	0.814	1.23
5910	1800	611	0.804	1.24
6240	1900	603	0.794	1.26
6560	2000	596	0.785	1.27
6900	2100	589	0.775	1.29
7220	2200	582	0.765	1.31
7550	2300	574	0.756	1.32
7880	2400	567	0.746	1.34
8200	2500	560	0.737	1.36
9842	3000	526	0.692	1.45

Effect of Temperature

With p held constant in the solubility formula as temperature is altered, a typical feature of gas solubility is observed. The solubility decreases as the temperature rises. This inverse relationship permits the generalization that cold water can hold more gas in solution than warm water. Water is surpassed by some other liquids in its capacity to dissolve gases. Furthermore, the relationship between temperature and gas solubility in water is complex, evidenced by the presence of low points in gas solubility curves plotted against temperatures. The resultant, somewhat parabolic, curves are of little concern to the limnologist, however, because within normal temperature ranges of natural waters the inverse relationship is nearly linear.

Effect of Salinity

The occurrence of various minerals in solution lowers the solubility of gas, but it has been customary in limnologic theory to overlook this. Usually, when saturations are defined, inland waters are considered to be pure with 0% salinity. The reduction of saturation values of gases in seawater, when compared with distilled water, is on the order of 20%. Seawater is variable, but a typical range is about 3–4% salinity (often expressed as parts per thousand, 30–40 ‰); this converted to parts per million or milligrams per liter is 35,000, far above most natural inland waters. Limnologists working in arid regions, however, are advised to keep this in mind, because they may work with saline pools and lakes containing 5 or 6 times the dissolved minerals that seawater holds.

Relative Saturation

Gas saturation is quantified on the basis of equilibria at the boundary of the water surface and the atmosphere. The usual definition of gas solubility is presented as the ratio of its concentration in the solution to its concentration above the solution. Relative saturation is the relation of existing solubility (amount of gas present) to the equilibrium content expected at the same temperature and partial pressure. It is expressed as a percentage. From Table 12-3 it can be seen that a concentration of gas at sea level, representing 100% saturation, would signify 124% of saturation at 1,800 m, other factors being equal. Similarly, a saturated situation at 1,800 m would amount to only 80.4% of sea-level saturation.

Other Sources of Gas

The important gases entering water at the air-water interface may have additional origins within the lake. Obviously CO_2 is produced through respiration and decay, whereas O_2 appears as a by-product of photosynthesis. The other gases common in some aquatic habitats are formed almost wholly within the habitat itself. For example, the aptly named marsh gas, methane (CH_4), is a prominent constituent of some waters but owes its origin to the anaerobic decomposition of plant and animal material rather than to atmospheric exchange. It enters the atmosphere from such sources as northern freshwater wetlands and from the artificially created wetland known as the rice paddy. Another source, perhaps one-fourth as important as the two types of wetland, is intestinal fermentation in domestic cattle and their relatives. Methane is increasing in the atmospheric due to human activities; it is more abundant in the Earth's atmosphere today than it has been in at least the last 400,000 years based on analyses of ice cores (Farmer and Cook, 2013).

Another consequential gas in some aquatic situations is hydrogen sulfide. It is formed within the lake by chemical and bacterial transformations. Although extremely soluble in water, it occurs only rarely in ordinary atmosphere. Therefore, it does not enter via the lake surface.

An important nitrogenous end product of heterotrophic bacterial breakdown of organic substances is ammonia. Aquatic invertebrates also release this gas as a major excretory product. Being extremely soluble, it is swept away from the organisms before toxic effects develop. Ammonia is prominent in the summer hypolimnion of eutrophic lakes, where it may occur as the gas NH_3, the ion, or in undissociated states such as NH_4OH. Its atmospheric store is exceedingly small, although it is believed that it was abundant in the earth's early atmosphere.

Oxygen: Introduction and Methodology

For most organisms, oxygen in the environment is a requisite for life. It is represented abundantly in the atmosphere (Table 12-1) and dissolves readily in water. There is a wealth of data on its occurrence in the seas, lakes, and streams; and a knowledge of the oxygen content and dispersion within a body of water reveals much about the nature of that habitat. Because of efficient methodology and the importance of oxygen, analysis of this gas is one of the first measurements made in lake and stream surveys.

It is quite common now to determine dissolved oxygen (DO) in the field or lab by using an electronic instrument. Electric current passes through a salt solution between metallic leads inside a probe, separated from the sampled water by a membrane. A meter reads the electric current, which is proportional to the DO in the water.

A chemical assay determining the dissolved oxygen in aqueous solutions was introduced many decades ago by Winkler and is still widely used. Modified through ensuing years, the method has proved relatively easy. The theoretical basis for oxygen determination by the Winkler method includes an alkaline phase and an acidic phase in the methodology. The important reactions depend on two facts: first, manganous hydroxide is easily oxidized to manganic hydroxide; and second, manganic salts are unstable in acid solutions with an iodide and revert to manganous salts, the acid radical combining with the iodide and freeing iodine. The sequence is (1) production of manganous hydroxide in the water sample to which manganous sulfate was introduced when KOH plus KI are added; (2) oxidation of manganous hydroxide to manganic hydroxide by the dissolved oxygen in the sample; (3) conversion of manganic hydroxide to manganic sulfate when concentrated sulfuric acid is added; (4) replacement of iodine in an iodide (KI) by sulfate, releasing free iodine; and (5) titration of the iodine solution with sodium thiosulfate until all free iodine has combined into sodium iodide. The end point, marked by the disappearance of the brown iodine color, is made sharper by addition of a starch indicator. The equations follow (reactions 3 and 4 proceed simultaneously):

(1) $MnSO_4 + 2KOH \rightarrow 4\ Mn(OH)_2 + K_2SO_4$

(2) $2Mn(OH)_2 + O_2 + 2H_2O \rightarrow 2Mn(OH)_4$

(3) $2Mn(OH)_4 + 4H_2SO_4 \rightarrow 2Mn(SO_4)_2 + 8H_2O$

(4) $2Mn(SO_4)_2 + 4KI \rightarrow 2MnSO_4 + 2K_2SO_4 + 2I_2$

(5) $4Na_2S_2O_3 + 2I_2 \rightarrow 2Na_2S_4O_6 + 4NaI$

Sources of Oxygen

Atmosphere and Solubility

When the atmospheric mixture of gases is in contact with water, some oxygen goes into solution if the water is undersaturated. It is about one-fourth as abundant in the air as nitrogen but is more than twice as soluble. The amount of oxygen absorbed depends on temperature, salinity, and pressure. Briefly, cold water absorbs more oxygen than does warm water, salinity decreases solubility, and pressure increases it.

Earlier work was reviewed by Benson and Krause (1980), who presented solubility values based on improved Henry coefficients of oxygen, or Henry's law constants (K in the equation $c = Kp$). Their values, given originally to four decimal places, are rounded off and presented in Table 12-4. The numbers show concentrations per volume (liter or dm^3) at various temperatures in equilibrium with a standard atmosphere saturated with water vapor at a total pressure of one atmosphere, including the pressure of the water vapor. At 20° the water vapor pressure is about 2.3% of an atmosphere.

At mean sea level beneath 1 atmosphere of pressure, pure water containing 10 mg O_2 per liter would be only 78% saturated if the temperature were 5° C. At 20° C this amount of oxygen would represent about 110% of saturation. In pure water at 15.38° C, allowed to come into equilibrium with sea-level atmosphere, 10 mg O_2 per liter would represent saturation.

From Henry's law it is evident that saturation, a function of pressure, would decrease with ascent above sea level. Fac-

Table 12-4   Solubility of oxygen in pure water at equilibrium with saturated air at one atmosphere*

°C	mg/liter or mg/dm³	°C	mg/liter or mg/dm³
0	14.62	21	8.91
1	14.22	22	8.74
2	13.83	23	8.58
3	13.46	24	8.42
4	13.11	25	8.26
5	12.77	26	8.11
6	12.45	27	7.97
7	12.14	28	7.83
8	11.84	29	7.69
9	11.56	30	7.56
10	11.29	31	7.43
11	11.03	32	7.30
12	10.78	33	7.18
13	10.54	34	7.06
14	10.31	35	6.95
15	10.08	36	6.84
16	9.87	37	6.73
17	9.66	38	6.62
18	9.47	39	6.51
19	9.28	40	6.41
20	9.09		

*To convert to:
  µg atoms/liter, multiply by 62.50
  ml or cm³/liter, multiply by 0.70
  µ mol/liter, multiply by 31.25
  mg/kg or ppm, divide by the density of the solution
Data from Benson and Krause (1980).

tors to correct for this phenomenon take into account the fact that saturation equilibria decrease with elevation. An example illustrates this. Assume there is a pond lying 1,500 m above sea level with water at 14° C and containing 5.45 mg O_2 per liter. Table 12-4 shows that 10.31 mg/liter is the saturation value for pure water at 1 atmosphere and 14° C. The amount 5.45 is 52.86% of 10.31; this calculation neglects a correction for reduced pressure, which according to Table 12-3 requires a factor of 1.2 for an altitude of 1,500 m. Thus, the product of 5.45 and 1.2 gives a relative saturation of 6.54 mg/liter, and it shows to what sea-level oxygen content 5.45 mg/liter at 1,500 m is comparable. Compared with 10.31, then, 6.54 mg/liter is 63.43%. A simpler step to show that oxygen in the pond water is closer to equilibrium than might be inferred at first glance is 52.86% × 1.2 = 63.43%.

The various minerals dissolved in water lower its ability to absorb and hold oxygen. Therefore, saturation-equilibrium values are greatest in dilute water. Limnologists rarely take this into account, and indeed it is usually insignificant. Compared with distilled water, seawater has its oxygen-dissolving capacity reduced about 18% because of the 35 g of salts dissolved in every kilogram.

Despite the gaseous nature of oxygen, it is rarely expressed as milliliters; commonly the data are set down as milligrams of oxygen per liter, greater by a factor of 1.4.[1]

The addition of atmospheric oxygen to a lake involves two processes. First, there must be a suitable gradient based on partial pressure differences of oxygen between the atmosphere and the water. This is the driving force, accounting for the flux of gas across a quiet boundary layer, a film of stagnant water. Although the direction and rate of transfer depend on the concentration gradient, the thickness of the film is important also. This is because only molecular diffusion would move oxygen across the boundary layer. High temperatures would speed the process, but even at 24° C the rate of molecular diffusion for oxygen is only 2.3×10^{-5} cm^2 per second. Obviously in an anaerobic body of water, if oxygen were distributed by molecular diffusion alone, it would require years for traces to reach 5 m below the surface. This is not the case, however, for a second process comes into play and spreads the gas to deeper layers. Turbulence carries the absorbed oxygen to lower levels, serving meanwhile to maintain a gradient at the surface film so that gaseous molecules continue to enter from the air.

Wind-driven waves and spray increase absorptive surfaces at the air-water interface and promote the eddies and currents that move the absorbed oxygen downward. Conversely, such agitation can bring about a loss of gas from the water, depending on the concentration gradient. Much of these increased rates of exchange is owed to reducing the thickness of the stagnant boundary layer through which gas molecules slowly diffuse. Broecker (1974), in a valuable discussion of oceanic-atmospheric gas exchange, stated that the mean thickness of the boundary layer in the sea is only 17 μm. In a small Canadian lake the layer was estimated to be, on the average, 295 μm thick (Schindler et al., 1972). Tsivoglou (1972), however, minimized the importance of, and even questioned the existence of, the stagnant layer in this discussion of gas transfer in well-mixed streams. He was especially concerned with the problem of reaeration of anaerobic running water, theorizing that parcels of anoxic water were constantly arriving at the surface to absorb oxygen and were being replaced immedi-

---

[1] The density of oxygen is 1.4276 mg/ml at 0° C and 760 mm Hg (101.325 kPa).

ately. Many stream ecologists follow a procedure outlined by Owens (1969) for calculating oxygen diffusion in or out of a lotic stretch. The method emphasizes the importance of flow rate and mean depth but is not concerned with a stagnant layer.

It is not unusual for people to think of a cold, cascading brook as containing the largest amount of oxygen of all aquatic habitats. This has some basis because of the increased solubility of oxygen in cold water and the aerating effect of tumbling waters. On the other hand, extreme agitation promotes gaseous loss, and in instances where values above saturation occur, a natural degasification follows.

There are a few ways by which dissolved gases can occur naturally at hyperbaric pressures, as Bouck (1976) termed supersaturation. He listed five methods, and at least one more can be added to his list. One of the most dramatic of these is the entrainment of air by water cascading over an escarpment or the spillway of a dam, the high pressure of the impacting water driving atmospheric gases into solution. Values up to 150% of saturation have been reported from the waters below Bonneville Dam on the Columbia River. This is similar to high velocity and turbulence in a stream, phenomena that can increase the gas content of the water and may lead to supersaturation, as was pointed out more than 50 years ago by Lindroth (1957).

If a body of water warms, the solubility percentage levels of its dissolved gases automatically rise; supersaturation and subsequent degasification might follow. Thus, diurnal solar heating and seasonal warming are sources of at least temporary supersaturation.

In some regions, cold rainwater and meltwater percolating into soil are warmed geothermally. Thus, cold groundwater with dissolved gases in equilibrium becomes oversaturated following subterranean heating.

Photosynthesis accounts for the high oxygen tensions found in algal-rich ponds during sunny, windless weather or, less commonly, in relatively undisturbed lake strata below the surface where oxygen produced by concentrations of algae accumulate (see Fig. 12-1, *C*).

A sixth method of creating supersaturated gas is exemplified by the emergence of subterranean waters with extremely high concentrations of CO_2, products of respiration, both aerobic and anaerobic decay, and the release from dissolved limestone. Thus, in the limnocrene Montezuma Well the influent water accounts for carbon dioxide concentrations 1,000 times the expected saturation values.

Excess gases, including oxygen, can create problems for fisheries' managers. The condition "gas bubble disease" has been known for many years (Fickeisen and Schneider, 1976). Lethal conditions are approached when fishes are subjected for a few hours to dissolved total gas pressures in excess of 115%. Bubbles begin to form within the tissues, and eventually gas emboli accumulate in the capillaries of the gills, leading to anoxia and death. This is especially true in very shallow waters; with greater depth the dangerous bubbles are kept in solution by hydrostatic pressure.

Invertebrate animals are also prone to gas bubble stress, although some (for example, crayfish, stoneflies) are hardier than fish (Nebeker, 1976). *Daphnia,* however, is as susceptible as fish, although the bubble effects differ. This may explain the puzzling absence of cladocerans and perhaps calanoid copepods and rotifers from the planktonic assemblage of Montezuma Well. The lack of a fish fauna in Montezuma Well is probably not a function of hyperbaric CO_2 pressure (more than 15 mmol/liter in the Well). (Natural concentrations of this gas rarely reach 30 mg/liter or 0.68 mmol/

liter). The tremendous amounts of CO_2 interfere with respiratory exchanges at the gill surface, and this could prove lethal before the appearance of gas bubble stress. The effects of high CO_2 are difficult to isolate, however, for in situations such as Montezuma Well the gas is accompanied by remarkable levels of bicarbonate (682 mg/liter; 11.2 mmol/liter). Studies on fishes' branchial carbonic anhydrase systems suggest

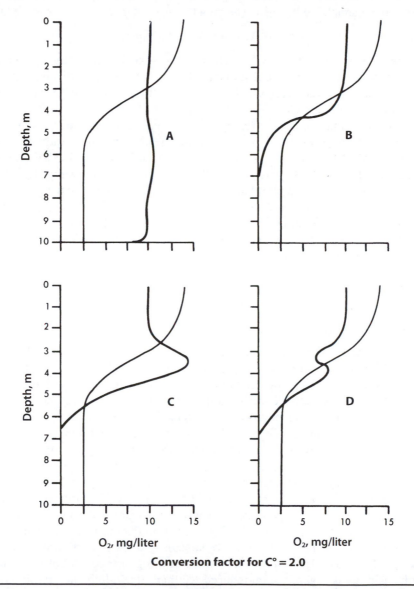

Figure 12-1 Oxygen profiles in thermally stratified lakes. **A**, orthograde oxygen profile; **B**, clinograde oxygen profile; **C**, positive heterograde oxygen profile; **D**, negative heterograde oxygen profile. Oxygen plotted with thicker lines; temperature plotted uniformly and with thinner lines; figures on x-axis times 2.0 = °C.

relationships among bicarbonate, salt transport, and acid-base exchange (Haswell et al., 1980). Normal osmoregulation and pH adjustment might be inhibited by enormous ambient bicarbonate concentrations.

Oxygen from Photosynthesis

Most lacustrine oxygen originates as a by-product of photosynthesis. The general and much simplified equation showing the process in green plants is the familiar

$$6CO_2 + 6H_2O \rightarrow C_6H_{12}O_6 + 6O_2$$

This reaction is powered by light, requiring 674 kcal/mol of sugar produced, as CO_2 is reduced to a unit of carbohydrate (CH_2O) and water is oxidized to O_2 by dehydrogenation. The oxygen resulting from photosynthesis comes, therefore, from the water rather than from the CO_2.

If only simple carbohydrate were produced, each mole of CO_2 absorbed would account for the release of 1 mol of oxygen. The photosynthetic quotient, PQ (the ratio $+\Delta O_2/-\Delta CO_2$), would be 1.0. Many observations have shown that this is rarely the case, and limnologists assume that a PQ of 1.2 is a realistic average figure.

In most lakes the phytoplankton contribute the bulk of the oxygen supply because tremendous amounts of chlorophyll are present in epilimnion algal populations. In shallow waters the limnetic phototrophs may be overshadowed by littoral species—the macrophytes, the attached algae, and benthic algal mats being the chief producers. In small brooks the periphyton algae account for most autochthonous production, although organic detritus imported from outside may be more important in the economy of the stream.

Loss of Oxygen

Decreases in oxygen can be attributed mostly to the respiration of plants, animals, and the aerobic bacteria of decay. Purely chemical oxidation can occur, but many oxidative processes in aquatic habitats are probably mediated through bacterial action.

There is considerable loss of O_2 at the interface of organic lake sediment and the overlying water. In this microzone where much of the organic decomposition occurs, the oxygen content may be much less than it is a few centimeters above. For this reason the morphology of a lake can influence the vertical oxygen curve. A certain region of water may have an unusually great area of sediment in contact with it compared with similar regions having a deeper water column.

Gas bubbles rising from the sediments remove oxygen from the water. This brings to mind a method of producing anaerobic water by bubbling nitrogen through it. Many years ago Lindeman (1942a) needed anaerobic lake water for some experiments on the tolerance of benthic organisms to anoxic conditions beneath the ice in Cedar Bog Lake. He toyed with the idea of using nitrogen bubbles but finally settled on shaking the organic, semi-reduced sediment (including bacteria) in bottles of lake water and allowing it to settle. When tested a day later, the supernatant water was found to be anaerobic and, therefore, sufficed for his experiments. Bacterial uptake may have been most important, but purely chemical oxidation unmediated by microorganisms can occur, especially in the sediments of dystrophic lakes. The difference between oxygen uptake in unmodified sediments and in those poisoned by formalin reveals the part played by bacteria in depleting the gas.

On a purely physical basis, the warming of a summer epilimnion could account for oxygen decrease. This is, of course, a function of lowered solubility of gas as water temperature rises.

The massive fish kills that occur sometimes in shallow, ice-covered lakes are the result of several factors. There is no gaseous exchange with the atmosphere, and if enough snow is present to prohibit light penetration, photosynthesis ceases. Respiration predominates, and in a shallow dimictic or cold monomictic lake with a small initial oxygen content, the level of this gas may fall to lethal thresholds for fish.

In some instances, attempts to alleviate low oxygen tensions by pumping air through holes in the ice have proved disastrous, for the organic sediments were stirred up, creating a situation reminiscent of Lindeman's bottles. An old article by Greenbank (1945) is still a valuable reference for the effects of snow and ice cover on aquatic fauna.

The amount of oxygen required by fishes is to a great extent a function of temperature. Moore (1942) found that during summer months when water temperatures range above 20° C, northern fishes usually die if confined to waters containing no more than 3.5 mg O_2 per liter, but this level is not lethal during winter when metabolism is low. Extreme and unusual adaptations are seen in some fish of desert thermal springs. Hubbs and Hettler (1964) reported that *Crenichthys* from a Nevada hot spring could survive less than 1.0 mg O_2 per liter at 30° C.

Catastrophic drops in summertime oxygen have been reported on several occasions despite the absence of an ice cover. These have been brought about by the concurrence of calm, hot weather with the decomposition of a massive organic aggregate derived from the death of an immense algal bloom. Barica (1975) reported the ability to predict summer fish kills in certain Manitoban prairie ponds. Kill occurred only in ponds with salinities in excess of 500 mg/liter and especially in those that had shown high winter concentrations of ammonia beneath the ice. The collapse of an enormous plankton bloom (usually the cyanobacterium *Aphanizomenon*), represented by more than 100 mg chlorophyll *a* per liter, brought about oxygen depletion and fish suffocation. A summer fish kill in a small Ontario lake was associated with an artificial destratification project. This brought about a sudden collapse of a tremendous population of the dinoflagellate, *Ceratium hirundinella*, rather than the typical breakdown of a cyanobacterial bloom (Nicholls et al., 1980).

In tropical regions a fish kill might be correlated with circulation periods. The oligomictic Ranu Lamongan, one of the Indonesian crater lakes studied by the German Sunda expedition in 1928 (Ruttner, 1931), turned over in July of 1974, following a long period of stagnation (Green et al., 1976). The result was a temporary deoxygenation of the small lake and the death of many fish.

The enrichment and pollution of the western arm of Lake Erie that had been going on for decades was brought suddenly to light in the summer of 1953 when a period of windless days permitted thermal stratification (Britt, 1955). The lower waters became anaerobic and destroyed the mayfly naiads (Ephemeroptera) that comprised an important element of the bottom fauna.

Vertical Distribution of Oxygen in Lakes

The sources of dissolved oxygen in a lake include that released during photosynthesis, and exchange with the atmosphere at the surface. A deep stratum is removed from both these sources if light does not penetrate to it. As a result, respiration and decomposition prevail there; oxygen is consumed rather than produced.

At times of circulation—during the spring and autumnal overturns in dimictic lakes, for example—oxygen is distributed more or less uniformly from top to bottom. If one should plot a curve based on oxygen values in relation to depth, the line would be nearly straight. This is an **orthograde** curve.

When thermal stratification occurs during summer months and the lake is no longer homogeneous throughout, the tropholytic zone becomes isolated from the upper waters. Now oxygen begins to be consumed there. In lakes with large hypolimnion volumes and relatively little production of organic matter in the epilimnion above, the demands on the oxygen in the tropholytic zone are so small that it shows no appreciable decline. The summertime oxygen profile, therefore, is orthograde despite thermal stratification. This is characteristic of oligotrophy (Fig. 12-1, *A*). The biomass, the ratio of epilimnion volume to hypolimnion volume, and the hypolimnion temperature interact to produce vertical oxygen curves.

If environmental factors favor the production of a large epilimnion biomass, the situation will be quite different during summer stratification. Great quantities of dead and dying organic matter cause a severe drain on the oxygen in hypolimnion waters. A decrease in oxygen occurs, and the vertical curve is now termed **clinograde** (Fig. 12-1, *B*). Clinograde oxygen distribution characterizes stratified eutrophic lakes.

In some lakes unusual oxygen distribution is observed. In one example the vertical profile shows a peak in the thermocline region; this is called the "metalimnetic oxygen maximum" by many authors. It is *positive heterograde distribution* (Fig. 12-1, *C*). The peak may be caused by gas well above saturation, in some instances over 400%. In many lakes dense layers of *Planktothrix agardhii* are responsible for unusual oxygen concentrations that persist throughout the summer. This cyanobacterium thrives in the dim light of the metalimnion; much of the oxygen it produces accumulates because photosynthesis exceeds respiration there, and turbulence is low at that depth. Many of these lakes have remarkably high relative depths (z_r). Eberly (1964) summarized what was known then about occurrence of positive heterograde oxygen distribution in the world's lakes. Since then another cause for the phenomenon in Mountain Lake, Virginia (z_r, 6.4%), was reported by Dubay and Simmons (1979). Neither phytoplankton nor epiphytic algae make important contributions to the maximum; instead, 39 species of macrophytes, growing down the slopes to a depth of 11 m, produce oxygen that moves horizontally to attain supersaturated levels in the late summer metalimnion (Fig. 10-9).

Persistent oxygen minima within vertical profiles are harder to explain. Such distribution is *negative heterograde* (Fig. 12-1, *D*). Some metalimnion oxygen minima are explained by respiration of a marked concentration of nonmigratory animals. Other possible causes were summarized by Shapiro (1960): oxygen consumption by decaying seston, having been slowed in its descent by colder, denser water; unusual morphologic features of the lake basin and horizontal movement of water from regions where organic sediments have lowered oxygen levels; and the phenomenon occasionally seen in artificial lakes where density currents result in water masses with low concentrations of dissolved oxygen becoming interpositioned between well oxygenated layers.

It is not unusual for a lake to show at least three types of oxygen curves during the year. Little Tom Wallace Lake is orthograde from late October to early April, then a positive heterograde distribution persists a while, but by midsummer there is a definite clinograde profile that remains until the autumnal overturn.

Measurement of Community Metabolism

The Light-Dark Bottle Technique

Community metabolism refers to the balance of production (evidenced by increase in oxygen, or removal of carbon dioxide from the water column by plants) and respiration (CO_2 added/O_2 removed). Metabolism is an important integrator of a variety of ecosystem processes, and as such indicates the overall condition of a particular lake; it is also of global concern, since respiration in lakes is estimated as 62–76 Tmol of carbon annually in the world's lakes, a significant global flux. Pace and Prairie (2005) present an overview of relevant data and also review methods used to determine respiration in lakes.

The estimation of total photosynthesis beneath a unit of water surface can be estimated in more than one approach. Although preceded by workers in the late nineteenth and early twentieth centuries, Gaarder and Gran (1927) are often credited with early attempts that led to development of one technique known as the light-dark bottle method.

The procedure begins by making a subsurface collection with an appropriate water sampler. After the sample is thoroughly mixed, three glass-stoppered reagent or BOD (biochemical oxygen demand) bottles are filled with what should be exactly equal quantities of water and its contained phytoplankton. One of the bottles is light-tight; this is the dark bottle, wrapped and blackened with electrician's tape or some other material that allows no light to strike the algae within. The dark bottle and a clear bottle are stoppered, immediately returned to the level from which the sample was taken, and suspended for a period of time. The third portion of the sample is analyzed quickly for dissolved oxygen. It is called the initial bottle, and its oxygen content represents the oxygen at the beginning of the experiment. IB, which equals the amount of oxygen in the initial bottle, is the base from which production will be calculated.

Later the subsurface bottles are brought up, and the oxygen they contain is assayed. The oxygen in the clear, light bottle is designated LB. Ideally, it has increased since it was first collected. LB – IB = net gain in oxygen, or net primary production expressed in terms of oxygen released. Now it is obvious that some oxygen could have been produced and subsequently consumed by respiratory activities in the bottle and, therefore, would not be available for testing. Activity in the dark bottle is assumed to be purely respiratory, and DB represents the oxygen remaining in the dark bottle after a period of total respiration. Theoretically, respiration (R) would have been similar in both the light and dark bottles and is shown by the decrease in oxygen content in the latter, so that IB – DB = R, the amount of oxygen that was respired in both bottles.

Net production plus respiration equals gross production, GP, the total amount of oxygen produced by photosynthesis during the time in question. Because the net gain plus respiration, yielding GP, is (LB – IB) + (IB – DB), cancellation leads to LB – DB = GP. Thus, gross primary production could be ascertained with only the light and dark bottles' oxygen contents after a given time, but without the initial bottle test neither respiration nor net production can be learned.

In eutrophic waters a test running from dawn to noon or from noon to dusk is adequate. Doubling the results approximates diurnal production and respiration. Dividing by the number of hours of incubation gives mean hourly rates. In oligotrophic waters with low phytoplankton densities, longer periods of time are needed for detectable oxygen changes to occur. But this is not satisfactory because the sampled plankton community should be kept as short a time as possible in the unnatural confinement of bottles.

In polluted ponds where, perhaps, the phytoplankton cells are so numerous that they make a green, soupy suspension, care must be taken not to incubate too long. Photosynthesis may proceed so rapidly that the water will become more than saturated and bubbles of oxygen will accumulate, or respiration may be so intense that the dark bottle becomes anaerobic. In each instance the endpoint was overshot, so rate cannot be determined; one does not know when the light bottle first became saturated, nor when oxygen reached the zero level in the dark bottle.

In eutrophic waters the oxygen method is probably more reliable than other procedures, but it also has faults. In an often-quoted article, Pratt and Berkson (1959) reported on their experiments that underscored sources of error in the light-dark bottle method. In addition to showing that much of the respiration attributed to the phytoplankton was actually bacterial, they demonstrated that phytoplankton and bacterial populations in the two bottles changed differentially after two days of confinement and that conditions inside the bottles did not mirror the situation outside.

The net production concept is based on the notion that respiration is caused by plants metabolizing their own photosynthates and tissues. This is a somewhat idealized approach. Some of the new primary production could have been passed on to herbivorous animals during the incubation of light-dark bottles. In a sense, the problems are like those inherent in estimating net production by harvesting a crop at some given time and comparing its mass and energy content with what it had been earlier. The part nibbled away by herbivores in the meantime is not evident, and, therefore, the estimate of net production is too low. For this reason some limnologists filter out the larger zooplankters before incubating.

About the time Pratt and Berkson were experimenting with dark and clear bottles, a few plant physiologists were commencing work on a new area of research, the phenomenon of **photorespiration**. Research has shown that respiration in the dark bottle may be quite different from that in the light one because of this phenomenon.

Photorespiration, occurring only in the light, reduces the rate of photosynthesis in some plants as CO_2 is depleted, and an energy-wasting competing reaction occurs. Oxygen serves as the electron acceptor, and the process is sensitive to oxygen tension. However, other plants may remove practically all the CO_2 from the environment without marked photorespiration occurring. At least some freshwater algae, such as *Chlorella*, are the latter type. The phytoplankton is complex, and over geologic time the algal flora has evolved different photosynthesis strategies (Tortell, 2000). Still, because it is favored by high light intensities and low CO_2 values, photorespiration may be accented in algal assemblages incubated in light bottles near the water surface.

Dark respiration differs. Its rate is more or less insensitive to ambient oxygen but levels off when the gas concentration is very low. It is centered in the mitochondria, coupled to ATP (adenosine triphosphate) generation, and the speed at which ATP is utilized governs the rate—the uptake of oxygen and evolution of CO_2 in the dark. In addition, there have been reports of dark respiration being surpassed threefold by light respiration. Therefore, there has been serious undermining of the assumption that the value R derived from decrease of oxygen in the dark bottle is always the same as that in the light bottle. This has led to the technique of incubating the experimental bottles for 24 hours, lessening errors caused by difference in night and daytime respiration.

Closer to the point was Golterman's limnologic study (1971) using organic inhibitors[2] such as those employed by plant physiologists. He found that oxygen consumption in the light was usually greater than respiration in the dark. This means that gross primary production might surpass that calculated from the usual light-dark bottle studies.

Diel Oxygen Changes in Natural Waters

Estimates of diel (24-hour) production can be made in natural waters by considering night as the dark bottle and day as the clear bottle. The increase in oxygen from dawn to dusk reflects net primary productivity. The decrease from dusk until dawn represents half the diel respiration. Adding the oxygen that disappeared at night to the daytime gain gives a sum that is daily gross photosynthesis.

An error that is inherent in any such consideration of unconfined water results from the diffusion of oxygen across the air-water boundary, both in and out of lake or stream. Ideally, a study of daily photosynthetic rates would be carried on when wind-induced turbulence is at a minimum. Should an afternoon gale spring up, the data for the day must be discarded. Similarly, a stream with considerable turbulence would require measurement of the gas changes through aeration processes.

Odum (1956b) described a method of estimating community metabolism from oxygen variations in lotic environments. Much like the diel oxygen flux in a lake, stream water downstream gains oxygen when photosynthesis occurs in daylight and decreases at night. Benthic plants and phytoplankton release oxygen into the water during the lighted hours. Benthic and planktonic organisms take up oxygen continuously. Perhaps chemical oxidation, especially in the sediments, also constantly depletes oxygen; low-oxygen groundwater may enter through the streambed. Furthermore, there is an exchange of oxygen with the air, the direction depending on the saturation gradient. McCutchan and colleagues (1998) describe methods for estimating uncertainty in measurements and improving the accuracy of the procedure, while Hall and Tank (2005) specifically address the effects of groundwater inputs on metabolism estimation.

The overall approach involves establishing two stations marking a stretch of stream, determining the area in square meters of the bottom, and monitoring the flow

---

[2] DNP (2,4-dinitrophenol) inhibits bacterial action without killing the microflora, and DCMU (dichloro-phenyl-dimethyl-urea) stops photosynthesis. DCMU does not interfere with respiration in the dark, but typical photorespiration seems to be inhibited by it.

of water during the study period. The change in dissolved oxygen concentration from the upstream station to the downstream site is recorded, so rate of production in terms of oxygen per square meter per hour can be estimated.

Following is a very simple example of production rate, uncorrected for diffusion and respiration. One June day starting at about noon, there was a gain of 1.17 mg O_2 per liter in a stretch of the stream leaving Montezuma Well, Arizona (Cole and Batchelder, 1969). The gain occurred during a flow of 40 minutes in a section of the stream with a total area of 1,053 m^2. The outflow from Montezuma Well was 256.8 m^3 per hour, or 171.2 m^3 per 40 minutes. The net production rate for 40 minutes was then

$$171.2 \text{ m}^3 \times 1.17 \text{ g O}_2/\text{m}^3 \times 1/1,053 \text{ m}^2 = 0.190 \text{ g O}_2/\text{m}^2$$

The hourly production rate was 0.285 g O_2 per square meter in that segment of the Arizona stream. As will be shown later, this represented the fixation of about 0.09 g C per square meter per hour or the net production of organic substances equaling 1.0 kcal/m^2. There were no data on respiration, so gross production for the 40-minute flow cannot be stated.

Profiles of Production Rates

As described in Chapter 10, the water column of a lake can include distinctive layers, with little or no vertical mixing. We may need to incubate water from several depths in light and dark bottles. From such experiments vertical profiles of photosynthetic rates can be plotted. The results give at least four types of curves (Fig. 12-2).

Findenegg (1964) clearly showed three of these curves, which he termed classes, in lakes of the European Alps, and in all three of his classes the rate of production is low at the surface. Within the upper meter the inhibitory effects of intense light, and perhaps specifically ultraviolet, are evident. The photoinhibition shown so clearly in Figure 12-2 is probably emphasized by the in situ light-dark bottle method. Phytoplankton assemblages trapped and held near the surface in glass containers are very different from those circulating in a turbulent epilimnion. Comparative studies by Harris and Piccinin (1977) imply that vertical mixing in the water column is sufficient to prevent photoinhibition. As a result, production rates in a well-mixed layer may be greater than would be inferred from light-dark bottle incubation experiments. This was substantiated by Marra (1978), who raised and lowered light-dark bottles in an attempt to mimic the motion provided by Langmuir circulation. Comparisons with concurrent carbon fixation in stationary bottles, suspended at different depths in the traditional manner, revealed a range of higher photosynthetic rates (19% to 87% greater) in the moveable bottles that simulated the transport of phytoplankton in a mixed layer.

Findenegg's first class is a production-depth curve in which there is a maximum rate in the upper epilimnion and a rapid decline deeper. This class is typical of lakes rich in phytoplankton and with a resultant low light penetration. This type of curve characterizes, for example, Lake Minnetonka, Minnesota, a very rich lake with cyanobacteria common in the phytoplankton. An article concerning photosynthesis in this lake (Megard, 1972) is a valuable reference to the interaction of factors contributing to planktonic productivity.

The second class contains no distinct maximum within the depth-production curve. It is more or less orthograde, although production is low in the surface stratum. Light penetrates deeply, and its lack is not an immediate limiting factor as in the first class. Nutrients are relatively low, and phytoplankton populations are impoverished. Photosynthesis occurs at a low rate throughout the euphotic zone.

The third class shows two peaks, one epilimnetic, the other metalimnetic. This brings to mind the heterograde positive oxygen curve, either temporary or prolonged.

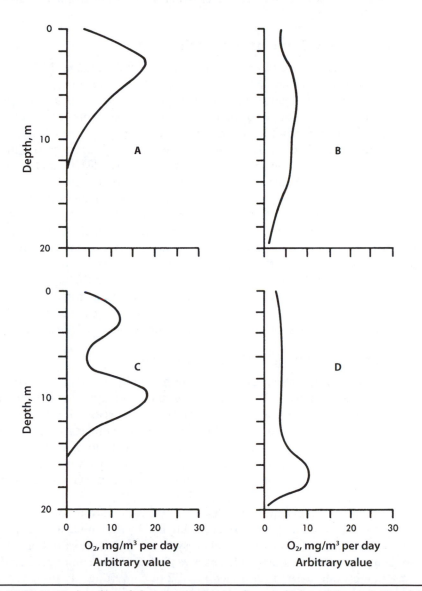

Figure 12-2 Vertical profiles of photosynthetic rates discussed in text. (Classes **A**, **B**, and **C** adapted from Findenegg, 1964; Class **D** from Rodhe et al., 1966.)

Perhaps the upper peak is due to optimum light conditions, the lower to favorable nutrient accumulation in addition to light penetration. In such cases, the euphotic zone extends into the metalimnion or even into the upper hypolimnion, rich in nutrients.

Another type of curve exists in some extremely oligotrophic high mountain lakes containing very clear water with great light penetration. Rodhe and coworkers (1966) explained that the trophogenic zone is far removed from the surface as a result of the inhibiting effects of ultraviolet light penetrating farther than in most lakes. Below the influence of ultraviolet, in a deep layer where visible light is still adequate, the phototrophic algae thrive.

An Integrated Curve of Production

If one plots the rate of oxygen increase against depth (Fig. 12-2), a curve is produced that may represent either net or gross production in terms of some unit of time—an hour or a day, for example. The titrations were probably done in terms of milligrams of O_2 per liter, which could be expressed as grams per cubic meter with no change. The area beneath the curve represents, then, grams of O_2 per cubic meter × meter (Fig. 12-3). By cancellation this becomes grams of O_2 per square meter in terms of some unit of time. The area of the curve can be established, and the primary productivity (expressed as oxygen) under 1 m^2 of lake surface can be found. If, however, you are a stickler for details, you will not accept this as a measure of productivity for the entire lake. Morphologic details were not taken into account. When the volume of each stratum is multiplied by the grams of O_2 per cubic meter produced within it and when the products for all strata are summed, the total production is revealed. Dividing the total oxygen by the surface area gives a depth-productivity figure for the entire lake that differs from that below the square meter where the oxygen production was determined. A modification of this procedure is shown in Table 12-5, where oxygen, relative areas, and their products are shown. When plotted against depth (Fig. 12-4), the curve encompasses total production per square meter of lake surface.

Conversion of Oxygen Data to Carbon

In many studies, researchers express primary production in terms of carbon fixed rather than oxygen evolved. Oxygen values, therefore, are often converted to carbon. One method would assume that 1 mol of oxygen is released for each mole of carbon dioxide that is fixed, as implied in the simple photosynthetic formula. The molecular weights, 44 for CO_2 and 32 for O_2. permit use of the factor $44/32 = 1.375$ to convert oxygen evolved to CO_2 consumed. Similarly, carbon, with an atomic weight of 12 compared with oxygen's molecular weight of 32, allows the factor $12/32 = 0.375$ to be used for conversions (Table 12-6 on p. 272).

The photosynthetic quotient (PQ), however, is not always one-to-one: if only a 6-Carbon sugar were being synthesized, as implied by the simple formula, the ratio O_2 / CO_2 would yield a PQ of 1.0; but photosynthetic quotients vary. For example, if fats are being synthesized to a significant extent, the ratio becomes greater than one. At present it is customary to use a factor of 1.2 in converting oxygen released by photosynthesis to carbon simultaneously fixed. On the average, oxygen is 1.2 times as great on a molar basis. For example, if one assumes that the annual production established

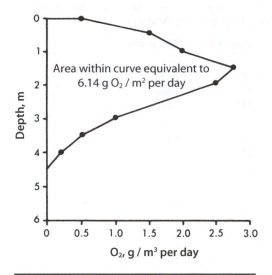

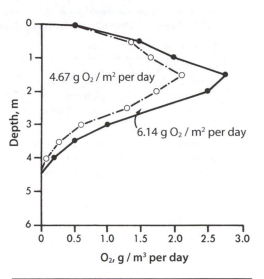

Figure 12-3 Integrated curve of oxygen production beneath 1 m² and uncorrected for morphologic factors (data from Table 12-5).

Figure 12-4 Two integrated curves of oxygen production: the solid-line curve (from Fig. 12-3) shows production beneath a square meter of the lake, uncorrected for morphologic details; the dotted-line curve, which shows production beneath same area, corrected for morphology, is only 0.76 the rate derived from the uncorrected curve (data from Table 12-5).

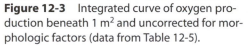

Table 12-5 Data from Tom Wallace Lake, Kentucky, used to compute primary production in two ways*

I Depth (m)	II Oxygen (g/m³ per day)	III Area$_z$ (ha)	IV A_z/A_0	V Column II × column IV (g O_2 /m³ per day)
0	0.50	2.34	1.0	0.50
0.5	1.50	2.14	0.914	1.37
1.0	2.00	1.96	0.837	1.67
1.5	2.75	1.80	0.769	2.11
2.0	2.50	1.64	0.701	1.75
2.5	2.00	1.49	0.637	1.27
3.0	1.00	1.36	0.581	0.58
3.5	0.50	1.22	0.521	0.26
4.0	0.20	1.11	0.474	0.09
4.5	0.00	0.85	0.460	0.00

*Column II plotted against depth gives unweighted production per square meter as in Fig. 12-3.
Column V plotted against depth gives primary production for the entire lake referred to unit area, square meter, as shown in Figure 12-4.

Table 12-6 Some conversions pertaining to primary production

To convert	To	Multiply by	Reciprocal
Assuming PQ = 1.0			
O_2 released	CO_2 absorbed	1.375	0.727
O_2 released	C fixed	0.375	2.667
O_2 released	(CH_2O) produced*	0.937	1.067
Assuming PQ = 1.2			
O_2 released	CO_2 absorbed	1.146	0.873
O_2 released	C fixed	0.312	3.205
O_2 released	(CH_2O) produced	0.889	1.125
Whatever the PQ			
g O_2 released	kcal	3.510	0.285
g CO_2 absorbed	kcal	2.553	0.392
g C fixed	kcal	9.361	0.107
g (CH_2O) produced	kcal	3.744	0.267
kcal	kJ	4.187	0.239
kcal/sec	kW	0.239	4.187

*(CH_2O) equals a unit of a carbohydrate molecule, one sixth of glucose, for example.

by Winkler analyses is 1.0 kg O_2 per square meter and that PQ is unity, 0.375 kg of carbon per square meter is the equivalent. If, however, a PQ of 1.2 is employed, the calculations become

$$1.0 \text{ kg } O_2 \times 0.375 \times 1/1.2 = 0.313 \text{ kg C}$$

It is conventional to phrase primary productivity in terms of the area on which radiant energy falls. The photosynthesis occurring beneath a square meter of a shallow, eutrophic pond may be intense, but the water column is short. For this reason the production per square meter in such a pond may be surpassed by that in a deep oligotrophic lake with marked light penetration. In volumetric terms of carbon fixed or oxygen released per cubic meter, however, the eutrophic pond's photosynthetic activity may be far greater.

An example of this is shown by a comparison of some data from transparent Lake Baikal with that of an imaginary small pond. The average Secchi disc transparency is 13 m, although 41 m has been recorded from Baikal (Kozhov, 1963), and photosynthesis occurs in the upper 50 m.

Data from Winberg (1963), Kozhov (1963), and Moskalenko (1972) suggest that 2.0 g O_2 per square meter could represent a typical summer day's production. This means that the average production per cubic meter in the 0 to 50 m stratum would be 0.04 g O_2. A pond with a mean depth of 2 m and with an identical intensity of photosynthesis (0.04 g O_2 per cubic meter per day) would have a daily production, on an areal basis, of only 0.08 g O_2 per square meter. To match Baikal's production of 2.0 g/m^2, the pond's volumetric daily rate would have to increase 25 times to 1.0 g/m^3.

The production of a transparent oligotrophic lake can be compared with that of a eutrophic pond in another manner. If the maximum volumetric rate is divided by the integrated production per square meter, the result is oxygen per meter, with no reference to the shape of the production curve or to the depth of the trophogenic zone.

Conversion of Oxygen Data to Calories

The conversion of all units of production to calories is recommended in an energy approach to what Winberg (1963) termed biotic circulation. The simple photosynthetic formula for the manufacture of glucose from CO_2 and H_2O with an attendant release of oxygen tells us that for every mole (180 g) of glucose, 6 mol (192 g) of oxygen are released. In caloric terms, about 674 kcal are released when 1 mol of glucose is burned. This means that every mole of oxygen evolved represents one sixth of 674 kcal, or 112.3 kcal.

Oxycaloric coefficients established by experimentation on mixtures of plant and animal proteins, fats, and carbohydrates indicate an average composition of plant and animal tissue of 3.5 cal/mg O_2. If 112.3 kcal is divided by 1 mol of oxygen, 32 g, the result is 3.5 kcal/g O_2, which is identical to 3.5 cal/mg O_2. Thus, the oxygen released in primary production tells much of the caloric value of newly produced organic molecules. Rates of production, biomass, and respiration can all be unified and reduced to caloric energy units, or to joules.

An Index of Productivity

Apparent Oxygen Deficits

If a stratified lake is visited on a summer day and data are collected on the vertical temperature profile and dissolved oxygen at various levels, important information can be inferred. The oxygen present at any depth can be compared with the amount that would be present at the existing temperature and pressure if it were saturated. This comparison gives the *actual deficit*. The *relative deficit* can be computed if we compare the summertime oxygen content with the amount of oxygen at the very end of the spring overturn when the lake was still uniform throughout. The *absolute deficit* compares the summertime oxygen content with the amount that would be present at 4° C if saturated.

Examples of these terms are presented for a hypothetical lake 500 m above sea level, with a certain hypolimnetic stratum at 7.0° C and containing 4.70 mg O_2 per liter. The saturation would be 12.14 mg/liter for that temperature at sea level (Table 12-4) and 11.44 when corrected with a factor of 0.942 for the reduced partial pressure at that altitude (Table 12-3). The actual deficit is 11.44 − 4.70 = 6.74 mg/ liter. At the close of the spring overturn the water at the level was known to have contained 11.03 mg/liter; the relative deficit is 11.03 − 4.70 = 6.33 mg/liter. The absolute deficit, based on saturation at 4° C at 500 m, is 12.35 − 4.70 = 7.65 mg/liter.

The Real Oxygen Deficit

These *apparent deficits* may be close to reality, but if oxygen were very scarce or absent, one would have to go beyond calculating the difference between 100% saturation and zero. Anoxic water implies more than just lack of oxygen, for there are substances present that occur only in the absence of oxygen and that were involved in its decline. For example, every milliliter of methane (CH_4) in anaerobic water represents the consumption of about 2.87 mg O_2. To determine the *real deficit*, therefore, one must take such substances and relationships into account when calculating actual, relative, or absolute deficits.

The Hypolimnetic Areal Deficit and Its Rate

The next step unifies the various deficits by relating them to the area of the hypolimnion. Visualizing organic material raining through the plane that is the roof of the hypolimnion and imagining some amount that passes through the average square centimeter of this plane leads to the concept of the hypolimnetic areal deficit. This concept involves the difference between total amount of oxygen beneath each square centimeter of the hypolimnetic area on the summer day when it was sampled and what could have been there at the observed temperature (saturation), the oxygen at the end of the vernal overturn, or saturated in 4° C water.

A knowledge of the lake's morphology is needed to find the total amount of oxygen present in the hypolimnion as well as the hypolimnion area. Samples must be taken from various depths, and the concentration of oxygen (milligrams per liter) for each stratum of the hypolimnion must be multiplied by volume (liter) of that stratum. For example, if oxygen amounts to 5 mg/liter in a certain stratum that is composed of a billion liters, the total oxygen present is 5 mg/liter $\times$ 10^9 liters = 5×10^9 mg O_2. Summing the amounts of oxygen concentrated in the various strata yields the total oxygen content of the hypolimnion.

Using volume data and theoretical saturation amounts, one can then find the total oxygen that might be there, that was present during spring circulation, or that would represent saturation at 4° C. The difference between these amounts and the actual summertime total divided by the area of the hypolimnion in square centimeters gives the *hypolimnetic areal deficits* in terms of milligrams of O_2 per square centimeter.

The discussion shows that the summertime hypolimnetic oxygen per unit area allows the designation of actual, relative, or absolute hypolimnetic areal deficits (either real or apparent) and permits comparison of different lakes. It does not indicate, however, what oxygen totals had been a week before or predict what they will be a week hence. A further dimension is needed: the *rate* at which the hypolimnetic oxygen disappeared per unit area of the hypolimnion. Oxygen assays must be made at intervals, preferably from the spring overturn well into the summer, but even incomplete data are instructive. Thus, if on May 25 the hypolimnion contains 0.83 mg O_2 per square centimeter of its surface and 10 days later on June 4 there is 0.33 mg/cm^2, calculations show that oxygen disappeared at the rate of 0.05 mg/cm^2 per day during that period.

The rate of the hypolimnetic areal oxygen deficit is an index of the intensity of decomposition in waters lying below the trophogenic zone. Hutchinson (1957) suggested figures that define the limits and allow interpretation of deficit rate as an index. A broad boundary between oligotrophic and eutrophic lakes is 0.025 to 0.055 mg O_2 per square centimeter per day. A lake exhibiting a deficit rate within this range might be termed mesotrophic.

In recent years additional attention has been directed toward the areal hypolimnetic oxygen deficit and it has acquired an acronym, **AHOD**. Cornett and Rigler

(1980) discussed the history of the concept, analyzing their own data and those of others, in search of a satisfactory model that would eliminate the effects of lake morphology. Of course, this is what conventional referring deficits to unit hypolimnetic area should eliminate or at least reduce. They did not claim to have achieved this, but their correlations of AHOD rates with other indices of productivity were not surprising. There were positive correlations with total phosphorus concentration and with pelagic primary productivity, and an inverse correlation with the average summer Secchi disc depth.

Factors Invalidating the Index

Unfortunately, there are factors that can render data on oxygen deficits useless for comparative purposes. Lakes shallower than 20 m may give spurious results. Thus, summer data from productive Lake Itasca falsely suggest oligotrophy because light penetrates the shallow hypolimnion, bringing about some photosynthesis and oxygen production. There is more going on, then, than just the day-to-day disappearance of oxygen that would prevail in deeper lakes with dark, undisturbed hypolimnia. Here is an instance where Austausch data are especially useful. Coefficients of eddy diffusivity provide information for calculating oxygen-deficit rates; the O_2 diffusing into the hypolimnion from upper waters should be taken into account if possible.

The value of the hypolimnetic deficit as an index depends on the assumption that the decay of autochthonous organic material is involved in the dwindling oxygen supply. Two lakes with identical productivity could show quite different deficits if the hypolimnion of one were receiving significant quantities of organic matter from without. The natural input of leaves and pollen and the unnatural introduction of cannery effluent and raw sewage place a burden on the hypolimnion's oxygen supply that is unrelated to epilimnetic production.

The principle of Van't Hoff, that metabolism roughly doubles with each 10° C rise in temperature, introduces another factor in hypolimnion oxygen deficit. If hypolimnetic temperatures are relatively low (at 10° C, for example), bacterial metabolism is somewhat depressed, and the quantity of decaying organic matter governs the distribution and diminution of oxygen. There is a pitfall in comparing the dynamics of oxygen consumption in this lake with consumption in a lake having summer hypolimnetic temperatures of 20° C. Microbial metabolism is accelerated to such an extent in the warmer hypolimnion that it assumes the primary position in determining the rate at which oxygen disappears. This bacterial factor precludes making comparisons of the hypolimnetic oxygen deficit as an index to productivity in a tropical lake and in a similar body of water from the temperate zone. Although they might have identical epilimnetic productivities, the oxygen would disappear more rapidly in the warm hypolimnion of the tropical lake.

In addition, to avoid oversimplification, one must realize that the disappearance of oxygen in the deeps is not always brought about by the decay of newly formed organic matter. Researchers have shown that, in some lakes, the sediments consume this gas to a significant degree; this means that the production of previous years can be involved in oxygen's hypolimnetic decrease. Moreover, much of the epilimnetic organic production can be mineralized before it reaches the hypolimnion. Despite these phenomena, comparing rates of hypolimnetic oxygen diminution is a good way of estimating the relative trophic status of lakes.

In a meromictic lake the real oxygen deficit in monimolimnetic water is great. Time is the major factor here. Many decades of anoxic stagnation contribute to accumulation of reduced substances. The annual production need not have been high through the years.

Under certain conditions, on a regional basis, comparisons can be made of lakes of varied depths and sizes by studying winter deficits without considering maximum depths and hypolimnetic disturbances. Schindler (1971a) compared a cluster of Canadian lakes, sharing climatic and edaphic environments but differing morphologically, by finding the rate at which oxygen was depleted beneath the ice. In early winter, supersaturation of oxygen was found in the upper water because of the freezing-out effect and because of the transparency of new, clear ice. Soon, however, snow lay on the ice of all the lakes, effectively blocking light penetration. It was possible to calculate total oxygen content in each lake at different times during the winter and to compute rates of depletion in terms of milligrams of O_2 per cubic meter per day.

The Isotopes of Oxygen

There are two stable isotopes of oxygen in addition to the common ^{16}O. The relative abundance of the three in ocean water are ^{16}O, 99.763%; ^{17}O, 0.0372%; and ^{18}O, 0.1995%. The $^{18}O/^{16}O$ ratio in materials is usually compared with a seawater standard, SMOW, the acronym for Standard Mean Ocean Water. The oxygen isotope composition of a substance reflects, among other things, the temperature at which it was formed or deposited. This has proved useful in studying oceanic ecosystems and in gaining information about past world climates. The temperatures at which marine sediments were laid down and at which $CaCO_3$ was formed in mollusc and foraminiferan shells is interesting, particularly to oceanographers. The analysis of ^{18}O in lake environments has more recently assumed importance, after marine research led the way: Stuiver (1970) demonstrated thermal effects in the isotope ratios of freshwater carbonates and discussed their value as indicators of past climates. Part of the delay in use of these analyses was the difficulty in obtaining sufficient volume in samples, but recent advancements have overcome the difficulty and in fact automated the process (Barth et al., 2004). Analysis of isotopic ratios is important to accurately accounting for community metabolism (Venkiteswaran et al., 2007).

Several types of limnologic information can be gained from studying the oxygen/isotope ratios. For example, a continental event, of consequence to limnologists about 12,000 years later, can be inferred from an abrupt decrease in the $^{18}O/^{16}O$ ratio in old sediments from the Gulf of Mexico. An influx of glacial meltwater, relatively high in ^{16}O, left its mark at the time the Wisconsin glaciers were disintegrating 2,000 km north up the Mississippi Valley (Kennett and Shackleton, 1975).

Covich and Stuiver (1974) made use of the oxygen isotopes to infer past hydrologic conditions in Laguna Chichancanab, a large closed basin in northern Yucatán. Because the common ^{16}O isotope has a lighter mass, it evaporates faster than ^{18}O, and the vapor from evaporating water contains relatively more ^{16}O than the water source. As evaporation proceeds, therefore, the remaining water becomes enriched with ^{16}O.

Covich and Stuiver assumed that, because of the lake's tropical setting, mean annual temperatures had remained stable in comparison with evaporation and inflow rates. They studied the oxygen isotope ratios in the shells of an aquatic snail taken from cores of lake sediment. Changes in the relative abundance of ^{16}O in the shells were correlated with other evidence and pointed to major fluctuations in the lake level during the past 28,000 years.

Because the heavier ^{16}O isotope enters plants less readily than the ^{16}O, one might expect organic lake sediments to be somewhat richer in ^{16}O than ^{18}O. This seems to be shown in isotope ratios in sediments from Lake Dweru (Lake George), Uganda, where changes have occurred during the 3,600 years of the lake's life. A relative increase of ^{18}O in the inorganic fraction of the bottom deposits, when ^{16}O has risen comparatively in the organic fraction, implies an increase in the lake's fertility (Viner, 1977).

13

Carbon Dioxide, Alkalinity, and pH
The CO_2 System

Carbon Biogeochemistry: Atmosphere and Water

Carbon is often considered the most important element on Earth, due to its unique and influential biogeochemistry. As the backbone of all organic materials, it is central to all life. Carbon has significant storage in air, water, soil, and rock. Unfortunately, humans have significantly altered the carbon cycle, resulting in serious environmental consequences worldwide.

Anthropogenic activity adds carbon to the atmosphere at an ever-increasing rate. The Intergovernmental Panel on Climate Change points out that atmospheric CO_2 levels exceed pre-industrial levels by almost 40%, most of the increase due to fossil fuel combustion, although other activities are also significant. Much of the CO_2 enters the oceans, lowering pH (IPCC, 2013).

Atmospheric carbon is important in limnology in two directions: waters both affect and are affected by carbon levels in the air. Streams and rivers in the United States alone emit nearly 100 Tg of carbon to the atmosphere every year, and all temperate-zone rivers of the Northern Hemisphere combined emit 0.5 Pg of carbon annually (Butman and Raymond, 2011). Rivers bury twice the terrestrial carbon that is buried in the oceans (Aufdenkampe et al., 2011). Lakes process terrestrial carbon dissolved in inflowing waters; the resulting flux of carbon to the air is 1.4 Pg annually, similar to the carbon released by fossil fuel combustion (Tranvik et al., 2009). Clearly, limnology is important to understanding global carbon dynamics.

The current atmospheric CO_2 level is 400 ppm (0.04%), a small amount, but it is nevertheless one of the major dissolved gases in lakes due its high solubility (see Table 12-2). The solubility of any gas is dependent on temperature, as described in the previous chapter; Table 13-1 provides values of the solubility factor (K) across a range of temperatures. If we consider these figures as Henry's law constants (K in the formula $c = Kp$), we can determine the amount of CO_2 that would be dissolved when in equilibrium with the atmosphere. By substituting the partial pressure, p, of the gas and solving the formula, we get the following at 20° C:

If $p = 0.03\%$
$\quad c = 1689 \times 0.0003 = 0.507$ mg CO_2 per liter
$\quad c = 38.39 \times 0.0003 = 0.0115$ mmol CO_2 per liter
$\quad c = 878 \times 0.0003 = 0.263$ ml CO_2 per liter
If $p = 0.044\%$
$\quad c = 1689 \times 0.00044 = 0.743$ mg CO_2 per liter

Table 13-1   Factors (K) for calculating solubilities of carbon dioxide in water at different temperatures according to Henry's law ($c = Kp$)*

°C	mg/liter	mmol/liter	ml/liter
0	3347	76.07	1713
5	2782	63.24	1424
10	2319	52.70	1194
15	1979	44.98	1019
20	1689	38.39	878
25	1430	32.50	759
30	1250	28.41	665
35	1106	25.14	592
40	970	22.05	519

* Factors should be multiplied by partial pressure (p) to give solubility (c) in units indicated. Mmols are derived by dividing mg by 44, the molecular mass of CO_2.

Inputs of Dissolved CO_2

Rainwater is charged with CO_2 as it falls toward earth, and could be introduced directly at the water surface. Water trickling through organic soil may become further loaded with products of decomposition and later enter a stream or lake from a subterranean source, introducing gaseous CO_2 in solution.

Subterranean water rich in CO_2, and hence (as will be shown) containing some carbonic acid, may dissolve carbonates and bring them into solution as bicarbonates. The bicarbonates are later introduced into the aquatic environment and made available to most aquatic plants as a source of carbon for photosynthesis.

The respiration of plants, animals, and aerobic bacteria of decay add CO_2 to the environment; anaerobic decomposition of carbohydrates in bottom sediments is another important source of CO_2 gas. Such free CO_2 produced within a body of water can cause the dissolution of $CaCO_3$ lying within the sediments and put it in solution as $Ca(HCO_3)_2$.

Isotopes of Carbon Found in CO_2

Carbon 13

The common stable carbon is ^{12}C, and the overwhelming majority (about 98.9%) of CO_2 molecules contain this isotope. Another nonradioactive isotope of carbon is the heavier ^{13}C. Photosynthetic organisms show a marked preference for ^{12}C and fractionate the two isotopes to produce organic material enriched in that common lighter isotope. The carbon of plant tissue is predominantly ^{12}C. The heavier isotopes have less chance of entering cells.

The value of isotopes for working out lacustrine carbon budgets was demonstrated in the 1960s in some Connecticut and New York lakes (Oana and Deevey, 1960; Deevey and Stuiver, 1964). The accumulation of metabolic CO_2 in the summer hypolimnion includes both free (gaseous) and half-bound (bicarbonate) portions. Some of it is produced aerobically by the oxidizing of organic seston particles in the water; another part is a result of anaerobic events, formed along with CH_4 and revealing another fractionation process. The sediments and the seston sinking to augment them are relatively enriched in ^{12}C through the action of green autotrophs in the epilimnion, as would be expected.

Hypolimnetic and profundal microorganisms responsible for the anaerobic metabolism utilize the parent material preferentially, incorporating the ^{12}C and respiring relatively greater amounts of ^{13}C. Thus they produce CO_2 and CH_4, which are comparatively high in heavy carbon. The summer increase of ^{13}C, then, reflects the intensity of fermentation processes in the profundal ooze—the *pelometabolism*.

Carbon 14

Radioactive carbon, ^{14}C, is formed at high altitudes where primary cosmic radiation enters the upper atmosphere and produces neutrons. Some neutrons bombard atmospheric nitrogen to form ^{14}C and hydrogen, as shown by the notation

$$^{14}N + n \rightarrow {}^{14}C + H$$

The newly formed ^{14}C atoms combine with oxygen to produce carbon dioxide that reaches the earth's surface via turbulent mixing and convection. Here it makes up a tiny fraction of the atmospheric CO_2 The gas containing the stable ^{12}C is about 1.2×10^{12} times more abundant than the CO_2 made up of the heavier radioactive isotope.

pH and the Hydrogen Ion

Definitions

We review the concept of pH here, because the form of dissolved inorganic carbon present in water is pH-dependent. The abbreviation pH refers to the "power of hydrogen," with "power" indicating an exponent of the base 10:

$$pH = \log \frac{1}{H^+} = -\log\left[H^+\right]$$

Since water is a weak electrolyte, by definition a small fraction of it dissociates into the ions that compose its molecule. The obvious ions in pure water would be H^+ and OH^-, H^+ making for acidity and OH^- typical of a base. The following dissociation equilibrium applies:

$$H_2O \leftrightarrow H^+ + OH^-$$

The prevailing practice treats the dissociation of water as a function of the hydrogen ions and hydroxyl ions alone, although the hydronium ion (H_3O^+) far surpasses the former in abundance. It forms when a hydrogen ion is hydrated with at least one molecule of water.

$$H_2O + H^+ + OH^- \rightarrow H_3O^+ + OH^-$$

The dissociation constant of pure water at a given temperature is expressed as

$$K_w = \frac{\left[H^+\right]\left[OH^-\right]}{H_2O} \quad or \quad \frac{\left[H_3O^+\right]\left[OH^-\right]}{H_2O}$$

The ionizing fraction is so small that the undissociated water approximates 1.0, and K_w is simply the product of H^+ and OH^-, the numerator alone. The product is the tiny value 10^{-14} at 24° C. In absolutely pure water the ratio of the number of hydrogen ions to hydroxyl ions is 1.0, the overall reaction being neutral. Each occurs, therefore, in a concentration of 10^{-7}. These concentrations are expressed as moles (gram ions) per liter. The effect of temperature on the dissociation constant, K_w, is set forth in Table 13-2.

Table 13-2   Effect of temperature on K_w, pK_w, (negative logarithm of K_w), and pH

°C	$K_w \times 10^{14}$	pK_w	pH
0	0.115	14.94	7.47
5	0.185	14.73	7.37
10	0.292	14.53	7.27
15	0.450	14.35	7.17
20	0.681	14.17	7.08
24	1.000	14.00	7.00
25	1.008	13.99	6.99
30	1.469	13.83	6.92
35	2.089	13.68	6.84
40	2.919	13.54	6.77

Water with neutral pH has a molar concentration of $H^+ = 7.0$. (Although not commonly used, pOH would then be 7.0 as well.) An increase in H^+ is reflected in a lowered pH; below 7.0 designates a surplus of H^+ and an acid reaction.

In a neutral solution the number of hydrogen ions would equal the hydroxyl ions; with any departure from neutrality, one ion increases as the other decreases (Table 13-3). A gain in H^+ to 10^{-6} mol/liter would signify a decrease in hydroxyl ions to 10^{-8}; their product, $10^{-6} \times 10^{-8}$, still equals 10^{-14}, or K_w. This constant relationship permits the notation of just one of the ions to describe whether a solution behaves as an acid or as a base. The hydrogen ion has been chosen for this purpose, and the logarithm of its concentration is used to describe the ratio of hydroxyl to hydrogen ions. Conventionally, the negative sign is dropped, and the exponent is designated pH. There are other ways of stating this. For example, pH is the logarithm to the base 10 of the reciprocal of the hydrogen ion molarity, or its cologarithm. Or, it may be stated that pH is the negative logarithm of the concentration of hydrogen ions in moles per liter.

At least two different methods are used to measure pH. First, indicators of various types are colored differently by acid and alkaline solutions. Color comparators that permit the matching of indicators (added to natural waters) to colored standards are accurate to ±0.1 pH units. More than 60 acid-base indicators ranging from pH 0 to 14 are available. A second method involves electrodes and either battery-operated

Table 13-3 Relations among pH, the seldom-used pOH, and equivalent acid and base normalities

pH	Acid normality	Base normality	pOH
0	1.0	0.00000000000001	14
1	0.1	0.0000000000001	13
2	0.01	0.000000000001	12
3	0.001	0.00000000001	11
4	0.0001	0.0000000001	10
5	0.00001	0.000000001	9
6	0.000001	0.00000001	8
7	0.0000001	0.0000001	7
8	0.00000001	0.000001	6
9	0.000000001	0.00001	5
10	0.0000000001	0.0001	4
11	0.00000000001	0.001	3
12	0.000000000001	0.01	2
13	0.0000000000001	0.1	1
14	0.00000000000001	1.0	0

meters in the field or line-operated instruments in the laboratory. This electrometric procedure has the advantage of accuracy, but the simplicity of comparing the hues of indicator solutions is attractive.

Sources of Hydrogen Ions

The main source of hydrogen ions within natural waters is carbonic acid in its various forms, which account for pH values somewhat below 6.0, approaching 5.0. One can generalize, then, that the providers of CO_2 in inland waters also supply hydrogen ions.

Atmosphere. Rainwater in equilibrium with atmospheric CO_2 would have a pH of about 5.6, if only carbonic acid were involved. Analyses have shown, however, that lower values prevail in many parts of the world. Acid deposition, also known as acid precipitation or "acid rain," has become a subject of international concern. There is good evidence that the increase in acidity is anthropogenic and relatively recent. The studies of Swedish and Norwegian workers, who documented the pH of precipitation for many decades, show that the temporal and geographic variation of pH and the associated strong acids are correlated with industrial sources. An early report discussed the effect of sulfurous fumes from the smelters at Sudbury, Ontario, on the poorly buffered La Cloche Mountain lakes. Records of 150 lakes show that in some there has been a hundredfold increase in hydrogen ion concentration. One result has been a high rate of fish mortality. Where soils are poorly buffered, much of the fish damage results from the arrival of acid-mobilized aluminates and other metal salts from the watershed. Baker and associates (1991) reported on 6,850 acidic lakes and streams with pH values below 5.0–5.5 in the United States. The atmosphere was the dominant source of acidity in most of the lakes and almost half the streams. Many streams owe their acidity to edaphic factors brought about by human mining activity (discussed in the next section).

Although the problems of this atmospheric pollution are owed primarily to human activity, there are some natural sources of strong acids. Volcanic emissions contaminate the atmosphere with SO_2, which accounts in part for acid rain and snow. Sulfur dioxide is an extremely soluble gas, and it reacts readily with water to form sulfurous acid, which is rapidly oxidized to sulfuric acid. Similarly, water solutions of the gas H_2S are slightly acidic, and subsequent oxidation can lead to the formation of H_2SO_4. The appreciable amount of sulfate ion that occurs in acid precipitation is largely in the form of H_2SO_4.

In addition to H_2SO_4, hydrochloric acid amounting to 1.8 mEq/liter has been found in rainwater near Mount Vesuvius. It owes its existence to that famous cone.

Another strong acid, HNO_3, occurs regionally in precipitation, but is probably a product of industrial activity. In the mid-1950s less than 2 kg/ha was the annual fall-out of nitric acid in central Europe; 10 years later it had increased to amounts of 6 to 8 kg/ha (Odén, 1976). The "baseline" coming from natural oxides of nitrogen, the precursors of HNO_3, in the atmosphere is estimated to have been 1 or less kg/ha annually.

The proton donors contributing to atmospheric acidity are of two main types: the strong acids that are highly or completely ionized and the weak acids that are undissociated. The protons in solution, hydronium or hydrogen ions, control the pH and are measured by standard electrometric pH procedure. The total concentration of protons (*total acidity*) includes the ionized component (*free acidity*) and those that are undissociated in the presence of the highly ionized types. Because carbonic acid is a weak acid, as pH values fall below 5.0, it starts to become neutralized; it is almost completely undissociated and donates few protons to the solution (none at about 4.3). The total acidity includes (1) the free protons that cause the ambient pH and (2) the bound protons. It is determined by NaOH titration past the phenolphthalein end point to pH 9.0. In a sense this is the reverse of an alkalinity titration and reveals the poorly dissociated proton donors in acidity. Carbonic acid is partially dissociated as the pH rises above 5.0, contributing to both free and bound acidity. Other weak proton donors include more than half-a-dozen organic acids in trace amounts, donating their hydrogen ions to the strong-base titrant. Some metals—aluminum, iron, and manganese—effectively consume hydroxide as the titration proceeds but contribute nothing to the free acidity. Above pH 8.0, NH_4^+ begins to release protons, and almost 50% of it will have been converted to NH_3 by the time pH 9.0 is achieved.

$$NH_4^+ + OH^- \rightarrow NH_3 + H_2O$$

Edaphic factors. Strip mining (surface mining) for coal locally increases the acidity of streams and lakes. Coal is often associated with large amounts of FeS_2 (pyrite). When these reduced rocks are exposed to air, they begin to oxidize. Much of the oxidation is brought about by iron- and sulfur-oxidizing bacteria. In strip mining the so-called tailings are left scattered after the coal has been removed. For this reason the formation of H_2SO_4 continues for many years; in some instances the soil remains acid for more than a century. In other types of coal mining, a seam may be worked and then closed with the tailings, leaving relatively little reduced rock exposed to the air. Water flowing through such mines eventually may become acid, however. Iron-rich, acidic water has seriously affected more than 1,500 km of U.S. waterways, with the states of Pennsylvania, Virginia, West Virginia, Kentucky, and Indiana being especially damaged. Some streams are almost lifeless, and the strip-mined regions, especially, continue to yield acid water for many years.

Other edaphic sources of acidity are natural, yet as effective in lowering pH as human disturbance. Some volcanic lakes are acid because of nearby active sources of sulfur compounds. Armitage (1958) wrote of one that occupied a crater in El Salvador and was ringed by sulfurous fumaroles and springs along its shores; its water had a pH of 2.0! Many Japanese volcanic lakes, similarly, are very acid (reviewed by Satake, 1980). On the other hand, other volcanic lakes, and particularly those in arid regions, are characterized by unusually high pH values. They are soda lakes, with marked concentrations of Na_2CO_3 and $NaHCO_3$, the sodium having been provided by lava.

Pyrite, the so-called fool's gold (FeS_2), may be found within the drainage basin of a body of water or be present in anaerobic peats or bottom sediments. Oxidation of this mineral leads to the formation of sulfuric acid in the following manner:

$$4FeS_2 + 15O_2 + H_2O \rightarrow 2Fe_2(SO_4)_3 + 2H_2SO_4$$

In anaerobic water, the gas H_2S may be evidenced by the rotten-egg odor it presents. Both phototrophic and chemotrophic bacteria can oxidize this compound to elemental sulfur and later to H_2SO_4. These events may be effective at the boundary of anaerobic and aerobic water in a stratified lake and, of course, are common in anoxic oozes.

Calcium sulfate is common in runoff waters and under some conditions can be present in rain. Any exchange between the bivalent cation Ca^{2+} and hydrogen ions would form the strong acid H_2SO_4, and this is does happen under certain conditions. A simple water softener based on ion exchange can be made from peat derived from *sphagnum* moss. As water trickles through the peat, calcium is adsorbed and the plant material concurrently yields hydrogen ions; as a result sulfuric acid is produced. This is why sphagnum bogs attain high acidity.

Experimental work by Bell (1959) underlined the fact that living sphagnum takes up cations differentially, the trivalent and bivalent ions being adsorbed much more readily than the monovalent. Dead sphagnum continues to act as an ion exchanger. The action of this moss is similar to that of the commercial resins and zeolites used in water deionizers. Zeolites make up a large family of natural hydrous aluminum silicates. They show molecular sieve effects and possess base exchange properties that make them valuable in water-softening procedures. Similarly, a complex system of minute pores in the cell walls of sphagnum may be adapted uniquely for such a role. Gorham (1957b) found that metallic ions adsorbed on peat surfaces were sometimes more than 100 times the amount present in the water.

Sulfuric acid is not the only agent lowering the pH of bog waters. Organic acids falling under the heading of humic substances are abundant in peaty materials. There are three groups of these substances: humic acids, fulvic acids, and humins. The first two are alkali extractable (0.5 N NaOH) and are distinguished by the fulvic acids remaining in solution when acidified, while the humic acids precipitate at pH 2. The humins are not alkali extractable. Humic materials can be found in soils as well as lake sediments. They arise from plant materials by a two-step process mediated by bacterial enzymes (Kononova, 1966). First, there is a synthesis of complex, high-molecular-weight yellow-brown substances. The second step is the conversion of open-chain compounds into aromatic compounds that undergo subsequent polymerization. The results are condensation products of phenols, quinines, and amino compounds. Some humic materials are found in soils and reach lakes via runoff. They are stronger than carbonic acid, bringing the pH down to about 4.0 or slightly lower. Shapiro (1957) reported that purified preparations of the yellow acids of lake water lowered the pH to 3.6 in aqueous solutions. Thus, humic acids can neutralize carbonic acid, preventing its dissociation at about pH 4.4. In turn, humic acids are neutralized by strong mineral acids. In many bog lakes, H_2SO_4 depresses the pH to 3.0 or below, where neither carbonic acid nor the humic acids dissociate; there is no alkalinity.

Humic compounds are rather poor buffers except between pH 4 and pH 5. This is the pH region near the endpoint in alkalinity titrations; here the buffering capacity of humic solutions surpasses that of the carbonate system. For this reason titrimetric determinations of total inorganic carbon are rendered less reliable in humic waters (Wilson, 1979).

Bayly (1964) published results of research on some acidic Australian lakes that owed their acidity (pH 4.0) to humic acids. Although the acidity of some North American bog lakes that have been studied is the result of H_2SO_4, there are thousands of humic lakes with natural pH values below 5.0; they are especially common in boreal and subarctic areas.

Edaphic factors are related to the acidifying effects of contaminated precipitation. In calcareous soils and limestone outcrops, where chemical weathering and ion exchange occur, much acidity is neutralized. By contrast, in thin soils overlying granitic rocks, the buffering capacity is weak, and the runoff to lakes and streams is acidic. Watershed vegetation adds to the problem, although trees generally improve local atmospheric conditions. Under the canopies of the European beech (*Fagus*), oak (*Quercus*), and various conifers, sulfur, nitrogen, and hydrogen ions collect. The foliage intercepts acidic dry fallout only to have it rinsed off during subsequent rains. The soil beneath the trees becomes markedly acid, and the mobility of soil ions increases; it is significant that aluminum, for example, is more abundant in acidified lakes than in waters that are uncontaminated.

Soil particles blown aloft neutralize the rain. Calcareous particles are especially effective (buffering at around pH 8.0), although dust composed of aluminosilicates and iron oxides can also combine with hydrogen ions. Prairie regions, where there is more natural airborne dust, have fewer problems with acidification of waterways than do forested regions. Volcanic particulates, sea spray, and fly ash from industrial sources have some buffering capacity and improve the atmospheric hydrogen-ion loading.

Ammonia serves to buffer the rains, but there are mixed effects. It neutralizes the sulfurous and sulfuric acids derived from SO_2, but this allows the oxidation of SO_2 to continue. Then, when the ammonium sulfates arrive at the ground or lake surface, plants take up the ammonium part of the salt, achieving a biologic acidification as effectively as the pure acid would have done.

Despite the paragraphs devoted to acidity derived from strong mineral acids and organic acids, it must be reiterated that the pH range in most inland waters extends from 6.0 to 9.0 and that the carbonic acid system controls it. Residual acidity following aeration and agitation is rare, except regionally where the precipitation is acid.

Limnologic Effects of Strong Acidity

The hydrogen-ion concentration of lakes does not fall smoothly as contamination proceeds. There is a slow decline to around pH 6.0, followed by a precipitous drop as the bicarbonate buffering capacity becomes saturated and ineffective. During the 1930s, the surface-water pH values in the Adirondacks of New York were mostly 6.0 to 7.5, with some ranging up to 8.5 and higher. Data from 1975 revealed, however, a bimodal pH distribution. There was a large group of waters below 5.0, and a second group still 6.0 to 7.5. A very small number of lakes fell in between 5.5 and 6.0, and the high pH waters were gone (Wright and Gjessing, 1976).

At the chemical level there is a shift from waters with bicarbonate as the major anion to those with a pH below 5.5, dominated by sulfate, and with a much lowered buffering capacity. In some Swedish lakes the Secchi disc transparency has increased from 3 m to 10 to 14 m since acidification, there being greatly reduced crops of phytoplankton and, hence, very little chlorophyll *a* in the open waters.

Much of the above suggests oligotrophy, and indeed, the events usually are described as oligotrophication. There are, however, overtones of dystrophy in the effects of acid precipitation. Biotic changes in the littoral zone are especially marked, and these may be accented by the early snowmelt each spring. There is a surge in shallow-water acidity as new cold water flows into the littoral area, for there has been an ionic concentration in the upper stratum of the winter snow; and this layer, therefore, is extremely acidic. Sphagnum moss invades, covering the substrate and replacing the tracheophytes *Isoetes* and *Lobelia*, two genera characteristic of the phytobenthos of poorly buffered waters. The moss takes up metallic cations and phosphate under acidic conditions, inhibiting their recycling and compounding what has been described as a self-accelerating oligotrophication. Invertebrate leaf-shredders are eliminated in acid lake shallows and streams, ciliates die, and there is a shift from bacteria to fungi as evidenced by accumulated fungal mats. Coarse organic detritus collects as the decomposition of cellulose is halved (shown by a 50% reduction of oxygen consumption) by the time pH diminishes from 7.0 to 5.2. The fact that lime applications can alleviate the situation recalls the experimental alkalization in dystrophic Wisconsin lakes (Hasler et al., 1951) and stresses the dystrophic nature of the so-called acidic oligotrophication.

The Fate of CO_2 in Water

The Two Dissociations of Carbonic Acid

When CO_2 enters pure water, a small proportion of it (well below 1%) is hydrated to form carbonic acid, as shown:

$$CO_2 + H_2O \leftrightarrow H_2CO_3 \tag{1}$$

Some of this carbonic acid dissociates into bicarbonate and hydrogen ions, bringing about a lowering of the pH, a typical occurrence when CO_2 is dissolved in water.

$$H_2CO_3 \leftrightarrow HCO_3 + H^+ \tag{2}$$

Compared with the total amount of CO_2 in the water, including both H_2CO_3 and CO_2, only a very small amount ionizes. The so-called first dissociation constant of carbonic acid is K_1'

$$K_1' = \frac{[HCO_3'][H^+]}{[CO_2 + H_2CO_3]}$$

Each bracket symbolizes molecular concentration. Thus, the ionization constant, K_1' above is the product of the concentration of the ions divided by the concentration of the non-ionized molecules, when they are at equilibrium. Table 13-4 shows the effect of temperature on the dissociation of carbonic acid, and Fig. 13-1 shows it as a function of pH. The relationships are valid in dilute waters but must be viewed with caution in concentrated, mineral waters, where activity is lowered.

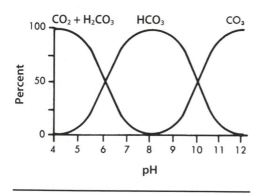

Figure 13-1 Theoretical relative proportions of the three forms of CO_2 (free CO_2, bicarbonate, and carbonate) in relation to pH.

From the curves in Fig. 13-1 it is obvious that by the time pH has dropped to about 4.5, nearing pH 4.3, reaction 2 has proceeded to the left, and almost no bicarbonate ion is present. For this reason, a common method of measuring bicarbonate ion concentration is based on titrating the unknown with a strong acid until the pH is lowered to about 4.3. Certain liquid indicators show a distinct color change near this pH, where CO_2 and H_2CO_3 are present, but the bicarbonate ion is nearly absent. Thus, the ionization of H_2CO_3 is completely reversed at about pH 4.3.

As pH rises, reaction 2 goes to the right, and dissociation of carbonic acid begins to assume importance at about pH 6.4. Gaseous CO_2 and H_2CO_3 decrease until they are no longer analytically present at approximately pH 8.3, the turning point of the indicator phenolphthalein. Below pH 8.3 phenolphthalein is colorless; it is pink at pH 8.3 and above.

From the curves in Fig. 13-1, we notice that relative HCO_3^- begins to decline above pH 8.3. The decline results from the following reaction (3) proceeding to the right:

$$HCO_3^- \leftrightarrow CO_3^{2-} + H^+ \tag{3}$$

Reaction 3 is the second dissociation of carbonic acid; its dissociation constant (Table 13-4) is based on the following ratio:

$$K_2' = \frac{[H^+][CO_3^{2-}]}{[HCO_3^-]}$$

From the preceding discussion it is obvious that carbonic acid dissociates twice, depending on pH. Similarly, it could be said that it is neutralized at two points on the pH scale. Above pH 8.3 the second dissociation prevails, resulting in the ions CO_3^{2-} and

Table 13-4 First (K'_1) and second (K'_2) dissociation constants of carbonic acid in relation to temperature, with co-logarithms pK'_1 and pK'_2

°C	$K'_1 \times 10^7$	pK'_1	$K'_2 \times 10^{11}$	ppK'_2
0	2.65	6.587	2.36	10.625
5	3.04	6.517	2.77	10.557
10	3.43	6.464	3.24	10.490
15	3.80	6.419	3.71	10.430
20	4.15	6.381	4.20	10.377
25	4.45	6.352	4.69	10.329
30	4.85	6.327	5.13	10.290
35	4.91	6.309	5.62	10.250
40	5.06	6.298	6.03	10.220

Data from Harned and Davis (1943) and Harned and Scholes (1941)

H^+, which are neutralized to HCO_3^- as pH is lowered to 8.3, the phenolphthalein end point. Below this pH, HCO_3^- and H^+ are the only dissociated ion pairs in an ideal carbonic acid system. They persist until about pH 4.4, the turning point of methyl orange, where most of these ions are neutralized to undissociated H_2CO_3. Putting the interactions together we get a stepwise picture of the forms of CO_2 and their dissociations:

$$CO_2 + H_2O \leftrightarrow H_2CO_3 \leftrightarrow HCO_3^- + H^+$$
$$\updownarrow$$
$$CO_3^{2-} + H^+$$

This discussion is relevant also to the dissociation of water (Table 13-2), buffer systems, and alkalinity. When the pH equals pK (Table 13-4), half the substance is ionized; changing the pH two units in one direction results in about 99% becoming ionized; moving pH two units the other way results, similarly, in almost all being unionized. This can be demonstrated in the curves in Fig. 13-1, where half the HCO_3, for example, is ionized near pH 6.3. Although the dissociation of the three forms of CO_2 (free CO_2, bicarbonate, and carbonate) and of water are treated separately on these pages, they co-occur as pH and temperatures change. These simultaneous shifts are accompanied by the dynamic balancing of ionic charges, the sums of positive and negative charges remaining equal to one another.

The constants K'_1 and K'_2 are modified by both temperature (Table 13-4) and ionic strength. Despite this fact, it is customary to make the following deductions after placing a few drops of phenolphthalein in a container of sampled water. With the appearance of a pink hue, the pH is 8.3 or above; there are probably carbonates present, and there is no carbonic acid or free CO_2 that can be analyzed, although theoretically it occurs in minute amounts up to a pH of 9.1. If colorless, the pH is below 8.3; free CO_2 and carbonic acid are present, and carbonate is virtually absent. Bicarbonate ions occur in both cases if the water is normal and lies somewhere within the pH range 4.3 to 12.6. Within limits, then, a few drops of phenolphthalein can reveal much about the nature of a water sample and the forms of CO_2 it contains.

Further examination of the symmetry of the three curves in Fig. 13-1 shows that pH 8.3 corresponds to the maximum in the HCO_3^- curve and is halfway between the points where concentrations of H_2CO_3 and free CO_2 nearly equal HCO_3 (about 6.4; Table 13-4) and where bicarbonate concentration approximates that of carbonate, about 10.2. These relationships permit the calculation of free CO_2 in some natural waters and are discussed again in later treatments of alkalinity and cations (such as calcium and sodium).

The Forms of CO₂ in Water and Total CO₂

In most natural waters gaseous CO_2 occurs in the presence of alkali metals or alkaline earth metals and combines with them to form the bicarbonates and carbonates discussed above. If calcium is used as an example, it is possible to describe an equilibrium relationship for its salts:

$$Ca(HCO_3)_2 \leftrightarrow CaCO_3 + CO_2 + H_2O$$

calcium	calcium	free	
bicarbonate	(mono) carbonate	CO_2	(4)

This reaction indicates that carbon dioxide has three important forms in water: the **half-bound** state, represented by the bicarbonate ion; the **bound** form, represented by monocarbonate; and the *free* dissolved gas. To these should be added the *hydrated* state, carbonic acid.

In another sense, it could have been stated that the equilibrium reaction 4 shows that carbonic acid is present in the water as an acid salt, as a neutral (or normal) salt, and as its anhydride, gaseous CO_2.

In a natural water very low in alkaline earth metals or in alkali metals, there would be relatively little bicarbonate formed. By contrast, when the metals are abundant, there is opportunity for the formation of much bicarbonate. In either case, after CO_2 is absorbed, it no longer obeys Henry's law and instead enters into chemical reac-

---

Case Study: Lake Nyos and Other "Killer Lakes"

Another limnologic aspect of gaseous CO_2 came to light in 1986 when the African Lake Nyos became known as the killer lake (Kerr, 1989). Lake Nyos is one of at least 40 lakes in Cameroon occupying volcanic craters. Its maximum depth is 208 m, the deepest 23-m layer making up a monimolimnion high in dissolved substances and gaseous CO_2 which slowly bubbles from the magma chamber beneath the crater. One night an "explosive" event caused the release of monimolimnetic CO_2 that rolled across the adjacent land killing 1,700 people. A nearby volcanic lake (Monoun) has similarly released a killing cloud of CO_2. No one knows the trigger for the deadly eruptions, although it is likely that a contributing factor was extended stormy weather that weakened the stratification in the lakes (Tassi and Rouwet, 2014).

Limnologists have studied the situation carefully, monitoring the lakes through regular sampling and installing an automated alarm system. Better still, both lakes now have long tubes extending from the surface and reaching to the hypolimnion. A continuous fountain of CO_2 is vented, stabilizing the gas level below, although additional vents must be installed to decrease the CO_2 level at Nyos and assure safety in the long term (Kusakabe et al., 2008).

Table 13-5   Equilibrium CO_2 in relation to total alkalinity

Total alkalinity as $CaCO_3$ (mg/liter)	Total alkalinity (mEq/liter)	Equilibrium CO_2 (mg/liter)
25	0.5	0.15
50	1.0	0.6
75	1.5	1.2
100	2.0	2.5
125	2.5	4.0
150	3.0	6.5
175	3.5	10.1
200	4.0	15.9
225	4.5	24.3
250	5.0	35.0
275	5.5	48.3
300	6.0	64.1

tions that result in its total content, in one form or another, being far above the saturation value for the gas.

Any equilibrium equation using certain calcium salts as examples should show a small amount of free CO_2 in the system with the bicarbonate; this gas is the **CO_2 of equilibrium**, sometimes called **attached CO_2** (not to be confused with bound or half-bound CO_2). The CO_2 of equilibrium is necessary for maintaining Ca $(HCO_3)_2$ in solution, but it will dissolve no more $CaCO_3$. If more gaseous CO_3 were added to the system, it would begin to dissolve $CaCO_3$, and reaction 4 would proceed to the left until an equilibrium were again attained. The gas in excess of the equilibrium level is termed **aggressive CO_2**. It would also be excess as gas in relation to the equilibrium between atmosphere and water. Total free CO_2 includes the aggressive and the equilibrium CO_2. The amounts of equilibrium CO_2 calculated as necessary to keep various amounts of $CaCO_3$ dissolved as Ca $(HCO_3)_2$ are shown in Table 13-5.

pH and Photosynthesis

There are interrelationships among various forms of CO_2, photosynthesis, and pH. If a water contains relatively great amounts of Ca $(HCO_3)_2$, with the equilibrium CO_2 that permits that solution, removal of the CO_2, as in photosynthesis, will disrupt the equilibrium. The reversible reaction below will proceed to the right with the precipitation of $CaCO_3$:

$$Ca(HCO_3)_2 \leftrightarrow \underset{\downarrow}{CaCO_3} + H_2O + \overset{\uparrow}{CO_2}$$

$$\overset{assimilated}{}$$

$$\underset{precipitated}{}$$

This reaction, resulting in a pH increase, is revealed sometimes by calcareous incrustations on plants and other submerged objects.

Further reactions are the hydrolysis of bicarbonate and carbonate and the appearance of hydroxyl ions with an accompanying increase of pH.

$$HCO_3^- + H_2O \leftrightarrow H_2CO_3 + OH^-$$

$$CO_3^{2-} + H_2O \leftrightarrow HCO_3^- + OH^-$$

The addition of free CO_2 to the system by respiration, for example, reverses the reactions and lowers the pH.

The pH of a typical calcareous water is the result, then, of the ratio of hydrogen ions (arising from the two dissociations of carbonic acid) to hydroxyl ions (provided by the hydrolysis of bicarbonate and carbonate). The importance of photosynthesis is obvious here, for plants can successively absorb CO_2, eliminate bicarbonates, precipitate carbonates, and form hydroxyl ions. All these events account for rises in pH.

Buffer Systems

Buffers are solutions that resist changes in hydrogen ion concentration when other solutions, acidic or basic, are added. A weak acid becomes a buffer when alkaline substances are added, and a weak base may become a buffer when acid is introduced. A dynamic equilibrium with a reversible reaction applies.

Mixing acidic and basic phosphates makes a common laboratory buffer. Because of the law of mass action, the pH of these buffers can be determined from the amounts of base and acid in the mixture. The association of a partially neutralized acid and its base has a certain pH, formulated as follows:

$$pH = pK' + \log \frac{C_B}{C_A}$$

In the preceding example, pK' is the negative logarithm of the ionization constant of the acid (A); C refers to the molar concentrations of the acid and its conjugate base (B). As an example, a buffer of about pH 6.8 is made by mixing a weak acid, Na_2HPO_4 (a slightly acidic primary salt of phosphoric acid) and a weak base, NaH_2PO_4 (an almost neutral secondary salt of phosphoric acid). This could be considered a mixture of a weak acid and its salt, or of two salts, one of which is less acid than the other. If the mixture includes a 0.06 molar concentration of NaH_2PO_4, and 0.04 of the acid Na_2HPO_4, then

$$pH = 6.8 + \log \frac{0.06}{0.04} = 6.8 + 0.18 = 6.98$$

Now, if 0.01 mol of HCl is added,

$$pH = 6.8 + \log \frac{0.05}{0.05} = 6.8$$

Or if 0.01 mol of NaOH had been added,

$$pH = 6.8 + \log \frac{0.07}{0.03} = 6.8 + 0.37 = 7.17$$

This buffer system, beginning at pH 6.98, varied no more than 0.19 pH units with the addition of considerable quantities of strong acid and base. By contrast, a liter of 0.1 molar solution of unbuffered NaCl would be changed from pH 7.0 to 2.0 by the addition of 0.01 mol of HCl!

Alkalinity

Very closely associated with the forms of CO_2 is the so-called **alkalinity** of the water. The use of the term is unfortunate, for it actually has little to do with pH terminology; waters on the acid side of the pH scale can rank high in alkalinity. Alkalinity is customarily expressed in terms of equivalent bicarbonate or carbonate, although other ions contribute to it. Additional names for alkalinity are titrable base, buffer capacity, excess base, acid-combining strength, and **SBV** (*Säuerbindungsvermögen*, as the Germans call it), loosely translated as acid capacity, or power to combine with acid. However termed, alkalinity is simply the property of resistance to acid.

A water containing some carbonic acid and one of its salts qualifies as a buffer solution. This is typical of most natural waters, and those with high total alkalinities are especially effective in resisting pH changes. When a strong base is added, it reacts with carbonic acid to form the bicarbonate salt and eventually carbonate, using up the base in the process. Likewise, when acid is added, it is used in the conversion of carbonate to bicarbonate and of bicarbonate to the undissociated H_2CO_3. These relationships explain why, in a natural body of water low in total alkalinity, the addition of respiratory CO_2 or the removal of CO_2 via photosynthesis results in far greater pH changes than in well-buffered water with a high total alkalinity. Temporary vernal pools that fill by snowmelt or rain contain notoriously poorly buffered water. In regions where precipitation is acid because of industrial pollution, many of these short-lived pools are no longer fit for embryologic development of the salamanders (*Ambystoma maculatum*) that require them for breeding sites (Pough, 1976).

Ordinarily, alkalinity is an index to the nature of the rocks within a drainage basin and to the degree to which they are weathered. Alkalinity commonly results from carbon dioxide and water attacking sedimentary carbonate rocks and dissolving out some of the carbonate to form bicarbonate solutions. If Me denotes an alkaline earth metal, such as calcium or magnesium, it is permissible to write

$$MeCO_3 + CO_2 + H_2O \rightarrow Me^{2+} + 2HCO_3^-$$

An assay of alkalinity is the common method for determining carbonate content. A given sample is titrated with a standard acid to the end point of methyl orange or bromocresol green, or with an electrode to about pH 4.5 or 4.4.[1] In North America, 0.02 N H_2SO_4 is frequently employed for this titration: 1 ml of this acid corresponds to 1 mg of $CaCO_3$. Therefore, if the water sample is 100 ml, the milliliters of acid titrant used to attain the end point are multiplied by 10 to arrive at total alkalinity expressed as milligrams of $CaCO_3$ per liter. Multiplying the total alkalinity by 0.599 then yields

---

[1] The exact end point depends on the total quantity of the three forms of CO_2. If total CO_2 is about 130 mg/liter, the end point is between 4.4 and 4.5. It is higher in more dilute concentrations and lower when greater amounts of CO_2 are present. For example, pH 5.2 is the end point for 4.4 mg/liter, and pH 4.2 applies for waters with 440 mg total CO_2 per liter.

CO_2^{3-} in milligrams per liter; the factor 1.219 converts total alkalinity to the HCO_3^- ion; and 0.02 times the alkalinity gives the alkalinity as milliequivalents per liter.

The rationale for these carbonate and bicarbonate evaluations is based upon the assumption that the titrations deal with salts of carbonic acid only. This leads to a reexamination of the curves in Fig. 13-1 to clarify the alkalinity procedure. Ideally, the aliquot to be assayed is tested first with phenolphthalein. If a pink color shows, one assumes the presence of carbonate, probably bicarbonate, and possibly OH^-. (Bicarbonates of alkali metals, such as $NaHCO_3$, cause the pink reaction even in the absence of carbonate.) The titration continues until the pink fades away, somewhere below but very near pH 8.3. This implies that the following steps could have taken place:

$$H_2SO_4 + Ca(OH)_2 \rightarrow CaSO_4 + 2H_2O$$

$$H_2SO_4 + 2CaCO_3 \rightarrow CaSO_4 + Ca(HCO_3)_2$$

Any hydroxide present, as in the first reaction above, is counted as an equivalent of normal carbonate. The second reaction shows that each carbonate ion in solution takes up one hydrogen ion to become a bicarbonate ion. When most of the carbonate ions have been converted, a small addition of acid produces a rapid lowering of pH, marked by the disappearance of the pink. This **phenolphthalein alkalinity** is only a fraction of the so-called carbonate alkalinity, including CO_3^{2-} and HCO_3^-, and is part of the total alkalinity. Continuing with the acid titration: in the presence of an indicator, such as methyl orange, that turns color near pH 4.4, the step, simplified, is

$$H_2SO_4 + Ca(HCO_3)_2 \rightarrow CaSO_4 + 2H_2CO_3$$

The titration leads to the point where carbonic acid no longer dissociates and alkalinity is zero (Fig. 13-1). Each bicarbonate ion took up one hydrogen ion to form a molecule of undissociated carbonic acid. The sum of the milliliters of acid titrant used to convert carbonate to carbonic acid by two steps is used in computing the total alkalinity.

It may be obvious now that alkalinity measures the buffering capacity of the water. If the assumption stands that only the carbonate-bicarbonate-carbonic acid buffering system is involved in the titration, then the total CO_2 content in its various forms can be stated with accuracy.

Factors Contributing to Alkalinity

Unfortunately, natural waters contain additional negative ions that react with hydrogen ions; therefore, an alkalinity titration may deal with buffer systems other than the salts of carbonic acid. For example, about 5% of the alkalinity in seawater comes from borate ($H_4BO_4^-$), and this anion could assume importance in inland waters of certain arid regions. Some saline desert lakes of California, such, as Little Borax and Mono Lake, contain great quantities of boron.

In some dystrophic bogs and other waters rich in humic acids, the occurrence of humates precludes the valid conversion of alkalinity titrations to carbonate. In polluted lakes and rivers, organic anions may become a part of the total alkalinity.

Phosphate alkalinity must be considered in some situations. The ions PO_4^{2-}, HPO_4^{2-}, and $H_2PO_4^-$ may all combine with H^+ to increase the titer. Silicates and, to a lesser extent, arsenates and aluminates could also be involved as buffers in alkalinity titrations.

Water specialists may convert total alkalinity to the bicarbonate ion and list it specifically as such in their reports; others report results as calcium carbonate equivalents. In some German and Austrian journals, however, it is common to find the major cations and anions reported in milligrams per liter, while the alkalinity titration is expressed in milliequivalents per liter, with no mention of carbonate or bicarbonate. This is a safe way to express alkalinity because a mixture of buffer systems might prevail. Accuracy is sacrificed by converting the results to carbonate or bicarbonate.

In 1939 The International Association of Physical Oceanography defined alkalinity as the number of milliequivalents of hydrogen ions neutralized by 1 liter of seawater. This somewhat unspecific approach recognizes that despite the major influence of HCO_3^- and CO_3^{2-}, borates account for about 5% of the alkalinity in marine samples.

Carbonate alkalinity can be calculated from determinations of pH, temperature, and total inorganic carbon dioxide. The following formula applies:

$$\text{Carbonate alkalinity} = \frac{\left[(H^+)+2K_2'\right] \times K_1'}{(H^+)^2+(H^+)K_1'+K_1'K_2'} \times \sum C \text{ in mmol/liter}$$
$$\text{(mEq/liter)}$$

The hydrogen ion concentration in the formula is derived from the pH in Table 13-6. The first and second dissociation constants of carbonic acid at different temperatures can be found in Table 13-4. The total inorganic CO_2, designated as ΣC, is

$$(CO_2) + (H_2CO_3) + (HCO_3) + (CO_3^{2-})$$

An excess in the titration (above the calculated carbonate alkalinity) reveals the presence of other buffers in the system that would have invalidated the use of acid titration for precisely assaying carbonate ion and other forms of carbon dioxide. For example, a detailed study of waters from the Columbia River and some of its tributaries disclosed that, on the average, only 94% of the total alkalinity is due to carbonate alkalinity (Park et al., 1969).

Diel Changes in Alkalinity

Usually, daily fluctuations in alkalinity are negligible. In dilute waters the effects of photosynthesis and respiration on the carbonic acid system are marked mostly by pH changes; despite changes in the proportion of bicarbonate to car-

Table 13-6   Conversions of pH values to hydrogen ion concentrations

pH	Molarity of hydrogen Ions (nol/liter)
6.0	1.0×10^{-6}
6.05	8.913×10^{-7}
6.10	7.943×10^{-7}
6.15	7.097×10^{-7}
6.20	6.310×10^{-7}
6.25	5.623×10^{-7}
6.30	5.012×10^{-7}
6.35	4.467×10^{-7}
6.40	3.981×10^{-7}
6.45	3.548×10^{-7}
6.50	3.162×10^{-7}
6.55	2.818×10^{-7}
6.60	2.512×10^{-7}
6.65	2.239×10^{-7}
6.70	1.995×10^{-7}
6.75	1.778×10^{-7}
6.80	1.585×10^{-7}
6.85	1.413×10^{-7}
6.90	1.259×10^{-7}
6.95	1.122×10^{-7}

The same basic values, with different exponents, apply in other ranges of pH; for example:
pH 5.8 is equivalent to 1.585×10^{-6} mol/liter
pH 7.8 is equivalent to 1.585×10^{-8} mol/liter

bonate, alkalinity titrations yield about the same result throughout the day and from day to day. The addition of CO_2 lowers pH without much change in alkalinity. Carpelan (1957) reported an exception to this where impressive diel fluctuations occurred in small pools within a desert canyon in California. Photosynthetic precipitation of $CaCO_3$ took place during the daylight hours, diminishing total alkalinity. This was accented by rising water temperatures that lowered the solubility of $CaCO_3$. The following night, **aggressive CO_2** from excess respiration redissolved the calcite to raise the alkalinity to a high dawn level. One sunny day in late May during the first 189 minutes after dawn, the pH rose from 7.1 to 7.6, and the loss of free CO_2 was matched by an increase in dissolved oxygen. During the next 40 minutes, however, the pH soared to 8.4 while alkalinity simultaneously decreased from 4.6 to 3.28 mEq/liter.

Often in hard carbonate lakes there is a decrease in epilimnion alkalinity during the growing season as photosynthetic uptake of CO_2 results in the precipitation of $CaCO_3$. The sinking carbonate is redissolved in the hypolimnion by the carbonic acid formed from hydration of the CO_2 of decay and is returned to the epilimnion at the next overturn. Seasonal heightening and ebbing of alkalinity values occur in such cases.

Methods of Analyzing for Free CO_2

Titration

Limnologic field tests for free CO_2 are somewhat unsophisticated and have been replaced to a great extent by calculation. A freshly collected water sample is measured out in a Nessler tube or graduated cylinder; a few drops of phenolphthalein are added, and a rapid titration with $NaOH$ or $NaHCO_3$ follows until the pink end of phenolphthalein is attained. The normality of the base solution is usually adjusted to facilitate ensuing calculations. For example, if 0.0227 N (1/44) $NaOH$ is used and 100 ml of water is tested, the milliliters of titrant multiplied by 10 gives free analytic CO_2 in milligrams per liter. Despite the small surface exposed in using an elongate, narrow cylinder to hold the sample and despite the gentle stirring that must be performed during the titration, there is probably some escape of CO_2 into the air. Imagine how this would apply to summer samples brought from the deeps and subjected to a sudden decrease in pressure and an increase in temperature during the titration. To compound the problem, the pink end point is not easily judged; pink may flash through the sample only to disappear as the worker prepares to record the results. This fading may be repeated a time or two before a pink lingers for a minute or so and is then termed "permanent." Because of this delay, considerable free CO_2 may have escaped the final reckoning.

An example of how CO_2 can be lost from a sample is illustrated by a simple test with the issue from a spring. A sample of water from an underground source, placed in a beaker and tested with phenolphthalein, will probably be colorless. After having been occupied with other tasks for a few minutes, the limnologist may return to find the container of spring water a reddish hue. Free CO_2 escaped from the water and the pH rose to 8.3 or above. Agitation or stirring would have hastened the process. This reveals something about the spring water as it first emerges: it contains gaseous CO_2; its pH is somewhere below 8.3; and there is no residual acidity, its hydrogen ion content being due to carbonic acid alone, rather than to mineral acids such as H_2SO_4.

Because of the interrelated chemistry of the CO_2 system just discussed, a better approach, one measuring the dissolved gas directly, is needed, and may at last have been recently found. Johnson and colleagues (2010) describe the use of a rugged, field-based infrared gas analyzer (IRGA) housed in a polytetrafluoroethylene (PTFE) tube or sleeve that is highly permeable to CO_2 but impermeable to water. This IRGA is submerged in the water and attached by a cable to a meter above. The apparatus was tested in a variety of settings and found accuracy within 2% across a wide range of concentrations and temperatures. More recently, Zosel and associates (2011) published a review of a variety of analytical approaches, some of which may yet be developed into meters available for the field limnologist. In the meantime, we still use the standard chemical analyses.

Calculation

Rough approximation. The relationships among bicarbonate, pH, and free CO_2 (including carbonic acid) shown in Fig. 13-1 permit calculation of any one when the other two are known. At the time of sampling, the pH of the water should be determined without delay and preferably in situ. If it is above pH 8.3, one assumes there is no free analytic CO_2; but if it is below this value, the assumption is that gaseous CO_2 occurs together with bicarbonate ions, and computing its strength is possible. The bicarbonate determination can be delayed in most instances with little harm, although it is not difficult to measure it in the field.

A rough approximation ignoring temperature and ionic strength of the water being tested for CO_2 is shown first. The pH is evaluated and then compared with total alkalinity expressed as $CaCO_3$ in mg per liter. Table 13-7 (on the following page) and Fig. 13-2 (on p. 299) can be used to make a crude approximation of the CO_2 to be expected. If pH were 7.2 and total alkalinity were found to be 185 mg/liter (Table 13-7), then the following equation would be closely applicable, even though it assumes a linear relationship between these factors. Such is not truly the case, as indicated by the straight lines on the semilog plot of Fig. 13-2.

$$\frac{200}{24.5} = \frac{185}{x \text{ mg CO}_2 \text{ per liter}}$$

Solving for x yields about 22.7 mg uncombined CO_2 per liter. Using the graph (Fig. 13-2) and interpolating to judge the point where total alkalinity of 185 mg/liter would fall at a pH of 7.2, one can estimate about 22 mg free CO_2 per liter.

Table 13-7 and Fig. 13-2 are based on data and computations presented by Moore (1939) and are familiar to North American limnologists because of their incorporation into *Standard Methods for the Examination of Water and Wastewater* (American Public Health Association, 2012). Later editions of this manual contain nomograms prepared by Dye (1952) that permit CO_2 approximations with corrections for the effects of temperature and other material dissolved in the water.

Rainwater and Thatcher (1960) presented some tables, based on the relationships between pH, alkalinity, and free CO_2 that permit rough approximations of free CO_2 within pH ranges 6.0 to 9.0. Their data were based on the following equation:

mg of CO_2 per liter $= 1.589 \times 10^6 \, [H^+] \times$ mg of alkalinity per liter (as HCO_3^-)

Table 13-7   Relationships among pH, total alkalinity (mg/liter as CaCO$_3$), and free CO$_2$ (mg/liter)

pH	Total alkalinity	Free CO$_2$	pH	Total alkalinity	Free CO$_2$
5.0	0	9.7	6.8	10	3.1
	1	24.3		50	15.4
	2	48.5		100	30.7
5.2	0	4.9	7.0	50	9.7
	2	26.5		100	19.4
	5	66.2		200	38.7
5.4	0	1.5	7.2	50	6.1
	2	16.1		100	12.3
	5	40.3		200	24.5
5.6	0	0.6	7.4	50	3.9
	5	24.7		100	7.8
	10	49.3		200	15.6
5.8	0	0.2	7.6	50	2.4
	5	15.5		100	4.8
	10	30.9		200	9.7
6.0	10	19.5	7.8	50	1.5
	15	29.2		100	3.1
	20	28.9		200	6.1
6.2	10	12.3	8.0	100	1.9
	20	24.5		200	3.8
	30	36.8		300	5.7
6.4	10	7.7	8.2	100	1.2
	30	23.2		200	2.4
	50	38.7		300	3.6
6.6	10	4.9			
	50	24.4			
	100	48.8			

The value of the term 1.589×10^6 [H$^+$] has been modified in Table 13-8 and shown for each 0.1 pH unit from 6.0 to 9.0; unlike Rainwater and Thatcher's presentation, these data apply to alkalinity expressed as CaCO$_3$, not as bicarbonate.

Using the factor for pH 7.2, which is 0.122 (Table 13-8), and multiplying by 185 mg/liter results in 22.6 mg CO$_2$ per liter, in close agreement with the other rough approximations.

The effect of temperature.   To correct for temperature's effect, the first dissociation constant (Table 13-4) can be used to write

$$[CO_2 + H_2CO_3] = \frac{\left[HCO_3^-\right]\left[H^+\right]}{K_1'}$$

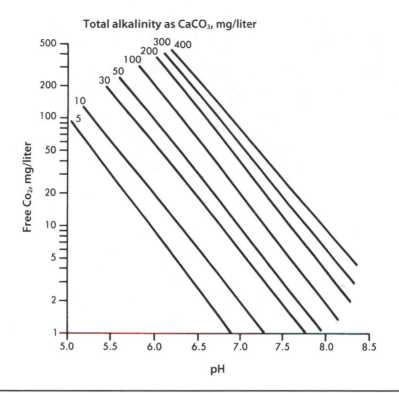

Total alkalinity as CaCO₃, mg/liter

Figure 13-2 Free CO_2 in relation to pH and total alkalinity. Based on data from E. W. Moore, 1939.

The bicarbonate (HCO_3^-) is to be expressed in moles per liter. (Dividing the alkalinity by 0.8202 gives alkalinity expressed as milligrams of HCO_3^- per liter. Shifting the decimal point three places to the left yields bicarbonate ions as grams per liter, and dividing by 61.018, the ionic weight of HCO_3^-, gives the molar concentration of bicarbonate.)

The [H^+] in the equation, of course, refers to moles of hydrogen ions per liter. Because pH, the conventional way of expressing hydrogen ions, is the negative log of the hydrogen ion concentration, we know that a pH of 5.0, for example, means that the molar concentration of H^+ is 10^{-5}. Other values are shown in Table 13-6.

Table 13-8 Factors useful in approximating free CO_2 in milligrams per liter at various pH levels*

pH	Factor	pH	Factor	pH	Factor
6.0	1.937	7.0	0.194	8.0	0.019
6.1	1.539	7.1	0.154	8.1	0.015
6.2	1.223	7.2	0.122	8.2	0.012
6.3	0.970	7.3	0.097	8.3	0.010
6.4	0.772	7.4	0.077	8.4	0.008
6.5	0.613	7.5	0.061	8.5	0.006
6.6	0.486	7.6	0.049	8.6	0.005
6.7	0.386	7.7	0.039	8.7	0.004
6.8	0.307	7.8	0.031	8.8	0.003
6.9	0.244	7.9	0.024	8.9	0.002

*Product of factors and total alkalinity, expressed as milligrams of $CaCO_3$ per liter, yields free CO_2.

Modified from Rainwater and Thatcher (1960).

The K_1' values are those presented by Harned and Davis (1943). Their apparent first ionization constants of carbonic acid as affected by temperature are shown in Table 13-4.

The equation yields CO_2 in moles per liter. Moles of CO_2 per liter × 44.01 (grams per mole) equal grams of CO_2 per liter, easily converted to mg per liter, a conventional way of expressing the results. The values derived from this equation, assuming a water temperature of 20° C, pH 7.2, and total alkalinity 185 mg/liter (as $CaCO_3$) are 225.6 mg or 0.2256 g HCO^{-3} per liter, which is 0.00369 mol/liter; molar concentration of H^+ at pH 7.2, from Table 13-6, is $6.31 × 10^{-8}$; and K_1' at 20° C is $4.15 × 10^{-7}$. Thus, corrected for the effect of temperature, free CO_2 is

$$\frac{(0.00369) × (6.31 × 10^{-8})}{4.15 × 10^{-7}} = 0.00056 \text{ mol } CO_2 \text{ per liter}$$

Continuing, $0.00056 × 44.01 = 0.0246$ g, or 24.6 mg of free CO_2 per liter.

Correcting for activity coefficient. A more accurate method of computing free CO_2 would include correcting for the *activity coefficient* (γ) of HCO_3^-. This is a function of the ionic strength of the solution: the higher the ionic strength of the water, the more interference there is with the activity of bicarbonate and the lower is γ.

The ionic strength of the water is a mathematic quantity that is used to describe the intensity of the forces restricting the freedom of ions in solution, causing them to depart from ideal thermodynamic activity. Ionic strength is defined as μ, or half the sum of the products of each ion's molar concentration and the square of its charge. The more concentrated a solution, the higher its ionic strength, and the greater the restriction in activity of any given ion. Its relation to the activity of HCO_3^- is shown in Table 13-9 and Fig. 13-3. Different ions contribute different increments of ionic strength (Mg^{2+} contributes about 1.6 times as much as an equal weight of Ca^{2+}, for example). For this reason it is dangerous to generalize, but in most fresh waters multiplying the number that quantifies the total dissolved solids (milligrams per liter) by 2.50 to $2.55 × 10^{-5}$ approximates the ionic strength.

Table 13-9 Approximate activity coefficients (γ) of HCO_3, CO_3^- and Ca^{2+} in relation to ionic strength of the water (μ)

Ionic strength (μ)	γHCO_3^{2-}	γCO_3^-	γCa^{2+}
0.0001	0.990	0.945	0.950
0.0005	0.975	0.910	0.920
0.0010	0.965	0.865	0.875
0.0050	0.930	0.730	0.740
0.0100	0.905	0.605	0.680
0.0500	0.815	0.450	0.495
0.1000	0.770	0.365	0.405

If, in the above example, the ionic strength had been 0.001 and the HCO_3 molarity had been corrected (a) for a reduced activity of $\gamma = 0.965$, the ratio would yield ultimately 23.8 mg free CO_2 per liter.

$$\frac{[aHCO_3^-][H^+]}{K_1'}$$

The titrations and calculations for free CO_2 are based on the assumption that the pH is essentially determined by carbonic acid relations. In waters containing strong mineral acids or organic humic acids, the relationships do not apply. Less obvious systems that introduce hydrogen ions could also invalidate calculations. The common utilization of ammonium ions by algae could set hydrogen ions free, along with ammonia as shown below:

$$NH_4^+ \leftrightarrow NH_3 + H^+$$

In most instances, where the pH of natural water is no lower than 6.0, there is no reason to suspect a source of acidity other than the ionization of carbonic acid. By bubbling CO_2 into a water sample, one can reduce the pH to values below 6.0, but waters below pH 5.0 probably owe their acidity to strong mineral acids or possibly to humic acids.

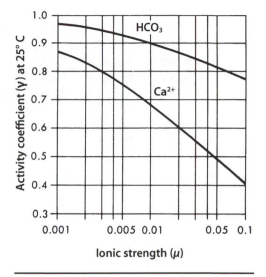

Figure 13-3 Activity coefficients (γ) for HCO_3 and Ca^{2+} in relation to ionic strength (μ) of solutions.

CO_2, pH, and Primary Production

In the final analysis, all carbon in living matter had its origin in CO_2 of the atmosphere, with photosynthesis as the basis of all plant and animal tissue. Primary productivity, the process wherein green organisms utilize a hydrogen donor to reduce CO_2, construct carbohydrate, and release O_2 as a by-product, has been the focus of attention for many ecologic workers over many years. Indeed, Rodhe (1969) emphasized the importance of the rate of carbon fixation in aquatic habitats to denote their relative place on a trophic scale. The rate at which autotrophic organisms, in the presence of light energy, produce organic compounds from inorganic CO_2 and water (the usual hydrogen donor in green-plant photosynthesis) can be estimated indirectly from the oxygen released or directly by assaying the rate at which CO_2 is absorbed.

In 1939 Rawson wrote that carbon dioxide "stands at the threshold of all production." Since then several methods for measuring the uptake of this gas in aquatic communities have been developed.

Assay by the Radiocarbon Method

Steemann Nielsen (1952), who measured carbon fixation in the oceans while serving on the Danish oceanographic vessel *Galathea*, introduced the radiocarbon method of measuring productivity. This method hinges on use of a radioactive isotope of carbon, ^{14}C, that has proved to be a relatively safe and effective tool because it is characterized by weak beta radiation and a half-life of 5,730 years. Several types of Geiger-Muller detectors and more modern scintillators can be used to "count" the radioactivity of ^{14}C.

The ^{14}C method for studying photosynthetic rates depends on taking samples, including the phytoplankton, from various depths and replacing them or incubating them under controlled light and temperature conditions in transparent bottles to which measured amounts of ^{14}C have been added. A small quantity of a dilute solution of NaH^{14}CO$_3$, amounting to 1 to 3 μCi per 125 ml of sample, is injected into the bottles.[2] This addition is followed by thorough shaking and about 4 hours of incubation. At the end of the chosen span of time the entire sample is filtered through a membrane to concentrate the algal cells that are then desiccated in preparation for future assay of radioactivity. The theory is that the assimilated ^{14}C in relation to the total ^{14}C available to the algae shows the ratio of the far more abundant nonradioactive ^{12}C assimilated to the total ^{12}C available. Thus

$$^{12}\text{C assimilated} = \frac{^{14}\text{C assimilated}}{^{14}\text{C available}} \times ^{12}\text{C available}$$

The total carbon dioxide in the system must be calculated first and converted to carbon before the formula is useful and before the radioactive counts from the filtered residual plant cells become meaningful. A correction factor of about 6% compensates for the slower rate at which algae take up the radioactive isotope. Further correction is effected by subtracting appropriate background counts, and additional refinement is possible by use of a dark-bottle control so that any algal uptake of carbon in the dark is also taken into account.

The ^{14}C procedure, briefly described here, approximates net production because it measures the radioactive carbon remaining *within* the plant cells at the end of the photosynthetic incubation period. The ^{14}C that was taken in by the cells and respired during the same time is part of the filtrate, along with the added radioactive material not taken up by the algal cells that are present on the membrane filter. Even this explanation oversimplifies the picture because there may be a loss of organic material containing ^{14}C that had been assimilated during the incubation period in addition to the respired inorganic ^{14}CO$_2$. Furthermore, the uptake of ^{14}C supplied as bicarbonate may result from reactions unrelated to photosynthesis.

The ^{14}C method of arriving at primary production is far more sensitive than the oxygen method. It is, therefore, superior in extremely oligotrophic waters because results may be obtained in 6 to 24 hours, while experiments based on oxygen changes might require several days. Where photosynthetic activity is great, the ^{14}C and oxygen methods are comparable, and satisfactory measurements can be made within a few hours using either technique.

Twenty-six years after his first report of the ^{14}C technique, Steemann Nielsen (1977) discussed the current state of our knowledge and some of the factors that can lessen the carbon test's dependability when not taken into account. Some sources of error are species dependent: certain algal cells are apt to rupture during filtration, allowing radioactive cell contents to pass into or through the filter pores, thereby being lost to the final assay; *Chlorella* is, for example, more fragile than *Scenedesmus*. Origi-

---

[2] The curie (Ci) is widely used although the SI unit of nuclear activity is the becquerel, amounting to one nuclear disintegration per second. The Ci unit of activity is the number of nuclear disintegrations per second from a gram of radium, 3.7×10^{10}.

nally, the ^{14}C technique was used in ocean water, where the total carbon content is about the same as the radioactive material in the standard ampules now available for testing photosynthetic rates. In oligotrophic lake water the CO_2 content may be only 1% of the $NaH^{14}CO_3$, in the ampules, and the uptake of isotopic carbon may be pronounced in such CO_2-poor water. Furthermore, the NaCl content of an entire ampule might raise the salinity of the incubating medium unnaturally if added in its entirety to an oligotrophic sample. In general, when the light-dark bottle method is used in any manner, care must be taken to prevent bacterial contamination, and toxicity to algae—from glass-cleaning agents or from the brass or rubber sampling gear itself.

About three decades after the ^{14}C-CO_2 method was first used, Peterson (1980) comprehensively reviewed its history and the alternate ways of measuring primary productivity. When compared with other techniques, including relatively new methods based on ATP changes and increases in particle numbers and volumes and the old light-dark bottle oxygen technique, it appears that the ^{14}C method underestimates the rate of carbon fixation. It is not clearly understood, but part of the difficulty has nothing to do with cell damage during filtration processes. The loss of $DO^{14}C$ from algal cells in addition to further losses of ^{14}C via dark respiration that may continue in light and the oxidation of glycolate that is not excreted as DOC contribute to underestimating ^{14}C that had been taken in.

Some biologic facts should be kept in mind while studying the uptake of CO_2. For example, certain aquatic plants cannot use the CO_2 in bicarbonate for photosynthesis. In this respect they resemble terrestrial plants, which depend on gaseous CO_2. Aquatic bryophytes are limited to regions where abundant free CO_2 prevails. For this reason one finds them near the orifices of springs or at impressive depths in clear lakes where light penetrates to strata moderately rich in dissolved CO_2, as in Thoreau's Walden Pond, Crater Lake (Oregon), and Lake Tahoe. The spectacular freshwater red alga *Batrachospermum* is another classic example. It occurs in spring brooks and especially near subterranean sources, but farther downstream in reaches of the outflow where crucial amounts of CO_2 have been lost to the air, the alga is no longer present (Minckley and Tindall, 1963). Hutchinson (1970b) analyzed the chemistry of habitats of some species of the aquatic angiosperm *Myriophyllum* and came to the conclusion that one species, *M. verticillatum*, is restricted to regions where free CO_2 abounds; two other species, possessing the typical versatility of higher aquatic plants, find bicarbonate an adequate source of carbon.

Estimate of Primary Production by pH Changes

Relationships between hydrogen ion and CO_2 have led to a method of estimating primary productivity by studying pH changes in aquatic habitats. At dawn, for example, one might expect the lowest pH of the 24-hour period because the CO_2 of respiration and decomposition would have accumulated since the preceding dusk. With the morning sun, photosynthesis commences, and CO_2 is taken up by the green autotrophic organisms of the community. As CO_2 is absorbed, the pH rises at a rate that can be translated to the rate at which carbon is being fixed by plants. Theoretical equations and nomographs based on them afford a rough index of the effects of CO_2 changes on pH, if the bicarbonate content is also known. Verduin (1951) used the equilibrium equations of E. W. Moore (1939) to study CO_2 removal during photosynthesis and CO_2 evolution by respiration. This procedure had been attempted many years before by Osterhout and Haas (1918), who used colored indicators rather than a precise pH meter.

Other substances in natural waters, however, could invalidate inferences based on theoretical carbonic acid equilibria, and these must be considered. Verduin (1956) was one of the first to realize that the theoretical curves, equations, and nomograms of E. W. Moore (1939), Dye (1944), and others do not apply to every type of water. To set the stage for studying photosynthetic rates of phytoplankton in a given body of water, he studied the effects of titrating with NaOH. Using filtered lake water, he would first lower the pH by introducing bubbles of CO_2. A differential titration with 0.01N NaOH followed, the rise in pH being recorded after the addition of an increment of 0.1 to 0.5 ml of the standard base. The rationale is that each milliliter of 0.01N NaOH will absorb 10 μmol of H_2CO_3 and will convert it to $NaHCO_3$, or at higher pH levels will convert 10 μmol of HCO_3^- to CO_3^{2-}. These steps approximate what occurs when plants take up free CO_2 and the so-called half-bound CO_2 in HCO_3^- as they reduce inorganic CO_2 to carbohydrate. Verduin had to consider another problem: titrating with a base (or acid) changes the alkalinity of the water. Therefore, after recording the pH change caused by adding a small amount of the standard base, he selected a new sample of the water being tested, brought it to some other pH level with CO_2 bubbles and performed another minititration. Combining the slopes of pH change for many of these small tests yielded a master curve for the particular water under investigation. Curves of this type, where pH is plotted against the added NaOH, may deviate markedly from the theoretical curve. Nevertheless, they can be used to match pH changes with carbon dioxide withdrawn from the aqueous system while carbon is photosynthetically fixed. Similarly, 1 ml of 0.01N HCl is equivalent to 10 μmol of CO_2. Many titrations, starting from various pH levels and employing HCl, yield data from which a curve depicting the addition of respiratory CO_2 and its effect on pH can be plotted. It is nearly identical to the NaOH curve, and either one can be used to find the results of removing or adding known quantities of CO_2. With this technique, standard light-dark bottle experiments can be performed in the same manner as the oxygen method but based on the rate at which hydrogen ion concentrations change.

Hypolimnetic CO_2 Increase as an Index of Productivity

Methods for estimating the amount of inorganic carbon fixed per unit of time have been discussed. These, as in the case of O_2 production, can be referred to activities within volumes of water or related to what occurs beneath the average square meter or square centimeter of water surface.

An indirect index of productivity within a lake, a value that can be used to compare one lake with another, was proposed by Ohle (1952). This index more nearly approximates the total amount of organic material decaying in the hypolimnion than does the rate at which oxygen disappears. It is the rate at which hypolimnetic CO_2 accumulates.

If all the CO_2 were produced by the aerobic decay of simple carbohydrates, this gas would appear in amounts equaling the disappearance of oxygen. The respiratory quotient ($RQ = CO_2/O_2$) would be 1.0. The materials that decay and are respired, however, include lipids and proteins, as well as other substances in various proportions. Because of these substances, the ratio of CO_2 produced to oxygen consumed is less than 1.0. An average figure of 0.85 seems to be a reasonable approximation of the RQ involving a mixture of organic compounds.

The moles of oxygen that disappear during a given time multiplied by 0.85 should approximate the CO_2 simultaneously produced by aerobic decay and respiration. Thus, if, 1,000 molecules of oxygen were used by the bacteria, 850 molecules of CO_2 would be liberated.

Although theoretical RQs fall in the range of 0.5 to 1.0, the actual values found empirically often surpass unity. The production of more CO_2 per unit of time than would be expected on theoretical grounds has been termed the Ohle anomaly. Actually, it is not especially anomalous: during late winter the RQ amounts to 9.0 under the ice in Lake Wingra, Wisconsin. This extra CO_2 results from anaerobic decay and is not, therefore, related to oxygen consumption. Alternate electron acceptors replace oxygen after it is depleted. Decomposition, then, continues to be limited only by the supply of electron acceptors or by the stores of organic matter that serve as electron donors. Certain bacteria that attack the celluloses of plant origin and some lipolytic microbes can function under anaerobic conditions. The appearance of CO_2 generated by their actions must be reckoned in computing the organic matter that is decaying in the hypolimnion. In addition, it is assumed that only about half the organic matter broken down by anaerobic events is reflected in this CO_2 production; the other half is represented by the marsh gas CH_4, methane. Methane is strictly a product of anaerobic decomposition of plant and animal remains.

Methanogenic prokaryotes of the Archaea type are represented by immotile, obligatory anaerobes belonging to such genera as *Methanobacterium*, *Methanococcus*, and *Methanosaeta*. They give rise to CH_4 at the expense of fatty acids, ketones, alcohols, and carbohydrates. They gain energy from an oxidative decomposition of organic compounds accompanied by the reduction of CO_2 to CH_4. Thus, the excess of free CO_2 that comes from anaerobic decay contains about half the carbon involved in that decay; the other half is represented by methane. The carbon in the two gases together indicates more nearly the total organic material that underwent decomposition within the hypolimnion than does the oxygen that disappeared.

Just as the rate of a hypolimnetic areal oxygen deficit can be calculated to derive an index of lake productivity, so the rate of CO_2 increase beneath unit area of hypolimnion can be computed for comparative purposes. The validity of both indices can be influenced adversely by organic allochthonous material, temperature, and morphologic relations.

The literature on the subject reveals there has been much more interest in the uptake of oxygen than in the accumulation of CO_2 when benthic and hypolimnetic metabolic events have been estimated. Also, the areal deficit rates for oxygen tend to underestimate lake bioenergetics when compared with daily increments of inorganic carbon. The increments of carbon are owed to anaerobic respiration and fermentation as well as oxidation (where O_2 plays the role of electron acceptor). Thus, as oxygen disappears in the hypolimnion in early summer, Dunham Pond, Connecticut, is oligotrophic on the basis of its mean hypolimnetic oxygen deficit corrected for eddy diffusion. It is mesotrophic, however, in terms of CO_2 accumulation converted to oxygen equivalents (Rich, 1978). Beneath the ice during the first half of winter the RQ is about 1.0, and the metabolic rates of oxygen dwindling and carbon dioxide increasing are at an oligotrophic level; during the latter half of winter the RQ rises as inorganic carbon accumulates at a mesotrophic rate while oxygen's deficit continues at an oligotrophic pace.

In later years, Rich focused his research on hypolimnetic and benthic events in relation to whole-lake metabolism (Rich, 1978; Rich and Wetzel, 1978). Much energy that once was considered lost or stored as detritus—nonpredatory loss of organic carbon—is in reality part of a substantial anaerobic metabolism that is revealed sometimes by high RQs. In anaerobic environments, reserves of electrons useful for reducing CO_2 to organic molecules are accumulated, for example, as H_2S. Anaerobic detrital metabolism and these stores of electrons power chemosynthesis and bacterial photosynthesis. Bacterial photosynthesis, using H_2S as a reductant, is about 9.4 times less difficult (thermodynamically) than the eukaryotes' and blue-greens' phytotrophic method, where water is the reductant. (Bacterial photosynthesis proceeds, therefore, relatively efficiently in dim light.) Later at the overturn, many reduced, energy-rich metabolic intermediates can be oxidized in an aerobic phase of the deep-water metabolism, oxygen being the terminal receptor. Thus, aerobic organisms gain energy with high oxygen uptake and low CO_2 production, which makes for temporarily low RQs.

Carbon as a Factor Limiting Primary Productivity and Its Role in Eutrophication

A limiting factor is an essential nutrient or other requirement, but this is only a partial definition. A necessary element that is always in adequate supply is not considered to be a limiting factor. If demand becomes greater than supply of such a needed element, it becomes limiting and governs growth or production despite the presence of all other requisites in adequate amounts. When an essential factor is in such short supply that it is limiting, its addition to the environment will cause a spurt in growth. Phosphorus, nitrogen, and silicon are the elements usually implicated when phytoplankton populations are growth limited.

Carbon dioxide is so common that it has rarely been considered a limiting factor in primary productivity, even though its quantities in carbonate and bicarbonate alkalinity have long been theorized to account in part for the trophic nature of lakes. You will recall that some of the highest photosynthetic productivity in natural water occurs in the soda lakes of Africa, where carbonate is combined with sodium. Talling and others (1973) reported alkalinities of 51 to 67 mEq/liter (2,550 to 3,350 mg/liter expressed as $CaCO_3$) in some Ethiopian lakes. There the standing stocks of *Spirulina* amounted to 10 to 20 g chlorophyll *a* per m^2, and 13.4 to 17.8 g of carbon were fixed each day beneath the average square meter of lake surface. In other instances in dense algal populations, there are brief effective limitations of growth because of carbon shortages. This is evidenced when additions of nutrients have no effect until supplemental carbon dioxide stimulates a burst of new growth.

Numerous observations have shown photosynthesis proceeding in epilimnetic strata where no free CO_2 could be demonstrated. This absence might lead to the erroneous conclusion that carbon dioxide is not necessary for photosynthesis! Closer to fact, however, is that in such instances, the green autotrophs utilize CO_2 as rapidly as it is produced; some temporary limiting effect might be inferred from this.

Lately, nutrient supply has received much attention because of society's acceleration of eutrophication processes that bring about nuisance blooms of cyanobacteria. These blooms seem to be the key to most of the events that diminish lake quality. For this reason, the search for the most important element inducing blooms has been the aim of much research, because with this information, effective purification of the sewage entering lakes can be achieved. The rationale, in simplistic terms, is that the culprit (element, ion, or compound), normally in limited supply, can be removed or intercepted before it reaches the lake.

A readily available source of information on these matters is the volume on eutrophication (National Academy of Sciences, 1969), and more to the point are the collected papers, *Nutrients and Eutrophication*, from the symposium of the American Society of Limnology and Oceanography (Likens, 1972). Evidence for the triggering of eutrophication by abrupt and massive addition of carbon via human activity is lacking; the finger points with a greater degree of accuracy to phosphorus as the element that can initiate such an event, and there are no published examples of algal nuisances being generated by such elements as silicon, manganese, or iron.

Phosphorus is an abundant and necessary element in plant and animal tissues. This explains why assays for ATP and DNA are sometimes employed to estimate biomass in aquatic habitats. It is rare in almost all ecosystems. Because of its scarcity, there is little doubt that phosphorus is often the master factor. Its meager supply limits the growth of freshwater phytoplankton. There are, however, other requisite elements that can hold back algal growth by their scarcity. Furthermore, at times cyanobacteria do not predominate in situations where the phosphorus supply is extraordinarily great.

Nitrogen, although an essential element, is often less limiting than phosphorus. It is more abundant than phosphorus, and, in addition to edaphic sources, the rich atmospheric store of nitrogen is made available in part by some nitrogen-fixing microbes. Oceanographers have found, however, that in coastal marine waters the nitrogen supply is often more limiting to algal growth than phosphorus. Early biologic tests performed by Potash (1956) showed that the algae of two Cornell ponds found phosphorus in short supply at one season and nitrogen compounds critical at another time of year. In addition, many experiments have shown that plant growth is stimulated best by the addition of a *mixture* of nitrogen and phosphorus compounds. Also, there are times when a marked response comes about from the addition of one of many trace elements, such as cobalt and molybdenum, or even common ions and compounds, such as SiO_2 (Goldman, 1972).

In Shriner's Pond, Georgia, Kerr and coworkers (1972) found evidence that the addition of the nutrients nitrogen and phosphorus caused a rapid increase in the heterotrophic bacteria before the larger, autotrophic algae responded. Increased bacterial decomposition of organic material made CO_2 available to the algae and stimulated their growth, an indirect effect of adding nitrogen and phosphorus.

Allen (1972) studied little Star Lake in Vermont, where total inorganic carbon supply was very low. A pronounced diel cycle in the waning and waxing of carbon took place. Free CO_2, present in the morning and largely a product of nocturnal respiration, was depleted by algal populations by noontime, and the bicarbonate level also fell. Allen theorized that, in poorly buffered Star Lake after the morning uptake of CO_2, a second source of carbon for bacteria and phytoplankton served to maintain a rather high production rate during the remainder of the daylight hours. This second supply of carbon was the release of organic compounds that algae and benthic macrophytes had produced during morning hours.

Schindler (1971c) and Schindler and associates (1972) reported the results of experiments on a Canadian lake, extremely impoverished in total CO_2 (about 1.1 mg/liter, with alkalinity less than 0.04 mEq/liter), that led them to conclude that CO_2 supply rarely limits phytoplankton growth. The addition of nitrate and phosphate salts to the lake increased the standing crop of planktonic algae thirtyfold, measured in terms of chlorophyll a, despite the initial low stock of carbon. Invasion of gaseous CO_2 from the air supplied the carbon needed for the excess productivity. Thus, the gradient set up by the stimulated algal population removing CO_2 allowed more than 700 mg of CO_2 to move across an average square meter of epilimnion surface each day. Schindler's report leads to speculating that in sunny epilimnion waters, where no free CO_2 can be demonstrated, there may be more photosynthesis proceeding than that which utilizes HCO_3^- as a carbon source. Gaseous CO_2 may still be important, although it is used up as rapidly as it invades from the atmosphere or is produced by respiration.

In poorly buffered waters the flux of CO_2 is greater than one might expect on purely physical grounds. As amounts of CO_2 and carbonic acid are lowered, the hydroxide concentration rises and there is a chemical enhancement of the CO_2 invasion across the air-water interface (Wood, 1977). At a pH of 10 the flux might exceed the theoretical rate by four times.

On the other hand, Foster (1973), studying two ponds containing tertiary-treated waste water from the city of Phoenix, came to the conclusion that CO_2 supply sometimes limited photosynthetic rate. The water entering the ponds carried a mean of 14.1 mg phosphate per liter; the algae utilized it to such an extent that on the average only 2.2 mg soluble PO_4 per liter was present in pondwater samples. The latter is a tremendous amount when compared to that in natural lakes and is far above a limiting threshold. Nitrogen also was abundant, averaging 1.2 mg/liter throughout the year. Foster found that, despite the plentiful stores of nitrogen and phosphorus, the rate of production often decreased spectacularly by noon. In some instances inhibition by intense incoming solar radiation was involved, but at other times free CO_2 seemed to be the limiting factor. The possibility of atmospheric CO_2 keeping pace with productivity in the Phoenix ponds seems remote; the amount of carbon fixed in 1 hour was usually greater than the entire daily input from the atmosphere over unit area of Schindler's Canadian lake. Several earlier studies on sewage lagoons had suggested, prior to Foster's work, that in such extremely fertile habitats carbon is often in short supply and is the limiting factor. This is probably rare in less polluted waters.

McIntire and Phinney (1965) studied experimental laboratory streams to which light and CO_2 input could be regulated. Some of their results (Fig. 13-4) show that primary productivity increased as a function of light intensity, up to about 0.005 cal/cm^2 per minute. After that, the carbon dioxide supply showed its effect; 45 mg CO_2 per liter resulted in 1.5 times the photosynthetic rate of 1.8 mg/liter. In this artificial system with phototrophic producers being periphytic algae, the limiting effect of CO_2 was obvious.

Wiegert and Fraleigh (1972) used a natural situation with little modification to demonstrate some effects of carbon limitation. In a stream charged with free CO_2 and issuing from a hot spring in Yellowstone National Park, they found diminished primary productivity of cyanobacteria at a downstream station. Solar radiation and phosphorus supply were unchanged from the source to the station in the lower

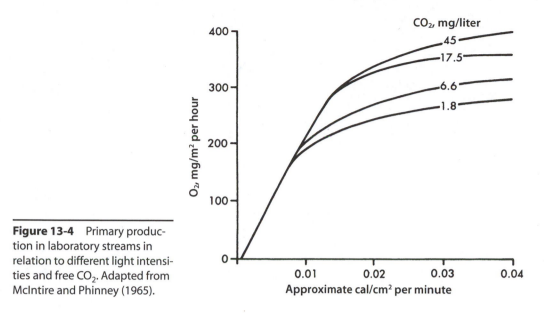

Figure 13-4 Primary production in laboratory streams in relation to different light intensities and free CO_2. Adapted from McIntire and Phinney (1965).

reaches, but much free CO_2 had been lost to the atmosphere by then. Its decrease was assumed to be responsible for the lessened productivity.

The relations of CO_2 and phosphorus to objectionable algal growth in lakes are complex. In the extremely polluted waters of sewage lagoons one would expect the cyanobacteria to be dominant. Often, however, this is not the case; mixed populations of small green algae occur, winning the competition with the cyanobacteria in these rich environments. Shapiro (1990) summarized the prevailing ideas concerned with factors in cyanobacteria dominance. In the presence of nutrients, high pH, and very little free CO_2, the cyanobacteria, efficient at utilizing CO_2 from low concentrations, predominate. When pH is lower and carbon dioxide tensions are higher, certain chlorophyta thrive and exclude their prokaryote competitors. In poorly buffered sewage lagoons, photosynthetic activity raises the pH to a high level, free CO_2 is at a premium, and cyanobacteria flourish just as they do in alkaline soda lakes. In well-buffered lakes, the pH may not rise to the threshold favoring cyanobacteria.

Brock (1973) presented evidence that cyanobacteria do not exist in habitats where the pH is below 4.0 (and rarely occur where pH is below 5.0), although the Archaea and some eukaryotic species representing many genera thrive.

In earlier experiments, cited in his 1990 paper, Shapiro had shown that natural populations of blue-green "algae" were replaced by chlorophytans when, after enrichment with phosphorus and nitrogen, carbon dioxide was added or the pH was brought down with acid, thereby making more CO_2 available from bicarbonate. The addition of phosphorus and nitrogen compounds alone stimulated the bluegreens to some extent, but the increments of CO_2 and hydrogen ions caused them to lose in competition with green algae.

The words of Golterman (1975: 383) serve to summarize this section. In his discussion of the role of carbon in hastening eutrophication, he states that there is no good evidence to bolster the notion that algal crops in an unfertilized lake are limited by carbon.

14

The Major Ions in Inland Waters

Sources of Ions

It is easy to speak of the chemical composition of the average fresh water of the world or of different continents, but at least two things must be kept in mind. First, there is no ideal toward which inland waters are evolving, and second, the average depends on what waters have been assayed and how the results have been weighted in the computations. Table 14-1 shows often-quoted data that were derived from Clarke (1924), Conway (1942), Livingstone (1963b), and Rodhe (1949). The first three authors were interested in the dissolved substances carried to the sea each year by the rivers of the world, and they estimated some data that describe the average chemical content of the world's rivers. Rodhe described a "standard" fresh water. The relative abundance of the major ions in lakes and rivers varies from continent to continent and region to region. Generally, researchers of dilute aquatic habitats assume proportions much like those shown in Table 14-1 on the next page.

Gorham's (1961) paper dealing with the major ions in inland waters is an excellent source of information. He underscores the factors operating to bring about and maintain the differences among fresh waters. Five obvious factors are climate, geography, topography, biotic activity, and time. These are not completely independent; they interact to produce variety.

The atmosphere—which receives materials from the sea, the soil, volcanoes, industrial and domestic pollution, and organic detritus blown aloft—contributes ion-laden rain, snow, gases, and dry fallout.

The soil and weathered rocks are sources of edaphic ions. The ions are released by these mechanisms: solution, including the complexities of solubility differences; oxidation and reduction; increasing or decreasing weathering and availability; chelation or other forms of complexing; and the action of hydrogen ions, supplied especially by carbonic acid but not limited to that source.

Total Dissolved Solids and Specific Conductance

A general measure of the solutes in water is Total Solids which is composed of two parts: the Total Suspended Solids (TSS) fraction is removed by a 2-μm filter, while

311

Table 14-1   Some estimates of the chemical composition of the world's rivers and of a "standard" fresh water compared with the principal ions of seawater

Ions	Clarke (1924) (% mEq)	Conway (1942) (mEq)	Rodhe (1949) (% mEq)	Livingstone (1963b) (% mEq)	Means (% mEq)	Livingstone (1963b) (mg/liter)	Seawater g/liter
Anions							
CO_3^{2-}	73.9	73.5	74.3	67.9	72.4	58.4	0.14
SO_4^{2-}	15.9	16.5	15.6	16.5	16.1	11.2	2.71
Cl^-	10.1	10.0	10.1	15.6	11.5	7.8	19.44
Totals					100.0	77.4	22.29
Cations							
Ca^{2+}	63.4	64.0	63.5	52.6	60.9	15.0	0.41
Mg^{2+}	17.5	17.0	17.5	24.0	19.0	4.1	1.30
Na^+	15.7	16.0	15.6	19.2	16.6	6.3	10.81
K^+	3.4	3.0	3.4	4.1	3.5	2.3	0.39
Totals					100.0	27.7	12.91

the Total Dissolved Solids (TDS) or filterable residue, passes through. TDS is simple to measure, but quite tedious. One filters a water sample (100 ml is convenient) and evaporates it at 105° ±2° C in a tared container until a constant weight is achieved. In samples from saline lakes, several weeks may pass before a constant weight is reached; for that reason a smaller aliquot, perhaps 25 ml, should be dried, although it could ultimately introduce a greater error. The TDS includes salts and organic residue. Further burning, this time at 500° to 550° C, would leave only the inorganic ash. The weight lost would be a measure of the organic content or may be best expressed simply as loss on ignition, with no mention of the specific organic designation. The residual ash represents the total salt content of the water. Usually, the filterable residue remaining after evaporation is all that is determined; this is **TDS**.

The mean of many analyses presented by Livingstone (1963b) suggests that the world's rivers contain an average of about 120 parts per million (ppm) of TDS. There is a remarkable range to be considered in lakes. For example, the ultraoligotrophic Waldo in Oregon has filterable residue from 1.0 to 5.5 ppm, averaging about 1.7 (Malueg et al., 1972). Eutrophic Lake Itasca waters contain about 185 ppm. In the sea, the TDS value is about 35,000 ppm; in the Great Salt Lake of Utah the value is roughly 220,000 ppm, representing an unusually concentrated solution.

Rawson (1951) showed something about the importance of TDS in lake productivity. He reviewed his earlier finding that the Slave River contributes about 54,000 metric tons of dissolved matter, along with 36,000 tons of suspended silt, per summer day to Great Slave Lake. Another great lake 400 km northeast, Great Bear, has no comparable influent, nor does it have such a large drainage area. Although the climatic environment of Great Bear is more severe, the two lakes are comparable in many ways and share a similar edaphic heritage. The TDS of the southern portion of Great Slave, where the Slave River enters, is about 160 ppm, compared with Great Bear's 98 ppm. Great Slave supports a limited commercial fishery; Great Bear, being more oligotrophic, cannot. The role of influent streams and the edaphic factor are illustrated further in Great Slave's eastern region, where streams draining the granitic pre-Cambrian Shield enter. Dissolved solids in that part of the lake range from 22 to 82 ppm.

Similarly, Hooper (1951) found that Demming Lake, receiving its water only via precipitation, seepage, and a very small drainage area, has a dissolved mineral content far less than that of nearby drainage lakes (those with influent streams). The result is a very low productivity in Demming, when compared with its neighbor lakes.

A shortcut in analyzing for TDS is testing for specific **conductance** (conductivity). A conductivity meter measures the flow of electrons through the water. The current in the presence of electrons is heightened; in distilled water, however, resistance to electron flow is very high because pure water is a poor conductor. Conductance is the reciprocal of resistance. Presently we find it expressed in two different ways as cgs (centimeter-gram-second) usage gives way to the International System. Formerly it was symbolized by reverse spelling of the name of the nineteenth century German physicist, George S. Ohm (1787–1854), whose well-known law states that an electric current is the quotient resulting from dividing potential difference, or electromotive force, by resistance, and whose name still stands for the unit of resistance, the *ohm*. Now, gaining very slowly, and especially so in North America, is the *siemens* unit, the equivalent of *mho,* and symbolized S. It is derived from and honors a German-born British engineer and inventor, William Siemens (1823–1883). The siemens unit expresses the conductance between two points; an electromotive force of 1 volt between these points in a conductor produces a current of 1 ampere. This is the reciprocal of its resistance expressed as ohms. In water the conductance is a measure of flow between two cm^2-electrodes set 1 centimeter apart. For this reason the conductance is expressed in units per centimeter at some specified temperature. In most fresh water it would be cumbersome to express the conductance in S units; usually use of μS units eliminates the need for decimal points. In saline water the mS unit is satisfactory, S × 0.001.

The flow between electrodes in a solution of electrolytes is increased with higher temperature; a compensating factor of 0.25 for every degree rise has been suggested. It is easy to establish an empirical factor in a given body of water by fixing upon the conductance of a single sample at various temperatures in the laboratory. The data acquired allow one to express conductance at any specified temperature. A favorite standard is 25° C, but many limnologists select 18° or 20° C, perhaps because they represent temperatures more often met in the field. Taking the TDS data from a given lake or lake district through the seasons and comparing them with concurrent conductivities reveals a relationship. A graphic plot of the affinities will show a line from which TDS can be inferred from conductance; this method precludes further filtering-evaporation procedures. The worker can establish a value for c in the formula $Kc = T$. In this, K is the conductivity in μS per centimeter at a specified temperature; c is a coefficient empirically determined to yield T, short for TDS. Thereafter, a simple assay of conductivity referred to a standard temperature could be used to find total dissolved solids.

Williams (1966) showed that in 62 samples from 41 Australian lakes, ranging from 0.05% to 20% salinity, there was excellent correlation between TDS and conductance. Caution should be observed, however, in transforming conductivity to TDS in samples above 5% salinity. In hypersaline waters there are inherent errors that are not significant in fresh water, because all ions do not contribute identical specific conductance. In the diversified ionic proportions of different hypersaline lakes, the sum of conductances could differ even when salinities are identical.

Because specific conductance does not refer to any particular chemical species, it has no direct ecological effect. However, it is useful as an easily-measured indicator of total chemical loading. Kimmel and Argent (2010) sampled nearly 11,000 fish in a Pennsylvania stream affected by mine pollution. They found a threshold of 3,000–3,500 µS per centimeter, above which the fish communities were significantly impaired.

Expression of Chemical Results

In most instances one can generalize that milligrams per liter or dm^3 and parts per million by weight are identical ways in which to express concentrations present in water. One ppm represents 1 mg of solute per kg of solution; 1,000 ppm is 1 part per 1,000 (‰) and 10,000 ppm is equivalent to 1%. Results reported in milligrams per liter are masses per volume, and the conversion to ppm is permissible if it is assumed that 1 liter of water has a mass of 1 kg, a density of 1.0. In desert limnology, an important error may be introduced by not correcting for dissolved minerals that raise the density of water substantially above unity. When mineral concentrations are greater than, roughly, 7,000 ppm, the conversion to or from milligrams per liter requires a correction factor to account for the change in solution density:

$$ppm = \frac{mg/liter}{density \text{ (or specific gravity)}}$$

Other volume concentration units are molarity and normality; the former is especially useful. Milliunits are appropriate. They describe typical freshwater concentrations without using values less than unity, such as milligrams per liter, and are preferred over grams per liter. Gases dissolved in water are often expressed as millimoles per liter, the quotient of the weight in milligrams divided by molecular weight. Twelve milligrams CO_2 per liter, then, equals 12/44, or 0.272 mmol/liter, or 272 µmol/liter.

Recently there has been a move to express water analyses in terms of chemical equivalents of the ions. This can be justified because it takes into account the combining or equivalent weights of ions (ionic weight/ionic charge), thereby permitting exploration of the cation-anion balance to see if an error has affected the chemical analyses. The equivalent sum of the cations should equal the equivalents due to the major anions. Usually ions are treated as milliequivalents per liter. The milliequivalent per liter value is the milligrams of an ion per liter divided by its equivalent weight. Thus, the equivalent weight of Ca^{2+} is 40.08/2 = 20.04. The equivalents in 1 g of calcium ion is 1/20.04 = 0.0499, or 49.9 mEq. Converting milligrams of Ca^{2+} per liter to milliequivalents per liter is brought about by multiplying by the factor, 0.0499. The reciprocal, 1/0.0499 = 20.04, the equivalent weight of calcium, and serves as a factor to convert calcium from milliequivalents per liter to milligrams per liter (Table 14-2).

Nonionized materials cannot be reported in mEq. This leaves total iron, silica, and oxygen, for example, to be expressed in terms of weight-based units, milligrams per liter or parts per million, or as molarity. The conglomeration called TDS must be reported as milligrams per liter or parts per million.

Salinity versus Chlorinity

Titration for chloride ions has been standard oceanographic procedure for many years. The results for seawater having around 535 mEq/ liter should be expressed as chlorinity rather than salinity. It is true that the assay of chlorinity in seawater gives a good index to whether salinity is low, as at a river's mouth, or relatively high, as in the Gulf of California, where evaporation is consequential. An empirical formula is

$$\text{Salinity} = 0.03 + 1.805 \text{ (chlorinity)}$$

This is expressed in grams of solid material dissolved in 1 kg of seawater, being parts per 1,000. Oceanographers define salinity as total grams of solid material in 1 kg of seawater after organic matter has been completely oxidized, all carbonate has been converted to oxide, and bromine and iodine have been transformed to chloride equivalent. In inland waters the chloride titration is not necessarily directly proportional to salinity, because there are numerous combinations of the various ions, while seawater and coastal brackish waters, by contrast, are rather stable with respect to ratios of major constituents.

Among limnologists the term salinity refers to the sum of anions and cations; its meaning is less restricted than that of chlorinity and perhaps less restricted than the word salinity as used by oceanographers. Sometimes it is equated with TDS, although they are not the same.

Salinity can be determined by measuring electrical conductivity. Among oceanographers this conversion is applicable universally; limnologists, by contrast, must make a few comparisons before they arrive at an empirical formula to convert conductivity to salinity in some body of water, and the formula may differ from lake to lake.

Salinity has ecological effects beyond the salty lakes or brackish coastal areas. Bos and others (1996) studied over 100 Canadian lakes and found that zooplankton communities differed based on salinity, with higher salinity associated with specific groups, and more dilute waters having less effect. Taxonomic variability among biota was highest in dilute waters.

Table 14-2   Factors to convert milligrams of common ions to milliequivalents and vice versa

Ions	mEq = mg ×	mg = mEq ×
Ca^{2+}	0.04990	20.04
Mg^{2+}	0.08224	12.16
Na^+	0.04350	22.99
K^+	0.02558	39.10
HCO_3^-	0.01639	61.02
CO_3^{2-}	0.03333	30.01
SO_4^{2-}	0.02082	48.03
Cl^-	0.02820	35.46

Carbonate

In most dilute freshwater in humid regions of the world, the principal anion is carbonate (see Table 14-1). Actually it occurs most often as bicarbonate ion, and usually with calcium. When it is compared by weight with other ions in solution, bicarbonate ion is customarily expressed as CO_3^{2-} because evaporation of a known amount of calcium bicarbonate solution leaves only the carbonate of calcium to be weighed. During evaporation, gaseous CO_2 and water are lost from bicarbonate ions, converting them to a lesser weight (factor 0.4917) of carbonate.

$$Ca(HCO_3)_2 \xrightarrow{\text{evaporation}} CaCO_3 + CO_2\uparrow + H_2O\uparrow$$

In the previous chapter much has been discussed about carbonate. It is unnatural to separate carbonate from discussions of CO_2, pH, alkalinity and even calcium in "typical" fresh water. Alkalinity is usually a measure of carbonate, and, despite pitfalls, the alkalinity titration usually quantifies this ion, the principal anion in most fresh waters. This is because the carbonate-bicarbonate complex is the common buffering system in fresh water.

Compounds of Carbonate

Indirectly, alkalinity measures cations that are balanced against carbonate and other anions that are important in buffering action. If the alkalinity were entirely the result of the carbonate buffering system in a pure calcium carbonate water and the results were expressed as mg/liter $CaCO_3$, the factor 0.4 would yield the Ca^{2+} portion. We must remember two things, however: a solution of pure $CaSO_4$ would have no buffering capacity and hence no alkalinity—the calcium content would be related in no way to alkalinity titrations, and a pure $NaCO_3$ solution might have a tremendous total alkalinity despite the absence of calcium! Some generalizations concerning alkalinity, calcium, and carbonate are made here because of the average freshwater chemical composition (Table 14-1).

Because the typical inland water can be described as a calcium bicarbonate solution, most carbonate is combined with calcium. This may be found in nature as *calcite* or, less commonly, as *aragonite*, which has the same chemical formula ($CaCO_3$) but is crystallized differently (in an orthorhombic system rather than hexagonal) and is unstable at ordinary temperatures; aragonite precipitates from thermal waters (usually above 30° C). Its presence in littoral sediments of freshwater lakes is owed to the accumulation of molluscan shells and especially those of the snails. *Magnesite*, the carbonate of magnesium ($MgCO_3$), and *dolomite*, a double carbonate of calcium and magnesium, $CaMg(CO_3)_2$, are relatively common. Carbonates of barium ($BaCO_3$) and strontium ($SrCO_3$) also occur.

In some arid regions precipitates of carbonate combined with sodium abound. They include *natron* ($Na_2CO_3 \cdot 10\ H_2O$) and *trona,* $Na_3(HCO_3)(CO_3) \cdot 2\ H_2O$. Ponds and lakes rich in Na_2CO_3 and the less soluble $NaHCO_3$ are called alkali lakes or soda lakes. They may be extremely high in total alkalinity but in no way can be termed "hard," as are the waters containing calcium and magnesium salts of carbonic acid. The potash ponds, occupying depressions among sand dunes in Nebraska, display noteworthy quantities of K_2CO_3. (The sodium borate mineral called borax, $Na_2B_4O_7 \cdot 10\ H_2O$, contributes to the alkalinity titrations of some desert lakes of the western United States but, of course, is not a measure of carbonate.)

In the stagnant, anoxic, summer hypolimnion waters of eutrophic lakes, ferrous bicarbonate—$Fe(HCO_3)_2$—is often plentiful, as is ammonium bicarbonate—NH_4HCO_3.

The high alkalinities from hypolimnetic samples even in calcareous regions thus indirectly reflect more than the abundance of alkaline earth or alkali metals. Conversely, large resources of those metals would go undetected in alkalinity determinations if they were present as sulfates, chlorides, or nitrates.

The carbonate of calcium is abundant in the earth's soils, although poorly soluble. In pure water, devoid of free CO_2, and at 1 atmosphere, only 6.2 mg $CaCO_3$ per liter would dissolve at 25° C (Schmalz, 1972). Its solubility decreases as temperature rises and increases in the presence of free CO_2. Thus, at equilibrium with normal atmospheric CO_2 the solubility of $CaCO_3$ increases to nearly 10 times that found in the absence of CO_2.

It is customary to say that $CaCO_3$ is virtually insoluble except in the presence of acid. With carbonic acid, it becomes $Ca(HCO_3)_2$. Because of this, it seems reasonable to express alkalinity titrations in terms of bicarbonate ions; but on the other hand, $Ca(HCO_3)_2$ is unstable to such an extent that when natural water is evaporated to determine its contained dissolved salts, the bicarbonate of calcium is destroyed and only carbonate remains.

Magnesium carbonate has about eight times the solubility of calcium carbonate (Fig. 14-1), but of the common carbonates in nature the only truly soluble forms are Na_2CO_3, K_2CO_3, and $(NH_4)_2CO_3$. The first of these, abundant in some lakes of arid regions in Africa and North America, is more soluble than its bicarbonate form, $NaHCO_3$. The nature of these carbonates is taken up in greater detail in the treatment of cations in inland waters.

Biota of Alkaline Carbonate Waters

In extremely saline environments, where relatively few organisms persist, there may be at least two distinct physiologic groups of salt-tolerant species. One of these inhabits chloride and sulfate waters. The second group includes the **natriophils** (L. *natrium*, sodium), typical of the alkali carbonate waters.

Certain cyanobacteria, such as *Arthrospira* and *Spirulina,* flourish in soda lakes, where the zooplankton may contain the rotifers *Hexarthra fennica* and *H. jenkinae,* the American cladoceran *Moina hutchinsoni,* or its Old-World counterpart, *M. mongolica. Diaptomus sicilis,* a North American calanoid copepod, may belong to such a group; it is significant that its synonymy includes an older name, *D. natriophilus. Diaptomus spinosus,* from saline ponds in the deserts of Asia and Africa, could be an ecologic equivalent of *D. sicilis.* Corixid bugs, especially those species referable to the genus *Corisella,* commonly appear in sodium carbonate pools. The larvae of one mosquito frequent alkali waters to such an extent that the species is appropriately named *Aedes natronius* (Gr., *aedes,* unpleasant).

There is danger in generalizing about these two salt-tolerant groups of organisms. Species that occur in all saline situations often blur distinctions. Moreover, different populations, assigned to the same species by taxonomists, may each represent distinct physiologic races or separate species. For example, *Artemia,* the brine shrimp, is extremely rare in soda lakes, and laboratory experiments have shown that some stocks cannot survive in high concentrations of carbonate. There are, however, a few populations of *Artemia* thriving in "lethal" waters, quite unaware of the danger and of the fact that they have never been classified with the natriophils (Cole and Brown, 1967). *Arte-*

mia monica is endemic to the waters of Mono Lake, which is a concentrated sodium lake, but not the NaCl (**saltern**) type (Fig. 14-2). It is unique among the species of the genus in producing benthic encysted embryos that overwinter beneath the water, diapausing six or seven months (Drinkwater and Crowe, 1991).

Diaptomus sicilis occurs in the Laurentian Great Lakes of North America as well as in alkali pools far to the west. At least three possible reasons account for this: that calanoid copepod is remarkably euryhaline; there are distinct physiologic races within the species; or there may be a complex of unrecognized, morphologically similar species lumped as one by taxonomists.

Sulfate

Some Forms of Sulfur

The sulfate ion, SO_4^{2-}, is usually second to carbonate as the principal anion in fresh waters, although chloride occasionally surpasses it.[1] Sulfate is discussed elsewhere in relation to sources of hydrogen ions and with respect to the cations with which it combines to form minerals. The minerals are treated in greater details when cations are discussed.

Free, or elemental, sulfur is inactive at ordinary temperatures. Therefore, beds of "brimstone" (sulfur in an uncombined state) are not rare. The beds are especially abundant in regions of recent volcanic activity. This element can combine with both metals and nonmetals to form many compounds. Beginning biology students are taught the element sulfur as an important constituent of protoplasm; it is in protein and specifically within those amino acids having sulfhydryl (SH) bonding; examples are cystine, cysteine, and methionine.

When sulfur is combined with hydrogen or oxygen, a series of reduced and oxidized forms of sulfur can be formed. The most reduced state is sulfide (S^{2-}), and the most important sulfides to limnology are the gas, hydrogen sulfide (H_2S), and ferrous sulfide (FeS). At the other end of the series, spanning SO_2 and SO_3, is the most oxidized form of sulfate SO_4^{2-}. Sulfate combined with hydrogen as sulfuric acid and with the alkaline earth and alkali metals is the most abundant form of sulfur in lakes and streams.

The evaluation of sulfate ions became relatively easy with the development of spectrophotometry. Older gravimetric procedures were replaced by measuring the turbidity of acidified samples treated with barium chloride. In most samples of fresh water, the only ion that precipitates with barium is sulfate; it forms crystals that approach uniformity in size and remain in suspension. In waters stained by humic acids and naturally turbid from particulate materials, the direct turbimetric method has drawbacks, and more sophisticated treatment is necessary to remove interfering substances. An atomic absorption method based on measuring barium before and after precipitation (by adding acidified $BaCl_2$) is extremely accurate.

Sources of Sulfate

Atmospheric sources of sulfate have increased since the onset of industrial activities, although volcanic emissions have added sulfur compounds to the air for eons.

---

[1] Silica often outranks sulfate, but very little is ionized in most fresh waters, so it is not considered here.

Our industries now contribute about 2.4 times more SO_2 than the annual contribution from volcanoes (about 29×10^6 tons) to the atmospheric load of this gas (Ivanov, 1981). Coal combustion is notorious, and copper smelting and paper manufacturing also produce gases and runoff rich in sulfur. Indeed, sometimes an important index of industrial water pollution is the presence of unusually high sulfate and sulfite levels.

Sulfate in seawater amounts to more than 2.7 g/liter. Intense evaporation in some arid-region lakes has concentrated sulfates well above that value, but fresh waters are usually 0.01 times or even less the marine level for the ion. Interestingly, those lakes nearest the sea, as is true for bodies of waters nearest industrial sites, show greater sulfate concentrations than comparable lakes farther inland. Sea spray swept aloft (with 60×10^6 tons of S/yr) can be carried inland, but evaporation probably occurs before much reaches the lake surface. Even so, salt particles serve as nuclei for cloud condensation, and ultimately raindrops carry the sea salts to inland lakes. The rain falling on sparsely populated islands in both the Atlantic and Pacific contains SO_4/Cl ratios approximating that of seawater, about 0.10, mEq/liter; elsewhere the oceanic rains contain excess sulfate, indicating pollution (Kroopnick, 1977).

The edaphic contribution is most important, however, and high-sulfate waters usually reflect the presence of old marine sediments. Calcium sulfate, especially, is to be considered in such instances, although sodium sulfate deposits can be of oceanic origin also. In addition, oxidation of sedimentary pyrites supplies sulfate.

The two most important stores of sedimentary sulfur are the minerals of $CaSO_4$ and the metal sulfides, especially pyrite (FeS_2). Those of the first type are essentially evaporites; the sulfides are largely a result of microbial action. There are four stable isotopes of sulfur, the commonest two being ^{32}S (95.1%) and ^{34}S (4.2%); ^{33}S and ^{36}S comprise less than 1% of the total. The proportions of ^{34}S and ^{32}S vary remarkably— up to 11%. This is probably because some sulfur bacteria fractionate the isotopes, selecting the lighter of the pair, ^{32}S, to enrich sedimentary sulfides, while the residual material is relatively enriched in ^{34}S. Evaporites tend to have a higher $^{34}S/^{32}S$ ratio than precipitated sulfides. There is a radioisotope useful in tracer studies, ^{35}S.

Moyle's (1945, 1956) studies of the regional limnology of Minnesota clearly show an interrelationship among geologic history, soils, forest types, water quality and lacustrine flora and fauna. There are three distinct soil types in the state. In the southwestern part, seas of Cretaceous age left sulfates which are now abundant in lake waters, ranging up to 500 mg/liter. The aquatic seed plants and the fish in this edaphic area differ markedly from those in the other two regions. *Ruppia*, a submerged pondweed called widgeon grass, flourishes in that region, recalling the stands of this species in the brackish waters of estuaries on the Atlantic coast and in some saline pools in western North America. It would not be noted, however, in other parts of Minnesota where terrestrial sulfate stores are minor.

Sulfate Lakes

Certain bodies of water can be termed sulfate lakes. In most instances these are concentrated waters, for bicarbonate is the principal anion of typical fresh water. Regional changes, however, have taken place in the past few decades as the SO_4^{2-} load has increased in precipitation downwind from industrial sources. Some unusual Nova Scotian lakes described by Gorham (1957a) contained Cl⁻ and SO_4^{2-} concentrations far higher than HCO_3, and in some cases sulfate surpassed chloride. Many of the lakes were acid, with pH values below 5.0. Gorham could demonstrate little or no increase in pH by aerating samples of these, their bicarbonate content being practically nil. In 14 of these more acidic lakes, the chloride/sulfate ratio differed from that found in sea spray, there being an excess of about 0.1 mEq/liter of sulfate. There was no satisfactory terminology to describe these lakes, although many resembled extremely dilute sea water except for the excess sulfate. We can now say that Gorham's discussion probably concerned one of the earliest examples of Canadian lakes acidified by precipitation; they typified a new kind of sulfate lake that has come into being since the early 1950s. Twenty-one years after Gorham had done his fieldwork, Watt and coworkers (1979) returned to the Nova Scotian lakes and found that 16 of Gorham's 23 higher-pH lakes (near neutral) showed remarkable increases in acidity. Local air pollution was identified as a factor in this, and a power station in Halifax seemed to be especially implicated.

In geologic tracts where rocks contain abundant $CaSO_4$, the water chemistry reflects this. Calcium sulfate is, however, the least soluble of the common sulfate compounds, and with concentration it precipitates from solution. For this reason it is uncommon in strongly arid regions, except when issuing from underground sources before evaporation has had its effect. Monkey Springs, Arizona, is a good example of one of these gypsum waters (see Fig. 14-2).

Sodium sulfate lakes are not rare in desert regions. Algerian *chotts*, which we would term playa lakes, contain this compound, as do the many saline bodies of water in southern and central Saskatchewan and the Bottomless Lakes of New Mexico. Magnesium sulfate waters are less common; Hot Lake, Washington, is one of these.

Sulfur Cycles and Productivity

Mann (1958) described an annual cycle in some British ponds that brings to mind the acid titration employed in alkalinity evaluation. The bicarbonate level was low in the spring, rising during the summer to an autumnal high that dwindled during the winter months. Sulfate concentrations, however, were high in spring, diminishing during the summer to a low point in fall. The decrease in sulfate was attributed to its reduction to sulfide, some of which entered the anaerobic sediment as the insoluble ferrous sulfide. Oxidation of sulfide to sulfate during winter circulation resulted in the formation of some H_2SO_4, which converted calcium bicarbonate to calcium sulfate. The action was in effect as though a partial titration, lowering the alkalinity, had been performed.

Similarly, indicator drops added to a measured water sample from streams receiving effluent from coal-mining operations often show the final end-point color before the titration has begun. There is no alkalinity; the "titration" has already been brought to completion by sulfuric acid from the mines, and the end point of pH 4.4 has been attained and passed.

Sulfate's effect on aquatic organisms is rarely obvious, although there are times when it may be a limiting factor in growth. The amino acids containing the sulfhydryl groups are normal in plant protein, and for this reason sulfur deficiency can inhibit algal populations, affecting both photosynthesis and also cell division processes (Antal et al., 2011). Probably in most instances, however, the sulfates are more than adequate for freshwater productivity.

Potash (1956) demonstrated that sulfate is adequate in waters by the commonly used technique of nutrient enrichment ("spiking"). He studied the response of algae to nutrients added to filtered pond water collected throughout the seasons. A green alga was added to the freshwater samples and allowed to grow for 10 to 12 days under constant conditions of light and temperature. In most instances the alga flourished as well in the unfertilized control water as in the media containing nutrients. On occasion nitrate or phosphate deficiencies were noted in the pond water, as evidenced by the greater growth in media containing these ions, but the alga always grew in the unmodified control water as well as it did in the water to which sulfate had been added. The inference is that adequate sulfate prevailed throughout the year in the rather ordinary ponds that Potash studied. Probably typical, if such a word can apply, are the data reported from nutrient enrichment experiments on the waters of 40 central European lakes belonging to five different nations (Thomas, 1973). Following addition of nitrate and phosphate all other elements were found to be adequate.

There has been a controversy in limnology concerning the effect of low levels of sulfate in Lake Victoria and some other large African lakes. It began when Beauchamp (1953) suggested that the low phytoplankton population of Lake Victoria is a function of a low level of sulfate, which he said ranged from 0.8 to 1.8 mg/liter. Lake Victoria and some other large African lakes, marked by poor SO_4^{2-} content, lie in regions where sedimentary rocks containing this ion are rare; yet the lakes contain less sulfate than their inflowing rivers. Sulfate is lowered to critical levels by at least three processes: protein synthesis locks up much of it, its recycling being dependent on decomposition rates that seem to be slow; adsorption on mineral particles such as calcite and diatomite; and much of the decaying protein releases H_2S, which forms less soluble metallic sulfides that precipitate and become buried in the sediments. Hesse (1957, 1958) believed the first two processes were effective in Lake Victoria, where he could find highly organic sulfur in the sediments with no evidence of sulfide. Shallow African Lake George, by contrast, is extremely rich in phytoplankton despite its poor sulfate content, 0.5 mg/liter or less. A high decomposition rate in Lake George, Beauchamp reasoned, accelerates all biologic cycles, and sulfate is regenerated so rapidly that it does not appear to be limiting.

Beauchamp's explanation is interesting because it outlines a mechanism whereby low concentrations of sulfate could limit phytoplankton crops, but later workers pointed out flaws in his argument. Fish (1956) showed that sulfate added to cultures of phytoplankton (not from Lake Victoria) in the lake's water, indeed, seemed to stimulate growth. Later, Evans (1961) reported that the predominant alga in the lake, a species of *Melosira*, showed no response to $MgSO_4$ added to cultures of lake water in which it grew, although phosphate additions stimulated its growth. The green alga *Ankistrodesmus falcatus* from Lake Victoria responded to enrichment with nitrate, sulfate, and phosphate, in that order, yet cyanobacteria from the lake showed no effects from the addition of any of the three nutrients. It seemed from these experiments that sulfate could be limiting at certain times but not always. Then Livingstone (1963b) questioned Beauchamp's quantitative data concerning the amounts of pointing out records of at least 3.4 mg/liter of sulfate from Lake Victoria's water. In 1965, Talling showed that on an areal basis the photosynthetic rate in Lake Victoria was high, amounting to a mean of approximately 7 g O_2 per square meter per day, although the phytoplankton, measured as chlorophyll *a*, was a "decidedly modest concentration." Later, Talling and Talling (1965) agreed that the sulfate level is unusually low in Lake Victoria and some other East African lakes but suggested that the analyses of Beauchamp and others had underestimated the amount. Subsequent analyses revealed two or three times more sulfate.

The Lake Victoria example of sulfate being so rare that it limits growth may be a myth, but there are probably some lakes where sulfate is critical. For example, algae from some lakes in New Zealand's South Island respond to sulfate enrichment of those lake waters (Goldman, 1972).

Hydrogen Sulfide

Hypolimnia of stratified eutrophic lakes or the monimolimnia of meromictic lakes may contain appreciable quantities of the very soluble gas H_2S. This is especially marked in lakes of regions high in edaphic sulfate, for that ion is the main source of hydrogen sulfide, and its concentration is directly related to H_2S. Meromictic Big Soda Lake, Nevada, and Soap Lake, Washington, are two outstanding North American examples, with recorded monimolimnetic amounts of 786 and 6,000 mg/liter, respectively. Some fjord lakes also contain great amounts. In recent years the Italian lakes Maggiore and Lugano have generated extremely high levels of sulfide, from 500 to 1,500 mg/liter, in their bottom waters and sediments (Sorokin, 1975). Increasing pollution associated with increased resident and tourist populations in the lakes' catchment areas plus contributions from paper mills have brought this about; Maggiore and Lugano may have the highest levels ever recorded from lakes that are at least occasionally holomictic. Hydrogen sulfide is poisonous to aerobic organisms because it inactivates the enzyme cytochrome oxidase.

Water samples from the anaerobic deep strata of eutrophic lakes may give off the characteristic rotten-egg odor of H_2S, while little or no sulfate can be detected because it has been chemically reduced. By contrast, epilimnetic waters may be almost odorless and contain significant quantities of sulfate. There may be, however, various types of volatile sulfur compounds in the upper waters and they are lost to the atmosphere during ice-free periods. Richards and coworkers (1991), who studied some lakes of the Canadian Shield and reviewed the existing literature, estimated that from 5 to 76 million kg of sulfur are lost from Canadian freshwater lakes each year. Moreover, collections from the stagnant hypolimnion may resemble dilute India-ink suspensions because of the presence of ferrous sulfide. The meromictic Black Sea owes its name to the dark waters lying 150 to 200 m below its surface. At least 99% of the H_2S in the Black Sea comes from the reduction of sulfates; the remainder is from decomposition of dead protein matter by putrefying bacteria.

The reduction of sulfate to sulfide probably does not occur until the redox potential falls, under anoxic conditions, below 0.1 V perhaps in the neighborhood of 0.06 (see Table 15-1). It is largely a phenomenon of anaerobic sediments, although occasionally it is reflected in overlying water or atmosphere. This leads into the field of bacteriology, following some generalizations about mineral cycles.

Sulfur Bacteria

Most students know the textbook example of carbon dioxide being reduced to an organic photosynthate with the evolution of oxygen as a by-product. This reduction process, in which solar radiation is the energy source and water serves as a hydrogen donor, is brought about by organisms—conspicuous green organisms. The reverse oxidation process, likewise, is well known. A similar cycle whereby sulfates are reduced or hydrogenated to form sulfhydryl groups of amino acids and, thereby, innumerable proteins, lacks the visibility of photosynthesis. The name *sulfuretum* has been suggested for an environment in which these reactions occur, and the organisms adapted for survival there comprise the **thiobios** (from the Greek *theion*, meaning brimstone). Typical of such habitats are the dim subsurface regions where pH is low, oxygen is absent, and sulfides are present: anaerobic marsh sediments; the organic oozes at the bottom of lakes and seas; and even the blackened anoxic layers a few centimeters below the surface of sandy beaches not far from the water's edge.[2] Organisms are involved in this reduction and in the reverse oxidation reaction, just as in the synthesis and destruction of carbohydrate, but most are not conspicuous. They are microscopic bacteria living in most instances in anaerobic habitats where they, lacking the enzyme catalase, escape the lethal effects of free oxygen. There is a varied array of metabolic processes by which bacteria deal with sulfur and its compounds. These are tightly coupled to the oxygen and iron cycles, but here we focus on sulfur alone.

Bacterial taxonomy and nomenclature are undergoing changes, but the sulfur bacteria fit three main groupings on the basis of morphology and physiology. There are pigmented species and large colorless forms, both filamentous and nonfilamentous. A physiologic or metabolic classification is convenient.

Sulfate reducers. When oxygen is absent and redox potentials fall below 0.1 V, the sulfate-reducing microbes begin their work. Indeed, in many such extreme habitats no other organisms can survive. *Desulfovibrio desulfuricans* is one of these; its metabolism is much like that of the denitrifying bacteria. This sulfate reducer, while transforming sulfate to sulfur and hydrogen sulfide, precipitates $CaCO_3$ at the same time. A simple way of showing this is in the two following reactions:

$$CaSO_4 + 2C \rightarrow CaS + 2CO_2$$
$$CaS + CO_2 + H_2O \rightarrow CaCO_3 + H_2S$$

The carbon shown in the first reaction is from decaying organic matter. Notice that organic compounds have been oxidized, not an unusual source of energy except that sulfate rather than oxygen was used as an electron acceptor. Because it derives its carbon from organic molecules, the sulfate respirer *Desulfovibrio* is essentially a heterotrophic organism. There is, however, some evidence that it can fix CO_2 in laboratory cultures; the energy source for this is the oxidation of hydrogen with the oxygen of sulfate. This chemolithotrophic ability underscores again the versatility of the prokaryotes.

Another group of reducing bacteria is those that attack organic material, liberating H_2S from the proteins with no accompanying calcite precipitation. These anaerobic bacteria of putrefaction degrade the sulfur-containing components of protein— cysteine, cystine, and methionine.

The value of anaerobic oozes and peats, where microbes can survive and carry on reducing activities, was elegantly expressed by Deevey (1970) in his essay "In Defense

---

[2] Green plants obtain their sulfur by similar metabolic processes, but this assimilatory sulfate reduction is neither so rapid nor on such a large scale as the bacterial reduction.

of Mud." In these dark anoxic sediments, sulfhydryl groups necessary for protein building are regenerated, and H_2S is formed to escape to the atmosphere. There it is oxidized and joins other oxides of sulfur, originating naturally from volcanoes and from combustion of coal and petroleum mixtures. Now the amount of sulfur derived from fossil fuels has overtaken the natural output to the air (Ivanov, 1981). In light of mounting industrial pollution that adds more than 100 million tons of H_2SO_4 to the atmosphere each year, any ecosystem where sulfuric acid can be chemically reduced is worth saving; these are among the many "ecosystem services" provided by wetlands, providing justification for their preservation.

Sulfur oxidizers. The sulfur-oxidizing microbes utilize the sulfide produced by *Desulfovibrio* as an electron donor, and they form free sulfur and sulfate. They contain at least five bacterial categories, some being anaerobic, pigmented, and photosynthetic. Those classified as the Chlorobiaceae are green species that oxidize the inorganic reductant H_2S to free sulfur. They contain bacteriochlorophyll *a*, which occupies a place in the absorption spectrum where the familiar green pigment of algae and higher plants is largely ineffective. This bacterial pigment is effective in wavelengths within the infrared range.

In anaerobic habitats where the level of H_2S attains 50 mg/liter or more, the nonmotile, green sulfur bacteria may be found. They are photolithotrophic, utilizing H_2S as a hydrogen donor for the reduction of CO_2 and regenerating elemental sulfur as a by-product of carbohydrate synthesis. The following simple reaction illustrates how similar their action is to the photosynthesis of higher green plants, which utilize water as a hydrogen donor to reduce CO_2 and release free oxygen as a by-product.

$$CO_2 + 2H_2S \xrightarrow{\text{light}} (CH_2O) + H_2O + 2S$$

Chlorobium is one of these autotrophic bacteria. Physiologically adapted to dim light, it often forms dense plates below the aerobic algae, and the blue-greens. One can generalize that these plates occur ordinarily near the contact layer between oxidative and reductive zones where light is less than 10% of that falling on the lake surface. *C. limicola* forms green layers because its carotenoids are yellowish and do not mask the green pigments. At least two other species have brown carotenoids that hide their green color. They form red-brown layers in subsurface anoxic strata where hydrogen sulfide prevails and where the absorption peak of their carotenoids (about 460 nm) is most efficient at absorbing wavelengths that reach those depths.

The purple sulfur bacteria Chromatiales, such as the motile *Chromatium* and *Thiopedia*, usually contain pigments with absorptive properties from 780 to 900 nm and another peak in the green-yellow wavelengths, about 500 to 600 nm. They form pink or red layers in meromictic lakes, saline lagoons, and in the sands of tropical beaches.

With the chlorophyll *a* of the cyanobacteria and higher plants considered, almost all wavelengths from 350 to 900 nm can serve as energy sources for pigmented prokaryotes and eukaryotic autotrophs.

Most purple sulfur bacteria, even more than their green relatives, are adapted for dim light and are inhibited photosynthetically by intense bright light. Furthermore, they function optimally at high temperatures. When purple bacteria form elemental sulfur, they keep the granules within their cells, rather than on the outside as the green bacteria do. The sulfur can be oxidized later to sulfuric acid with a gain of energy in a reaction such as the following:

$$2H_2O + 2S + 3O_2 \rightarrow 2H_2SO_4$$

In most ponds where this reaction is effected, there is little or no increase in hydrogen ion concentration because bases are abundant enough to neutralize the acid.

The Chromatiales include colorless species that also oxidize inorganic substances, such as H_2S, but are aerobic. They are autotrophic organisms but chemotrophic rather than phototrophic. The energy they require is attained by direct oxidation of H_2S and of elemental sulfur:

$$2H_2S + O_2 \rightarrow 2H_2O + 2S$$
$$2S + 2H_2O + 3O_2 \rightarrow 2H_2SO_4$$

Thiobacillus is one of these acid formers, and, as one might guess, it can thrive in extremely acid environments. At least one species uses nitrate as an electron acceptor rather than oxygen while it oxidizes sulfur.

The gliding, filamentous *Beggiatoa* is a spectacular member of the colorless sulfur bacteria. Its cells, stuffed with sulfur granules, are arranged in trichomes, reminiscent of the Cyanobacterium *Oscillatoria*. Indeed the entire group of gliding bacteria may have been derived from the cyanobacteria through loss of pigment and are assigned to a family different from that of *Thiobacillus*. Assemblages of *Beggiatoa* are found at the boundary of aerobic and anaerobic waters, appearing like wispy cotton strands. The two requirements, a low oxygen tension and H_2S, make it necessary for these colorless organotrophs to occupy this subsurface, watery interface. They are not good chemolithotrophic bacteria; they require an organic carbon source.

A fifth group, the purple nonsulfur bacteria (family Rhodospirillaceae), differs from the preceding four by not depositing sulfur. In the light most of them can use a variety of hydrogen donors, including reduced sulfur compounds. In the dark some species become heterotrophic, using organic compounds and oxygen. All require organic compounds for growth; this and the fact they do not deposit sulfur separate them from the purple sulfur bacteria.

Chloride

Authors who have summarized preexisting data to present the chemical composition of the rivers and freshwater lakes of the world have shown that chloride usually ranks third among the anions (see Table 14-1). Much depends, however, on the "world" sampled. An Australian limnologist finds the fresh water in the "down under" world to be quite different from "standard" waters. Although chloride's position may vary in some Australian states, in Victoria and Tasmania it usually takes precedence among the anions.

Chlorine is an element of the halogen group that includes also fluorine, iodine, and bromine. It is by far the most abundant of these, surpassing other members of this group in the sea, in inland saline and polluted waters, as well as in the purest of lakes or streams. Molecular chlorine (Cl_2) is a heavy, yellow, lethal gas; but in most natural waters it is dissociated as chloride ions, which combine with all common cations.

Where chloride is most abundant, as in the sea and in concentrated desert pools, most of it combines with sodium. The Dead Sea, representing an extreme in concentration of salts, is an exception; it contains some 35.6×10^9 tons of $MgCl_2$ and $NaCl$. (The former compound is 64% of the total.)

Chloride is the major **halide** stored in most freshwater algal cells, although the marine kelps and other brown algae are noteworthy for their uptake and concentration of iodide. In the American southwestern deserts an exotic tree has established itself along the few surviving water courses, competing with the native cottonwoods. This is the salt-cedar, *Tamarix*. It takes up sodium chloride and exudes it from its leaves to the ground below.

Sources of Chloride

Edaphic factors.   Igneous rocks have been considered a minor source of chloride even though a mineral called *sodalite*, $Na_8(AlSiO_4)_6 Cl_2$, is fairly common. The outflow from hot springs, if derived from molten igneous rock, is by definition *magmatic* water and is often rich in chloride. In some instances the heated water passing through other strata dissolves salt on its way, but Johnston (1980) has found that some magmas themselves may contain up to 1% chlorine, which exceeds earlier estimates by 40 times or more. Some of this water may be *juvenile*, never having circulated before and defined as newly released water that had been combined in minerals or dissolved in igneous rocks. It is not known how much juvenile chloride enters natural surface waters, but it is estimated to surpass a million metric tons each year.

Evaporites are very important as edaphic chloride sources. Extensive beds of salt derived from evaporation of former bodies of water are common. Chloride coming from such stores can be called *cyclic*; it has been in the oceans at least once before and will return.

Atmosphere.   The atmosphere carries windblown chloride inland from the seacoasts. Many measurements of this meteoric chloride have been made in rainwater. This falls under the classification of cyclic Cl^-, as does the salt in ancient sedimentary beds. Cyclic, meteoric chloride is pronounced in some surface waters of Australia, many of which resemble dilute seawater in chemical constitution (Williams, 1964).

Windblown chloride may be immediately derived from the sea or may have been part of seawater no more than a few months before. The classic example of the latter is the Rann of Cutch in India. During the annual dry season a great arm of the sea dries, and the new evaporite is transported inland via the winds.

Volcanic gases contribute chloride to the air, accounting for some atmospheric acidity. Hydrochloric acid is released in quantities far greater than previously estimated. Ordinarily it goes into solution and returns to earth with the rain. Perhaps once a year a volcanic explosion is of such magnitude that it injects ash, gas, and HCl directly into the stratosphere (about 11 to 16 km above the earth's surface) where free chlorine atoms are released to join anthropogenic fluorocarbons in destroying ozone (Johnston, 1980).

Pollution.   In certain lake districts contamination from domestic sewage can be monitored by chloride assays. Comparatively high concentrations imply that the lake in question may be receiving pollutants. This is because human and animal excretions contain, on the average, 5 g Cl per liter. Thus, subsurface seepage from septic tanks introduces soluble chloride salts to nearby lakes. High chloride levels are found in some desert pools in the American Southwest and in the so-called *r'dirs* of the North African desert; they are water holes, polluted by the wastes of domestic and wild animals.

A further source of chloride in northern United States and Canada is the application of NaCl and $CaCl_2$ to highways and city streets to clear them of ice. The amount used each winter can be estimated in millions of tons, and much of it eventually finds its way into lakes, streams, and the groundwater. The effects are more than just seasonal. Winter peaks of Ca^{2+}, Na^+, and Cl^- amounting to a sum of 16.2 g/liter were noted in an Ontario stream monitored by Crowther and Hynes (1977). Groundwater discharge during the summer, however, supplied saline mixtures that were derived originally from winter applications to roads. Direct runoff, later meltwater, or storm sewers entering directly into lakes all introduce dense solutions of $CaCl_2$ and NaCl during winter months, but the effects may be seen later. In some instances deep density layers formed during the de-icing season prohibit complete mixing at the spring overturn and incipient meromixis results (Judd, 1970). Thus, the lake-bottom fauna is subjected to more than increased salinity; longer anaerobic periods are induced and only the hardiest species survive. Concern is growing about this environmental threat, which may worsen with time.

Calcium

Sources of Calcium

The predominant compound in many interior waters, $CaCO_3$, is also one of the least soluble. Only a small amount can be dissolved in pure water,[3] but in the presence of carbonic acid it is represented abundantly as the soluble $Ca(HCO_3)_2$. Thus, a treatment of the silvery white alkaline earth metal named calcium can hardly be divorced from earlier discussions of alkalinity, the forms of carbonic acid or CO_2, pH, and the anion CO_3^{2-}.

The earth's crust contains an ample store of calcium as a constituent of certain silicates. For example, *anorthite* ($CaAl_2Si_2O_8$) is a common member of the feldspar group of silicates; feldspars are the most abundant of all minerals and make up about 60% of the earth's coating. In addition, over extended periods of time, immense deposits of sedimentary $CaCO_3$ have been laid down by living things. They are now a substantial part of the calcium reserve, awaiting the attack of rainwater, rich in CO_2, that will change them to the soluble bicarbonate that may enter aquatic environments.

The initial input of allochthonous calcium compounds to aquatic habitats comes about through various erosional phenomena. Particulate material can be washed almost directly into a lake basin from its banks or be carried from afar by influent streams or by wind. Usually the waterborne detritus surpasses that of aeolian origin,

---

[3] In distilled water at 25° C the solubility of $CaCO_3$ is on the order of only 15 to 20 mg/liter; in the presence of acid, however, it is converted to the soluble $Ca(HCO_3)_2$. According to Schmalz (1972, Fig. 3), pure water in contact with atmosphere containing a partial pressure of CO_2 equaling 0.035% could dissolve about 66 mg of $CaCO_3$ per liter or roughly 26 mg of calcium; the increased solubility is owed to the effect of carbonic acid.

but there are exceptions. Gorham (1958b) pointed out that some small lakes in the English Lake District, lying in noncalcareous soils and having no more than 0.2 mEq total salts per liter receive more than half their calcium and in excess of 80% of their magnesium from precipitation. Dust blown aloft from the ground is the main source of these two earth metals found in snow and rain.

Chemical weathering of calcium compounds can put them into solution, and they enter the aquatic environment in the dissolved state. Subsurface seeps or influent streams thus carry calcium bicarbonate to the lake basin.

Calcium carbonate is the main constituent of chalk, limestone, and the metamorphosed rock called marble. Its common form is the mineral *calcite*, the most stable type at normal temperatures and low pressures. Formed under a much narrower range of conditions than calcite is *aragonite*, a less prevalent type of $CaCO_3$. Its crystalline structure differs from calcite, and it has a higher specific gravity. Because it changes spontaneously into calcite, there are no ancient aragonites. It is represented in deposits from hot springs and geysers, cave stalactites, and in most molluscan shells.

Solubility of Calcium Compounds

Much interest has been aroused in the solubility of $CaCO_3$, and a great deal of research has been directed toward this phenomenon. Certain engineers, especially, have focused attention on this because precipitation of $CaCO_3$ in steam boilers and in water pipes has economic importance. The limnologist finds this subject engaging because of the biogenic precipitation of calcite by photosynthesizing organisms and by the events occurring at or slightly downstream from spring orifices, where conspicuous crusty deposits often form. In each instance the loss of CO_2 caused a shift to the left in the reaction $CaCO_3 \leftrightarrow Ca(HCO_3)_2$. Equilibrium that had been achieved beneath the ground was disturbed when the water emerged, and the pressure on it was suddenly released, permitting the escape of CO_2. Phototrophic plants also trigger the reaction to the left by absorbing CO_2.

At least four names apply to calcareous deposits in lakes, streams, and at spring sources. The differences among them are not always clear cut, and the use of terms may differ from one geographic region to another. Common usage denotes **marl** as the calcareous material in the littoral region of hard-water lakes. It is usually fine grained, although marl pebbles are common, and the coarse contributions from gastropod and clam shells may be important. Individuals of some species of *Chara* are encrusted with flaky limestone that they have precipitated from the water. This accounts for the application of the name stonewort to this large alga. Through the years the accumulation of dead *Chara* contributes to littoral shelves of marl. The photosynthesis of microscopic algae and cyanobacteria, both planktonic and benthic, also precipitates $CaCO_3$, but in small particles. In the deeps of the lake, most of the $CaCO_3$ may be dissolved to $Ca(HCO_3)_2$ by the acid nature of hypolimnion water. Because of this, marl deposits are especially prominent features of lake margins where pH values are high and photosynthesis is intense during sunny hours. Thus, biogenesis is important in marl formation, and the results are typically a soft, calcareous-clay sediment of whitish hue.

Sinter is a general term applied to chemical sediments deposited by mineral springs. Some are siliceous, building terraces of *geyserite*, such as those around the hot springs of Yellowstone National Park familiar to North Americans. Calcareous sinter is called *tufa* or *travertine*. It is deposited from solution in ground waters, as evidenced by stalactites and stalagmites in limestone caves, or from surface waters not long after they have emerged from subterranean sources. When travertine is extremely porous, it is termed calcareous tufa or *spring deposit*. The name tufa has also been applied to some siliceous sinters.

Despite the importance of organisms in the formation of marl, many lakes with rich organic sediments and suspended detritus are not active marl formers. This may be attributed in part to the CO_2, derived from decay, that dissolves marl and hinders its deposition. The littoral marl benches or shelves reported to modify the morphology of some Michigan lakes were not conspicuous in these waters until they had attained alkalinities of at least 105 mg/liter (Hooper, 1956). Although this concentration is above the level of calcium carbonate saturation, it is not extraordinarily so when compared with some lakes having higher rates of organic sedimentation.

The symmetrical curves that represent the dissociation of carbonic acid as a function of pH (see Fig. 13-1) imply that almost no carbonate occurs below a pH of 8.3. This suggests that $CaCO_3$ would not be precipitated below the point on the scale where phenolphthalein becomes colorless. However, other factors, such as temperature, may be involved. The solubility of $CaCO_3$ decreases as the temperature rises from 0° to 35° C (Table 14-3). Moreover, the nature of ions other than calcium and carbonate and their concentrations affects the ionic strength of the solution and thus the activity of individual ions. The activities of bicarbonate and calcium are especially important for any consideration of the solubility of calcium carbonate (Table 13-9 and Fig. 13-3). The total concentration of ions present may not reveal the amounts available to participate in chemical reactions because activity coefficients can be well below unity. Dissolved materials may thus remain in solution, though simple theory would predict their precipitation. In laboratory experiments sodium and potassium salts increase the solubility of $CaCO_3$ even in the absence of free CO_2.

Investigation of subterranean water at spring orifices reveals that $CaCO_3$ can be saturated and precipitate at pH levels well below the 8.3 that is shown on curves in the idealized graphs portraying solutions containing only species of carbonic acid (see Fig. 13-1). The point at which saturation occurs, therefore, may lie at a greater hydrogen ion concentration than one would expect for the conversion of bicarbonate to carbonate and its subsequent precipitation.

Table 14-3   The solubility product of calcite ($CaCO_3$) as a function of temperature. All K_{sp} values should be multiplied by 10^{-9}.[*]

C°	0	5	10	15	20	25	30
K_{sp}	4.467	4.457	4.385	4.236	4.406	3.802	3.524
pK_{sp}	8.350	8.351	8.358	8.373	8.393	8.420	8.453

[*] Shown are K_{sp} values $\times 10^9$, the products of Ca^{2+} and CO_3^{2-} in moles/liter which mark the solubilities in saturated solutions; and pK_{sp} the negative logarithms of these solubility products. Note that a $CaCO_3$ solution at 0° C would precipitate at all higher temperatures shown.

From Jacobson and Langmuir, 1974.

There are certain ways of expressing the degree of saturation in a body of water. Usually Langelier (1936) is credited with defining the pH of saturation (pH$_s$). In a given solution where the alkalinity and calcium contents are known, Langelier's pH is at a point where the water is neither undersaturated nor oversaturated with $CaCO_3$. An **index of saturation** can be set down if we know the difference between observed pH (pH$_{obs}$) and the computed pH$_s$. If the index is positive, equilibrium would be attained by precipitation of $CaCO_3$. If, however, pH$_{obs}$ – pH$_s$ is negative, the water is undersaturated and equilibrium could be established by the dissolution of $CaCO_3$ or, as the engineers might say, by "corrosion, rather than scaling." Thus, water with a calculated pH, of 7.8 and with an observed pH of 8.0 has a positive index, 0.2, and is obviously oversaturated; if the observed pH were 7.6, the index would be –0.2, indicating undersaturation. When water emerges from a subterranean origin, it is probably in equilibrium at pH$_s$, but the nature of the solution changes as free CO_2 escapes to the air. The index of saturation becomes positive, to be followed soon by precipitation of $CaCO_3$ to reestablish a condition of equilibrium in which pH$_{obs}$ – pH$_s$ = 0. Indices of saturation based on the differences between pH$_{obs}$ and pH$_s$ are not precisely quantitative. They do disclose, however, the directional tendency that must be followed to achieve equilibrium.

An entirely different way of expressing the degree of saturation would be the ratio of the thermodynamic ionic product of Ca^{2+}, and CO_3^{2-} (corrected for activity, a) to its solubility-product constant, K_{sp}.

$$\frac{a Ca^{2+} \times a CO_3^{2+}}{K_{sp}} = \text{relative saturation}$$

A quotient of 1.0 represents equilibrium; a figure less than 1.0 denotes undersaturation; and of course a figure greater than 1.0 reveals supersaturation.

A direct test to see whether water is saturated was explained by Weyl (1961), who devised a "carbonate saturometer." It is essentially a slightly modified electric pH meter. The water sample is tested immediately after collecting; crystalline calcite is sprinkled in, contacting the glass electrode, and any pH change is noted. If the water is saturated with calcium carbonate, no pH change occurs; pH$_{obs}$ is pH$_s$, and its degree of saturation is 1.0. Had the water been oversaturated, the added calcite would have provided nuclei for carbonate precipitation with release of hydrogen ions and falling pH. Had calcite been added to water undersaturated with $CaCO_3$, hydrogen ions would have been consumed with a rise in pH. To understand these events we recall the second dissociation of carbonic acid:

$$HCO_3^- \rightarrow CO_3^{2-} + H^+$$

The changes occur within 2 minutes after the addition of calcite crystals when a pH plateau is reached and held for about a minute before drifting on. The magnitude of the pH deflection to the plateau affords an index of the degree of undersaturation or oversaturation.

$CaCO_3$ is precipitated from water when various factors interact to bring it above the equilibrium level. This generalization is valid, but some others are not always applicable. A long-prevailing idea has been that marl accumulates only at the margins of lakes where littoral calcareous shelves are perceptible and that in the acid milieu of the hypolimnion, carbonate is dissolved and, therefore, cannot be deposited in profundal ooze. Megard (1968) disclosed that the calcareous nature of some profundal sediments need not be explained only by erosion of relatively large particles from marginal banks. He demonstrated that in some Minnesota lakes, precipitation of profundal calcium carbonate comes about during the spring overturn, even though the calcium concentration in hypolimnion waters had been higher during winter stagnation before the profundal marl was laid down and even though the pH was circumneutral rather than above 8.3. The vernal phytoplankton pulse, with its accompanying photosynthetic activity, removed enough CO_2 to disturb the equilibrium and effectuate the deposition of marl. Even more remarkable is the deposition of a layer of calcite each summer in the deeps of meromictic Fayetteville Green Lake; the sedimentation occurs beneath monimolimnic water of pH 7.1. Rising temperatures initiate the calcite precipitation (Brunskill, 1969).

After the silicate and carbonate minerals of calcium, the sulfates rank as its most abundant store. These are *gypsum* ($CaSO_4 \cdot 2H_2O$) and *anhydrite* ($CaSO_4$). Sedimentary beds containing these two minerals are usually of marine origin, but evaporation of saline inland lakes accounts for some terrestrial deposits. Anhydrite precipitates from seawater at high temperatures (about 42° C and above) and at lower temperatures in the presence of high salinity. At ordinary temperatures gypsum is usually the first mineral thrown down as seawater evaporates; later, when salinities have mounted, anhydrite is deposited. Anhydrite can be converted to gypsum by the action of rainwater, and, conversely, some anhydrite sediments are thought to have been formed by dehydration of earlier gypsum beds.

The solubility of the sulfate of calcium in pure water surpasses that of the carbonate by about 120 times. In normal seawater 970 mg of calcium sulfate per liter would represent approximate saturation, but this is because of the reduced activities in water of high ionic strength. Inland waters in contact with gypsum rocks become rich in $CaSO_4$ because of this increased solubility, which does not require carbonic acid. $CaSO_4$ must be considered a compound of fresh water, however, because it precipitates when concentrated in saline basins. It has about 0.02 times the solubility of sodium sulfate, the least soluble of the abundant compounds in the typical concentrated waters of arid regions.

To many Americans, the impressive dunes of White Sands National Monument in New Mexico typify gypsum. That state boasts an immense gypsum resource, and because of the solubility of the mineral some interesting lakes occur there. These are solution basins, circular in outline or in the design of a figure 8 where adjacent lakes have coalesced. They are protected now within the Bottomless Lakes State Park. Some have steep-walled sides, recalling the Yucatán cenotes far to the southeast. The cenotes, however, mostly occupy limestone pits, and carbonic acid contributed much to their genesis, although many have waters high in $CaSO_4$ (Covich and Stuiver, 1974).

The most soluble of the common compounds and sources of calcium is its chloride. It is 30,000 times more soluble than $CaCO_3$. There are only a few good examples of calcium chloride waters. For example, the flow from Clifton Hot Springs in Arizona contains much $CaCl_2$, although surpassed by NaCl. Also, there are 5.4×10^9 tons of $CaCl_2$ in the Dead Sea, though outstripped 6.5 times by NaCl and $MgCl_2$. A remarkable body of water, Don Juan Pond in Antarctica, definitely belongs to the $CaCl_2$ category. It is a shallow pool upon the bottom of which lie needlelike crystals. These are $CaCl_2 \cdot 6H_2O$ and were new to science when first described and named *Antarcticite* in 1965 by Torii and Ossaka.

Marl Lakes

In the glaciated region of the Laurentian Great Lakes there are many bodies of water called marl lakes. These are oligotrophic, distinguished by low primary productivity as well as scanty littoral periphyton and macrophyte growth. Certain diatoms flourish in marl lakes, but low levels of phosphorus, iron, calcium, and potassium serve to restrict other algae and the cyanobacteria. Their waters, overlying beds of calcareous sediments, are high in pH and saturated concentrations of bivalent cations. In extreme cases, the monovalent ions Na^+ and K^+ are surpassed tenfold by Ca^{2+} and Mg^{2+}. Nutrient elements such as iron, manganese, and especially phosphorus are scarce because they are bound in an unavailable form or have precipitated out with the cations. The coprecipitation of phosphorus and calcium is especially important in taking phosphate out of circulation. Phosphate adsorbs on calcium carbonate, and they precipitate concurrently when the pH rises to values near 9.0. Otsuki and Wetzel (1972) suggested that in the microenvironments at algal-cell and macrophyte surfaces, where CO_2 removal leads to high pH, the resultant coprecipitation of carbonate and phosphate limits plant growth, serving as a density-dependent control on populations in marl lakes. Also, the phosphate ion can react with Ca^{2+} to form calcium phosphate, a compound with low solubility in water. The study of marl lakes has contributed to the general concept that the speed of eutrophication, following the introduction of phosphate, is sometimes greater in low-calcium, soft-water lakes than in hard-water lakes.

Biota of Calcium Waters

Among plants and animals there are some that can be classed as *calciphiles* (calcium lovers) and others as *calciphobes* (haters of calcium). In certain taxonomic groups all species favor calcium, and species numbers decrease as that element becomes scarcer. Probably most of the Bivalvia and Gastropoda typify the strict calciphiles.

Other groups have both calciphilic and calciphobic species. The planktonic rotifer genus *Brachionus* includes common species, such as *B. calyciflorus*, *B. angularis*, and *B. quadridentatus*, said to be typical calciphiles, while other members of the genus are less restricted to high-calcium waters. Some other rotifers are restricted to acidic waters low in calcium and high in hydrogen ion concentration.

The family Desmidiaceae of the green algae is composed of hundreds of species, including many planktonic forms. The finest desmid floras flourish in acidic waters low in calcium, although certain species can be found in high-calcium habitats. (Moss, 1972) brought up the complexities of these relationships. In laboratory experiments even desmid species deemed oligotrophic do not behave like calciphobes.

An excellent example from the cladocerans is *Holopedium gibberum.* Typical of the calciphobes, it swims in the plankton of waters low in both calcium and pH. An upper limit of 20 mg per liter has long stood in the literature, but some populations have been found in water containing almost twice that amount. The low pH associated with some low Ca^{2+} waters may preclude fishes and account for *Holopedium's* "abhorrence" of calcium.

Though very important to higher plants, calcium is no more than a micronutrient to most algae; thus, relatively small quantities suffice for large phytoplankton populations at times. Calcium abundance, however, can mirror the presence of rarer and essential plant nutrients and a favorable edaphic heritage. Many calcareous soils are well endowed with phosphorus, for example. Dilute, soft water, poor in calcium, may also be low in most other ions. Lack of calcium may play a role in promoting dystrophy, decreasing the rate at which organic substances are precipitated, mineralized, and recycled for use by the primary producers.

Magnesium

Sources and Types of Magnesium Compounds

Magnesium is usually the second most abundant cation in inland waters of temperate regions. Its source is both silicate and nonsilicate minerals of the earth's crust. Also, near the seacoast some waters have high magnesium concentrations believed to represent aerial transport of ocean spray. A liter of seawater contains about 1.3 g of magnesium.

Magnesium is represented in most of the main subclasses of the silicates that compose continental rocks. One such as *forsterite* (Mg_2SiO_4) can be altered by carbonic acid to form silica, carbonate, and *serpentine* ($H_4Mg_3 Si_2O_9$) in the following manner:

$$5Mg_2SiO_4 + 4H_2O + 4CO_2 \rightarrow 2H_4Mg_3Si_2O_9 + 4MgCO_3 + SiO_2$$

The process of serpentinization produces $MgCO_3$ from several types of silicate minerals. This carbonate can be dissolved by carbonic acid in the same manner as limestone, although it is about eight times more soluble than calcite in pure water.

The magnesium carbonate *magnesite* and a double carbonate called *dolomite*, $CaMg(CO_3)_2$, are less common than calcite. Dolomite represents an intermediate step in the replacement of calcite by magnesium-bearing solutions and contains varying proportions of the two elements calcium and magnesium. It resists weathering more than calcite, and yet it releases magnesium and calcium in approximately equal ratios despite the greater solubility of most magnesium compounds.

Epsomite, known as Epsom salt, is a soft, whitish sulfate of magnesium ($MgSO_4 \cdot 7H_2O$). It is 150 times more soluble than gypsum. It occurs in some mineral spring deposits, in salt sediments of lacustrine and marine origins, and in caves. The best example of a body of water typified by magnesium sulfate is the small, saline Hot Lake in the arid eastern region of Washington. Hot Lake occupies a shallow depression that came about through mining operations for epsomite. Magnesium sulfate's source is directly edaphic in this lake, rather than climatic when relative enrichment is brought about by evaporation and the precipitation of less soluble compounds.

Magnesium chloride ($MgCl_2$), like calcium chloride, is extremely soluble. Its occurrence in marked concentrations is typical of the most condensed of desert waters. A well-known example is the Dead Sea.

When seawater is evaporated and most of its salts have crystallized and precipitated, the "mother liquor" that remains is called **bittern**. It contains bromides and magnesium salts. For this reason, the Dead Sea and other bodies of saline water, such as Lake MacDonnell in the Australian desert, can be termed specifically bittern lakes, signifying their high magnesium content.

Because most magnesium compounds surpass similar calcium compounds in their ability to remain in solution, they precipitate at different rates. Calcium usually surpasses magnesium in fresh waters because there is a preponderance of calcium over magnesium in sedimentary rocks. With evaporation and resultant concentration, however, magnesium may assume more importance. Ca/Mg ratios in waters change first when the solubility product of $CaCO_3$ is reached and calcite precipitates. For this reason the relative amounts of these two metals in lacustrine sediments or encrusted on aquatic plants usually favor calcium when compared with the adjacent lake water. A further example of the different solubilities of calcium and magnesium is provided by events in the stream leaving Montezuma Well. At the start the Ca/Mg ratio approximates 2.8 on the basis of milliequivalents per liter, but only 1 mile (1.6 km) downstream it has dropped to 2.2. This is because the loss of free CO_2 and the rise in pH, accompanied by some evaporation, cause precipitation of $CaCO_3$, while magnesium compounds remain in solution. Similarly, the Ca/Mg ratio is 2.2 in the world's rivers running to the sea, but it is reduced to about 0.2 in seawater (Table 14-1).

Magnesium lies at the heart of the chlorophyll molecule and seems to occupy, therefore, a central position in community ecology. Too much of this light metal, however, is detrimental. For example, magnesium salts, especially $MgSO_4$, produce anesthesia in both invertebrates and vertebrates. In laboratory experiments it has been demonstrated that pure 0.25 M solutions of $MgCl_2$ are slightly toxic to the hardy *Artemia*. Rawson and Moore (1944) found it puzzling that Last Mountain and Echo Lake were more productive than about 60 other somewhat saline lakes they had studied within the same district of Saskatchewan. Point after point were compared with no fundamental differences coming to light. The only thing that set these two lakes apart was the low magnesium content of their waters when compared with the others.

The Concept of Hardness

Cations that form insoluble compounds with soap contribute to what is termed "hardness" in discussions of water quality. An early method of quantifying hardness was based on titration with a standard soap. The soap-consuming power of the water is lowered by heavy metals and the alkaline earth metals. The two most common bivalent cations of lake water, calcium and magnesium, usually account for most of the hardness. At present, total hardness is determined by titration with ethylenediaminetetraacetic acid (EDTA) or some other compound that chelates magnesium and calcium. This is more nearly accurate than the older soap method, and furthermore, the titrant is usually standardized to give results equivalent to milligrams of $CaCO_3$ per liter just as in alkalinity titrations.

The carbonate hardness of water includes the portion equivalent to the bicarbonate and carbonate in the water. This has been called **temporary hardness**, for it is the part that disappears as water is softened by boiling and by the ensuing precipitation of calcium carbonate and magnesium carbonate. Noncarbonate or "permanent" hardness is the difference between total and carbonate hardness. Permanent hardness is caused by the sulfates and chlorides of bivalent cations. The SO_4^{2-} and Cl^- ions are not revealed by alkalinity titrations, and their compounds of calcium and magnesium do not precipitate with boiling. Silica, on the other hand, may be part of alkalinity titrations but does not contribute to hardness, even though it forms incrustations as does $CaCO_3$.

The generality that magnesium compounds are more soluble than molecules containing calcium does not apply in the case of the hydroxide, and this phenomenon is utilized in hardness assays. First, total hardness, mostly the sum of calcium and magnesium, is established. A dye that shows a reddish hue in the presence of the two cations is added to the sample, and an EDTA titration follows until the two cations are completely chelated and the dye no longer indicates them. A second titration is performed on a sample that is treated with strong KOH, which precipitates $Mg(OH)_2$. The result yields calcium hardness, expressed as $CaCO_3$. The difference between total and calcium hardness is magnesium hardness.[4]

In the upper midwestern portion of the United States and adjacent Canada, where North American limnology got its start, hardness is often equated erroneously with alkalinity. Calcium is the common cation associated with carbonate, and the acid titrations of alkalinity assays indirectly measure most of that earth metal. The monovalent alkaline metals such as sodium and potassium are used in the manufacture of soap, and obviously they do not contribute to hardness. For this reason, some concentrated waters of arid zones are extremely soft, although their alkalinities are high. All students of limnology should experience titrating and recording some 6,000 mg total alkalinity per liter in a sample that has practically no hardness. (The sample might be from a soda lake such as Moses Lake, Washington.) This exercise would serve as a reminder that the words "hardness" and "softness" cannot be used interchangeably with alkalinity.

Sodium

Sources and Types of Sodium Compounds

The monovalent cation sodium is the sixth most abundant element in the lithosphere. This alkali metal is very reactive and soluble; when leached from the rocks, its compounds tend to remain in solution. For this reason, it is at least the third most abundant metal in lakes and streams, and in many instances it ranks first. Among igneous rocks, the feldspars, aluminosilicates of alkali, and alkaline earth metals are the most abundant of all minerals. A representative of them, *albite*, is sodium feldspar ($NaAlSi_3O_8$). When attacked by carbonic acid, albite decomposes to yield sodium carbonate, silica, and clay minerals.

When alkali-rich sources of igneous rocks are deficient in silica, they form a type collectively called the feldspathoids. Such rocks react with silica to form feldspars, a reaction that explains their absence in the presence of quartz. At least three common types of feldspathoids contain sodium; of these, *nepheline* ($NaAlSiO_4$) is predominant. In East Africa, lake basins lying in humid, hot regions contain sodium carbonate-

---

[4] Interference from iron and manganese are bothersome in many instances, and this description of methodology may be too idealized. Despite this, if the titrations are referred to milligrams of $CaCO_3$ per liter, the calcium hardness $\times 0.4$ approximates the calcium ion content in milligrams per liter, and the magnesium hardness multiplied by 0.243 gives the value of Mg^{2+}.

bicarbonate waters because of the dissolution of soluble, feldspathoid, volcanic rocks in the drainage area. In more arid parts where chemical weathering is practically non-existent, the ionic composition of lake waters is determined by the occasional precipitation—sodium chloride is predominant. The drainage basin of Lake Nakuru lies across both climatic zones; as a result, its concentrated waters are high in Na_2CO_3, $NaHCO_3$, and NaCl (Hecky and Kilham, 1973).

The commonest water-soluble mineral is *halite* or simply NaCl. A liter of seawater contains about 30 g of this and, therefore, roughly 11 g of sodium. Extensive beds of rock salt, representing evaporation of seawater, are impure halite, a common source of sodium chloride in some springs. Ocean spray, swept up by the wind and carried inland, is believed to account for the high NaCl values in Australian (Williams, 1964) and Nova Scotian lakes (Gorham, 1957a). Solid salt, swept up from dried inland basins such as the flat floor of ancient Lake Bonneville, can be carried by winds to add to aquatic sodium concentrations far inland.

Mirabilite ($Na_2SO_4 \cdot 10H_2O$), another name for Glauber's salt,[5] may serve as a source of sodium in some instances. Other sources are natron and trona, deposits of hydrated Na_2CO_3. Such sedimentary accumulations of sodium compounds should be thought of, however, as having been derived from concentrated waters (by evaporation or freezing-out effects) rather than as being a common source of sodium for waters of humid regions. Agricultural activities involving irrigation in arid zones may put old salts back into solution. The Salton Sea in southern California is a good example of this. Its salts were deposited by pluvial Lake Cahuilla.

Types of Sodium Lakes and the Cyanobacteria

In arid tracts where closed basins hold concentrated waters, there are at least three types of sodium lakes: *salterns*, the commonest and much like concentrated seawater with a preponderance of NaCl; lakes typified by much Na_2SO_4, such as saline lakes of Saskatchewan (Rawson and Moore, 1944); and the soda lakes characterized by $NaHCO_3$ and Na_2CO_3, termed alkali waters because of their high pH.

Some of the most luxuriant populations of cyanobacteria in the world are known from the soda lakes, and there has been suspicion that sodium might be a minor factor in eutrophication. Provasoli (1969) brought together the experimental evidence founded on laboratory experiments that showed Na^+ and K^+ are necessary for cyanobacterial growth and that increasing concentrations of alkali metals leads to flourishing populations of these prokaryotes. As the enrichment of lakes via cultural eutrophication has proceeded, concomitant increases in sodium and potassium have scarcely been noted, despite the fact that these monovalent cations are essential nutrients for the cyanobacteria. The general abundance of these alkali metals, however, precludes their being limiting factors, except in rare situations. Furthermore, the pH and carbonate ion of soda lakes may be keys to the tremendous blooms of cyanobacteria they support. The concentrated salterns and sodium sulfate waters do not exhibit such large populations; this suggests that sodium *per se* does not always favor enormous blooms.

---

[5] *Glauberite* is more of a mixture—$Na_2Ca(SO_4)_2$.

Potassium

Sources and Types of Potassium Compounds

Potassium, a close relative of sodium, is usually the fourth ranking cation in lake water. In some closed basins it surpasses sodium, but this is unusual. It is weathered from various feldspars that have the formula $KAlSi_3O_8$ but does not remain in solution so well as sodium. It recombines easily with other products of weathering, being removed from solution by adsorption on clay, for example. Moreover, the feldspathoids containing potassium do not weather so readily as the sodium minerals. *Leucite* ($KAlSi_2O_6$), one of the potassium types, exists as crystals within volcanic rocks. Potassium also tends to form micas, which are insoluble and unavailable to aquatic ecosystems. All these phenomena interact to make potassium rarer in water than sodium is, but more common in the world's rocks.

In plants, both extracellular and intracellular fluids contain an excess of K^+ over Na^+. In animals, extracellular Na^+ often surpasses potassium. There is some evidence that highly concentrated waters with a pronounced potassium content are lethal to many aquatic animals, the Na/K ratio being less than ten. Some habitats are known, however, where K^+ surpasses Na^+, and euryhaline species, such as *Artemia*, thrive.

Plants take up potassium to such an extent that an important commercial source of the element has been wood ash. Conversely, 90% of the world's production of potassium salts is used for fertilizers, to be taken up by domestic plants.

The dung of herbivores reflects the high potassium content of their plant food. In the Arizona desert pond Little Triste, all the common ions become concentrated as the pond level falls because of intense evaporation. When a violent rainstorm fills or partially replenishes the water, the concentrations of all ions, except K^+, show dilutions' effect. Potassium concentration rises, apparently the result of rabbit and cattle droppings being swept from the desert floor into the pond.

Potash is the name for K_2CO_3, but the word has been used indiscriminately to refer to KOH and potassium oxide. The so-called potash lakes, occupying depressions among the sand dunes in Nebraska, serve as sources of potassium-rich brines. In many of these waters potassium surpasses the other common alkali metal, sodium.

Sylvite (*KCl*) is less common than halite but occurs with it and gypsum in old marine deposits.

Older analyses of water constituents lumped Na^+ and K^+ together because of the difficulty in the chemical procedures for measuring them. With the development of flame photometry in the early 1940s, the detection and evaluation of these two metals became simple. Sodium burns with a yellow flame, and potassium produces violet. With proper filters these two elements can be detected, and the intensity of the color can be related to a standard curve prepared by flaming known amounts. Today ratios between sodium and potassium are common in accounts of water quality. (Calcium, magnesium, and lithium are often assayed by flame photometry because filters have been developed to screen out all but the characteristic wavelength of each individual flame.) The use of atomic absorption spectrophotometry is an alternative to the older flame technique.

Some confusion exists in terminology where certain potassium and sodium compounds are concerned, but this has little limnologic importance. *Caliche* ($NaNO_3$) is also known as "Chile saltpeter." In the southwestern United States and in Mexico, caliche is the name applied to the hard-pan layer beneath the surface of desert soils where leached minerals have precipitated; much of this evaporite is $CaCO_3$. Also, KNO_3 is termed saltpeter but without the Chilean prefix.

Water Chemistry and Desert Limnology

Desert limnology is best defined in terms of hydrography and limited to regions where no runoff reaches the sea—the climes where closed basins occur, as evaporation substantially exceeds precipitation. Desert waters are generally high in electrolytes and are quite different from the dilute waters of humid regions. The chemical limnology of arid lands deals especially with evaporation, concentration, precipitation of compounds, and the relative changes in ionic abundance. A review of desert limnology was published by Cole in 1968; Bayly and Williams (1973) summarized further details, and the 30 papers assembled by Williams (1981) following an international symposium on salt lakes attest to the interest in the subject.

Fig. 14-1 reveals comparative solubilities of the common compounds of fresh water. The common compounds of dilute inland waters, the carbonates of the alkaline earth metals, are the first to achieve their solubility products and precipitate. Thus, $CaCO_3$ and $MgCO_3$ are diminished, with marked relative increases in SO_4 and Cl^-. The next precipitate is $CaSO_4$, leading to the comparative importance of Na^+ and Mg^{2+} among the cations. The subsequent precipitation of sodium sulfate leads to a water dominated by compounds of chloride. If all carbonate were not deposited with calcium and magnesium, that linked to sodium would precipitate next, followed by $MgSO_4$, and the sodium chloride saltern would result. These salterns, such as the Great Salt Lake of Utah, are fairly common in North America. The theoretical stages when only the extremely soluble chlorides of the alkaline earths persist are rare, but bitterns such as the Dead Sea do exist. Calcium chloride is the most soluble compound shown in Fig. 14-1, but one would not expect its occurrence at the termination of an evaporation sequence, because calcium precipitates earlier with carbonate and sulfate. The shallow Antarctic Don Juan Pond is a notable exception; it is a $CaCl_2$ pool. This is the saltiest water body on Earth and is located in an extreme desert. Recent work by Dickson and others (2013) using time-lapse photography solved the old mystery of the water source to the pond: the air. Water is absorbed directly from the air by the super-salty pond through a process called deliquescence; no groundwater enters, and evaporation exceeds precipitation.

However, many desert lakes are less concentrated than would be theorized on the basis of their age and the annual input of salt. Moreover, many are quite different in ionic composition from the salterns and bitterns toward which evaporating waters seem to be

	Cl^-	SO_4^{2-}	CO_3^{2-}
Ca^{2+}	1.5	0.006	0.00005
Mg^{2+}	1.3	0.9	0.0004
Na^+	1.0	0.3	0.4

Figure 14-1 Relative solubilities of some compounds in distilled water at 10° C; NaCl = 1.0. The three compounds bounded by dotted lines are typical of dilute water only.

proceeding. Such discrepancies are explained by two facts: first, there are seven or eight mechanisms by which salt can be lost from closed basins; and second, the dilute waters that initiate the evaporative series may be strikingly different ionic mixtures. Details of these methods were reviewed by Hutchinson (1957) and Cole (1968).

There are two main groupings of arid-land waters and a terminology to describe them. Sodium chloride waters, similar to concentrated ocean brine or brackish waters (best described as dilute sea water), are termed **thalassohaline**. Others quite different from seawater in relative ionic content are **athalassohaline**. Figure 14-2 was devised to portray the different types of water, including seawater. The examples chosen show a typical thalassohaline saltern, the Great Salt Lake and the Dead Sea bittern. Athalassohaline examples include: a "typical" dilute freshwater calcium carbonate lake and a gypsum spring (neither of which could exist in a concentrated state); a soda lake; two other types of sulfate lakes; and Mono Lake, California, as an example of "triple water" with the ions CO_3^{2-}, Cl^-, and SO_4^{2-} in nearly equal proportions.

Biota of Saline Waters

The biota of concentrated inland waters are typically hardy species with a freshwater ancestry. Bayly (1967) constructed a biologic classification of Australian aquatic environments, coming to the conclusion that marine and interior saline waters are at opposite ends of a spectrum. Bayly, a student of calanoid copepods, showed that Australian centropagids, characteristic of the highest salinities, seem to have had direct

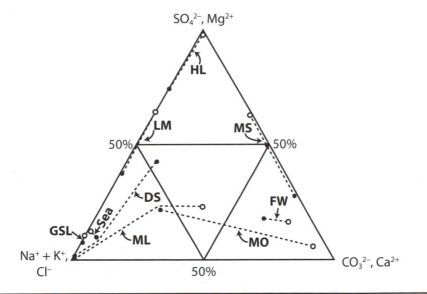

Figure 14-2 Triangular coordinate plot of different types of inland waters. Solid circles are cations; open circles are anions. **FW**, "Standard" fresh water, from Rodhe, 1949; **MO**, Moses Lake, Washington, soda lake; **MS**, Monkey Spring, Arizona, gypsum water; **ML**, Mono Lake, California, "triple" water; **DS**, Dead Sea, Israel, bittern; **GSL**, Great Salt Lake, Utah, saltern; **LM**, Little Manitou, Saskatchewan, sodium sulfate lake; **HL**, Hot Lake, Washington, magnesium sulfate lake; **Sea**, mean sea water.

freshwater ancestors and were less closely related to marine and brackish water forms. Later he reviewed our knowledge of salinity tolerances shown by animals in hypersaline waters and discussed their osmoregulatory behavior (Bayly, 1972).

In general, the diversity of species is inversely correlated with salinity. There is risk in setting down exact tolerance limits because relative proportions of major ions may be more important than total salt content. There is, however, only a meager list of species found in solutions in excess of 10% salinity. Usually, specific cyanobacteria and diatoms persist in concentrated water where green algae no longer survive. Notable exceptions are the branched green alga *Ctenocladus circinnatus* and the green flagellate *Dunaliella*. Cells of the former survive surrounded by crystallized salt, and the latter survives in a wide range of salinities by changing its cellular glycerol content. Because glycerol also protects against freezing, *Dunaliella* remains motile until surrounding water temperatures fall to about −18° C.

Euryhaline invertebrates include cosmopolitan species and others that are more local in distribution but represented from region to region by ecologic equivalents. The rotifer *Brachionus plicatilis* is an example of the first category. The calanoid copepod *Diaptomus nevadensis* of North America and *Arctodiaptomus salinus* of the Old World exemplify the latter. The Diptera have many hardy species, including ephydrids, culicids, and tabanids. To a lesser degree, certain species of the chironomids are capable of life in concentrated waters. A few Coleoptera and Hemiptera, especially corixids in the Hemiptera, are found in hypersaline environments.

Seemingly, an invertebrate adapted for terrestrial existence would be prepared for the osmotic hardships of hypersaline water. Both habitats are, in a sense, arid. Examples are scarce, although in Australia *Haloniscus*, a member of the terrestrial oniscid isopods, lives in concentrated pools.

Among the Vertebrata, some cyprinodont fishes are extraordinarily euryhaline. The southwestern pupminnows, *Cyprinodon*, have been found in pools of at least twice the concentration of seawater.

The species that occur in concentrated thalassohaline and athalassohaline waters are capable of withstanding major fluctuations in physicochemical features of the environment. They are found in lakes having a wide range of salinities. Marine species, by contrast, have evolved in a stable habitat, and most plants and animals from the sea that are found also in inland waters are not capable of any remarkable osmoregulation. They occur in waters differing little in salinity from seawater. A few marine species, such as the mullet *Mugil cephalus* and the edible blue crab *Callinectes*, are capable of progressing in land and move up rivers in gradients of decreasing salinity.

Researchers usually focus on saline lakes as habitats where physiologic stress on the biota is accented and where extreme adaptations are necessary for survival. Williams (1972) suggested they could go further and recognize the value of salt lakes for fundamental ecologic research. He pointed out that there are several factors making trophic-dynamic studies in saline lakes easier compared with investigations of freshwater ecosystems. These are (1) a far less complex biota with greatly decreased biotic diversity and, therefore, simplified trophic relations, (2) the discreteness of the salt-lake habitat that always lacks outflow and often permanent inflow and shows minimal contact with contiguous ecosystems, and (3) the relatively low habitat heterogeneity in shallow saline lakes that often lack littoral macrophytes, for example. Moreover, it is interesting that extremely high rates of primary production are known from condensed desert waters, and following this plant growth some unusual secondary production occurs. Perhaps this is partially due to the remarkable efficiencies of the animals, as studies on the typical saline-water grazer *Artemia* suggest. Also, organisms from the simple ecosystems represented by some desert waters are amenable to laboratory experimentation, contributing, therefore, to our understanding of the entire ecosystem. Data from desert lakes may be more rewarding for mathematical and computer simulation than are those from more complex ecosystems.

Keith Walker, a student of Williams', published results of a detailed study that supports Williams' ideas (Walker, 1973). The shallow, 21-ha Lake Werowrap in Australia exemplified the advantages offered by saline lakes for the holistic, ecosystem approach to limnology. The trophic structure of the shallow lake was simple and its primary productivity fairly high, the annual carbon fixation amounting to 435 g/m^2.

We state again that we must not define limnology narrowly as the study of fresh waters. Many limnologists are intrigued by saline bodies of water and devote much time in researching them. This was emphasized by the appearance of a new journal in 1992, the *International Journal of Salt Lake Research*, another part of the complex subject called limnology.

15

Redox, Metals, Nutrients, and Organic Substances

Oxidation and Reduction: Redox Potential

In the previous three chapters we considered the common chemicals of limnology as gases, solutions, and precipitates. Many of these chemicals undergo reactions involving the gain or loss of electrons (electrochemistry).

Long ago, "oxidation" was thought of as the combination of a substance with oxygen. "Reduction" implied removal of oxygen whose mass was therefore reduced. Later, chemists understood that these reactions were actually caused by transfers of electrons, so "oxidation" can occur without involvement of oxygen. Since electrons are transferred between atoms, these "redox reactions" are always coupled, the oxidizing agent gaining electrons and the reducing agent losing electrons.

One of Earth's most common elements, iron, undergoes a familiar redox reaction. Ferrous iron can be oxidized to the ferric state by giving up an electron, and ferric iron can be reduced by the addition of an electron. These events can occur without the participation of oxygen or hydrogen.

$$Fe^{2+} \quad \rightarrow \quad Fe^{3+} + \ e^-$$
$$\text{ferrous} \qquad \text{ferric}$$
$$\text{iron} \qquad \text{iron}$$

Both processes are changes in the availability of the outer orbital electrons of chemical elements. Some resist rearrangement of these electrons more than others do. Sodium, for example, is very resistant to reduction, normally existing in an oxidized state.

Reduction and oxidation occur simultaneously, for an ion can be reduced only by the gain of electrons that arrive from an outside source as another ion is oxidized by their loss. This permits us to think of oxidation-reduction pairs, such as ferric and ferrous iron. Also, inherent in the theme of oxidation-reduction is the relative nature of an element or ion. When compared with aluminum, sodium is a strong reducer, giving up electrons to become oxidized. However, compared with lithium, another alkali metal, sodium is a weaker reducer and more of an oxidant. Thus, ion A may reduce B but in turn is reduced by C. Or stated in another fashion, A oxidizes C but is oxidized by B.

Some of the above is reminiscent of the principle upon which a common battery, the voltaic cell, is produced. A copper sulfate solution and a zinc sulfate solution are joined by a wire and salt bridge. The copper solution, a stronger oxidant than the zinc, takes up electrons. As a result electrons flow through the wire, oxidizing Zn to Zn^{2+} as Cu^{2+} is reduced to Cu. The voltage is about 1.10 but would be different if some reductant other than zinc were compared with the copper oxidant. The tendency of a copper–copper sulfate mixture to oxidize by taking electrons from other pairs is relative.

The relative difference in potential that drives a current is expressed as voltage. It is analogous to the difference in hydraulic pressure that forces water through a pipe. The intensity of the current flowing in the wire that connects two contrasting mixtures is a function of their relative states of oxidation or reduction and is measurable. It is the oxidation-reduction potential, or simply the redox potential, the difference in voltage that can be quantified by the counter-potential needed to bring the current to a halt. The electromotive force is termed E_h.

Potential and voltage imply comparisons. A single solution, a mixture of a redox couple (an oxidant and its reductant), must be compared with some standard. The standard generally selected is the hydrogen electrode,[1] where the redox pair is molecular hydrogen (the reductant) and hydrogen ion (the oxidant). The electrode is a glass tube through which gaseous H_2 at a pressure of 1 atmosphere can be passed over an included foil of platinum. The platinum adsorbs hydrogen but, being less active, does not lose or gain electrons itself. The electrode is immersed in a 1N acid solution; the pH, therefore, is zero. The flow of electrons may be toward the electrode of the mixture being compared or toward the standard hydrogen electrode, which can act as an anode, with hydrogen molecules giving up electrons to form hydrogen ions:

$$H_2 \rightarrow 2H^+ + 2\ e^-$$

Or it can perform as a cathode, receiving electrons:

$$2H^+ + 2\ e^- \rightarrow H_2$$

The direction of flow indicates whether the mixture being tested tends to oxidize, taking up electrons readily, or whether it is a reducing solution, giving off electrons to the hydrogen system. The redox potential of the H^+–H_2 pair is arbitrarily defined as zero. Other redox couples and complex mixtures with potentials greater than zero are those that take up electrons and oxidize the hydrogen electrode; negative potentials are those shown by mixtures that yield electrons to the standard electrode. High positive voltage indicates oxidizing mixtures; decreasing voltage reflects an increase of reducing elements. A surplus of the reducing elements, when compared with the hydrogen standard, is shown by negative voltage. This scheme is the so-called European-sign convention. Confusingly, it is sometimes reversed, with positive values signifying that the reduced member of a redox pair is a more effective reducer than the H_2 in the H^+–H_2 couple.

Hydrogen ion concentration affects E_h to such an extent that it is necessary to correct voltage to some standard pH, usually 7.0; E_h is then expressed as E_7. At 18° C a lowering of 1 unit of pH is attended by an increased potential of 0.0577 V; at 20° C

---

[1] The calomel electrode of HgCl or Hg_2Cl_2 is another common arbitrary standard. It has a positive voltage of 0.242 compared with the hydrogen electrode at pH 0.

this is 0.0581. Temperatures in this range are considered normal, and customarily, solutions tested at any pH have their potentials referred to pH 7.0 by adding 0.058 V for every unit above neutral or by subtracting the same for each unit of the sample below pH 7.0. From data based on such manipulations, it appears that oxygenated epilimnion waters usually have an E_7 voltage near 0.5, although the oxidant–reductant couple, O_2–H_2O, theoretically have a potential of about 0.82 V.

As oxygen tensions decline, there may be little change in potential. In small fertile lakes, however, there is an increasing tendency for reduction to accompany the decrease of hypolimnetic oxygen, and negative voltages often accompany anaerobic water. At and just below the profundal sediment-water interface, sharp E_h changes are to be expected in rich, stratified bodies of water.

As certain voltages are reached in lake waters and sediments, limnologically important ions are reduced as shown in Table 15-1. Mortimer (1941–1942) has shown that the conversion of trivalent to bivalent iron, occurring at E_7 voltages between 0.3 and 0.2, is especially significant. This change can be witnessed in stratified eutrophic lakes when depletion of oxygen and accumulations of reduced substances bring about iron reduction (Fig. 15-1 on the next page).

In Table 15-1 a sequence of conversions from oxidized to reduced states can be seen. In a summer hypolimnion, nitrite, for example, begins to be reduced to ammonia as voltage falls to 0.4; sometime later, ferric iron may be reduced. It must be understood that the voltages shown in Table 15-1 are those at which the reactions have been found to occur; they are not the potentials of the redox pairs opposite them. In neutral solutions the potential *(E₇)* voltage of the trivalent-bivalent iron couple is about 0.36 V, and in tables of standard electrode potentials the potential is listed as 0.77 V at 25° C.

Table 15-1 Redox values of limnologic interest: voltages at which important reductions and oxidations occur and associated oxygen concentrations

Redox couples	E_7 (volts)	Dissolved O_2 (mg/liter)
NO_3 to NO_2^-	0.45 to 0.40	4.0
NO_2^- to NH_3	0.40 to 0.35	0.4
Fe^{3+} to Fe^{2+}	0.30 to 0.20	0.1
SO_4^{2-} to S^{2-}	0.10 to 0.06	0.0

Data from Mortimer (1941–1942).

Iron

Iron is an abundant and important element, unsurpassed by any other heavy metal in the earth's crust. In living systems it is associated with numerous enzymes—peroxidase, catalase, cytochrome oxidase, and nitrogenase. Iron is necessary to photosynthesizing plants, where it is the metal part of at least two plant cytochromes that function in the transfer of electrons during photosynthesis.

Iron is found in two states, the oxidized ferric (Fe^{3+}) and the reduced ferrous (Fe^{2+}). Reducing and acid conditions promote the solubility of iron. Most ferrous compounds are soluble; a noteworthy exception is FeS (Fig. 15-1). In aqueous environments the common ferric compounds are insoluble; as a result, iron precipitates in alkaline and oxidized conditions.

Many of the conversions reducing and oxidizing iron are mediated by microorganisms. Chemosynthetic bacteria belonging to the *Thiobacillus-Ferrobacillus* group possess enzyme systems that transfer electrons from ferrous iron to oxygen, and this transfer results in ferric iron, water, and some free energy that is used for synthesizing organic compounds from CO_2. Bacteria in this group work at a low pH (below 5.0) where the oxidation of ferrous iron would proceed very slowly without enzymatic aid. Other bacteria use the energy derived from oxidizing iron to assimilate, rather than create, organic material. Generally, the microbial iron reducers cause the acceptor to act in low pH and E_h environments but not necessarily in anoxic habitats.

Bacteria and plants can modify environments so that iron becomes either self-oxidizing or self-reducing. The elevation of oxygen values and the consumption of CO_2 promote oxidation and, hence, precipitation of iron. Some bacterial metabolic products, such as the H_2SO_4 produced by *Desulfovibrio*, promote iron solubilization. At pH values from 7.5 to 7.7, a threshold is reached where iron in the form of $Fe(OH)_3$ is precipitated automatically. This means that iron would not be found except in acid to neutral water that is very low in oxygen and with redox potentials of 0.3 to 0.2 V—such as in the hypolimnion of a stratified eutrophic lake. There it would be present in the soluble reduced state. With the introduction of oxygen at circulation, the iron would be oxidized and precipitated. A similar phenomenon is seen in streams contaminated by the acidic wastes from coal mines. Rusty colored masses of precipitated flocculent $Fe(OH)_3$ mark a recovery zone where pH has risen and oxygen increased. These flocculated accumulations of ferric hydroxide, called "yellow dog," often damage the fish by clogging their gills in a seemingly habitable stream region.

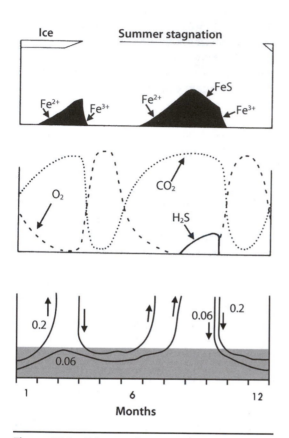

Figure 15-1 Schema of events caused by anoxic and intense reducing conditions in the hypolimnion of a dimictic, eutrophic lake. Lowest block shows the progress of two critical isovolts throughout the year within the sediment (shaded) and the overlying hypolimnetic water; 0.2 V line represents the redox potential at which ferric iron is reduced to ferrous iron; 0.06 V line represents the redox potential at which sulfate is reduced to sulfide. Top block shows the appearance of Fe^{2+} in the water beneath winter ice and subsequent precipitation of Fe^{3+} at the vernal overturn, the appearance of Fe^{2+} during the summer stagnation, the precipitation of some iron as FeS when the redox potential falls to 0.06 V, and the precipitation of Fe^{3+} at the fall overturn. Middle block shows the appearance and disappearance of the gases O_2, CO_2, and H_2S in hypolimnetic waters.

Chelating Agents

It is significant that some conversions of insoluble ferric iron to a soluble state do not require chemical reduction of the metal. Adding chelating agents results in an increase in solubility, and it is this available soluble iron, not total iron, that is important in algal nutrition.

Chelators are a class of compounds, organic and inorganic, that can hold metal atoms or ions. The metals are held between atoms of a single chelator molecule, frequently being pinched between two nitrogen, oxygen, or sulfur atoms. Chlorophyll is a chelate molecule with a magnesium atom held by nitrogen atoms. *Siderophore* is the specific term for a chelator that holds iron. The word is derived from the Greek words *sideros*, applying to iron, and *phoros*, meaning bearer. Many microorganisms, both prokaryotic and eukaryotic, aid in solubilization of ferric iron by releasing natural chelators that make assimilation possible. Other metals, including some trace elements, are solubilized by such chelators. Alanine and glycine are two amino acids with known chelating properties, and the porphyrins represent another common type of organic chelator.

In marl lakes the pH is on the alkaline side and soluble iron is scarce. The iron present in the water column is in the ferric state and is sestonic, in many instances adsorbed to particles en route to the bottom deposits. The increase in algal primary productivity following the addition of chelators to marl lake waters is a response to the solubilization and hence availability of iron (Wetzel, 1972). Some of the newly solubilized iron in the complexed, chelated molecule can be used by iron-limited cells as the molecule is degraded. In Wetzel's marl lakes, amino acids, citrate, and the synthetic chelators EDTA (ethylenediaminetetraacetic acid) and NTA (a trisodium salt of nitrilotriacetic acid) are effective. The EDTA is used to chelate calcium and magnesium in water hardness tests; NTA[2] achieved prominence because it was tested as a substitute for phosphorus compounds in laundry detergents in Canada.

In instances where iron and other heavy metals are excessively concentrated, chelation can have an opposite effect on algal populations. In a sense, the metal in the complexed state is taken out of the algal environment, reducing its toxic properties. Chelators are not equally biodegradable: EDTA, for example, is more stable than NTA and the amino acids, although iron-starved cells can degrade the complexed molecule.

The Iron Cycle in Lakes

A most important limnologic feature of iron is its seasonal behavior in the hypolimnion. We owe much of our knowledge of the iron cycle in lakes to Mortimer (1941–1942). In well-oxygenated waters, ferric iron occurs but is rare because of its insolubility. During the spring overturn, most of it is in the sediments. In the profundal ooze it exists as ferric hydroxide, ferric phosphate, and perhaps a ferric silicate and

---

[2] Hamilton (1972) reviewed much of what is known about this chelating compound. Although lacking the important nutrient phosphorus, NTA could possibly accelerate algal growth by making iron and other essential metals available. Moreover, NTA contains nitrogen and decomposes to form glycine and glycolic acid during biodegradation. These are probably useful to some algae, and certainly they may be deaminated to ammonia, which is readily available to phytoplankters. In any habitat where nitrogen rather than phosphorus is limiting, the substitution of NTA for phosphoric acid detergents will not prohibit eutrophication. The coastal seawaters, receiving tremendous quantities of sewage, are this type of habitat (Ryther and Dunstan, 1971).

ferric carbonate complex. An oxidized micro-zone of iron-containing molecules in a complex colloidal layer seals nutrient within the sediments, and little escapes to the overlying water. This would be the condition throughout the summer stagnation period in a stratified oligotrophic lake. In a eutrophic lake, by contrast, CO_2 collects, oxygen becomes scarce or absent, the pH falls, and the redox voltage drops to 0.3 or 0.2. Then conversion of ferric to ferrous iron commences (Fig. 15-1). Because this reduced iron is soluble, the oxidized seal disappears. The various substances mobilized with ferrous iron—including phosphorus and silica—then become abundant in hypolimnetic waters. Ammonia also may be profuse, having been produced at higher redox potentials by the reduction of nitrite (Table 15-1) and having arisen as a result of microbial action on decaying protein. Much of the well-known increase in summertime hypolimnion alkalinity, unaccounted for in terms of $Ca(HCO_3)_2$, is due to ammonium bicarbonate and ferrous bicarbonate.

The disappearance of the oxidized barrier and the release of nutrients to the supernatant water lead to the consumption of more oxygen and perhaps to the escape of even more nutrients. Thus, the first appearance of a reduced microzone at the water-sediment interface during the course of a lake's history must be a momentous event. The surface-oxidized sediments of oligotrophic lakes serve to keep elements of fertility from being recycled and to maintain the trophic state at an oligotrophic equilibrium.

The hypolimnetic agents of fertility, however, are not completely mixed throughout the lake, because oxygenated conditions at the autumnal overturn cause rapid conversion of Fe^{2+} to Fe^{3+} and the immediate precipitation of $FePO_4$ and other iron compounds (Fig. 15-1). The hypolimnion has been likened to an iron trap: most of the iron that arrives there is retained, alternating between mobile soluble and immobile insoluble states. This explains a minor source of the world's iron ore, the so-called bog iron—*siderite*, $FeCO_3$; *hematite*, Fe_2O_3; "ochre," $Fe(OH)_3$—occupying concavities that were former lake basins.

In large lakes, where seiches are effective, internal waves might bring hypolimnetic waters close to the surface at the windward end of the lake. This upwelling could be followed by some mixing and transport of the rich water into the trophogenic zone.

Iron is not abundant in waters of meromictic lakes. The reason for its scarcity is the same as that which accounts for a decline in iron as summer stagnation proceeds (but *before* the overturn) in some polluted eutrophic lakes (Fig. 15-1). In both instances intense anaerobiosis and reducing conditions (as are met in the monimolimnia of meromictic lakes) bring redox potentials below 0.1 V, and sulfate is reduced to sulfide (Table 15-1). Much of the sulfide, in addition to forming H_2S, combines with iron as FeS. Ferrous sulfide is insoluble and precipitates to contribute to the shiny black luster of sapropel. The precipitation of FeS represents the final stage in summer stagnation, although in many eutrophic lakes the overturn terminates stagnation before this stage is reached.

Manganese

Very close to iron in the economy of lakes and behaving in much the same manner is the heavy metal manganese. Manganese has four valence states, and it alternates between reduced soluble and oxidized (less soluble) conditions.

Manganese is a necessary nutrient for plants and animals. It stimulates plankton growth, perhaps by activating enzyme systems and by having at least some effect on the vitamin thiamine. Its abundance in the igneous rocks and soils is perhaps no greater than 0.5%, but this amount seems adequate, and manganese deficiencies are rare.

Manganese, being especially soluble in the bivalent state, is reduced and mobilized at a higher (almost two times) redox voltage than iron. Under strong oxidizing conditions, manganese is part of the colloidal microzone seal and serves with iron as a barrier between deeper sediment and supernatant water. At this time manganese is probably in the tetravalent or trivalent condition.

During summer stagnation, manganese goes into solution earlier than iron, but it is precipitated later than iron when the overturn occurs. Manganese, then, is more apt to be lost in lake outflows than iron is. The hypolimnion is not so effective a trap for manganese as it is for iron. Because of this difference, past oxidizing-reducing conditions have been inferred from Fe/Mn ratios changing in the sediments. Mackereth (1966) published some instructive data based on analyses of cores from the eutrophic English lake, Esthwaite Water. Changing Fe/Mn ratios imply fluctuations in reducing oxidizing environments during the postglacial history of the lake. When oxidizing conditions began to be replaced by a reducing environment, there was an increase in the Fe/Mn ratio, signifying chemical reduction and solubilization of manganese, some of which was lost. Later, as redox conditions brought about the mobilization of ferrous iron, the ratio diminished. Similarly, the advent of oxidizing situations led to initial increases in the ratio because iron precipitated first; the later arrival of manganese in the sediments lowered the Fe/Mn value again.

Phosphorus

Phosphorus is absolutely necessary to all life; it functions in the storage and transfer of a cell's energy and in genetic systems. The universality of ATP (adenosine triphosphate) as an energy carrier and the presence of phosphate groups in nucleotides, and hence nucleic acids, underscores living organisms' need for phosphorus. Found in meteorites, rocks, soils, and even in the sun's atmosphere, it is not one of the rarest elements. Phosphorus is much scarcer, however, than the other principle atoms of living organisms (carbon, hydrogen, oxygen, nitrogen, and sulfur). Its abundance at the surface of the earth is about one tenth of 1% by weight. It is taken up rapidly and concentrated by living organisms.

Sources and Nature of Phosphorus

Phosphorus oxidizes very readily and occurs in the earth's rocks principally as **orthophosphate** PO_4^{3-}. The main source of this ion is igneous rocks containing the phosphatic mineral *apatite*, $Ca_3(PO_4)_3^+$ united with either OH^-, Cl^-, or fluoride. The fluoroapatite is the chief mineral source in igneous material, but there are at least 205 minerals, many very rare, that contain phosphorus (Fisher, 1973).

The reduced, gaseous phosphine (PH_3) is but a fleeting state because it oxidizes so rapidly. It is thought that the eerie lights sometimes flickering over marshes, will-o'-the-wisps, could be phosphine and perhaps other gases produced by anaerobic decomposition burning on contact with air. There is little other phosphorus in the earth's atmosphere save what adheres to dust and debris. The five other major elements of tissue are represented, sometimes abundantly, in air and rain.

Vallentyne (1974) detailed the case of mysterious toxins causing extensive fish and invertebrate kills in Placentia Bay, Newfoundland, a story that teaches us that, among other things, we still have much to learn. The danger of elemental phosphorus as an industrial pollutant was underestimated when a shoreside factory was built to produce it from phosphatic rocks shipped from Florida. It was assumed that phosphorus escaping into the natural environment would be oxidized rapidly and rendered harmless, as is the nature of the element. Finally it came to light that elemental phosphorus in the colloidal state does not easily oxidize in water and that this was the cause of the disaster. The fish deaths actually were an important alarm signal, for as Vallentyne noted, debilitating hepatitis comes from as little as 0.6 mg, and 100 mg of P is lethal to a human.

Igneous rock was the original source of phosphorus on earth. Subsequent weathering, probably aided by carbonic acid, liberated much of the phosphorus, which was redeposited or carried to the seas. The flow from continents to seas was the major movement. Phosphorus accumulations of biotic origin formed near the oceans when upwelled phosphate was taken up by planktonic organisms and concentrated by marine fish that served as food for guano-producing sea birds. Great accumulations of guano were built up on dry cliffs and islands, especially off the west coast of South America. In addition, phosphorus is concentrated in the skeletons of vertebrate animals as an important constituent of bone. Phosphate accumulations are reminders of the major role played by organisms in concentrating this essential element. Other biotic sources of phosphate concentration are human waste and the release of laundry detergents into the environment. These accounted, in part, for the 300% increase in the phosphate content of Switzerland's Lake of Zurich since World War II (Thomas, 1973).

With industrial and agricultural use of guano beds there has been a reverse flow from the sea to the continents. Since the phosphorus from rocks, both igneous and sedimentary, is also mined and used, phosphorus has been spread throughout a large part of the globe. It is now being washed seaward, some of it for the second time, with much of it reaching lacustrine basins en route.

The decay and mineralization of plant and animal corpses is a source of phosphorus to the living components of ecosystems; the bacteria change it from organic molecular phosphorus to inorganic orthophosphate, which can be utilized by plants. This recycling does not represent a steady income of phosphorus from without. Only the rocks and the material from the seas can supply that.

Analysis of Phosphorus

There are some troublesome areas in analyzing and presenting data for phosphorus. The usual method of testing for it is based on the reaction of acidified ammonium molybdate with inorganic phosphate (orthophosphate). The reaction forms a molybdophosphoric acid that turns blue when reduced. Stannous chloride, the usual reducing agent, causes the development of molybdenum blue when soluble orthophosphate is in the sample. (Contamination of glassware, being more critical than it is in many water-chemistry routines, may be a source of error here.) Organic phosphates, known as metaphosphates or polyphosphates, are not shown by this test. They are revealed after being hydrolyzed to orthophosphate by heat and acid. The molybdate test on such a hydrolyzed sample gives total phosphate, and by subtraction (total phosphate minus orthophosphate) the organic phosphate can be rated. Testing filtered and unfiltered samples adds further dimensions; the categories of soluble and particulate phosphate are specified.

The reactive soluble orthophosphate is the fraction that is immediately useful for autotrophic plants. The tests for it, however, overestimate levels because of some hydrolysis of soluble polyphosphates in the acid medium during the time it takes the blue color to appear. Furthermore, when lake water is filtered in preparation for determining soluble reactive phosphate, algal cells may be injured. This might cause release of easily hydrolyzed phosphorus compounds to the filtrate.

There are pitfalls to watch for when one reads reports on phosphorous tests. Briefly, some authors have been careless in reporting their data and have confused orthophosphate with orthophosphate-P, and total phosphate with total phosphorus. Phosphate amounts are 3.07 times the phosphorus atom in the ion. Also, phosphorus can be limiting even at extremely low levels, and special techniques are necessary in extremely oligotrophic lakes to avoid overestimation (Hudson et al., 2000).

Phosphorus in Lakes

In 1957 Hutchinson presented a table that showed the mean of samples from several lake districts was about 20.8 mg total P per cubic meter (or that same amount in micrograms per liter); this equals 0.0208 mg total P per liter or 0.0638 mg total phosphate per liter. Far greater amounts have been reported from saline basins of arid climates. Hutchinson listed Goodenough Lake, British Columbia, with 208 g/m^3, which is 10,000 times the mean from dilute lakes in humid regions. The concentrated waters of Red Pond at the Long-H Ranch, Arizona, had more than 150 g/m^3 one summer following the sudden crash of a flourishing *Artemia* population.

In many instances light penetration as measured with a Secchi disc is reduced in ponds or lakes with high phosphorus content. In such cases the poor transparency may be a result of one of two phenomena. First, there may be enormous numbers of algal cells and the Z_{SD} is strongly correlated, although inversely, with chlorophyll *a* also. Second, in argillotrophic bodies of water, especially in arid regions, there is a rapid attenuation of vertical light penetration although there are few algal cells and little chlorophyll *a*. In such waters the Secchi disc depths are still negatively correlated with total phosphorus. because most of the P is adsorbed to or associated in some way (bacterial attachment) with the particles that absorb and scatter light.

Phosphorus cycles have engaged limnologists for a long time because the element is often a limiting factor for primary production and, therefore, for fish yield. The picture has changed, however—now there is concern about too much phosphorus!

Early investigations were restricted to measuring changes over periods of days or monitoring algal response after addition of phosphate. Commonly, only traces of orthophosphate are found in the water when photosynthesis is proceeding at a good rate. This situation resembles the apparent absence of free CO_2 on sunny days in rich waters while primary production is considerable. The phosphate, of course—as is true for CO_2—has been taken up by the plants, and its scarcity in the water in no way implies unimportance. The rate at which it is taken up and recirculated could only be guessed by early investigators, although they presumed rapid cycling.

With the availability of radioisotopes after World War II, there arose the possibility of quantifying phosphorus dynamics by using ^{32}P in a phosphate. Hutchinson and Bowen (1947, 1950) at Linsley Pond and Coffin and coworkers (1949) in Canada were the first to add radio-phosphorus directly to lakes and follow it as best they could. There was good evidence that phosphate was absorbed rapidly by the plankton and transferred, presumably by sedimentation, to the hypolimnion. Subsequently, inorganic phosphorus was released from the sediments. A few years later when ^{32}P was available in greater quantities, Hayes (1955), Hayes and Phillips (1958), and Rigler (1956, 1964) made some important contributions in Canada. Rigler (1964) published the results of tracer studies in bog lakes that revealed very rapid cycling of phosphorus. He added radiophosphate to the epilimnion, and the results agreed with earlier reports: it had but a brief time in the upper waters.

More important, Rigler discovered that low levels of phosphate, remaining constant for hours, did not reflect the inability of plankton cells to take up the ion. Rather, a metabolic loss matched the uptake, and soluble inorganic phosphate was replaced many times during an hour. By separating phytoplankters from bacterial cells, Rigler demonstrated that the bacteria were more responsible for the rapid cycling of phosphorus. The turnover time, which was approximately 5 minutes, suggests that aquatic bacteria are formidable competitors with the algae where this scarce element is concerned. Subsequent studies with labeled phosphate-P in aquatic environments quite different on a trophic scale show that the turnover time is from 1 to 8 minutes in a summer epilimnion (Rigler, 1973). The turnover rates reflect the amount of phosphate in relation to algal demand according to a study of some East African lakes (Peters and MacIntyre, 1976). In the concentrated soda waters of Lake Nakuru (about 15 mS/cm at 20° C), with about 0.023 mg/liter of PO_4-P, the turnover rate is relatively low compared with freshwater Lake Naivasha, where 50% of the experimental ^{32}P was taken up in less than 1 minute and where the turnover rate compares with the fastest reported from North American lakes.

Rigler found in his tracer study that 77% of the radiophosphate added to the water was gone within 4 weeks, but only 3% ended up in the sediments, and only 2% left via the lake's outlet. Theoretically, the remainder had been taken up largely by the littoral organisms. Uptake of phosphate through the leaves as well as the root system may be possible for many aquatic macrophytes. Later, Rigler (1973) pointed out some inadequacies in those early studies with ^{32}P. He cited the work of one of his students, who added large amounts of tracer to a Canadian lake and analyzed its fate more thoroughly (Chamberlain, 1968). From this study arose the concept of a less simplistic compartmentalization in the lake system. At least two conspicuous compartments exist in the epilimnion: the radioactive P is taken up rapidly by particles of less than 70 μm and sedimented rapidly; sestonic particles greater than 70 μm are much slower in reaching tracer equilibrium, and they remain in the epilimnion a longer time.

A common concept of the phosphorus cycle assumes that a dead organism, starting as particulate organic phosphorus, becomes soluble organic phosphorus and, eventually, soluble inorganic phosphorus. (The changes are mediated by bacteria.) This is probably a little too simple. Living plankton organisms and littoral plants excrete soluble organic phosphorus compounds (broadly, DOM), much of which aggregates to become colloidal and hence POM. Another fraction becomes particulate by bacterial incorporation. LaRow and McNaught (1978) calculated that the zooplankters in Lake George, New York, excreted about 20% of the phosphate required by the phytoplankton. Lehman (1980) showed that the herbivore *Daphnia pulex* excreted soluble P equivalent to 1.5% to 2.5% of its total phosphorus each hour, and this was taken up rapidly by algae. From 35% to 60% of the total P in this herbivore was turned over every day as egested and excreted matter. There are difficulties in distinguishing zooplankton-P from phytoplankton-P and bacterial-P in samples from the epilimnion, but it is accepted that phosphorus exists in different compartments and interdependent pools.

Although new data continuously accumulate on the complexities of the phosphorus cycle in lake and stream waters, some generalities can be made. First, the speed with which phosphorus is whisked back and forth between the abiotic and biotic worlds emphasizes the usefulness of the ecosystem concept and leads to the conclusion that the best figure to use in quantifying phosphorus in a body of water is *total* phosphorus, including sestonic as well as dissolved values. Second, there is a mechanism whereby the normal accumulation of phosphorus in a lacustrine system does not lead rapidly to a condition favoring the production of a soupy mass of cyanobacteria. This is because of rapid sedimentation, followed by little return of phosphorus in oligotrophic lakes and only more or less seasonal releases in eutrophic lakes. According to Megard (1972), the total phosphorus at any one time in the waters of Lake Minnetonka is about equal to the annual increment from outside sources.

Jones (1972) examined 16 lakes in the English Lake District in search of indices that best correlated with high summer standing crops and degree of eutrophication. As others have found, total phosphorus stood at the top of the list. Interestingly, the **alkaline phosphatase** activity of epilimnetic samples was also significant. (The enzyme is an excellent indicator of biomass in the English lakes.) Berman (1970) reported on the annual fluctuations of this enzyme in Israel's Lake Kinneret and showed its importance to the algae and dinoflagellates of that lake. Enzymatic hydrolysis, whether occurring freely in the water or at bacterial and algal cell surfaces, affords green phototrophs access to the total phosphorus of the ecosystem by making some of it available as soluble inorganic phosphate, the useful state of the element.

Eutrophication and Phosphorus

Unappealing eutrophication resulting from sewage and other pollutants associated with human activity has stimulated research on ways to keep phosphorus out of lakes. One approach is to reduce or eliminate the use of high-phosphate detergents. Another scheme is to modify sewage before it can enter a waterway. This method has been successful in Switzerland, where ferric chloride precipitates about 90% of the phosphate from the sewage as ferric phosphate (Thomas, 1969). (A similar phenomenon occurs in a eutrophic lake at the overturn.) Aluminum salts are also used to precipitate phosphorus from sewage effluents. A third method is to be certain that domestic wastes do not reach the lakes; the successful bypassing of Lake Washington (Edmondson, 1991) and some European lakes (Thomas, 1973) has been well documented.

Case Study: Lake Erie Eutrophication

For a century, the United States and Canada have both noted with concern the profound changes in Lake Erie. Driven by human population growth and nearby deforestation, followed later by industrial and municipal waste discharges, this Great Lake was famously declared "dead" in the late 1960s (Burlakova et al., 2014). The subsequent implementation of improved wastewater treatment and a ban on phosphate detergents turned the fortunes of the lake, and then invasion by exotic Dreissenid (Zebra and Quagga) mussels further reduced phosphorus levels, markedly increased water transparency, and even allowed resumption of game fishery. It appeared that the lake was recovered—a real success story.

Unfortunately, more recently we have noted otherwise. A study by 28 authors assessed the role of N, P, and climate on the "re-eutrophication" of Lake Erie (Scavia et al., 2014). To prevent blooms of toxic cyanobacteria such as *Microcystis*, or hypolimnetic hypoxia (dissolved oxygen < 2 mg/L), and to restore biotic communities will require a massive effort: reducing total P by 46% and reactive P by 78% from recent average levels. We know which watersheds are the important sources and that inputs are from non-point sources, in particular runoff from agricultural operations. Controlling these inputs would likely also result in lowered nitrogen loading, and that element may be limiting to productivity under certain circumstances (Chaffin et al., 2014).

Nutrient Loading

With widespread acceptance of the opinion that lake eutrophication is brought about by excessive amounts of phosphorus and nitrogen, the concept of *nutrient loading* arose. This idea holds that there is a relationship between a lake's trophic state and the nutrients entering the lake and that it can be quantified. Several authors have attempted various approaches to predict what load of nutrient would become critical. The critical level or load (L_c), if exceeded, would lead from oligotrophy to eutrophy.

Until fairly recently most authors have concentrated on phosphorus, assuming it to be the limiting factor in most aquatic ecosystems. More than a half-century ago the importance of relative proportions of nutrients to phytoplankton communities in the English Lake District was suggested by Pearsall (1932), and this idea of ratios has been stressed since (Kilham and Kilham, 1982; Tilman, 1982). Phytoplankton populations are structured, in part, by the relations among important resources, and many sequential changes in algal or cyanobacterial species can be attributed to specific and differing nutrient ratios needed for optimum growth. The subject of nutrient proportions (stoichiometry) will be addressed in the subsequent discussions of nitrogen and silica.

There has been an evolution of methods by which to estimate critical loading values. In the simplest model, mean depth was the sole factor to consider in relation to the annual P loading. Subsequently, shoreline, volume, area, water-retention time, and flushing rate were taken into account. Rigler (1975) and Vollenweider (1976), both of whom had contributed much to the development of nutrient-loading concepts, reviewed the subject. Vollenweider presented a refined equation, which seems widely applicable, to define the annual critical loading value in milligrams of phosphorus per square meter:

$$L_c = (10 \text{ to } 20) \times q_s \left(1 + \sqrt{\overline{z}/q_s}\right)$$

In the equation, q_s is the so-called hydraulic load, calculated by dividing the total discharge of water into the lake by lake area and, therefore, is expressed in terms of meters per year. The hydraulic load also equals the quotient of mean depth and the theoretical water-filling time in years. The water-filling time is, of course, the ratio of lake volume to the total annual discharge; its reciprocal is the flushing rate. Vollenweider's (1976) discussion addressed phosphorus sedimentation and the residence time of incoming phosphorus in relation to the hydraulic residence time.

Lake dimensions and the annual water budgets are important factors in eutrophication and are part of Vollenweider's improved equation. Big lakes are less susceptible to pollution than small bodies of water because of dilution's effect, but other factors modify this generality. A study of the phosphorus supply of some oligotrophic Nova Scotian lakes led to the conclusion that lakes with shorter water-renewal times were less vulnerable to pollution than lakes with lower flushing rates (Kerekes, 1975). Moreover, the unstratified lakes were less susceptible to the buildup of nutrient loads than were the deeper waters of stratified lakes during stagnant periods. These two points are emphasized strongly by further examples from Canada and from Italy. The small Canadian Cameron Lake receives annual P loadings (1.7 to 2.2 g/m^2) that, according to some estimates, surpass by ten times critical thresholds for much larger lakes. The average total P in the lake, however, is only 0.01 mg/liter. Very high flushing rates—the annual outflow averages more than 16 times the lake's volume—counteract the nutrient loading, and Cameron Lake does not show the classic reduced transparency and high chlorophyll content of eutrophy (Dillon, 1975).

A second example that contradicts the prevailing idea that large lakes are less susceptible than smaller ones comes from stratified lakes that rarely circulate. In oligomictic lakes the actual time of water renewal may be far different from that calculated as the ratio of lake volume to outflowing water. Lake Maggiore, for example, theoretically renews itself every 4 years (a flushing rate of 0.25 per year), but the short, irregular periods of overturn, perhaps once every 5 to 7 years, led Tonolli and associates (1975) to conclude that it would take 14½ years to renew 50% of the water and that some original water would remain after 90 years! The hydraulic residence time is far longer than the theoretical 4 years, derived from volume equals 37.7 km^3 and discharge equals 9.4 km^3 per year. The mean P loading (L_c) of the different tolerances proposed by Vollenweider (1976) is about 1 g/m^2 per year for Maggiore. Even if the hydraulic residence time were only 16 years, the L_c estimate, founded on the assumption that annual holomixy is the rule in Lake Maggiore, would be four times too high.

Another type of loading is shown clearly in polluted shallow lakes, when external sources of phosphorus (sewage for example) are reduced or eliminated (Ryding and Forsberg, 1976). This is an internal loading derived from the sediments, which hinders the recovery that a deeper lake might exhibit following the abatement of allochthonous nutrient input. In one shallow, nonstratified Danish lake the annual release of phosphorus from the sediments was slightly greater than the amount received from external sources. This **internal loading**—the release of phosphorus from the sediments—occurred despite dissolved oxygen tensions of at least 60% of saturation in the bottom water (Riemann, 1977). In two eutrophic lakes in Ohio, their maximum depths about 12 m, 65% to 100% of the phosphorus increase during summer stratification could be attributed to internal loading (Cooke et al., 1977). Carey and Rydin

(2011) compared P in eutrophic and oligotrophic lakes (94 in all), finding that enriched systems have much P at the surface of sediments, easily recirculated in the water column, while oligotrophic lakes effectively bury P. So, a "tipping point" exists where mesotrophic systems are most vulnerable to P additions which will cause a long-term change in the nutrients throughout the system. This, however, is only ascertained through careful study of P deep in the sediments.

Wetlands, like lakes, can be sensitive to phosphorus enrichment. For example, the Everglades of south Florida has been seriously impacted by P additions, in some cases inputs three times or more background levels. Since the system is naturally oligotrophic, the soil, water, and biota all are effected, with Sawgrass (*Cladium*) replacement by Cattail (*Typha*) and loss of periphyton mats, which support productive consumer communities of fish, shrimp, and numerous invertebrates (Noe et al., 2001).

Nitrogen

The versatile element nitrogen is found in four recognized spheres of the Earth. The data presented by Sweeney and others (1978) imply more than 6.12×10^{16} metric tons as the total, with 93.75% tied up in the lithosphere and not taking part in geochemical cycles. About 6.2% is atmospheric and the tiny remainder belongs to the hydrosphere and biosphere. Perhaps only 9×10^{11} tons are incorporated in living biomass. Despite the small portion in the biosphere, most nitrogen cycling is brought about by organisms.

The free, inert nitrogen so abundant in the atmosphere and pressing down on us at about 755 g/cm^2 is denied to higher plants and animals without the aid of other organisms. Only a few prokaryotes can convert the strongly bonded molecules of gaseous nitrogen to useful compounds.

Nitrogen, although absolutely necessary for life as we know it, does not command the attention that phosphorus does, because nitrogen is more abundant and has more sources for living organisms. The tremendous atmospheric stores contribute in more than one manner. Rainwater contains various forms of nitrogen, some of which the rain is returning to the lithosphere. Nitrate and perhaps other oxides of nitrogen in the rain may have been formed photochemically or to a lesser extent by electrical discharge. Perhaps the oxidation of ammonia, common in rainwater, is also a factor in the origin of the nitrate.

Nitrogen Fixation

Molecular nitrogen dissolves readily (see Table 12-2) and enters the hydrosphere or the soil where a few organisms can convert it to useful compounds. During industrial techniques or laboratory procedures, the covalent triple bonds of the N_2 molecule ($N \equiv N$) can be broken only at high pressures and temperatures. A few bacteria, including some cyanobacteria, have the remarkable ability to break this bond at ordinary temperatures and pressures, a biologic process that requires more energy than any other. In a review article on the subject of nitrogen reduction, Brill (1979) wrote that it requires as many as 36 ATP molecules to fix a molecule of N_2.

Stewart (1977) summarized our knowledge of nitrogen-fixing organisms, listing 28 bacterial genera and 21 genera of the cyanobacteria. In addition, the paper includes the known symbiotic associations where one member is a prokaryotic organism capable of breaking the bonds of N_2. There are two main groups of organisms that crack the molecular bonds of nitrogen, bringing about the reduction of dinitrogen. They are the symbiotic fixers and the nonsymbiotic type. The symbiotic fixers, or reducers, reduce N_2 in association with eukaryotic plants and have long been known. They are made up of a half dozen species of *Rhizobium* living in tumorlike root nodules of leguminous plants. The legumes use the ammonia fixed by *Rhizobium* to synthesize amino acids and proteins. The agricultural technique of planting alfalfa and clover to improve soil is based on this bacterial-seed plant symbiosis.

The rather high nitrogen content of desert waters has been attributed to the abundance of wild legumes in arid lands. Many species of cat-claw (*Acacia*) characterize the Australian deserts and have been suspected of being responsible. In the American Southwest, *Acacia*, mesquite (*Prosopis*), and ironwood (*Olneya*) have been assumed to be the source of edaphic nitrates.

Many nonlegumes have root nodules harboring colonies of nitrogen-fixing microorganisms. At least 15 species of alders (*Alnus*) are known to have such root nodules, and many grow in moist soil near water. Goldman (1961) showed that alder leaves falling in oligotrophic Castle Lake, California, accounted for one third of the lake's nitrogen supply.

The nonsymbiotic N_2 reducers include, first, free-living bacteria, such as the aerobic *Azotobacter* and members of about a dozen genera of anaerobic bacteria, of which *Clostridium pasteurianum* may be the best known. These bacteria work in the dark, utilizing organic matter as an energy source for reduction. The second group of nonsymbiotic nitrogen fixers are photosynthetic organisms relying ultimately on light for energy. One of these is the purple non-sulfur bacterium, *Rhodospirillum*. Because it is an obligate anaerobe, its distribution is limited. Often it occurs in anoxic zones beneath algal plates, its bacteriochlorophyll utilizing wavelengths (800 to 890 nm) not absorbed by algal pigments. Simple alcohols and organic acids serve as electron donors for *Rhodospirillum*.

Aerobic photosynthetic nitrogen reducers are represented by roughly 60 species of blue-greens, especially those of the family Nostocaceae. They are widely distributed when compared with *Rhodospirillum;* they contain chlorophyll *a* and rely on water as an electron donor. All photosynthetic fixers, anaerobic and aerobic, convert radiant energy to ATP to power the reduction process.

Historically, Drewes (1928) demonstrated that pure cultures of cyanobacteria fixed nitrogen. One of the first reports of blue-green nitrogen fixers in nature concerned *Anabaena* in rice paddies (De, 1939). Soon thereafter a soil-dwelling *Nostoc* capable of reducing gaseous N_2 was discovered.

A well-known symbiotic relationship between a plant and a cyanobacterium takes place in an aquatic setting. *Azolla is* a small floating fern that sometimes blankets small ponds. Its fronds are porous, and within the pores lives a species of *Anabaena,* a member of the Nostocaceae and capable of fixing nitrogen. *Azolla* thrives in nitrogen-poor waters thanks to its prokaryote partner. The use of *Azolla* in Asian rice culture is centuries old. *Azolla-Anabaena* populations are allowed to cover the flooded rice paddy before the rice is planted. The subsequent death of *Azolla,* brought about by shading from the growing rice plants, ensures the latter a rich nitrogen supply.

Some complexities of the symbiotic relations are revealed by studies on the macroscopic, siphonous marine alga *Codium*. In 1975 Head and Carpenter found that the aerobic bacterium *Azotobacter* on the surface of the alga fixed N_2. The fixation proceeded best in the light and with dissolved material supplied by the green alga. In 1981 Rosenberg and Paerl found representatives of three cyanobacterial genera, *Anabaena*, *Calothrix*, and *Phormidium*, living in a chemically reduced microzone within the algal thallus. Only the first two genera possess heterocysts yet the cyanobacteria fixed 1.2 mg N_2/kg of dry mass of *Codium* each hour. The fixation reported earlier for the *Azotobacter* was six times this rate (Head and Carpenter, 1975).

Some of the dinitrogen-reducing, soil-dwelling cyanobacteria have remarkable colonizing abilities. Species of *Anabaena* and *Nostoc* became established on the cooled lava within 18 months after it had been extruded in Iceland and were fixing nitrogen at a remarkable rate (Englund, 1976). Cyanobacteria are abundant in desert soils (Cameron, 1963), and they may be implicated in the high nitrogen levels in desert waters that have been attributed to the legumes. *Nostoc* occurs symbiotically within the desert lichens *Peltigera* and *Dermatocarpon*, and it has been demonstrated that these lichens fix nitrogen (Snyder and Wullstein, 1973). Those authors also found *Azotobacter* so closely associated with the soil in which desert mosses grow that they assigned the bryophyte *Grimmia* a role in nitrogen fixation. There are more reasons for the abundance of nitrates in desert soils than can be attributed to legumes alone—the sources are varied. Certainly there are unusual opportunities for concentration of edaphic nitrogen compounds in climates where rainfall is rare. The high nitrate levels reported from flash-flooding desert waters (Fisher and Minckley, 1978) probably are the result of fixation by a host of organisms, not just *Rhizobium* species in leguminous root nodules. Despite this, Grimm and others (1982) found that nitrogen is unusually low in desert streams and may be a limiting nutrient factor. In most instances the ratio of N to P was less than 6 and it was always less than 15. Nitrate was highest at the sources of the desert streams, but it was taken up rapidly downstream. Nitrate's uptake was much greater than that of phosphate, and it was positively correlated with primary productivity and the abundance of periphyton.

The ability to fix nitrogen, limited to a few prokaryotes, is based on the unique possession of a complex enzyme, **nitrogenase**. Functional nitrogenase consists of two proteins, molybdoferredoxin and azoferredoxin. The former contains Mo and Fe; the latter contains only iron. This explains why legumes do poorly in soils deficient in molybdenum; their symbiotic prokaryotes are unable to synthesize one component of nitrogenase. With an electron donor lower on the redox scale, an energy source, and in the presence of magnesium and the absence of oxygen, the prokaryotes that possess nitrogenase can reduce N_2 to an ammonium ion that can be aminated later.

Early techniques used to study cyanobacterial fixation of N_2 were based on (1) culturing the prokaryotes in media deprived of combined nitrogen and determining gains in total N within the system, gains that could be attributed only to fixation of gaseous N, and (2) monitoring cellular uptake of the heavy isotope, ^{15}N, in cultures where no other nitrogen source was possible. Today, however, nitrogenase activity is usually measured by testing it with an electron acceptor. Nitrogenase is a versatile enzyme that reduces a variety of substrates that are analogues of N_2, having triple bonds between N or C atoms. For example, it reduces acetylene ($HC \equiv CH$) to ethylene. Organisms that lack nitrogenase cannot bring about the reduction of acetylene.

In the brine pond called Solar Lake (Sinai), remarkable for its high subsurface temperatures and the presence of a sulfide-dependent anoxic photosynthetic cyano-bacterium (*Oscillatoria limnetica*), there is another unusual feature: high levels of back-ground ethylene which interfere with acetylene reduction assays of nitrogen fixation. The source of the C_2H_4 is not known; it may be released from bottom sediments with some sort of biotic origin (Potts, 1979). The possibility of microbial origin seems rea-sonable because some marine algae produce ethylene enzymatically.

The ^{15}N isotope still is useful for other nitrogen studies, however. Using nitrate labeled with ^{15}N, one can test for the rate of nitrate uptake by algae. In laboratory cul-tures, the Michaelis-Menten kinetic expression is useful in enzyme-substrate studies and describes the results well. In this expression, V = velocity of uptake of nitrate-nitrogen per unit of particulate N in the algae; V_{max} = the maximum velocity of uptake; S = concentration of substrate, the nitrate nutrient in this case; and K_m, the Michaelis-Menten constant, is the value of S when $V = \frac{1}{2} V_{max}$.

$$V = \frac{V_{max}S}{K_m + S}$$

MacIsaac and Dugdale's (1969) experiments with natural populations of phyto-plankters have shown that K_m is rather high in eutrophic regions and lower in oligotro-phic areas. Because K_m is a figure that relates inversely to a phytoplankter's capability of taking up a nutrient, this suggests that there may be adaptations for taking nitrate up rapidly where it is scarce. This applies to other nutrients as well: where phosphate is abundant, it is taken up at a slower rate than in PO_4-poor environments.

The site of nitrogen fixation (reduction of gaseous N_2) is the **heterocyst** in the cyanobacteria; with few exceptions only species with heterocysts have the ability to fix nitrogen. All genera of the Nostocaceae have species possessing heterocysts and the family rates high as a reducer of inert N_2. *Aphanizomenon*, having a lower ratio of het-erocysts to vegetative cells than *Anabaena*, fixes less N_2 than does *Anabaena* (Horne and Fogg, 1970; Horne, 1979). Horne and others (1979) found the major factor in nitrogen fixation during a springtime bloom of *Aphanizomenon* was the number of het-erocysts. Also, as though there were some sort of negative feedback, they found het-erocyst formation was suppressed by low levels of ambient nitrate and by higher levels of ammonia. Representatives of some other families, such as *Gloeotrichia* of the Rivu-lariaceae, possess heterocysts and also fix molecular nitrogen.

Van Gorkom and Donze (1971) found that in aerobic situations nitrogen is fixed almost exclusively by the heterocysts, but vegetative cells can do this work anaerobi-cally. The absence of oxygen is probably the key. The aerobic forms such as *Azotobacter* and the blue-green "algae" sequester their nitrogenase in vesicles and heterocysts, respectively, creating minute anoxic environments for the enzyme. The obligate anaer-obe *Clostridium* does not partition its nitrogenase in any protective enclosures, and *Rhizobium* lives in anaerobic root nodules. Many *Rhizobium* cells live within an anaer-obic root nodule characterized by a bright red lining; the red color comes from a pro-tein, leghemoglobin, which binds stray O_2 and prevents it from reaching and inactivating the bacterial nitrogenase. Some other free-living, facultative anaerobic bacteria are known to reduce N_2, but do so only in anoxic environments. In those cyanobacteria that seem to fix nitrogen in vegetative cells, the phenomenon is much reduced and ambient O_2 is very low.

Years ago, Carpenter (1973) reported nitrogenase activity in a nonheterocystous cyanobacterium of the sea; this was a species of *Trichodesmium* (placed in *Oscillatoria* by some authors, including Carpenter). The filaments or trichomes of *Trichodesmium* occur in bundles (like those of *Aphanizomenon,* Fig. 2-1, *F*). When the trichomes are numerous and arranged in parallel alignment, the center of each bundle is anoxic, and acetylene reduction proceeds at a high rate (Bryceson and Fay, 1981). Agitation and water turbulence expose the inner region to oxygen; nitrogen fixation, therefore, as inferred from acetylene reduction, is impaired. In Clear Lake, California, there is early fixation of N_2 (before heterocysts have appeared) in flakelike colonies of *Aphanizomenon.* This may be possible because of an anaerobic central region in the flakes of trichomes (Horne et al., 1979). Nevertheless, freshwater *Oscillatoria* lack heterocysts, and their ability to reduce dinitrogen has not been substantiated. Interestingly, one of these, *O. rubescens,* is said to foretell the onset of cultural eutrophication in lakes; it is N-limited and does not appear until the nitrogen content of the water reaches a critical level.

Assimilation

There are three principal inorganic states and sources of nitrogen that plants can take up. These are nitrate-nitrogen (NO_3–N), ammonium-nitrogen (NH_4–N), and in those prokaryote species with nitrogenase, N_2–N. In a study of planktonic cyanobacteria, Ward and Wetzel (1980) found that the radiant energy necessary for *Aphanizomenon* to assimilate N_2–N was greater than that required for NO_3–N, which in turn was greater than that required to assimilate NH_4–N. As a result, the lowest light intensity at which growth would occur was determined by the state of the nitrogen source. The highest growth rates at all experimental light intensities occurred with ammonium nitrogen as a source; by contrast, *Aphanizomenon* would not grow at all when N_2–N was the only source of nitrogen and the light intensity was 150 to 200 lux (about 0.24 to 0.32 W/m^2).

Once ammonia (NH_3) has been formed by the reduction of N_2 it can be converted to nitrite or nitrate. Nitrate and ammonia can be taken up by other plants, but after assimilation, enzymes reduce nitrate to NH_3, which can then be aminated. The ammonia is built into aspartic and glutamic acids, the amino acids that serve as forerunners of all other organic nitrogenous compounds (reviewed by Rosswall, 1981).

Ammonification

Aquatic animals, in contrast to most terrestrial forms, commonly excrete ammonia as a waste product of metabolism. Jacobsen and Comita (1976) reported experimental results that imply 0.2 mg of nitrogen per day would be released by 1,000 individuals of *Daphnia pulex* or, on a dry-weight basis, 1 g of *Daphnia* excretes 5.1 mg of N each day. In some nitrogen-poor environments the excretory contribution from the zooplankton provides up to 90% of the nitrogen requirements of the primary producers (Jawed, 1973). Lehman (1980) underscores the difficulty in measuring the nutrients released by zooplankters: there is a simultaneous uptake by the phytoplankton. The rapid recycling of nutrients (such as NH_3 and dissolved phosphorus compounds) between grazers and algae makes possible high productivity and rapid cell division even when outside inputs and dissolved pools of the nutrients would hardly suffice for a day's support of the autotrophs. Furthermore, the rates of uptake are faster in nutrient-deficient cells. It is difficult to say precisely whether the main source of ammonia dissolved in epilimnion water comes from bacterial mineralization of dead plants and animals or whether it is excreted by living animals. In the lake's tropholytic hypolimnion, bacterial action and autolysis of dead material is the major source of NH_3 as the bacteria of decay assimilate some of the dead organic material to build their cells but convert the rest to by-products. The steps in ammonification are roughly the opposite of assimilation, amination, and protein synthesis, and both anaerobic and aerobic bacteria function in the process.

Nitrification

Some bacteria gain energy from oxidizing ammonia: both the newly derived gas from dinitrogen reduction and that produced by ammonification. This phenomenon is called **nitrification**, and the two kinds of organisms playing parts in the process are chemolithotrophic bacteria in the genus *Nitrosomonas* and a somewhat mixotrophic group belonging to *Nitrobacter*. They gain their energy from oxidizing ammonia to nitrite and nitrite to nitrate, respectively. Because only aerobic organisms participate in nitrification, ammonia accumulates in the anoxic hypolimnion but is scarce in the well-oxygenated epilimnion.

Denitrification

Working in the biogeochemical cycling of nitrogen is a group of mostly facultative anaerobic microbes that restore the gases N_2 and N_2O to the atmosphere. The process is called **denitrification**.

Before industrialization, the amount of nitrogen fixed by bacteria and cyanobacteria was probably balanced by natural denitrification. Thus, nitrogen was returned to the air in an inert state that could be converted to a useful condition only by the work of another set of prokaryotes. Our learning how to fix N_2 industrially (at $500°$ C and several hundred atmospheres) has changed the picture. The annual industrial production approximates 50 million tons, and the denitrifying bacteria are hard pressed to keep pace with this extra substrate. Because denitrification proceeds best in anaerobic environments, the preservation of swamps, marshlands, and tundra has even more to recommend it than the aesthetic worth of these habitats.

The process of denitrification can result in a significant loss in the combined nitrogen available to an ecosystem unless it is matched by fixation. Thus, denitrification, along with sedimentation, is considered one of the major sinks for nitrogen in an aquatic environment.

In Japanese Lake Tizaki there are 44 "strains" of denitrifying bacteria during the course of the year (Terai, 1979). During the early summer, denitrification occurs only in the hypolimnion, but by the end of the summer it is going on everywhere except at the very surface. Both anoxic and aerobic zones have their denitrifying bacteria.

Nitrogen and Eutrophication

Ryther and Dunstan (1971) came to the conclusion that nitrogen is the critical limiting factor in coastal marine waters. Their calculations show that the phosphorus received via sewage and terrestrial runoff in general is more than adequate for plankton growth. There are two reasons for the relative scarcity of nitrogen in coastal marine waters: (1) there is a low nitrogen/phosphorus (**N/P**) ratio in contributions from sewage and other land-derived wastes, and (2) phosphorus regenerates more rapidly than does ammonia from decomposing organic matter. There is a generalization to be derived here: when phosphorus is added to an aquatic system experimentally or by way of sewage or other terrestrial import, nitrogen often becomes a limiting factor even though it outranked phosphorus at first.

Schelske (1975) proposed a model to account in part for an increase of cyanobacteria following enrichment by phosphorus in the upper Laurentian Great Lakes. In the model, the addition of phosphorus accelerates algal activity and the uptake of the rather scarce nitrogen. This gives the advantage to organisms that can fix nitrogen and accounts for the shift in community composition from diatoms, in these lakes, to heterocystous cyanobacteria. Schelske's idea was strengthened by Schindler's (1977) experiments in an oligotrophic ELA lake in Ontario. Following the addition of fertilizer containing both N and P, there was a general increase in photosynthesis and the standing stock of chlorophyll a but no change in species composition. When fertilizers containing P but lacking N were added, the results were different. For the first time in years of observation blue-greens appeared abundantly, and they were nitrogen-fixing species. From this Schindler theorized that programs to reduce nitrogen loading might have adverse effects on water quality. In ELA lakes lowering the N/P atomic ratio to about 22 favors the nitrogen-fixing cyanobacteria, some of which are vacuolate and form objectionable nuisance blooms.

Kilham and Kilham (1982) and Tilman (1982) summarized their own work and that of others to show many examples of algal populations varying with changes in resource ratios; the different species within a larger taxonomic group are neither identical in their requirements and responses to nutrient ratios nor equal in ability to compete for nutrients. As the level of one nutrient rises or falls, its ratio to all other nutrients, of course, changes. A species that had been limited by a nutrient that is added will be limited by a different nutrient and will be forced to compete for it. N/P ratios are lowered when P is added and species that once were P limited become limited by N. Green algae, some diatoms, and cyanobacteria incapable of reducing N_2 may find high N/P ratios favorable. The addition of P changes the picture.

We now understand that humans are profoundly changing nitrogen biogeochemistry on a global scale, and once again limnologists provide evidence to help us understand the ecology. Holtgrieve and others (2011) found a clear anthropogenic signature of nitrogen deposits in 25 remote lakes, with an isotopically distinct signal starting around 1895, the beginning of the commercial process of N-fixation. We are coupling changing nitrogen cycling with changing climate to alter ecosystem functioning.

Silica

Silicon is the second most abundant element in the lithosphere. Its main source in fresh and sea waters is weathering of the extremely abundant feldspar rocks. In inland waters it ranges from 0.1 to 4,000 ppm, representing the extremes from snowmelt to hot mineral springs. In rivers and lakes silicon commonly ranges from 2 to 25 ppm and is usually expressed as silica (SiO_2) in water analyses.

Silica is an essential nutrient for diatoms, which build their frustules of this glassy material. It may also be essential for the growth of planktonic chrysophyceans such as *Dinobryon*, *Uroglena*, and *Mallomonas*, which construct silicified scales and spores. Among freshwater animals, the main users are the sponges, building their glassy spicules with silica. In the sea silica is scarce, averaging about 5 mg/liter; this has been attributed to the uptake by radiolarians, silicoflagellates, siliceous sponges, and diatoms.

The chemical test that is commonly employed for silica works only when this compound is in a soluble state; the particulate frustules of diatoms in a sample are not reactive. Because solubility of silica increases directly with temperature, high concentrations are found in geyser waters and hot springs. Concentrations of silica are reported in parts per million or milligrams per liter rather than in milliequivalents per liter because it is usually not ionized. Dissolved silica tends to complex with H_2O, perhaps as $SiO_2 \cdot 2H_2O$ or H_4SiO_4. The latter only dissociates above pH 9.0; thus, silicate ions become important in alkalinity titrations only at low hydrogen ion concentrations.

The relationships between silica and the biotic segment of the aquatic ecosystem have been clearly demonstrated. In an early study Birge and Juday (1911) described a Wisconsin lake with a metalimnetic stratum of diatoms that accounted for a positive heterograde oxygen curve and a striking silica minimum at the same level. Lund (1964) showed that the course of events through the years reveals an inverse ratio for the diatom crop and the soluble silica in the water (Fig. 15-2). Some diatom blooms have been shown to deplete silica concentrations to deficiency levels (Munawar and Munawar, 1975).

The crash of the diatom populations at the end of summer is accompanied by a slow rise in reactive soluble silica in the water samples. The decay is slower for silica than for organic compounds, and many frustules may be buried and lost. Because of this, the paleolimnologist can read the record of the past from the siliceous remains of diatoms in vertical cores of lake sediments. Although many sedimentary diatom frustules may be lost, there is an increase in hypolimnetic silica during stagnation. Presumably, much of this could be distributed to the epilimnion at the overturn. Some, however, may be precipitated with iron.

These sequential events have been quantified for Lake Ontario (Nriagu, 1978). About 30% of the silica that reaches the sediments via diatom frustules is retained there permanently. The remainder is regenerated slowly from the sediments and incorporated in some subsequent diatom bloom.

Kilham (1971) hypothesized that diatom sequences were governed by ambient silica. From a literature search he found that different planktonic species were not numerically dominant until silica concentrations were great enough to match their

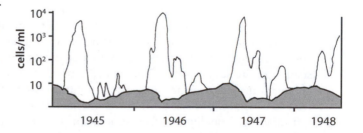

Figure 15-2 The relationship between soluble silica (dark area) and the waxing and waning of diatom populations (cell numbers shown by thin line). The greatest SiO_2 concentration shown here is about 2.5 mg/liter. Adapted from Lund (1964).

demands. For example, a variety of *Melosira granulata* did not bloom until SiO$_2$ was 13.4 mg/liter; *Stephanodiscus*, on the other hand, was predominant when SiO$_2$ levels were 0.05 of this. Subsequent studies (Kilham and Kilham, 1982; Tilman, 1982), demonstrated that the important datum to consider is the **Si/P** atomic ratio, not simply silica or the silicon atom alone. This idea was offered as a possible explanation for the appearance of cyanobacteria at the expense of diatom plankters in Lake Michigan (Schelske and Stoermer, 1972) and in the upper Great Lakes (Schelske, 1975). The addition of phosphorus stimulates growth of diatoms, which utilize more silica as their populations increase. After their growing season, the diatoms sink, and much of their siliceous material reaches the sediments before decaying. Some is buried by subsequent deposition of other material and is lost from the open-water silica cycling. Non-siliceous algae then compete successfully with the diatoms, eventually replacing them. The authors pointed out that over the course of 40 years silica had diminished steadily in the waters of Lake Michigan, and recently the dominance of the diatoms has lessened as cyanobacteria have become more abundant. Lake Superior, the most oligotrophic, has the highest concentration of silica, although it amounts to no more than 2.0 to 2.2 mg/liter. Silica depletion reflects eutrophication in these lakes because their normal phytoplankton assemblages are mostly diatoms. There is far less silica regeneration from the settling frustules in the deeps than there would be in a smaller lake with an anoxic summer hypolimnion; thus the enrichening of the trophogenic layers with SiO$_2$ in autumn is relatively less. To maintain the ability of diatoms to compete successfully with cyanobacteria, silica would have to be supplied in proportion to increased anthropogenic phosphorus. Assuming a silica/phosphorus ratio of 100 in diatoms (not unreasonable, according to Schelske), 1 g of SiO$_2$ per cubic meter would be taken up for an increase of 10 mg P per cubic meter.

Inorganic Micronutrients

Groups of elements that are needed in trace quantities are called micronutrients. They stand in contrast to other nutrient elements, most of which are abundant in nature and are required in greater quantities, according to the review by O'Kelley (1974). The abundant nutrient elements fall under the heading of macronutrients required by algae; they include calcium, magnesium, potassium, and sulfur—nitrogen and phosphorus not being considered here. The micronutrients can be divided into two groups: those apparently needed in small amounts by all algae and those that are required by members of a few algal groups. Of those needed by all algae, chloride, iron, and manganese have been mentioned already; the others are copper, molybdenum, and zinc. Inorganic micronutrients required by some (but not all) algae include boron, cobalt, iodine, silicon, sodium, and vanadium.

Copper and zinc probably behave much like iron and manganese; these are heavy metals (>5 g/cm^3). They are insoluble in oxidized states and in intensely reducing environments where they form sulfides. Other elements, such as vanadium, chromium, and selenium, are soluble when oxidized (the opposite is true of iron). Cobalt and nickel behave somewhat similarly, but their oxidation-solubility pattern may resemble that of iron to a degree. The list of minor elements in lake waters could include boron, cadmium, iodine, rubidium, and poisonous forms, such as lead, mercury, and arsenic.

Both essential and nonessential heavy metals become toxic at high concentrations. Of the latter group, mercury is the most poisonous, although cadmium, zinc, and lead must be considered also. The deleterious effects that high levels of metals have are based on their poisoning algal enzyme systems. Thus they may interfere with chlorophyll and protein synthesis as well as with photosynthesis and respiration. Factors contributing to reducing toxicity of a given concentration of metal are increasing magnesium-calcium hardness, rising pH, high phosphorus content of the water, and complexing agents that join the metals and render them less available. Of the last category, the best known chelators are EDTA, NTA, fulvic acids, humic acids, and amino acids. Metals simply adsorbed on negatively charged particles, however, enter the watery medium easily and become available as nutrients, or as toxins if in excess, after being desorbed.

The heavy metal *molybdenum* (Mo) occurs in seven stable and five radioactive isotopes, and its importance to the growth of higher plants has been known for decades. In 1960 Goldman demonstrated that its deficiency limits primary production in Castle Lake, California, where addition of this element stimulates algal growth. Apparently, the beneficial effect of molybdenum is based on its relation to the enzyme systems that reduce N_2 (the nitrogen-fixing enzymes) and to those that reduce nitrate. It is directly incorporated into nitrogenase, and specifically to the molybdoferredoxin component.

Molybdenum's lacustrine cycle is similar to the iron cycle because it decreases at the same time $Fe(OH)_3$ and FeS are removed by precipitation. As might be suspected, it is commoner in deep water and is concentrated in plankton cells and sediments. Dumont's (1972) report of the annual fluctuations of molybdenum in a eutrophic pond showed seasonal changes recalling the iron cycle. It was scarce in winter water samples when the sediments were oxidized; coinciding with the release of phosphorus, it appeared in abundance early in the summer; and it decreased during later summer when H_2S was obvious in the lower waters.

Copper is one of the few trace minerals that has been studied limnologically. Riley (1939) followed the annual copper cycle in some Connecticut lakes and found its abundance to vary more than twentyfold during the course of a year.

Copper in the form of $CuSO_4 \cdot 5H_2O$ has been used as an algaecide for many years in aquatic ecosystems. It has a special effect on diatoms when at toxic levels. In addition to inhibiting growth, it interferes with the uptake and proper utilization of silicic acid. The result is the occurrence of aberrant frustules (Rueter et al., 1981). In lakes where copper sulfate has been used to control algae, it accumulates as copper carbonate in the sediments, where it may be detrimental to the bottom fauna. Furthermore, as is true for the other metals, it may reenter the water column from the bottom deposits where it is stored—an internal loading effect that can be harmful or beneficial depending on concentration.

Copper carbonate has been used for decades for killing the snail hosts of the flukes that cause a dermatitis called "swimmers' itch." The cercarial larvae leave the snails and burrow into the skin of a wrong host—the human (McMullen, 1941). Recently, organic algicides and molluscicides have largely replaced copper.

Cobalt is a trace element essential to life; when lacking, it leads to the inactivation of certain enzymes. This is because it, as well as some other metal ions (copper, zinc, iron, and magnesium), serve as cofactors or coenzymes; they function in binding substrate molecules to the enzyme-active site. Cobalt is found in the unique vitamin B_{12}, properly termed cobalamin. No other known vitamin contains a metal ion. Most cyanobacteria need cobalamin and are able to synthesize it; this accounts for their dependence on an external source of cobalt.

Zinc is significant to living systems. A metal constituent of dehydrogenases, it is required in photosynthesis as an agent in hydrogen transfer. Another metabolic role of zinc is in the synthesis of protein. It is probably never in short supply because the rain carries from 2.5 to 12 mg/m^3. It precipitates in the lake sediments as a sulfide and coprecipitates with $CaCO_3$ and $Fe(OH)_3$. The radioisotope ^{65}Zn has served in studies of the uptake of this metal by organisms.

Vanadium has assumed importance in medical geology because it inhibits cholesterol synthesis; the incidence of vascular disease is said to be low where vanadium is present in soils and water. It increases the rate of photosynthesis in the green alga *Scenedesmus*, and substituting for molybdenum, it can be used by *Azotobacter* as a catalyst for dinitrogen reduction.

Dissolved Organic Substances

Data from various authors agree with the early report by Birge and Juday (1934) that freshwater lakes contain from 0.1 to 50 mg dissolved organic material (DOM) per liter. More modern analytical techniques have made it possible to identify specific organic compounds in natural waters and lacustrine sediments. Furthermore, the materials released by different organisms can be identified taxonomically in general terms to at least broad categories: bacteria, yeast, and fungi; cyanobacteria and dinoflagellates; mosses, ferns, and lycopods; gymnosperms and higher aquatic plants.

There are probably five sources of these dissolved materials: organic compounds of allochthonous origin, soluble organics from the decay of aquatic organisms, photosynthates or extracellular metabolites excreted from phytoplankton, extracellular metabolites or photosynthates secreted by littoral macrophytes, and excretions from the lake fauna. Particulate organic material (POM) is separated from DOM on the basis of the filter's pore size; for this reason it is possible that much that has been classified as DOM in the past was actually colloidal material.

Organic compounds in aquatic systems can be placed in five or six functional categories. In addition to serving directly as sources of energy, they are associated with nutrient (and toxic) trace elements. Indeed, most metals are transported downstream or exist in lake waters complexed in some manner with organic material, being adsorbed or occurring as metallic coatings on particulate detritus. This explains why some water analysts have gone astray by filtering samples before testing for metals. Other organic substances clearly are biologically active: toxins, vitamins, and metabolic products with pheromone-like properties.

Humic Substances

Humic material playing a role in acidifying some waters were discussed in detail in Chapter 13. They cause the yellow stain in coastal marine waters as well as those in some inland lakes, and they are derived from the decay of plant material in the soil and from autochthonous plant detritus. The materials are a complex mixture. Analyzing them physically and chemically is also complicated, and their properties still are somewhat obscure.

Their ecologic significance in aquatic habitats has not been explored thoroughly, even though they represent a large portion of the organic matter in both fresh water and seawater. As might be expected, the highest values, up to 50 mg/liter, are found in dystrophic lakes. Published papers from a symposium held in the Netherlands during May 1972, involving 41 authors, are a good source of information (Povoledo and Golterman, 1975).

Humic substances are, in one sense, a vast pool of dead material in the soil, water, and aquatic sediments, both freshwater and marine. They are polymeric mixtures derived mostly from plant matter—lignins, cellulose, proteins, and fats. Microbial degradation of lignins, for example, synthesizes phenols, which are oxidized to quinones and subsequently undergo polymerization to form complex molecules of high molecular weight. They are remarkably resistant to decay. Because the –COOH radical is abundant in the mixture, they are commonly called humic acids.

Since they are dead organic residues, derived from all sorts of ecosystems, there are different types of **humus** material. Shapiro (1957) published results of intensive studies on these yellow organic acids. He called them *humolimnic acids* to distinguish them from the humic acids of soils.

Povoledo (1964) stated that the colored material in the nondialyzable fraction of lake sediment and lake water differed from soil humic acid and found that 14 amino acids associated with the humic acids of Lake Maggiore differed somewhat from those reported from soil humic substances.

The contrast between humus of soils and aquatic sediments is emphasized further by the term **kerogen**, applied to lacustrine and marine deposits (Swain, 1975). There are some fundamental structural differences, and in general the carbon content of kerogen is lower and the H/C and N/C ratios are higher than those of soil humus. Perhaps the differences should not be stressed, because there is much overlap. Although one technique separates humic substances into two main fractions, one with a molecular weight of less than 10,000 and the other with a molecular weight greater than 100,000, there are mixtures of the two types in both lake sediments and soil humus. In general, kerogens are said to contain a greater fraction of the heavier component, but an observation by Povoledo (1975) is significant—another example of the ties linking terrestrial and aquatic environments. He described the experience of watching the dark-brown humic flow collected by small streams from the groundwater of the scanty soils of the Canadian Precambrian Shield and discharged, with its color and organic content, into lakes.

The humic materials in lake waters hold in suspension remarkable quantities of metallic ions compared with their solubilities in pure water. Shapiro (1966) reported high iron concentrations in membrane-filtered humic waters, suggesting that humic materials peptize (chelate) the metal, thereby keeping it in solution at pH and redox levels where it would normally precipitate. In the absence of complex-formers the solubility of ferric iron in the pH range of 6.0 to 9.0 does not exceed about 10^{-8} mol/liter, roughly 0.56 μg/liter (Stumm and Morgan, 1970). The chelating action of humic substances is a kind of biologic conditioning that has influence on aquatic productivity, as reviewed by Prakash and coauthors (1975). The humic compounds of less than 700 molecular weight bind two to six times more metals than the heavier weight fractions, and it is significant that the solubility decreases with molecular weight, the lighter molecules remaining in solution and representing much of the stain in waters. Through their chelating activity, essential trace metals are made available and the solubility behavior of important anions is modified. The precipitation of phosphates reacting with iron, calcium, aluminum, and other metals may be delayed in the presence of humic compounds, thereby increasing the residence time of nutrients in the water column. The inverse of this is, of course, the diminished toxicity of some metals (Ag, Cd, Cu, Hg, Pb, and so on) when they are complexed. The toxicity decreases from those organic-metal compounds with low stability to those with high complex stability, probably related to the degree of difficulty microbes have in degrading them. For example, cadmium ranks with mercury as a very dangerous contaminant in aquatic systems; it is lethal to salmon in concentrations of only 5 µg/liter and to algae at about the same level. Its toxicity is lessened significantly in the presence of aquatic humus, probably a function of reduced biologic availability (Gjessing, 1981).

The more strictly physical effects of humic materials are related to light penetration and heating. The name **gilvin** (from the Latin *gilvus*, pale yellow) was proposed for the dissolved yellow stain in lake and coastal marine waters by Kirk (1976). He showed that light's vertical penetration is reduced sharply where humic materials are abundant: from 70% to 80% of the attenuation can be attributed to absorption by the yellow-brown substances. Similarly, in "black-water lakes" of the North American Southeast a large part of the daily heat income is stored in a small volume of surface water (M. C. Waldron and J. E. Schindler, personal communication, 1981). The rapid absorption of solar radiation by the dissolved and colloidal organic acids, and the reduced evaporation rates caused by the contact of humid air and lake surface account for high temperatures in such lakes, compared with nearby transparent bodies of water, or with ponds and lakes in very arid zones.

Vitamins

It was once believed that higher plants have no need for vitamins. In this modern, oxygen-laden world, however, the plants, like us, need antioxidants to survive. We turn to plants as a source of such important antioxidants as vitamin C, which protects us all. By definition, a vitamin is an organic compound that an organism requires but cannot synthesize; it must get the substance, perhaps only in small quantities, from the environment in some way. There have been anthropocentric overtones to discussions of vitamins in freshwater research. Workers concentrate on the compounds that are human vitamins. (If we were members of the species *Rattus norvegicus*, there would be no such thing as vitamin C!) Conversely, there are probably vitamins in the lives of aquatic organisms about which we know nothing or which we have not classified as such. The tissues of some higher aquatic plants are rich in vitamins. Perhaps they produce them; if so, we are correct in saying they have no need for them from environmental sources. Kurata (1981) showed that the epiphytic microorganisms associated with the littoral macrophyte *Elodea densa* abundantly produce and release vitamins. The epiphytic forms are yeasts, actinomycetes, fungi, and bacteria; the bacteria exceed the others in abundance and in production and release of the vitamins thiamine (B_1), biotin, and B_{12} (cobalamin). Kurata's experimental work showed that the epiphytic producers relied on organic substances released by *Elodea*. Possibly *Elodea* gained B vitamins in turn from the epiphytic synthesizers.

Only three vitamins are essential to the algae and they are common in lake waters: biotin, thiamine, and B_{12}. Biotin is required by chrysomonads, dinoflagellates, and at least one euglenoid. Thiamine is less often required by algae than is vitamin B_{12} (Provasoli and Carlucci, 1974).

The importance of vitamin B_{12} (cobalamin) as an algal growth factor was explained in part by Gleason and Wood (1977). In cyanobacteria, some anaerobic bacteria (e.g., *Clostridium, Rhizobium*), and the Euglenophyta, the vitamin serves as a coenzyme for a unique type of ribonucleotide reductase. The enzyme is crucial to cell division, supplying the precursors for DNA synthesis. Presumably most other organisms possess an iron-containing reductase that is not dependent on vitamin B_{12} and, hence, cobalt. Gleason and Wood suggested that the discovery of a specific inhibitor for the cyanobacterial reductase might serve to control them without affecting coexisting organisms that feature the iron-containing enzyme.

The large rotifer *Asplanchna brightwelli* makes clear-cut responses to a-tocopherol (vitamin E), which it receives secondhand from herbivorous prey that feed on the algal source. Vitamin E causes amictic females to produce mictic offspring, whose haploid eggs develop into males if unfertilized and into diploid, resistant eggs when fertilized. This kind of production signals the end of the population, for the fertilized eggs will lie dormant for some time. The advantages of such a scheme whereby the population disappears while primary production is high are obscure. Apparently the males require a-tocopherol for fertility, receiving the vitamin from their mothers. Perhaps the benefits from sexuality and genetic recombination outweigh other considerations, and the population, therefore, responds to vitamin E even though it means the end of actively feeding individuals when food resources are abundant (Gilbert, 1972).

In 1980, Gilbert summarized much of his work on the induction of three distinct female body forms in *Asplanchna sieboldi* by different levels of dietary tocopherol. The responses are complex, but in this species tocopherol can stimulate postnatal growth as well as affect embryogenesis. One result is that a molecular signal gained from eating herbivorous prey species stimulates the growth necessary for capturing them more efficiently.

Nutrient Stoichiometry

As mentioned in previous discussions about individual elements (e.g., nitrogen or silica) it is important to realize that the amounts of any one of them can affect the status and ecological interactions of others. In recent years, researchers have focused on these chemical limitations and interactions, the subject of *stoichiometry*.

The famous Redfield Ratio of C:N:P in oceans of 106:16:1 simply doesn't apply to inland waters, which are much too variable for such a constant stoichiometry (Hecky et al., 1993). Instead, the relative amounts of an element in a particular lake may be an order of magnitude different than this formula.

Moreover, we are changing the natural stoichiometry. In particular, we have enriched both carbon and biologically available nitrogen in the air through modern industrial processes. As is predicted by previous discussions, additions of N to the air (as much as doubling of natural levels) shifts phytoplankton to P-limited communities. This is observed in thousands of lakes throughout the northern hemisphere (Elser et al., 2009). In addition, the algae respond by increasing the ratio of N to P in their tissues, an adaptation to unbalanced eutrophication (Peñuelas et al., 2012).

As humans continue to interact with the waters of our planet, we will rely on limnologists to help us understand this chemistry and the other functioning of these ecosystems, and hopefully preserve their integrity well into the future.

Glossary

A   area

A_d area of drainage basin

A_o area at depth zero; surface area

A_z area at depth z

aggradation   the process wherein a stream deposits its excess load to its channel

aggressive CO_2   gas in excess of the CO_2 needed to dissolve $CaCO_3$; gaseous CO_2 in excess of the equilibrium level needed to dissolve $CaCO_3$ to $Ca(HCO_3)_2$

AHOD   the areal hypolimnetic oxygen deficit, usually expressed as a rate per day, referring to the decrease in O_2 beneath unit area of hypolimnion surface; mg O_2/cm^2 per day or mg O_2/m^2 per day

alkaline phosphatase   an enzyme that hydrolizes organic phosphorus compounds, making some P available as soluble inorganic phosphate

alkalinity   the buffering system or titratable base in water; the milliequivalents of hydrogen ions neutralized by a liter of water; often expressed as $CaCO_3$ in mg/liter

allochthony   referring to materials being formed elsewhere and transported to the site in question; allochthonous, *adj.*

allotrophy   the main source of chemical energy in an ecosystem coming from outside

alluvium   materials transported by a stream or river

amixis   the absence of circulation periods, as in permanently ice-covered lakes

ammonification   production of ammonia (NH_3) from organic nitrogenous compounds by decay of dead material and metabolism in living organisms

anaerobiosis   the development of oxygen-free conditions in a lake

anemotrophy   a process in which most of the chemical energy in a system arrives by wind

anoxygenic photosynthesis   autotrophy based on a solar energy source, but producing sulfur as a by-product rather than O_2 as typifies oxygenic photosynthesis

anthropogenic   having its origin in the activities of humans

argillotrophy   referring to systems in which the main source of nutrition for the animals is in clayey particles and their attached organic matter

assimilation number   the ratio of carbon fixed (or oxygen released) photosynthetically per hour or day to the chlorophyll a present

astatic   unstable, in limnology usually referring to water level

astrobleme   terrestrial configuration caused by meteoritic impact

atelomixis   an incomplete vertical mixing of stratified water masses, homogenizing layers of different chemical properties without disturbing the underlying hypolimnion; in stratified tropical lakes

athalassohaline referring to waters with ionic proportions quite different from the dissolved salts in seawater

aufwuchs *see* periphyton

Austausch coefficients coefficients that quantify turbulent movements of various types in excess of the normal diffusion and conductivity caused solely by molecular action

autochthonous referring to material formed in situ, or on the spot; authigenic

autochthony the state of being or having been formed in situ

autotrophic referring to a lake whose nutrition is derived from the synthesis of high-energy organic compounds from inorganic substances; photosynthesis; chemosynthesis; primary productivity in general; autotrophic, *adj,* pertaining to the ability to perform primary production

b breadth or width of a body of water

$\bar{b}$ mean breadth of a body of water; area divided by length

bathylimnion the deepest part of a stratified lake, removed from wind action and below the turbulent currents that characterize the clinolimnion

bed load the part of the stream's load that is rolling and sliding along because it is too heavy to be carried by suspension

benthic referring to bottom zones or bottom-dwelling forms

benthos bottom-dwelling biota; the benthic community

biogenic meromixis meromixis (partial circulation) owed to the accumulation of substances of biologic origin in the monimolimnion

biotype an assemblage of organisms sharing genetic characteristics

bittern "mother liquor," the bromide-magnesium mixture that remains when seawater is nearly evaporated; bittern lakes are especially high in $MgCl_2$

bound CO_2 an older term for CO in the carbonate form, $CO_3{}^{2-}$

braided stream a stream with complex, anastamosing multiple channels rather than a single larger channel

caldera a collapsed volcanic crater formed when molten rock flows out from beneath; caldera lakes occupy such concavities

capacity the theoretical load that a stream can transport

chemocline a density gradient, or pycnocline, owed to changing salt concentrations

chemolithotrophy autotrophy in which inorganic chemical bonds are the source of energy and inorganic substances serve as electron donors; metabolism based on chemosynthesis

chemosynthesis the metabolism used by certain autotrophic prokaryotes to power cells using inorganic oxidation, independent of light

cline any sort of gradient or continuum

clinograde referring to distribution that shows a gradient, often applied to temperature or oxygen curves

clinolimnion that portion of a lake where turbulence prevails so that temperature distribution and diffusivity decrease exponentially from high points in the epilimnion to the bathylimnion

cold monomixis complete circulation once a year when ice cover is absent; comprised of cold thereimictic and warm thereimictic lakes

competence a stream's ability to move a particle of a given size

conductance conductivity, the indirect measure of electrolytes in water; the reciprocal of resistance; an electromotive force of 1 volt between two points is 1 mho or 1 siemens(S); specific conductance, usually the electron flow between two cm^2-electrodes, set 1 cm apart

copro- a prefix referring to feces

copropel *see* gyttja

CPOM coarse particulate organic matter, greater than 1 mm diameter

-crene or **creno-** suffix or prefix designating a spring or seep

crenogenic meromixis   meromixis in which the density contrast between the waters of the monimolimnion and the mixolimnion is caused by subsurface flows of saline water from springs or seeps

cryo-   referring to ice

cryogenic lake   a lake occupying a thaw basin in a region of permanently frozen ground

cryptodepression   the portion of a lake basin that is below sea level; z_c

cryptorheism   hidden drainage where stream flow is below the surface, usually in limestone, karstic regions

cyclic succession   pattern occurring in shallow wetlands over many years, of a productive plant community alternating with open water

cyclomorphosis   the sequential appearance of con-specific different morphs, usually seen in micro-crustaceans and rotifers

degradation   erosion of a stream's channel; the inverse of aggradation

delineation   determination of a wetland boundary, usually because of a legal requirement

denitrification   the microbial production of N_2 and N_2O from nitrites and nitrates, mostly anaerobic, but not always

density current   a current moving along the bottom slope or cutting through other water masses, its greater density owed to salinity, turbidity, or temperature

detritus   dead organic matter and its associated microbial elements, particulate (POM) or dissolved (DOM)

diapause   a quiescent, physiologically inactive stage that is regularly a part of the life cycle

dichothermy   a phenomenon in which a vertical temperature profile contains an inflection, a low temperature point bounded by higher temperatures; seen in some meromictic lakes

dimixis   two circulation or mixing periods, vernal and autumnal, per year; dimictic lakes circulate after the ice melts before summer stratification and after temperature stratification is destroyed and before the ice forms

direct stratification   condition in which dense water lies below lighter water, usually referring to a temperature profile

discharge   the volume of stream flow passing a point during some period of time; often expressed as cfs or cubic feet/second

distributary   one of the branches formed by a major stream as it flows across its delta to the sea

D_L shoreline development index

dolina   a depression owed to dissolution and collapse in limestone substrata; a sink or swallow hole (plural *doline*)

DOM   dissolved organic material not retained by membrane filters with pore size 0.45 μm

D_v volume development index

dystrophy   the condition in water in which decay is hindered and recycling of nutrients is slowed

ecologic efficiency   the amount of energy, expressed as a per cent, captured by one trophic level from the lower or preceding level

ecosystem   the total living and nonliving components of a community

ecotone   a transitional area between two communities

ectogenic meromixis   partial circulation in which the density contrast between the waters of the monimolimnion and the mixolimnion is caused by waters from outside sources, saline water being introduced into a dilute lake or dilute water being introduced from elsewhere to a saline lake

edaphic   referring to the ground or soil, especially with reference to materials derived from them or their influence

eddy   a current or small whirlpool running contrary to the main flow

endorheic   lake having no surface outflows

epeirogeny   the tectonic events that raise large crustal blocks

epigean   living above ground; surface dwelling

epilimnion   the upper, well-mixed, well-illuminated, nearly isothermal region of a stratified holomictic lake

epilithic   growing on a stony substratum

epipelic   growing on a mud or ooze surface

epiphytic   adjective applied to organisms living on or attached to a plant surface

epizoic   applies to growth on an animal's body surface

equilibrium CO$_2$ the CO$_2$ gas necessary for maintaining CaCO$_3$ in solution as Ca(HCO$_3$)$_2$

euphotic zone   water depth receiving at least 1% of the light intensity present at the surface

eutrophy   the condition of water being rich in plant nutrients, and the results of this condition

evorsion   basin caused by erosion from falling water

exorheic   lake drained by surface outflow

femtoplankton   composed of plankters sized from 0.02 to 0.2 μm

fluviatile   referring to streams; fluvial

FPOM   fine particulate organic matter, with diameter between 0.45 μm and 1 mm

gilvin   a name proposed for the yellow humic molecules that stain natural waters

Gleasonian (cyclical) succession   change in wetland vegetation over time due to clearing by animals followed by drawdown of water and revegetation from buried seeds

graben   lake formed in the rift between continental plates

graded stream   a stream in dynamic equilibrium between erosional and depositional forces, flowing over a bed composed of alluvium brought and deposited by its own flow

gross production   total organic production including losses to respiration and other metabolic processes; net production plus respiration

gyttja   semireduced, fine-grained, organic, profundal sediments of eutrophic lakes; copropel

half-bound CO$_2$ an old term for CO$_2$ in the bicarbonate form, $-HCO_3^-$

halide   a binary compound of one of the halogen elements, bromine, chlorine, fluorine, iodine

hardness   the quality of water that prevents soap from dissolving; mostly caused by Ca^{2+} and Mg^{2+}

heat budget   the heat absorbed by a body of water during any specified period of time, expressed as cal/cm^2 of water surface

heterocyst   a modified cell in the "blue-green algae" (Cyanobacteria); the site of the enzyme nitrogenase

heterotrophy   referring to the gaining of energy via the degradation of the organic molecules originally produced by autotrophic organisms; herbivory, omnivory, detritivory, carnivory, microbial decay

HGM   Hydrogeomorphic system for assessing wetland function in comparison to idealized "reference" systems

holomixis   complete circulation or mixing throughout as in an holomictic lake

homoiothermy   *see* isothermy

humus   high-molecular-weight organic molecules, polymeric, derived mostly from plant decay; humic acids with the $-COOH$ radical; humolimnic acids in lake waters and sediments

hydrarch succession   orderly, predictable change in a plant community over many years from aquatic to wetland to terrestrial

hydric soils   mucks and other soils formed during prolonged submersion under water; indicator of wetland conditions

hydrograph   figure recording water surface elevation over time

hydrophyte   any aquatic plant, usually the larger rooted type; macrophyte

hydropsammon   interstitial sand-dwelling community below shallow water

hypolimnion   the poorly illuminated lower region of a directly stratified lake; denser, colder water protected from wind action; lies below the metalimnion; overlies the profundal zone

hypsographic   referring to measurement and mapping of elevations and contours

index of saturation   difference in pH from neutral; this measures the potential calcium carbonate dissolution into, or precipitation from, solution

interstitial   spaces between soil particles containing unique water chemistry and organisms

inverse stratification   condition where warm water lies beneath colder water in a vertical temperature profile; winter stratification beneath ice cover

isobath   a subsurface contour line

isothermy/isothermal   having the same temperature; homoiothermal

kerogen   humic substances of marine and lacustrine deposits as opposed to soil humus

L length of shoreline

l maximum length of a lake or pond

laminar flow   smooth, slow flow when all units are moving at the same speed and parallel to each other with no mixing

Langmuir cells   cylindrical helices of spiraling water moving parallel to the wind's direction

lentic   referring to standing water, as in ponds and lakes

limnocrene   a pool formed by the ponding of the water issuing from a spring or artesian well

lithotrophy   primary production, or autotrophy, in which inorganic substances are electron donors; chemolithotrophy; photolithotrophy

littoral zone   the marginal region of a body of water; the shallow, near-shore region; often defined by the band from zero depth to the outer edge of the rooted plants

loading   addition of nutrients or pollutants to water from sources either internal or external to the system

lotic   referring to running water, as in streams

maar   a volcanic explosion crater

macrobenthic   referring to bottom-dwelling organisms retained by screens with interstices from 1.00 to 0.425 mm (arbitrary)

marl   calcareous sediments, usually soft and littoral, mostly $CaCO_3$

meander   a stream sinuosity (curve, bend)

MEI   morphoedaphic index to fish production; TDS of water/$\bar{z}$ of lake, expressed as mg/liter divided by m

meiobenthos   bottom-dwelling organisms that are retained by screens with interstices about 0.05 to 0.045 mm

meio-microbenthos   the smallest animals living at the bottom of lakes

meromixis   partial circulation, the lower denser layers never mixing with the upper

mesolimnion   the central stratum between the epilimnion and hypolimnion in a stratified lake; metalimnion; the region occupied by the thermocline

mesothermy   phenomenon in which vertical temperature profile contains an inflection—a warm temperature point bounded by lower temperatures; seen in some meromictic lakes

mesotrophy   the condition of water being only moderately rich in plant nutrients, and the results

metagenomics   study of organisms in the environment by collecting and analyzing genetic material

metalimnion   *see* mesolimnion

methanogenic   referring to an organism that produces methane gas

microbenthos   bottom-dwelling organisms that pass through a screen with interstices 0.045 to 0.05 mm; mostly ciliates, gastrotrichs, rotifers, and so on

microbial loop   cycling of nutrients (in particular, dissolved carbon) in the water column by microscopic and very small planktonic organisms

microplankton   composed of plankters sized from 20 to 200 μm

minerotrophic   wetlands such as bogs and fens, with mineral-rich water

mixolimnion   the upper layer that occasionally mixes in a meromictic lake

mixotrophy   referring to the ability to use both organic and inorganic carbon sources in nutrition

monimolimnion   the permanently stagnant layer in meromictic lakes, below the chemocline

monomixis   referring to one regular period of circulation or mixing per year; includes cold monomixis and warm monomixis

nanoplankton   composed of plankters sized from 2 to 20 μm

natriophils   organisms living in sodium-rich alkaline waters

nekton   large invertebrates and fish occurring in limnetic zone and capable of swimming independently of turbulence

net plankton   the portion of the plankton community retained by a net with interstices of 50 to 63 μm

neuston   the community of organisms associated with the film produced by surface tension; includes hyponeuston and epineuston, living at the under-surface and the upper surface, respectively

nitrification   the process of oxidizing NH_3 to NO_2, and NO_2 to NO_3

nitrogenase   an enzyme that reduces N_2 to NH_3, possessed only by a few prokaryotes

nutrient spiraling   an imaginary spiral followed by a cycling nutrient as it flows downstream

N/P   the ratio of nitrogen to phosphorus; best expressed as an atomic ratio

oligohaline   coastal wetlands with water having relatively low salt concentration

oligomixis   rare or occasional circulation of water masses; oligomictic lakes rarely or irregularly mix

oligotrophy   the condition of water being poor in plant nutrients, and the results

ombrotrophic   referring to a system receiving all its water and minerals via precipitation; ombrotrophic bog organic drift the mostly nocturnal downstream floating of stream organisms

organotrophy   nutrition in which energy is gained from the fermentation or oxidation of organic compounds

orthograde   referring to straight distribution, as in a vertical oxygen or temperature profile

orthophosphate   inorganic phosphate

P/A   the ratio of organic production to assimilated food; net growth efficiency expressed as a percent; K_2

paleolimnology   the study of lacustrine sediments and relics preserved in them

palynology   the study of pollen grains and spores

PAR   photosynthetically active radiation, 350 or 400 to 700 nm

P/B   the ratio of the annual production to the mean annual standing stock, or biomass, of an organism; the annual turnover or replacement number, times per year

P/C   the ratio of organic production to food consumed; gross growth efficiency when expressed as a percent; K_1

peat   layer of accumulated undecomposed plant matter, often *Sphagnum* moss

pel-   referring to mud, clay, ooze; from the Greek word *pelos*

pelagic   referring to open-water regions not directly influenced by the shore and bottom; limnetic

periphyton   the biota attached to submersed surfaces; community of sessile organisms on lake and stream substrata

phagotrophy   the method by which some organisms or cells acquire nutrition, using pseudopodia to engulf their prey or other particles

phenolphthalein   alkalinity part of the water's total alkalinity based on the amount of standard acid used to bring a sample from a pH above 8.3 (where phenolphthalein is pink) to the colorless turning point; expressed as mEq/liter, mg/liter, or ppm; caused especially by OH^- and CO_3^{2-}

photoautotrophs   photosynthetic organisms

photolithotrophic   photosynthetic organisms living in water or mud

photorespiration   a process in plant metabolism which attempts to reduce the consequences of a wasteful oxygenation reaction by the enzyme RuBisCO

phytobenthos   plants and algae growing at the bottom of a lake

phytoplankton   that portion of the plankton community composed of algae and cyanobacteria

phytoplankton succession   pattern of seasonal changes in algae populations

phytotrophy   photosynthesis or photoautotrophy in organisms that possess chlorophyll *a* and that use water as an electron donor in reducing carbon

picoplankton   composed of plankters sized from 0.2 to 2 μm

plankter   an individual of the plankton community

playa   an extremely shallow, flat, closed basin that holds water occasionally

pluvial lakes   formed in the past when melting glaciers made water more plentiful

poikilothermy   phenomenon in which vertical temperature profile is complex, including at least one low point and one high point bounded by warmer and colder water, respectively

polar lake   a lake or pond in which the surface water temperature does not rise above 4° C during the warm season

polymixis   many circulation or mixing periods per year or nearly continuous circulation

POM   all particulate organic material retained by a membrane filter with pore size of about 0.45 μm

P/R   the ratio of gross primary productivity to respiration in a community; or, gross photosynthesis/total respiration

protists   eukaryotic, unicellular or colonial organisms, either plant-like algae or animal-like protozoans

psammon   the community in the interstitial water among beach sand grains

pycnocline   general term for a density gradient whatever its cause

redox (reduction) potential   affinity of a chemical to electrons, and therefore its relative oxidation state; measured in Volts

Reynolds number   a dimensionless number marking the boundary between laminar and turbulent flow; diameter of the flow divided by viscosity

rheo-   prefix referring to current, flow

riparian   referring to the streamside, the land bordering streams

salinity   the total inorganic salts dissolved in water, expressed in various ways: %, ‰, g/kg, ppm, mg/liter, mg/dm^3; in seawater it is closely related to Cl$^-$ concentration and conductance

saltern   a concentrated saline, its relative ionic composition much like seawater; thalassic

saprobity   *see* heterotrophy

sapropel   reduced, amorphous sediment of polluted or hypereutrophic lakes, glossy black with FeS, often odorous with H$_2$S

SBV   *sauerbindungsvermögen* (German); the water's buffering capacity, titratable base, alkalinity, acid-combining strength, usually expressed as mEq/ liter

Secchi disc   a white, or black and white disc about 20 cm in diameter, lowered into water to measure transparency on the basis of visibility

seiche   an oscillation of water reestablishing equilibrium after having been disturbed

seres   temporary ecological community found during succession

seston   organic particulate matter in water including living and nonliving fractions

sinter   a general term for sediments deposited by mineral springs including the siliceous geyserite and the calcareous tufa or travertine

Si/P   the ratio of silicon to phosphorus in water or tissue

spiraling length   the distance downstream a nutrient atom travels to complete its cycle in the aqueous, sestonic, and consumer compartments

stability of stratification (*S*)   the amount of work necessary to mix a body of water to uniform density without adding or subtracting heat in the process, usually expressed as g-cm/cm^2 of lake surface

stoichiometry   the relative quantities of nutrients such as nitrogen and phosphorus in water

stream order   the position a section of a stream occupies in relation to the tributaries contributing to it; the higher the order the more tributaries it has

sublittoral zone   the bottom region lying between the littoral and the profundal zones

tarn   a small mountain lake, especially one that occupies a cirque

TDS   total dissolved solids; filtrable residue; usually expressed as g/liter or mg/liter following evaporation of a measured sample of filtered water

tectonic   referring to all types of activities in the earth's crust

temporary hardness   the water hardness removed by the precipitation of $CaCO_3$ and $MgCO_3$ following boiling

thalassohaline   referring to water with ionic proportions similar to those of seawater

thalweg, talweg   German for "the valley path"; the longitudinal deepest channel in a stream bed

thermal bar   a vertical 4° isotherm or a wall of 4° C water separating two masses of less dense water

thermocline   a density gradient or pycnocline owed to changing temperatures; the planar thermocline is the imaginary plane at the depth where the rate of temperature change is the greatest in a vertical temperature profile

thiobios   organisms adapted for life in anaerobic low-pH habitats where sulfides abound

tripton   old word for nonliving organic particles in water; particulate detritus; POM

troglobitic   referring to obligatory cave-dwelling organisms or to obligatory hypogean organisms

trophogenic zone   a region in a body of water where synthesis of organic compounds is predominant; usually referring to the photosynthetic region

tropholytic zone   a region in a lake or stream where the destruction of organic compounds is predominant; usually referring to region of decay

turbidity   measure of cloudiness of water

turbulence   erratic, mixing, sinuous flow marked by eddies and units moving at different speeds and different directions from the main flow

turbulent flow   movement within water column causing mixing, adding oxygen and churning dissolved materials

univoltine   pertaining to organisms that have but one generation per year

UPOM   smallest particulate organic matter; particles between 0.45 and 0.5 μm

V volume of a body of water

varve   an annual layer of sediment marked by distinct seasonal laminae

warm monomixis   one period of complete circulation during the cold time of year in an ice-free lake; temperature stratification occurs in summer

z referring to depth; vertical distance from the surface

$\bar{z}$ mean depth of a body of water, V/A

z_c cryptodepression; the depth of the basin portion below sea level

z_g the depth designated as the center of gravity of a body of water; the depth that separates the mass of the lake into two equal parts

z_m maximum depth of a body of water

z_r relative depth, the ratio of z_m to the mean diameter of lake surface, expressed as percent

Z_{SD} the maximum depth of Secchi-disc visibility, usually in meters

zoobenthos   animals living at the bottom of a lake

zooplankton   the fraction of the plankton community composed of animals; the individual is the zooplankter

Literature Cited

Agrawal, M., and M. K. Agrawal. 2011. Cyanobacteria-herbivore interaction in freshwater ecosystem. J. Microbiol. Biotechn. Res. **1**:52–66.

Alatalo, R., and R. Alatalo. 1977. Components of diversity: multivariate analysis with interaction. Ecology **58**:900–906.

Allen, H. L. 1971. Primary productivity, chemoorganotrophy, and nutritional interactions of epiphytic algae and bacteria on macrophytes in the littoral of a lake. Ecol. Monogr. **41**:91–127.

Allen, H. L. 1972. Phytoplankton photosynthesis, micro-nutrient interactions, and inorganic carbon availability in a soft-water Vermont lake. Pp. 63–83 in G. E. Likens, ed., Nutrients and eutrophication: The limiting-nutrient controversy. ASLO Special Symposia 1. Lawrence, KS: American Society of Limnology and Oceanography.

Allen, K. R. 1951. The Horokiwi stream: a study of a trout population. New Zealand Marine Dept., Fish. Bull. **10**:1–239.

American Public Health Association. 2012. Standard methods for the examination of water and wastewater, 20th ed. New York: Author.

Amoros, C., and G. Bornette. 2002. Connectivity and biocomplexity in waterbodies of riverine floodplains. Freshwater Biol. **47**(4):761–776.

Anderson, D. M., P. M. Glibert, and J. M. Burkholder. 2002. Harmful algal blooms and eutrophication: Nutrient sources, composition, and consequences. Estuaries **25**(4):704–726.

Anderson, E. R. 1952. Water-loss investigations: vol. 1, Lake Hefner studies. Tech. Rept. U.S. Geol. Surv. Circ. **229**:71–119.

Anderson, E. R., and D. W. Pritchard. 1951. Physical limnology of Lake Mead. San Diego, CA: U.S. Navy Electron. Lab. Rept. 258, Probl. NEL 2J1.

Anderson, G. C. 1958a. Seasonal characteristics of two saline lakes in Washington. Limnol. Oceanogr. **3**:51–68.

Anderson, G. C. 1958b. Some limnological features of a shallow saline meromictic lake. Limnol. Oceanogr. **3**:259–270.

Anderson, L. W. J., and B. M. Sweeney. 1977. Diel changes in sedimentation characteristics of *Ditylum brightwelli:* Changes in cellular lipid and effects of respiratory inhibitors and ion-transport modifiers. Limnol. Oceanogr. **22**:539–552.

Anderson, R. O., and F. F. Hooper. 1956. Seasonal abundance and production of littoral bottom fauna in a southern Michigan lake. Trans. Amer. Microsc. Soc. **75**:259–270.

Anderson, R. S. 1967. Diaptomid copepods from two mountain ponds in Alberta. Canadian J. Zool. **45**:1043–1047.

Anderson, R. S., and A-M. De Henau. 1980. An assessment of the meiobenthos from nine mountain lakes in western Canada. Hydrobiologia **70**:257–264.

Angelier, E. 2003. Ecology of streams and rivers. Enfield, NH: Science Publishers.

Angino, E. E., K. B. Armitage, and J. C. Tash. 1965. A chemical and limnological study of Lake Vanda, Victoria Land, Antarctica. Univ. Kansas Sci. Bull. **45**:1097–1118.

Anholt, B. R. 1995. Density dependence resolves the stream drift paradox. Ecology **76**:2235–2239.

Antal, T. K., T. E. Krendeleva, and A. B. Rubin. 2011. Acclimation of green algae to sulfur deficiency: Underlying mechanisms and application for hydrogen production. Appl. Microbiol. Biot. **89**(1):3–15.

Arhonditsis, G. B., M. T. Brett, C. L. DeGasperi, and D. E. Schindler. 2004. Effects of climatic variability on the thermal properties of Lake Washington. Limnol. Oceanogr. **49**(1):256–270.

Armitage, K. B. 1958. Lagos volcánicos de El Salvador. Comun. Inst. Trop. Invest. Cient. **7**:39–48; 18 figs.

Armitage, K. B., and H. B. House. 1962. A limnological reconnaissance in the area of McMurdo Sound, Antarctica. Limnol. Oceanogr. **7**:36–41.

Arnold, D. E. 1971. Ingestion, assimilation, survival, and reproduction by *Daphnia pulex* fed seven species of blue-green algae. Limnol. Oceanogr. **16**:906–920.

Audzijonytė, A., J. Damgaard, S. L. Varvio, J. K. Vainio, and R. Väinölä. 2005. Phylogeny of *Mysis* (Crustacea, Mysida): History of continental invasions inferred from molecular and morphological data. Cladistics **21**(6):575–596.

Audzijonytė, A., M. E. Daneliya, N. Mugue, and R. Väinölä. 2008. Phylogeny of *Paramysis* (Crustacea: Mysida) and the origin of Ponto-Caspian endemic diversity: Resolving power from nuclear protein-coding genes. Mol. Phylogenet. Evol. **46**(2):738–759.

Aufdenkampe, A. K., E. Mayorga, P. A. Raymond, J. M. Melack, S. C. Doney, S. R. Alin, … and K. Yoo. 2011. Riverine coupling of biogeochemical cycles between land, oceans, and atmosphere. Front. Ecol. Environ. **9**(1):53–60.

Auguet, J. C., A. Barberan, and E. O. Casamayor. 2009. Global ecological patterns in uncultured Archaea. ISME J **4**:182–190.

Avila, T. R., A. A. D. S. Machado, and A. Bianchini. 2012. Estimation of zooplankton secondary production in estuarine waters: Comparison between the enzymatic (chitobiase) method and mathematical models using crustaceans. J. Exp. Mar. Biol. Ecol. **416**:144–152.

Azam, F., T. Fenchel, J. G. Field, J. S. Gray, L. A. MeyerReil, and F. Thingstad. 1983. The ecological role of water column microbes in the sea. Mar. Ecol. Prog. Ser. **10**:257–263.

Babitski, V. A. 1980. The microzoobenthos of three lakes of different types. Hydrobiol. J. **16**:27–34.

Baker, A. L., and A. J. Brook. 1971. Optical density profiles as an aid to the study of microstratified phytoplankton populations in lakes. Arch. Hydrobiol. **69**:214–233.

Baker, L. A., A. T. Herliky, P. R. Kaufmann, and J. M. Eilers. 1991. Acidic lakes and streams in the United States: The role of acidic deposition. Science **252**:1151–1154.

Barica, J. 1975. Summerkill risk in prairie ponds and possibilities of its prediction. J. Fish. Res. Bd. Canada **32**:1283–1288.

Barland, K. 1991. Trapped seawater in two Norwegian lakes: Kilevannet, a "new" lake with old trapped seawater, and Rorholtfjorden. Aquatic Sci. **53**:90–98.

Bärlocher, F., and B. Kendrick. 1973. Fungi in the diet of *Gammarus pseudolimnaeus* (Amphipoda). Oikos **24**:295–300.

Barnes, D. F. 1960. An investigation of a perennially frozen lake. Bedford, MA: U.S. Air Force Survey in Geophysics (ARDC) 129.

Barns, S. M., R. E. Fundyga, M. W. Jeffries, and N. R. Pace. 1994. Remarkable archaeal diversity detected in a Yellowstone National Park hot spring environment. P. Natl. Acad. Sci. **91**(5):1609–1613.

Barringer, R. W. 1967. World's meteorite craters ("Astroblemes"). Version VII, Feb. 1967. Meteoritics **3**:151–157.

Bartelt-Hunt, S. L., D. D. Snow, T. Damon, J. Shockley, and K. Hoagland. 2009. The occurrence of illicit and therapeutic pharmaceuticals in wastewater effluent and surface waters in Nebraska. Environ. Pollut. 157(3):786–791.

Barth, J. A., A. Tait, and M. Bolshaw. 2004. Automated analyses of 18O/16O ratios in dissolved oxygen from 12-mL water samples. Limnol. Oceanogr-Meth. 2(2):35–41.

Bartsch, A. F., and M. S. Allum. 1957. Biological factors in treatment of raw sewage in artificial ponds. Limnol. Oceanogr. 2:77–84.

Batzer, D. P., and R. R. Sharitz, eds. 2006. Ecology of freshwater and estuarine wetlands. Oakland: University of California Press.

Bayley, P. B. 1995. Understanding large river-floodplain ecosystems. BioScience 45:153–158.

Bayley, S. E., and C. M. Prather. 2003. Do wetland lakes exhibit alternative stable states? Submersed aquatic vegetation and chlorophyll in western boreal shallow lakes. Limnol. Oceanogr. 48(6):2335–2345.

Baylor, E. R., and W. H. Sutcliffe, Jr. 1963. Dissolved organic matter in seawater as a source of particulate food. Limnol. Oceanogr. 8:369–371.

Bayly, I. A. E. 1964. Chemical and biological studies on some acidic lakes of east Australian sandy coastal lowlands. Australian J. Mar. Freshwat. Res. 15:56–72.

Bayly, I. A. E. 1967. The general biological classification of aquatic environments with special reference to those of Australia. Pp. 78–104 in A. H. Weatherby, ed., Australian inland waters and their fauna: Eleven studies. Canberra, NSW: Australian National University Press.

Bayly, I. A. E. 1972. Salinity tolerance and osmotic behavior of animals in athalassic saline and hypersaline waters. Ann. Rev. Ecol. Syst. 3:233–268.

Bayly, I. A. E., and W. D. Williams. 1973. Inland waters and their ecology. Brisbane: Longman Australia Pty Ltd.

Beadle, L. C. 1981. The inland waters of tropical Africa. An introduction to tropical limnology, 2nd ed. London: Longman Group Ltd.

Beauchamp, P. de. 1952. Un facteur de la variabilité chez les rotiferes du genre Brachionus. Compt. Rend. Acad. Sci. Paris 234:573–575.

Beauchamp, R. S. A. 1953. Sulphates in African inland waters. Nature 171:769–771.

Beauchamp, R. S. A. 1964. The Rift Valley lakes of Africa. Verh. Internat. Verein. Limnol. 15:91–99.

Beeton, A. M. 1962. Light penetration in the Great Lakes. Univ. Michigan, Great Lakes Res. Div., Inst. Sci. Tech. 9:68–76.

Belk, D. 1972. The biology and ecology of Eulimnadia antlei Mackin (Conchostraca). Southwest. Nat. 16:297–305.

Belk, D. 1977a. Zoogeography of the Arizona fairy shrimps (Crustacea: Anostraca). J. Arizona Acad. Sci. 12:70–78.

Belk, D. 1977b. Evolution of egg size strategies in fairy shrimps. Southwest. Nat. 22:99–105.

Belk, D., and G. A. Cole. 1975. Adaptational biology of desert temporary-pond inhabitants. Pp. 207–226 in N. F. Hadley, ed., Environmental physiology of desert organisms. Stroundsburg, PA: Dowden, Hutchinson and Ross.

Bell, H. L. 1971. Effect of low pH on the survival and emergence of aquatic insects. Water Res. 5(6):313–319.

Bell, P. R. 1959. The ability of sphagnum to absorb cations differentially from dilute solutions resembling natural waters. J. Ecol. 47:351–355.

Benson, B. B., and D. Krause, Jr. 1980. The concentration and isotopic fractionation of gases dissolved in freshwater in equilibrium with the atmosphere: I. Oxygen. Limnol. Oceanogr. 25:662–671.

Bent, A. M. 1960. Pollen analysis of Deadman Lake, Chuska Mountains, New Mexico, M.S. Thesis, University of Minnesota, Minneapolis.

Berg, C. O. 1950. Biology of certain Chironomidae reared from *Potamogeton*. Ecol. Monogr. **20**:83–101.

Berg, C.O. 1963. Middle Atlantic States. Pp. 191–237 in D. G. Frey, ed., Limnology in North America. Madison: University of Wisconsin Press.

Berggren, M., L. Ström, H. Laudon, J. Karlsson, A. Jonsson, R. Giesler, ... and M. Jansson. 2010. Lake secondary production fueled by rapid transfer of low molecular weight organic carbon from terrestrial sources to aquatic consumers. Ecol. Lett. **13**(7):870–880.

Berman, T. 1970. Alkaline phosphatases and phosphorus availability in Lake Kinneret. Limnol. Oceanogr. **15**:663–674.

Berninger, U.-G., B. J. Finlay, and P. Kuuppo-Leinikki. 1991. Protozoan control of bacterial abundances in freshwater. Limnol. Oceanogr. **36**:139–147.

Berra, T. M. 2007. Freshwater fish distribution. Chicago: University of Chicago Press.

Bieler, R., and Mikkelsen, P. M. 2006. Bivalvia—a look at the branches. Zool. J. Linn. Soc-Lond. **148**(3):223–235.

Bilby, R. E., and G. E. Likens. 1981. Importance of organic debris dams in the structure and function of stream ecosystems. Ecology **61**:1107–1113.

Bindloss, M. E. 1976. The light climate of Loch Leven, a shallow Scottish lake, in relation to primary production by phytoplankton. Freshwat. Biol. **6**:501–518.

Birge, E. A. 1897. Plankton studies on Lake Mendota: II. The Crustacea from the plankton from July, 1894, to December, 1896. Trans. Wisconsin Acad. Sci. Arts Lett. **11**:274–448.

Birge, E. A. 1916. The work of the wind in warming a lake. Trans. Wisconsin Acad. Sci. Arts Lett. **18**:341–391.

Birge, E. A., and C. Juday. 1911. The inland lakes of Wisconsin: The dissolved gases of the water and their biological significance. Wisconsin Geol. Nat. Hist. Surv. Bull. 22.

Birge, E. A., and C. Juday. 1934. Particulate and dissolved organic matter in inland lakes. Ecol. Monogr. **4**:440–474.

Birge, E. A., C. Juday, and H. W. March. 1928. The temperature of the bottom deposits of Lake Mendota: A chapter in the heat exchanges of the lake. Trans. Wisconsin Acad. Sci. Arts Lett. **23**:187–231.

Blackburn, W. M., and T. Petr. 1979. Forest litter decomposition and benthos in a mountain stream in Victoria, Australia. Arch. Hydrobiol. **86**:453–498.

Blinn, D. W., M. Hurley, and L. Brokaw. 1981. The effect of saline seeps and restricted light upon the seasonal dynamics of phytoplankton communities within a southwestern (USA) desert canyon stream. Arch. Hydrobiol. **92**:287–305.

Blinn, D. W., and D. B. Johnson. 1982. Filter-feeding of *Hyalella montezuma*, an unusual behavior for a freshwater amphipod. Freshwat. Invert. Biol. **1**:48–52.

Blueweiss, L., H. Fox, V. Kudzma, D. Nakashima, R. Peters, and S. Sams. 1978. Relationships between body size and some life history parameters. Oecologia **37**:257–272.

Bonomi, G., and D. Ruggiu. 1966. II macrobenton profondo del Lago di Mergozzo. Mem. 1st. Ital. Idrobiol. **20**:153–200.

Bos, D. G., B. F. Cumming, C. E. Watters, and J. P. Smol. 1996. The relationship between zooplankton, conductivity and lake-water ionic composition in 111 lakes from the Interior Plateau of British Columbia, Canada Int. J. Salt Lake Res. **5**(1):1–15.

Boschetti, C., A. Carr, A. Crisp, I. Eyres, Y. Wang-Koh, E. Lubzens, ... and A. Tunnacliffe. 2012. Biochemical diversification through foreign gene expression in bdelloid rotifers. PLoS Genet. **8**(11):e1003035.

Bouck, G. R. 1976. Supersaturation and fishery observations in selected alpine Oregon streams. Pp. 37–40 in D. H. Fickeisen and M. J. Schneider, eds., Gas bubble disease. Washington, DC: Technical Information Center, Office of Public Affairs, Energy Research and Development Administration.

Boulton, A. J. 2007. Hyporheic rehabilitation in rivers: restoring vertical connectivity. Freshwater Biol. **52**(4):632–650.

Boulton, A. J., H. M. Valett, and S. G. Fisher. 1992. Spatial distribution and taxonomic composition of the hyporheos of several Sonoran Desert streams. Arch. Hydrobiol. **125**:37–61.

Bousfield, E. L. 1989. Revised morphological relationships within the amphipod genera *Pontoporeia* and *Gammaracanthus* and the "glacial relict" significance of their postglacial distributions. Canad. J. Fish. Aquat. Sci. **46**:1714–1725.

Bradbury, J. P. 1971. Limnology of Zuni Salt Lake, New Mexico. Bull. Geol. Soc. Amer. **82**:379–398.

Bradbury, J. P. 1975. Diatom stratigraphy and human settlement in Minnesota. Geol. Soc. Amer., Special Papers **171**:1–74.

Bradbury, J. P. 1988. Fossil diatoms and neogene paleolimnology. Palaeogeogr., Palaeoclimatol., Palaeoecol. **62**:299–316.

Brand, C., and M. L. Miserendino. 2011. Life history strategies and production of caddisflies in a perennial headwater stream in Patagonia. Hydrobiologia **673**(1):137–151.

Brett, M. T., and C. R. Goldman. 1996. A meta-analysis of the freshwater trophic cascade. P. Natl. A. Sci. **93**(15):7723–7726.

Brewer, M. C. 1958. The thermal regime of an Arctic lake. Trans. Amer. Geophysics Union **39**:278–284.

Bridge, J. S. 2003. Rivers and floodplains: Forms, processes, and sedimentary record. Oxford, UK: Wiley-Blackwell.

Bridgeman, T. B., D. W. Schloesser, and A. E. Krause. 2006. Recruitment of Hexagenia mayfly nymphs in western Lake Erie linked to environmental variability. Ecol. Appl. **16**(2):601–611.

Bridgham, S. D., H. Cadillo-Quiroz, J. K. Keller, and Q. Zhuang. 2013. Methane emissions from wetlands: biogeochemical, microbial, and modeling perspectives from local to global scales. Global Change Biol. **19**(5):1325–1346.

Brill, W. J. 1979. Nitrogen fixation: Basic to applied. Amer. Scientist **67**:458–466.

Brinkhurst, R. O. 1966. The Tubificidae (Oligochaeta) of polluted waters. Verh. Internat. Verein. Limnol. **16**:858–859.

Brinkhurst, R. O. 1974. The benthos of lakes. New York: St. Martin's Press.

Britt, N. W. 1955. Stratification in western Lake Erie in summer of 1953: Effects on the *Hexagenia* (Ephemeroptera) populations. Ecology **36**:239–244.

Brock, T. D. 1973. Lower pH limit for the existence of blue-green algae: Evolutionary and ecological implications. Science **179**:480–483.

Broda, E. 1975. The evolution of the bioenergetic process. Oxford, UK: Pergamon Press.

Broecker, W. S. 1974. Chemical oceanography. New York: Harcourt Brace Jovanovich.

Brönsted, J. N., and C. Wesenberg-Lund. 1911. Chemische-physikalische Untersuchungen der dänischen Gewässer nebst Bemerkungen über ihre Bedeutung fur userere Aufassung der Temporalvariationen. Internat. Rev. Hydrobiol. **4**:251–290, 437–492.

Brook, A. J. 1965. Planktonic algae as indicators of lake types, with special reference to the Desmidiaceae. Limnol. Oceanogr. **10**:403–411.

Brook, A. J., A. L. Baker, and A. R. Klemer. 1971. The use of turbimetry in studies of the population dynamics of phytoplankton populations with special reference to *Oscillatoria aghardii* var. *isothrix*. Mitt. Internat. Verein. Limnol. **19**:244–252.

Brooker, M. P. 1985. The ecological effects of channelization. The Geog. J. **151**:63–69.

Brooks, J. L., and E. S. Deevey, Jr. 1963. New England. Pp. 117–162 in D. G. Frey, ed., Limnology in North America. Madison: University of Wisconsin Press.

Brooks, J. L., and S. I. Dodson, 1965. Predation, body size, and composition of plankton. Science **150**:28–35.

Brown, K. L., and C. Lydeard. 2010. Mollusca: Gastropoda. In J. H. Thorp and A. P. Covich, eds., Ecology and classification of North American freshwater invertebrates. London: Academic Press.

Brown, L. A. 1929. The natural history of cladocerans in relation to temperature. II. Temperature coefficients for development. Amer. Nat. **63**:346–352.

Brunskill, G. J. 1969. Fayetteville Green Lake, New York. II. Precipitation and sedimentation of calcite in a meromictic lake with laminated sediments. Limnol. Oceanogr. **14**:830–847.

Brunskill, G. J., and D. W. Schindler. 1971. Geography and bathymetry of selected lake basins. Experimental Lakes Area, northwestern Ontario. J. Fish. Res. Bd. Canada **28**:139–155.

Bryceson, I., and P. Fay. 1981. Nitrogen fixation in *Oscillatoria (Trichodesmium) erythraea* in relation to bundle formation and trichome differentiation. Marine Biol. **61**:159–166.

Bryson, R. A., and V. E. Suomi. 1951. Midsummer renewal of oxygen within the hypolimnion. J. Mar. Res. **10**:263–269.

Buckland, S. T., A. E. Magurran, R. E. Green, and R. M. Fewster. 2005. Monitoring change in biodiversity through composite indices. Philos. T. R. Soc. B. **360**(1454):243–254.

Burgis, M. J., and P. Morris. 1987. The natural history of lakes. Cambridge: Cambridge University Press.

Burlakova, L. E., A. Y. Karatayev, C. Pennuto, and C. Mayer. 2014. Changes in Lake Erie benthos over the last 50 years: Historical perspectives, current status, and main drivers. J. Great Lakes Res **40**(3):560–573.

Busch, D. E., and S. G. Fisher. 1981. Metabolism of a desert stream. Freshwat. Biol. **11**:301–308.

Butman, D., and Raymond, P. A. 2011. Significant efflux of carbon dioxide from streams and rivers in the United States. Nat. Geosci. **4**(12):839–842.

Cameron, R. E. 1963. Algae of southern Arizona: I. Introduction-blue-green algae. Rev. Algol. **6**:282–318.

Cannell, M. G. R., and Thornley, J. H. M. 2000. Modelling the components of plant respiration: Some guiding principles. Ann. Bot-London **85**(1):45–54.

Caraco, N. F., J. J. Cole, and D. L. Strayer. 2006. Top down control from the bottom: Regulation of eutrophication in a large river by benthic grazing. Limnol. Oceanogr. **51**(1):664–670.

Carey, C. C., and E. Rydin. 2011. Lake trophic status can be determined by the depth distribution of sediment phosphorus. Limnol. Oceanogr. **56**(6):2051–2063.

Carlson, R. E. 1977. A trophic state index for lakes. Limnol. Oceanogr. **22**(2):361–369.

Carpelan, L. H. 1957. Hydrobiology of the Alviso salt ponds. Ecology **38**:375–390.

Carpelan, L. H. 1958. The Salton Sea: Physical and chemical characteristics. Limnol. Oceanogr. **3**:373–386.

Carpenter, E. J. 1973. Nitrogen fixation by *Oscillatoria (Trichodesmium) thiebautii* in the southwestern Sargasso Sea. Deep-Sea Res. **20**:285–288.

Carpenter, S. R., ed. 1988. Complex interactions in lake communities. New York: Springer-Verlag.

Carpenter, S. R., J. J. Cole, M. L. Pace, R. Batt, W. A. Brock, T. Cline, ... and B. Weidel. 2011. Early warnings of regime shifts: A whole-ecosystem experiment. *Science* **332**(6033):1079–1082.

Carpenter, S. R., and J. F. Kitchell, eds. 1996. The trophic cascade in lakes. Cambridge: Cambridge University Press.

Carter, J. C. H. 1965. The ecology of the calanoid copepod *Pseudocalanus* minutus Krøyer in Tessiarsuk, a coastal meromictic lake of northern Labrador. Limnol. Oceanogr. **10**:345–353.

Casamitjana, X., and E. Roget. 1990. The thermal structure of Lake Banyoles. Verh. Internat. Verein. Limnol. **24**:88–91.

Casamitjana, X., and E. Roget. 1993. Resuspension of sediment by focused groundwater in Lake Banyoles. Limnol. Oceanogr. **38**:643–656.

Casanova, M. T., and M. A. Brock. 2000. How do depth, duration and frequency of flooding influence the establishment of wetland plant communities? Plant Ecol. **147**(2):237–250.

Caspers, H., and L. Karbe. 1967. Vorschlage für eine saprobiologische Typisierung der Gewässer. Int. Rev. Hydrobiol. **52**:145–162.

Castenholz, R. W. 1960. Seasonal changes in the attached algae of freshwater and saline lakes in the Lower Grand Coulee, Washington. Limnol. Oceanogr. **5**:1–28.

Cattaneo, A., and J. Kalff. 1979. Primary production of algae growing on natural and artificial aquatic plants: A study of interactions between epiphytes and their substrate. Limnol. Oceanogr. **24**:1031–1037.

Cavicchioli, R. 2002. Extremophiles and the search for extraterrestrial life. Astrobiology **2**(3):281–292.

Chaffin, J. D., T. B. Bridgeman, D. L. Bade, and C. N. Mobilian. 2014. Summer phytoplankton nutrient limitation in Maumee Bay of Lake Erie during high-flow and low-flow years. J. Great Lakes Res. **40**(3):524–531.

Chamberlain, W. M. 1968. A preliminary investigation of the nature and importance of soluble organic phosphorus in the phosphorus cycle of lakes. Ph.D. thesis, University of Toronto.

Clarke, F. W. 1924. The data of geochemistry, 5th ed. Bull. U.S. Geol. Surv. 770, U.S. Washington, DC: U.S. Government Printing Office.

Clarke, G. L. 1939. The utilization of solar energy by aquatic organisms. Pp. 27–38 in E. R. Moulton, ed., Problems of lake biology. Lancaster, PA: American Association for the Advancement of Science.

Cochran-Biederman, J. L., and K. O. Winemiller. 2010. Relationships among habitat, ecomorphology and diets of cichlids in the Bladen River, Belize. Environ. Biol. Fish. **88**(2):143–152.

Coffin, C. C., F. R. Hayes, L. H. Jodrey, and S. G. Whiteway. 1949. Exchange of materials in a lake as studied by the addition of radioactive phosphorus. Canadian J. Res. **7**:207–222.

Coffman, W. P. 1978. Chironomidae. Pp. 345–376 in R. W. Merritt and K. W. Cummins, eds., An introduction to the aquatic insects of North America. Dubuque, IA: Kendall/Hunt.

Cohen, A. S. 2003. Paleolimnology: The history and evolution of lake systems. New York: Oxford University Press, USA.

Cohen, Y., E. Padan, and M. Shilo. 1975. Facultative anoxygenic photosynthesis in the cyanobacterium *Oscillatoria limnetica*. J. Bacteriol. **123**:855–861.

Cohen, Y., W. E. Krumbein, M. Goldbery, and M. Shilo. 1977. Solar Lake (Sinai) physical and chemical limnology. Limnol. Oceanogr. **22**:597–608.

Cole, G. A. 1954. Studies on a Kentucky Knobs Lake: I. Some environmental factors. Trans. Kentucky Acad. Sci. **15**:31–47.

Cole, G. A. 1955. An ecological study of the microbenthic fauna of two Minnesota lakes. Amer. Midl. Nat. **53**:213–230.

Cole, G. A. 1968. Desert limnology. Pp. 423–486 in G. W. Brown, Jr., ed., Desert biology. New York: Academic Press.

Cole, G. A., and W. T. Barry. 1973. Montezuma Well, Arizona, as a habitat. J. Arizona Acad. Sci. **8**:7–13.

Cole, G. A., and G. L. Batchelder. 1969. Dynamics of an Arizona travertine-forming stream. J. Arizona Acad. Sci. **5**:271–283.

Cole, G. A., and R. J. Brown. 1967. The chemistry of Artemia habitats. Ecology **48**:858–861.

Cole, G. A., and W. L. Minckley. 1968. Anomalous thermal conditions in a hypersaline inland pond. J. Arizona Acad. Sci. **5**:105–107.

Cole, G. A., and R. L. Watkins. 1977. *Hyalella montezuma*, a new species (Crustacea: Amphipoda) from Montezuma Well, Arizona. Hydrobiologia **52**:175–184.

Cole, G. A., M. C. Whiteside, and R. J. Brown. 1967. Unusual monomixis in two saline Arizona ponds. Limnol. Oceanogr. **12**:584–591.

Comita, G. W. 1964. Energy budget of *Diaptomus siciloides*. Verh. Internat. Verein. Limnol. **15**:646–653.

Comita, G. W. 1972. The seasonal zooplankton cycles, production and transformation of energy in Severson Lake, Minnesota. Arch. Hydrobiol. **70**:14–66.

Contos, J., and N. Tripcevich. 2014. Correct placement of the most distant source of the Amazon River in the Mantaro River drainage. Area **46**.1:27–39.

Conway, E. J. 1942. Mean geochemical data in relation to oceanic evolution. Proc. Roy. Irish Acad. (B) **48**:119–159.

Conway, K., and F. R. Trainor. 1972. *Scenedesmus* morphology and flotation. J. Phycol. **8**:138–143.

Cook, R. E. 1977. Raymond Lindeman and the trophicdynamic concept in ecology. Science **198**:22–26.

Cooke, G. D., M. R. McComas, D. W. Walker, and R. H. Kennedy. 1977. The occurrence of internal phosphorus loading in two small, eutrophic glacial lakes in northeastern Ohio. Hydrobiologia **56**:129–135.

Corliss, J. O. 2002. Biodiversity and biocomplexity of the protists and an overview of their significant roles in maintenance of our biosphere. Acta Protozool. **41**(3):199–220.

Cornett, R. J., and F. H. Rigler. 1980. The areal hypolimnetic oxygen deficit: An empirical test of the model. Limnol. Oceanogr. **25**:672–679.

Costerton, J. W., K. J. Cheng, G G. Geesey, T. I. Ladd, J. C. Nickel, M. Dasgupta, and T. J. Marrie. 1987. Bacterial biofilms in nature and disease. Ann. Rev. Microbiol. **41**(1):435–464.

Coulter, G. W., ed. 1991. Lake Tanganyika and its life. Natural History Museum Publications and Oxford/New York: Oxford University Press.

Covich, A., and M. Stuiver, 1974. Changes in oxygen 18 as a measure of long-term fluctuations in tropical lake levels and molluscan populations. Limnol. Oceanogr. **19**:682–691.

Covich, A. P., J. H. Thorp, and D. C. Rogers. 2010. Introduction to the subphylum Crustacea. Ch. 18 in J. H. Thorp and A. P. Covich, eds., Ecology and classification of North American freshwater invertebrates. San Diego, CA: Academic Press.

Cowles, H. C. 1899. The ecological relations of the vegetation on the sand dunes of Lake Michigan. Bot Gaz. **27**:95–117, 167–175, 281–308, 361–391.

Crawford, J. T., N. R. Lottig, E. H. Stanley, J. F. Walker, P. C. Hanson, J. C. Finlay, and R. G. Striegl. 2014. CO_2 and CH_4 emissions from streams in a lake-rich landscape: Patterns, controls, and regional significance. Global Biogeochem. Cy. **28**(3):197–210.

Crow, G. E., and C. B. Hellquist. 2006. Aquatic and wetland plants of northeastern North America, Vol. II: A revised and enlarged edition of Norman C. Fassett's A Manual of Aquatic Plants, Vol. II: Angiosperms: Monocotyledons. Madison: University of Wisconsin Press.

Crowther, R. A., and H. B. N. Hynes. 1977. The effect of road deicing salt on the drift of stream benthos. Environ. Pollut. **14**:113–126.

Culver, D. A. 1977. Biogenic meromixis and stability in a soft-water lake. Limnol. Oceanogr. **22**:667–686.

Cummings, K. S., and D. L. Graf. 2010. Mollusca: Bivalvia. In J. H. Thorp and A. P. Covich, eds., Ecology and classification of North American freshwater invertebrates. New York: Academic Press.

Cummins, K. W. 1978. Ecology and distribution of aquatic insects. Pp. 29–32 in R. W. Merritt and K. W. Cummins, eds., An introduction to the aquatic insects of North America. Dubuque, IA: Kendall/Hunt.

Cummins, K. W., and M. J. Klug. 1979. Feeding ecology of stream invertebrates. Ann. Rev. Ecol. Syst. **10**:127–172.

Cummins, K. W., and J. C. Wuycheck. 1971. Caloric equivalents for investigations in ecological energetics. Mitt. Internat. Verein. Limnol. **18**:1–158.

Curl, H., Jr., J. T. Hardy, and R. Ellermeier. 1972. Spectral absorption of solar radiation in alpine snowfields. Ecology **53**:1189–1194.

Daborn, G. R. 1975. The argillotrophic lake system. Verh. Internat. Verein. Limnol. **19**:580–588.

Dai, A., T. Qian, K. E. Trenberth, and J. Milliman. 2009. Changes in continental freshwater discharge from 1948 to 2004. J. Climate 22:2773–2792.

Darnell, R. M. 1961. Trophic spectrum of an estuarine community, based on studies in Lake Pontchartrain, Louisiana. Ecology 42:553–568.

De, P. K. 1939. The role of blue-green algae in nitrogen fixation in rice-fields. Proc. Roy. Soc. London (B) 127:121–139.

Debroas, D., F. Enault, I. Jouan-Dufournel, G. Bronner, and J-F. Humbert. 2011. Metagenomic approach studying the taxonomic and functional diversity of the bacterial community in a lacustrine ecosystem. Pp. 287–293 in Frans J. de Bruijn, ed., Handbook of molecular microbial ecology II: Metagenomics in different habitats. Hoboken, NJ: John Wiley and Sons.

Deevey, E. S., Jr. 1941. Limnological studies in Connecticut: VI. The quantity and composition of the bottom fauna of 36 Connecticut and New York lakes. Ecol. Monogr. 11:413–455.

Deevey, E. S., Jr. 1942. A re-examination of Thoreau's "Walden." Quart. Rev. Biol. 17:1–11.

Deevey, E. S., Jr. 1955. The obliteration of the hypolimnion. Mem. 1st. Ital. Idrobiol. Suppl. 8:9–38.

Deevey, E. S., Jr. 1957. Limnologic studies in Middle America with a chapter on Aztec limnology. Trans. Connecticut Acad. Arts Sci. 39:213–328; 4 plates.

Deevey, E. S., Jr. 1970. In defense of mud. Bull. Ecol. Soc. Amer. 51:5–8.

Deevey, E. S., Jr., G. B. Deevey, and M. Brenner. 1980. Structure of zooplankton communities in the Petén Lake District, Guatemala. Pp. 669–678 in W.C. Kerfoot, ed., Evolution and ecology of zooplankton communities. Hanover, NH: University Press of New England.

DeLong, J. P. 2008. The maximum power principle predicts the outcomes of two-species competition experiments. Oikos, 117(9):1329–1336.

Denman, K. L., and Gargett, A. E. 1983. Time and space scales of vertical mixing and advection of phytoplankton in the upper ocean. Oceanography 28(5):801–815.

Dickson, J. L., J. W. Head, J. S. Levy, and D. R. Marchant. 2013. Don Juan Pond, Antarctica: Near-surface $CaCl_2$-brine feeding Earth's most saline lake and implications for Mars. Scientific Reports, 3. doi:10.1038/srep01166.

Dietz, R. S., and J. F. McHone. 1972. Laguna Quatavita: Not meteoritic, probable salt collapse crater. Meteoritics 7:303–307.

Dijkstra, K. D. B., M. T. Monaghan, and S. U. Pauls. 2014. Freshwater biodiversity and aquatic insect diversification. Annu. Rev. Entomol. 59:143–163.

Dillon, P. J. 1975. The phosphorus budget of Cameron Lake, Ontario: The importance of flushing rate to the degree of eutrophy of lakes. Limnol. Oceanogr. 20:28–39.

Dillon, T. M., T. M. Powell, and L. O. Myrup. 1975. Low frequency turbulence and vertical temperature microstructure in Lake Tahoe, California-Nevada. Verh. Internat. Verein. Limnol. 19:110–115.

Dodds, W. K. 2006. Eutrophication and trophic state in rivers and streams. Limnol. Oceanogr. 51(1):671–680.

Dodson, S. I., and D. L. Egger. 1980. Selective feeding of red phalaropes on zooplankton of Arctic ponds. Ecology 61:755–763.

Dodson, S. I., C. E. Cáceres, and D. C. Rogers. 2010. Cladocera and other Brachiopoda. Pp. 774–828 in J. H. Thorp and A. P. Covich, eds., Ecology and classification of North American freshwater invertebrates, 3rd ed. London: Academic Press/Elsevier.

Dolman, H. 2008. The role of the hydrological cycle in the climate system. In M. Bierkens, H. Dolman, and P. Troch, eds., Climate and the hydrological cycle. London: International Association of Hydrological Sciences.

Drewes, K. 1928. Ober die Assimilation des Luftstickstoffe durch Blaualgen. Zentbl. Bakt. Parastke Abt. II 76:88–101.

Drinkwater, L. E., and J. H. Crowe. 1991. Hydration state, metabolism and hatching of Mono Lake Artemia cysts. Biol. Bull. 180:432–439.

Dubay, C. I., and G. M. Simmons, Jr. 1979. The contribution of macrophytes to the metalimnetic oxygen maximum in a montane, oligotrophic lake. Amer. Midl. Nat. **101**:108–117.

Dumont, H. J. 1972. The biological cycle of molybdenum in relation to primary production and waterbloom formation in a eutrophic pond. Verh. Internat. Verein. Limnol. **18**:84–92.

Dunn, D. B., G. K. Sharma, and C. C. Campbell. 1965. Stomatal patterns of dicotyledons and monocotyledons. Am. Midl. Nat.**74**(1):185–195.

Dunson, W. A., and G. W. Ehlert. 1971. Effects of temperature and salinity and surface water flow on distribution of the sea snake *Pelamis*. Limnol. Oceanogr. **16**:845–853.

Dussart, B. 1966. Limnologie: L'Etude des eaux continentales. Paris: Gauthier-Villars.

Dye, J. F. 1944. The calculation of alkalinities and free carbon dioxide in water by use of nomographs. J. Amer. W. W. Assoc. **36**:895–900.

Dye, J. F. 1952. Calculation of the effect of temperature on the three forms of alkalinity. J. Amer. W. W. Assoc. **44**:356–372.

Eberly, W. R. 1964. Further studies on the metalimnetic oxygen maximum with special reference to its occurrence throughout the world. Invest. Indiana Lakes Streams **6**(3):103–139.

Eddy, S. 1963. Minnesota and the Dakotas. Pp. 301–315 in D. G. Frey, ed., Limnology in North America. Madison: University of Wisconsin Press.

Edmonds, J. S., and J. V. Ward. 1979. Profundal benthos of a multibasin foothills reservoir in Colorado, U.S.A. Hydrobiologia **63**:199–208.

Edmondson, W. T. 1960. Reproduction rates of rotifers in natural populations. Mem. Ist. Ital. Idrobiol. **12**:21–77.

Edmondson, W. T. 1963. Pacific Coast and Great Basin. Pp. 371–392 in D. G. Frey, ed., Limnology in North America. Madison: University of Wisconsin Press.

Edmondson, W. T. 1980. Secchi disk and chlorophyll. Limnol. Oceanogr. **25**:378–379.

Edmondson, W. T. 1991. The uses of ecology: Lake Washington and beyond. Seattle: The University of Washington Press.

Edwards, C. 1980. The anatomy of Daphnia mandibles. Trans. Amer. Microsc. Soc. **99**:2–24.

Eggleton, F. E. 1931. A limnological study of the profundal bottom fauna of certain freshwater lakes. Ecol. Monogr. **1**:231–332.

Eggleton, F. E. 1956. Limnology of a meromictic interglacial plunge-basin lake. Trans. Amer. Microsc. Soc. **75**:334–378.

Eichhorn, R. 1957. Zur Populationsdynamik der Calaniden Copepoden in Titisee und Feldsee. Arch. Hydrobiol. Suppl. **24**:186–246.

Elgmork, K. 1980. Evolutionary aspects of diapause in freshwater copepods. Pp. 411–417 in W. C. Kerfoot, ed., Evolution and ecology of zooplankton communities. Hanover, NH: The University Press of New England.

Ellis, B. K., J. A. Stanford, D. Goodman, C. P. Stafford, D. L. Gustafson, D. A. Beauchamp, ... and B. S. Hansen. 2011. Long-term effects of a trophic cascade in a large lake ecosystem. P. Natl. A. Sci. **108**(3):1070–1075.

Elser, J. J., T. Andersen, J. S. Baron, A. K. Bergström, M. Jansson, M. Kyle, ... and D. O. Hessen. 2009. Shifts in lake N: P stoichiometry and nutrient limitation driven by atmospheric nitrogen deposition. Science, **326**(5954), 835–837.

Elster, H. I. 1954. Über die Populationsdynamik von *Eudiaptomus gracilis* Sars und *Heterocope borealis* Fischer in Bodensee-Obersee. Arch. Hydrobiol. Suppl. **20**:546–614.

Elton, C. 1927. Animal ecology. London: Sidgwick and Jackson Ltd.

Engel, R. 1962. *Eurytemora affinis,* A calanoid copepod new to Lake Erie, Ohio J. Sci. **62**:252.

Englund, B. 1976. Nitrogen fixation by free-living microorganisms on the lava field of Heimaey, Iceland. Oikos **27**:428–432.

Eppley, R. W., and F. M. MaciasR. 1963. Role of the *alga Chlamydomonas mundana* in anaerobic waste stabilization lagoons. Limnol. Oceanogr. **8**:411–415.

Evans, J. H. 1961. Growth of Lake Victoria phytoplankton in enriched cultures. Nature **189**:417.

Evans, M. S., and J. A. Stewart. 1977. Epibenthic and benthic microcrustaceans (copepods, cladocerans, ostracods) from a nearshore area in southeastern Lake Michigan. Limnol. Oceanogr. **22**:1059–1066.

Ewing, M., F. Press, and W. L. Donn. 1954. An explanation of the Lake Michigan wave of 26 June 1954. Science **120**:684–686.

Farmer, G. T., and J. Cook. 2013. Carbon dioxide, other greenhouse gases, and the carbon cycle. Pp. 199–215 in Climate change science: A modern synthesis. New York/London: Springer-Verlag.

Fenchel, T. M. 1978. The ecology of micro- and meiobenthos. Ann. Rev. Ecol. Syst. **9**:99–121.

Fenchel, T. M. 2008. The microbial loop—25 years later. J. Exp. Mar. Biol. Ecol. **366**(1), 99–103.

Fickeisen, D. H., and M. J. Schneider. 1976. Gas bubble disease. Oak Ridge, TN: Technical Information Center, Office of Public Affairs, Energy Research and Development Administration.

Findenegg, I. 1935. Limnologische Untersuchungen im Kärtner Seengebiet. Internat. Rev. Hydrobiol. **32**:369–423.

Findenegg, I. 1937. Holomiktische und meromiktische Seen. Internat. Rev. Hydrobiol. **35**:586–610.

Findenegg, I. 1964. Types of planktonic primary production in the lakes of the Eastern Alps as found by the radioactive carbon method. Verh. Internat. Verein. Limnol. **15**:352–359.

Findenegg, I. 1971. Die Produktionsleistungen einiger planktischer Algenarten in ihrem natürlichen Milieu. Arch. Hydrobiol. **69**:273–293.

Fish, G. R. 1956. Chemical factors limiting growth of phytoplankton in Lake Victoria. East Afric. Agric. J. **21**:152–158.

Fisher, D. J. 1973. Geochemistry of minerals containing phosphorus. Pp.141–152 in Griffith et al., eds., Environmental phosphorus handbook. New York: John Wiley and Sons.

Fisher, S. G. 1977. Organic matter processing by a stream-segment ecosystem: Fort River, Massachusetts, U.S.A. Internat. Rev. Hydrobiol. **62**:701–727.

Fisher, S. G., and G. E. Likens. 1973. Energy flow in Bear Brook, New Hampshire: An integrative approach to stream ecosystem metabolism. Ecol. Monogr. **43**:421–439.

Fisher, S. G., and W. L. Minckley. 1978. Chemical characteristics of a desert stream in flash flood. J. Arid Environ.**1**:25–33.

Fitzpatrick F. A., I. R. Waite, P. J. D'Arconte, M. R. Meador, M. A. Maupin, and M. E. Gurtz. 1998. Revised methods for characterizing stream habitat in the National Water-Quality Assessment Program. U.S. Geological Survey, Water-Resources Investigations Report 98-4052.

Florin, M. J., and H. E. Wright, Jr. 1969. Diatom evidence for the persistence of stagnant glacial ice in Minnesota. Bull. Geol. Soc. Amer. **80**:295–704.

Forbes, S. A. 1887. The lake as a microcosm. Bull. Peoria Sci. Assoc. **1887**:77–87.

Forel, F. A. 1892, 1895, 1904. Le Léman: Monographie limnologique, 3 vols. Lausanne: F. Rouge.

Foster, J. M. 1973. Limnology of two desert recharged-groundwater ponds. Ph.D. thesis, Arizona State University, Tempe.

Francoeur, S. N., M. Schaecher, R. K. Neely, and K. A. Kuehn. 2006. Periphytic photosynthetic stimulation of extracellular enzyme activity in aquatic microbial communities associated with decaying Typha litter. Microb. Ecol. **52**(4), 662–669.

Frantz, T. C., and A. J. Cordone. 1967. Observations on deepwater plants in Lake Tahoe, California and Nevada. Ecology **48**:709–714.

Frey, D. G. 1955. Längsee: A history of mermomixis. Mem. Ist. Ital. Idrobiol. Suppl. **8**:141–161.

Frey, D. G. 1959. The taxonomic phylogenetic significance of the head pores of the Chydoridae (Cladocera). Internat. Rev. Hydrobiol. **44**:28–50.

Frey, D. G. 1982. Questions concerning cosmopolitanism in Cladocera. Arch. Hydrobiol. **93**:484–502.

Frey, H., W. Haeberli, A. Linsbauer, C. Huggel, and F. Paul. 2010. A multi-level strategy for anticipating future glacier lake formation and associated hazard potentials. Nat. Hazard Earth Sys., **10**(2):339–352.

Fryer, G. 1968. Evolution and adaptive radiation in the Chydoridae (Crustacea: Cladocera): A study in comparative morphology and ecology. Phil. Trans. Roy. Soc. London (B) **254**:221–385.

Fryer, G. 1987. A new classification of the branchiopod Crustacea. Zool. J. Linnean Soc. **91**:357–383.

Gaarder, T., and H. H. Gran. 1927. Investigations of the production of plankton in the Oslo Fjord. Rapports et procs-verbaux des réunions, Cons. Internat. Explor. Mer. **42**:1–48.

Gabriels, W., K. Lock, N. De Pauw, and P. L. Goethals. 2010. Multimetric Macroinvertebrate Index Flanders (MMIF) for biological assessment of rivers and lakes in Flanders (Belgium). Limnologica: Ecol. and Mgmt. of Inland Waters **40**(3):199–207.

Gallardo, B., M. García, Á.Cabezas, E. González, M. González, C. Ciancarelli, and F. A. Comín. 2008. Macroinvertebrate patterns along environmental gradients and hydrological connectivity within a regulated river-floodplain. Aquat. Sci. **70**(3):248–258.

Gallardo, B., S. Gascon, M. Gonzalez-Sanchis, A. Cabezas, and F. A. Comín. 2009. Modelling the response of floodplain aquatic assemblages across the lateral hydrological connectivity gradient. Mar. Freshwater Res. **60**(9):924–935.

Gause, G. P. 1934. The struggle for existence. Baltimore: Williams and Wilkins.

Gebhardt, W. 1986. Photosynthetic efficiency. Radiat. Environ. Bioph. **25**(4):275–288.

Geiling, W. T., and R. S. Campbell. 1972. The effect of temperature on the development rate of the major life stages of *Diaptomus pallidus* Herrick. Limnol. Oceanogr. **17**:304–307.

Genin, A., J. S. Jaffe, R. Reef, C. er, and P. J. Franks. 2005. Swimming against the flow: A mechanism of zooplankton aggregation. *Science* **308**(5723):860–862.

Gianniou, S. K., and V. Z. Antonopoulos. 2007. Evaporation and energy budget in Lake Vegoritis, Greece. J. Hydrology **345**:212–223.

Gilbert, J. J. 1966. Rotifer ecology and embryological induction. Science **151**:1234–1237.

Gilbert, J. J. 1972. a-Tocopherol in males of the rotifer *Asplanchna sieboldi:* Its metabolism and its distribution in the testes and rudimentary gut. J. Exper. Zool. **181**:117–128.

Gjessing, E. T. 1981. The effect of aquatic humus on the biological availability of cadmium. Arch. Hydrobiol. **91**:141–149.

Gleason, F. K., and J. M. Wood. 1977. Ribonucleotide re-ductase in blue-green algae: dependence on adenosylcobalamin. Science **192**:1343–1344.

Gliwicz, Z. M., and E. Biesiadka. 1975. Pelagic water mites (Hydracarina) and their effect on the plankton community in a neotropical man-made lake. Arch. Hydrobiol. **76**:65–88.

Goldman, C. R. 1960. Molybdenum as a factor limiting primary productivity in Castle Lake, California. Science **132**:1016–1017.

Goldman, C. R. 1961. The contribution of alder trees (*Alnus tenuifolia*) *to* the primary productivity of Castle Lake, California. Ecology **42**:282–288.

Goldman, C. R. 1970. Antarctic freshwater ecosystems. Pp. 609–627 in M. W. Holdgate, ed., Antarctic ecology, Vol. 2. New York: Academic Press.

Goldman, C. R. 1972. The role of minor nutrients in limiting the productivity of aquatic ecosystems. Pp. 21–38 in G. E. Likens, ed., Nutrients and eutrophication: The limiting-nutrient controversy. ASLO Special Symposia 1. Lawrence, KS: American Society of Limnology and Oceanography.

Goldman, C. R. 1981. Lake Tahoe: Two decades of change in a nitrogen deficient oligotrophic lake. Verh. Internat. Verein. Limnol. **21**:45–70.

Goldman, C. R., D. T. Mason, and J. E. Hobbie. 1967. Two Antarctic desert lakes. Limnol. Oceanogr. **12**:295–310.

Goldspink, C. R., and D. B. C. Scott. 1971. Vertical migration of *Chaoborus flavicans* in a Scottish loch. Freshwat. Biol. **1**:411–421.

Golterman, H. L. 1971. The determination of mineralization losses in correlation with the estimation of net primary production with the oxygen method and chemical inhibitors. Freshwat. Biol. 1:249–256.

Golterman, H. L. 1975. Physiological limnology. Amsterdam: Elsevier.

Golubic, S. 1969. Cyclic and noncyclic mechanisms in the formation of travertine. Verh. Internat. Verein. Limnol. 17:956–961.

Golubic, S. 1973. The relationship between blue-green algae and carbonate deposits. Pp. 434–472 in N. G. Carr and B. A. Whitton, eds., The biology of the blue-green algae. Bot. Monogr. 9. Berkeley: University of California Press.

Goodman, D. 1975. The theory of diversity-stability relationships in ecology. Quart. Rev. Biol. 50:237–266.

Gorham, E. 1957a. The chemical composition of lake waters in Halifax County, Nova Scotia. Limnol. Oceanogr. 2:1221.

Gorham, E. 1957b. The development of peat lands. Quart. Rev. Biol. 32:145–166.

Gorham, E. 1958a. The physical limnology of Northern Britain: An epitome of the bathymetrical survey of the Scottish freshwater lochs, 1897–1909. Limnol. Oceanogr. 3:40–50.

Gorham, E. 1958b. The influence and importance of daily weather conditions in the supply of chloride, sulphate and other ions to fresh waters from atmospheric precipitation. Phil. Trans. Roy. Soc. London (B) 247:147–178.

Gorham, E. 1961. Factors influencing supply of major ions to inland waters, with special reference to the atmosphere. Bull. Geol. Soc. Amer. 72:795–840.

Gorham, E. 1964. Morphometric control of annual heat budgets in temperate lakes. Limnol. Oceanogr. 9:525–529.

Goulden, C. E. 1968. The systematics and evolution of the Moinidae. Trans. Amer. Philos. Soc. 58(6):1–101.

Govedich, F. R., B. A. Bain, W. E. Moser, S. R. Gelder, R. W. Davies, and R. O. Brinkhurst. 2010. Annelida (Clitellata): Oligochaeta, Branchiobdellida, Hirudinida, and Acanthobdellida. Ecol. and Classif. of N. Amer. Freshwater Invert. 3:385–436.

Graf, W. L. 1999. Dam nation: A geographic census of American dams and their large-scale hydrologic impacts. Water Resour. Res. 35:1305–1311.

Grattan, L. M., D. Oldach, T. M. Perl, M. H. Lowitt, D. L. Matuszak, C. Dickson, ... and J. G. Morris Jr. 1998. Learning and memory difficulties after environmental exposure to waterways containing toxin-producing Pfiesteria or Pfiesteria-like dinoflagellates. Lancet, 352(9127):532–539.

Gray, J. S., and M. Elliott. 2009. Ecology of marine sediments: From science to management. Oxford, UK: Oxford University Press.

Green, J., S. A. Corbet, E. Watts (nee Betney), and O. B. Lan. 1976. Ecological studies on Indonesian lakes: Overturn and restratification of Ranu Lamongan. J. Zool. 180:315–354.

Green, M. A., Emery, K., Hishikawa, Y., Warta, W., and Dunlop, E. D. 2012. Solar cell efficiency tables (version 39). Prog. in Photovoltaics: Res. and Apps. 20(1):12–20.

Greenbank, J. T. 1945. Limnological conditions in ice-covered lakes, especially as related to winter-kill of fish. Ecol. Monogr. 15:343–392.

Gregory, K. J. 2006. The human role in changing river channels. Geomorphology 79:172–191.

Grimaldi, S., and A. Petroselli. 2014. Do we still need the Rational Formula? An alternative empirical procedure for peak discharge estimation in small and ungauged basins. Hydrological Sciences J. doi: 10.1080/02626667.2014.880546.

Grimm, N. B., S. G. Fisher, and W. L. Minckley. 1982. Nitrogen and phosphorous dynamics in hot desert streams of southwestern U.S.A. Hydrobiologia 83:303–312.

Grubaugh, J., B. Wallace, and E. Houston. 1997. Production of benthic macroinvertebrate communities along a southern Appalachian river continuum. Freshwater Biol. 37(3):581–596.

Gumprecht, B. 1999. The Los Angeles River. Baltimore, MD: Johns Hopkins University Press.

Gupta, H. K. 2002. A review of recent studies of triggered earthquakes by artificial water reservoirs with special emphasis on earthquakes in Koyna, India. Earth-Sci. Rev. **58**(3):279–310.

Hadzisce, S. D. 1966. Das Mixiphänomen im Ohridsee im Laufe der Jahre 1941/42–1964/65. Verh. Internat. Verein. Limnol. **16**:134–138.

Hairston, N. G., Jr. 1979a. The adaptive significance of color polymorphism in two species of *Diaptomus* (Copepoda). Limnol. Oceanogr. **24**:15–37.

Hairston, N. G., Jr. 1979b. The relationship between pigmentation and reproduction in two species of *Diaptomus* (Copepoda). Limnol. Oceanogr. **24**:38–44.

Hairston, N. G., Smith, F. E., and Slobodkin, L. B. 1960. Community structure, population control, and competition. Am. Nat. **94**(879):421–425.

Hall, C. A. S. 1972. Migration and metabolism in a temperate stream ecosystem. Ecology **53**:585–604.

Hall, R. O., and J. L. Tank. 2005. Correcting whole-stream estimates of metabolism for groundwater input. Limnol. Oceanogr-Meth. **3**:222–229.

Haltuch, M. A., P. A. Berkman, and D. W. Garton. 2000. Geographic information system (GIS) analysis of ecosystem invasion: Exotic mussels in Lake Erie. Limnol. Oceanogr. **45**(8):1778–1787.

Harned, H. S., and R. Davis. 1943. The ionization constant of carbonic acid in water, and the solubility of CO_2 in water and aqueous salt solutions from 0 to 50°. J. Amer. Chem. Soc. **65**:2030–2037.

Harned, H. S., and S. R. Scholes, Jr. 1941. The ionization constant of HCO_3^- from 0 to 50°. J. Amer. Chem. Soc. **63**:1706–1709.

Harris, G. P., and B. B. Piccinin. 1977. Photosynthesis by natural phytoplankton populations. Arch. Hydrobiol. **80**:405–457.

Harris, G. P., and J. N. A. Lott. 1973. Observations on Langmuir circulations in Lake Ontario. Limnol. Oceanogr. **18**:584–589.

Hart, D. D. 1981. Foraging and resource patchiness: Field experiments with a grazing stream insect. Oikos **37**:46–62.

Hartland-Rowe, R. 1972. The limnology of temporary waters and the ecology of Euphyllopoda. Pp. 15–31 in R. B. Clark and R. J. Wootton, eds., Essays in hydrobiology. Exeter: University of Exeter Press.

Hartman, R. T., and D. L. Brown. 1967. Changes in internal atmosphere of submersed vascular hydrophytes in relation to photosynthesis. Ecology **48**:252–258.

Harvey, P. H., and L. Partridge. 1998. Evolutionary ecology: Different routes to similar ends. Nature **392**(6676):552–553.

Hasler, A. D. 1947. Eutrophication of lakes by domestic drainage. Ecology **28**:383–395.

Hasler, A. D., O. M. Brynildson, and W. T. Helm. 1951. Improving conditions for fish in brown-water bog lakes by alkalization. J. Wildl. Mgt. **15**:347–352.

Haswell, M. S., D. G. Randall, and S. F. Perry. 1980. Fish gill carbonic anhydrase: Acid-base regulation or salt transport? Am. J. Physiol. **238**(3):R240–R245.

Hayes, F. R. 1955. The effect of bacteria on the exchange of radiophosphorus at the mud-water interface. Verh. Internat. Verein. Limnol. **12**:111–116.

Hayes, F. R. 1957. On the variation in bottom fauna and fish yield in relation to trophic level and lake dimensions. J. Fish. Res. Bd. Canada **14**:1–32.

Hayes, F. R., and J. E. Phillips. 1958. Lake water and sediment: IV. Radiophosphorus equilibrium with mud, plants, and bacteria under oxidized and reduced conditions. Limnol. Oceanogr. **3**:459–480.

Head, W. D., and E. J. Carpenter. 1975. Nitrogen fixation associated with the marine macroalga *Codium fragile*. Limnol. Oceanogr. **20**:815–823.

Hecky, R. E., and P. Kilham. 1973. Diatoms in alkaline, saline lakes: Ecology and geochemical implications. Limnol Oceanogr. **18**:53–71.

Hecky, R. E., P. Campbell, and L. L. Hendzel. 1993. The stoichiometry of carbon, nitrogen, and phosphorus in particulate matter of lakes and oceans. Limnol. Oceanogr. **38**:709–724.

Heiri, O., and A. F. Lotter. 2003. 9000 years of chironomid assemblage dynamics in an Alpine lake: Long-term trends, sensitivity to disturbance, and resilience of the fauna. J. Paleolimnology **30**: 273–289.

Heiri, O., and A. F. Lotter. 2005. Holocene and Lateglacial summer temperature reconstruction in the Swiss Alps based on fossil assemblages of aquatic organisms: A review. Boreas **34**:506–516

Heisler, J., P. M. Glibert, J. M. Burkholder, D. Maerson, W. Cochlan, W. C. Dennison, ... and M. Suddleson. 2008. Eutrophication and harmful algal blooms: A scientific consensus. Harmful algae **8**(1)3–13.

Hensen, V. 1887. Über die Bestimmung des Planktons oder des im Meere treibenden Materials an Pflanzen and Thieren. Ber. Komm. Wiss, Unters. Deutschen Meere, Kiel **5**:1–107.

Henson, E. B., A. S. Bradshaw, and D. C. Chandler. 1961. The physical limnology of Cayuga Lake, New York. Cornell Univ. Agric. Exper. Sta. Memoir **378**:1–63.

Hershey, A. E., J. Pastor, B. J. Peterson, and G. W. Kling. 1993. Stable isotopes resolve the drift paradox for Baetis mayflies in an arctic river. Ecology **74**(8):2315–2325.

Hesse, P. R. 1957. The distribution of sulphur in the muds, water and vegetation of Lake Victoria. Hydrobiologia **11**:29–39.

Hesse, P. R. 1958. Fixation of sulphur in the muds of Lake Victoria. Hydrobiologia **11**:171–181.

Hillebrand, H. 2002. Top-down versus bottom-up control of autotrophic biomass—a meta-analysis on experiments with periphyton. J. N. Am. Benthol. Soc. **21**(3):349–369.

Hillman, T. J. 1986. Billabongs. Pp. 457–470 in P. DeDeckker and W. D. Williams, eds., Limnology in Australia ((Monographiae Biologicae). Melbourne: CSIRO; Melbourne: CSIRO; Dordrecht, Netherlands: Dr. W. Junk b.v.

Hobbie, J. E. 1964. Carbon 14 measurements of primary production in two Arctic Alaskan lakes. Verh. Internat. Verein. Limnol. **15**:360–364.

Hoenke, K. M., M. Kumar, and L. Batt. 2014. A GIS based approach for prioritizing dams for potential removal. Ecol. Eng. **64**:27–36.

Hohman R., Kipfer, F. Peeters, G. Piepke, and D. M. Imboden. 1997. Processes of deep-water renewal in Lake Baikal. Limnol. Oceanogr. **42**(5):841–855.

Holliday, V. T. 1997. Origin and evolution of lunettes on the high plains of Texas and New Mexico. Quaternary Research **47**:54–69.

Holtgrieve, G. W., D. E. Shindler, W. O. Hobbs, et al. 2011. A coherent signature of anthropogenic nitrogen deposition to remote watersheds of the northern hemisphere. Science **334**(6062):1545–1548. doi: 10.1126/science.1212267.

Holzner, C. P., W. Aeschbach-Hertig, M. Simona, M. Veronesi, D. M. Imboden, and R. Kipfer. 2009. Exceptional mixing events in meromictic Lake Lugano (Switzerland/Italy), studied using environmental tracers. Limnol. Oceanogr. **54**(4):1113–1124.

Hooper, F. F. 1951. Limnological features of a Minnesota seepage lake. Amer. Midl. Nat. **46**:462–481.

Horne, A. J. 1979. Nitrogen fixation in Clear Lake, California: 4. Diel studies on *Aphanizomenon* and *Anabaena* blooms. Limnol. Oceanogr. **24**:329–341.

Horne, A. J., and G. E. Fogg. 1970. Nitrogen fixation in some English lakes. Proc. Roy. Soc. (B) **175**:351–361.

Horne, A. J., J. C. Sandusky, and C. J. W. Carmiggelt. 1979. Nitrogen fixation in Clear Lake, California: Repetitive synoptic sampling of the spring *Aphanizomenon* blooms. Limnol. Oceanogr. **24**:316–328.

Hrbácek, J. 1962. Species composition and the amount of zooplankton in relation to the fish stock. Naklad. Cesk. Akad. Ved. **72**:1–116.

Hubbs, C., and W. F. Hettler. 1964. Observations on the toleration of high temperatures and low dissolved oxygen in natural waters by *Crenichthys baileyi*. Southwest. Nat. **9**:245–248.

Hudson, J. J., W. D. Taylor, and D. W. Schindler. 2000. Phosphate concentrations in lakes. Nature **406**:54–56.

Hughes, R. N. 1970. An energy budget for a tidal-flat population of the bivalve *Scrobicularia plana* (DaCosta). J. An. Ecol. **39**:357–381.

Hurlbert, S. H. 1971. The non concept of species diversity: A critique and alternative parameters. Ecology **52**:577–586.

Hutchinson, G. E. 1937. A contribution to the limnology of arid regions primarily founded on observations made in the Lahontan Basin. Trans. Connecticut Acad. Arts Sci. **33**:47–132.

Hutchinson, G. E. 1941. Limnological studies in Connecticut: IV. The mechanisms of intermediary metabolism in stratified lakes. Ecol. Monogr. **11**:21–60.

Hutchinson, G. E. 1957. A treatise on limnology: Vol. I. Geography, physics, and chemistry. New York: John Wiley and Sons.

Hutchinson, G. E. 1959. Homage to Santa Rosalia or why are there so many kinds of animals? Am. Nat. **93**(870):145–159.

Hutchinson, G. E. 1967. A treatise on limnology: Vol. II. Introduction to lake biology and the limnoplankton. New York: John Wiley and Sons.

Hutchinson, G. E. 1970a. The biosphere. Sci. Amer. **223**(3):44–53.

Hutchinson, G. E. 1970b. The chemical ecology of three species of *Myriophyllum* (Angiospermae, Haloragaceae). Limnol. Oceanogr. **15**:1–5.

Hutchinson, G. E. 1975. A treatise on limnology: Vol. III. Limnological botany. New York: John Wiley and Sons.

Hutchinson, G. E., and V. T. Bowen. 1947. A direct demonstration of the phosphorus cycle in a small lake. Washington, DC: Proc. Nat. Acad. Sci. **33**:148–153.

Hutchinson, G. E., and V. T. Bowen. 1950. Limnological studies in Connecticut: IX, A quantitative radiochemical study of the phosphorus cycle in Linsley Pond. Ecology **31**:194–203.

Hutchinson, G. E., and H. Löffler. 1956. The thermal classification of lakes. P. Natl. Acad. Sci. USA **42**(2):84–86.

Hutchinson, G. E., and G. E. Pickford. 1932. Limnological observations on Mountain Lake, Virginia. Internat. Rev. Hydrobiol. Hydrogr. **27**:252–264.

Hutchinson, G. E., G. E. Pickford, and J. F. M. Schuurman. 1932. A contribution to the hydrobiology of pans and other inland waters of South Africa. Arch. Hydrobiol. **24**:1–154; tables 1–6, plates 1–8.

Hynes, H. B. N. 1969. The ecology of flowing water in relation to management. J. Wat. Poll. Contr. Fed. **42**:418–424.

Hynes, H. B. N. 1970. The ecology of running waters. Toronto: University of Toronto Press.

Idso, S. B. 1973. On the concept of lake stability. Limnol. Oceanogr. **18**:681–683.

Idso, S. B., and G. A. Cole. 1973. Studies on a Kentucky Knobs Lake: V. Some aspects of the vertical transport of heat in the hypolimnion. J. Ecol. **61**:413–420.

Imboden, D. M., and S. Emerson. 1978. Natural radon and phosphorus as limnologic tracers: Horizontal and vertical eddy diffusion in Greifensee. Limnol. Oceanogr. **23**:77–90.

Iovino, A. J., and W. H. Bradley. 1969. The role of larval Chironomidae in the production of lacustrine copropel in Mud Lake, Marion County, Florida. Limnol. Oceanogr. **14**:898–905.

IPCC. 2013. Summary for policymakers. In T. F. Stocker, D. Qin, G.-K. Plattner, et al., eds., Climate change 2013: The physical science basis. Contribution of Working Group I to the Fifth Assessment Report of the Intergovernmental Panel on Climate Change. Cambridge, UK/New York: Cambridge University Press.

Ivanov, M. V. 1981. The global biogeochemical sulphur cycle. Pp. 61–78 in G. E. Likens, ed., Some perspectives of the major biogeochemical cycles. SCOPE 17. New York: John Wiley and Sons.

Jackson, C. R., P. F. Churchill, and E. E. Roden. 2001. Successional changes in bacterial assemblage structure during epilithic biofilm development. Ecology 82(2):555–566.

Jacobsen, R. J., and G. W. Comita. 1976. Ammonia-nitrogen excretion in *Daphnia pulex*. Hydrobiologia 51:195–200.

Jacobson, R. L., and D. Langmuir. 1974. Controls on the quality variations of some carbonate spring waters. J. Hydrol. 23:247–265.

Jankowski, T., D. M. Livingstone, H. Bührer, R. Forster, and P. Niederhauser. 2006. Consequences of the 2003 European heat wave for lake temperature profiles, thermal stability, and hypolimnetic oxygen depletion: Implications for a warmer world. Limnol. Oceanogr. 51(2):815–819.

Jawed, M. 1973. Ammonia excretion by zooplankton and its significance to primary productivity during summer. Marine Biol. 23:115–120.

Jenkin, P. M. 1957. The filter-feeding and food of flamingoes (Phoenicopteri). Phil. Trans. Roy. Soc. London (B) 240:401–493.

Jewett, S. G., Jr. 1963. A stonefly aquatic in the adult stage. Science 139:484–485.

Johnson, B. D., and R. C. Cooke. 1979. Bubble populations and spectra in coastal waters: A photographic approach. J. Geophys. Res-Oceans (1978–2012) 84(C7):3761–3766.

Johnson, E. A., and K. Miyanishi. 2008. Testing the assumptions of chronosequences in succession. Ecol. Lett. 11(5):419–431.

Johnson, M. S., M. F. Billett, K. J. Dinsmore, M. Wallin, K. E. Dyson, and R. S. Jassal. 2010. Direct and continuous measurement of dissolved carbon dioxide in freshwater aquatic systems—method and applications. Ecohydrol. 3(1):68–78.

Johnson, R. H., E. DeWitt, and L. R. Arnold. 2012. Using hydrogeology to identify the source of groundwater to Montezuma Well, a natural spring in Central Arizona, USA: Part 1. Environ. Earth Sci. 67(6):1821–1835.

Johnston, D. A. 1980. Volcanic contribution of chlorine to the stratosphere: More significant ozone than previously estimated. Science 209:491–493.

Jónosson, P. M., ed. 1992. Ecology of oligotrophic, subarctic Thingvallavatn. Odense, Denmark: Oikos.

Jonasson, P. M., and J. Kristiansen. 1967. Primary and secondary production in Lake Esrom: Growth of *Chironomus anthracinus* in relation to seasonal cycles of phytoplankton and dissolved oxygen. Internat. Rev. Hydrobiol. 52:163–217.

Jones, J. G. 1972. Studies on freshwater microorganisms: Phosphatase activity in lakes of differing degrees of eutrophication. J. Ecol. 60:777–791.

Juday, C. 1916. Limnological studies on some lakes in Central America. Trans. Wisconsin Acad. Sci. Arts Lett. 18:214–250.

Juday, C., and E. A. Birge. 1932. Dissolved oxygen and oxygen consumed in the lake waters of northeastern Wisconsin. Trans. Wisconsin Acad. Sci. Arts Lett. 27:415–486.

Judd, J. 1970. Lake stratification caused by runoff from street deicing. Water Res. 8:521–532.

Junk, W. J., P. B. Bayley, and R. E. Sparks. 1989. The flood pulse concept in river-floodplain systems. Canad. Spec. Publ. Fisheries Aquat. Sci. 106(1):110–127.

Junk, W. J., S. An, C. M. Finlayson, B. Gopal, J. Květ, S. A. Mitchell, ... and R. D. Robarts. 2013. Current state of knowledge regarding the world's wetlands and their future under global climate change: a synthesis. Aquat. Sci. 75(1):151–167.

Kajak, Z., and J. I. Rybak. 1966. Production and some trophic dependencies in benthos against primary production and zooplankton production of several Masurian lakes. Verh. Internat. Verein. Limnol. 16:441–451.

Kalyuzhnaya, M. G., A. Lapidus, N. Ivanova, A. C. Copeland, A. C. McHardy, E. Szeto, ... and L. Chistoserdova. 2008. High-resolution metagenomics targets specific functional types in complex microbial communities. Nat. Biotechnol. 26(9):1029–1034.

Kambesis, P. N., and J. G. Coke. 2013. Overview of the controls on eogenetic cave and karst development in Quintana Roo, Mexico. In M. J. Lace and J. E. Mylroie, eds., Coastal karst landforms, Coastal Research Library 5. doi: 10.1007/978-94-007-5016-6 16.

Karaus, U., S. Larsen, H. Guillong, and K. Tockner. 2013. The contribution of lateral aquatic habitats to insect diversity along river corridors in the Alps. Landscape Ecol. **28**. doi: 10.1007/s10980-013-9918-5.

Karl, D. M., C. O. Wirsen, and H. W. Jannasch. 1980. Deep-sea primary production at the Galapagos hydro-thermal vents. Science **207**:1345–1347.

Karr, J. R. 1991. Biological Integrity: A long-neglected aspect of water resource management. Ecol. Appl. **1**:66–84.

Keddy, P. A. 1992. Assembly and response rules: Two goals for predictive community ecology. J. Veg. Sci. **3**:157–164.

Kennett, J. P., and N. J. Shackleton. 1975. Laurentide ice sheet meltwater recorded in Gulf of Mexico deep-sea cores. Science **188**:147–150.

Kerekes, J. 1975. Phosphorus supply in undisturbed lakes in Kejimkujik National Park, Nova Scotia (Canada). Verh. Internat. Verein. Limnol. **19**:349–357.

Kerfoot, W. C. 1977. Implications of copepod predation. Limnol. Oceanogr. **22**:316–325.

Kerr, P. C., D. L. Brockway, D. F. Paris, and J. T. Barnett, Jr. 1972. The interrelation of carbon and phosphorous in regulating heterotrophic and autotrophic populations in an aquatic ecosystem, Shriner's Pond. Pp. 41–62 in G. E. Likens, ed., Nutrients and eutrophication: The limiting-nutrient controversy. ASLO Special Symposia 1. Lawrence, KS: American Society of Limnology and Oceanography.

Kilham, P. 1971. A hypothesis concerning silica and the freshwater planktonic diatoms. Limnol. Oceanogr. **16(1)**:10–18.

Kilham, P., and J. M. Melak. 1972. Primary northupite deposition in Lake Mahega, Uganda. Nature Phys. Sci. **238**:123.

Kilham, S. S., and P. Kilham. 1982. The importance of resource supply rates in determining phytoplankton community structure. In D. G. Meyers and J. R. Strickler, eds., Trophic dynamics of aquatic ecosystems. Washington, DC: AAAS Symposium.

Kilham, S. S., and P. Kilham. 1990. Tropical limnology: Do African lakes violate the "first law" of limnology? Verh. Internat. Verein. Limnol. **24**:68–72.

Kimmel, W. G., and Argent, D. G. 2010. Stream fish community responses to a gradient of specific conductance. Water, Air, and Soil Pollution **206**(1-4):49–56.

Kirk, J. T. O. 1976. Yellow substance (gelbstoff) and its contribution to the attenuation of photosynthetically active radiation in some inland and coastal south-eastern Australian waters. Austral. J. Mar. Freshwat. Res. **27**:61–71.

Klemer, A. R., J. Feuillade, and M. Feuillade. 1982. Cyanobacterial blooms: Carbon and nitrogen limitation have opposite effects on the buoyancy of *Oscillatoria*. Science **216**:1629–1631.

Kolpin, D. W., E. T. Furlong, M. T. Meyer, E. M. Thurman, S. D. Zaugg, L. B. Barber, and H. T. Buxton. 2002. Pharmaceuticals, hormones, and other organic wastewater contaminants in US streams, 1999–2000: A national reconnaissance. Envir. Sci. Tech. Lib. **36**(6):1202–1211.

Kononova, M. M. 1966. Soil organic matter. Oxford, UK: Pergamon Press.

Kopp, M., J. M. Jeschke, and W. Gabriel. 2001. Exact compensation of stream drift as an evolutionarily stable strategy. Oikos **92**:522–530.

Koshinsky, G. D. 1970. The morphometry of shield lakes in Saskatchewan. Limnol. Oceanogr. **15**:695–701.

Kostalos, M., and R. L. Seymour. 1976. Role of microbial enriched detritus in the nutrition of *Gammarus minus* (Amphipoda). Oikos **27**:512–516.

Kozhov, M. 1963. Lake Baikal and its life. The Hague: Dr. W. Junk N.V.

Kozlovsky, D. G. 1968. A critical evaluation of the trophic level concept: I. Ecological efficiencies. Ecology **49**:48–60.

Kroopnick, P. 1977. The SO_4:Cl ratio in oceanic rainwater. Pacific Sci. **31**:91–106.

Krueger, D. A., and S. I. Dodson. 1981. Embryological induction and predator ecology in *Daphnia pulex*. Limnol. Oceanogr. **26**:219–223.

Krumholz, L. A., and G. A. Cole. 1959. Studies on a Kentucky Knobs Lake: IV. Some limnological conditions during an unusually cold winter. Limnol. Oceanogr. **4**:367–385.

Kubly, D. M. 1982. Physical and chemical features of playa lakes in southeastern California, USA. Arch. Hydrobiol. Suppl. **62**:491–525.

Kubly, D. M. 1992. Aquatic invertebrates in desert mountain rock pools: The White Tank Mountains, Maricopa County, Arizona. J. Arizona-Nevada Acad. Sci. **26**:55–69.

Kuehn, K. A., B. M. Ohsowski, S. N. Francoeur, and R. K. Neely. 2011. Contributions of fungi to carbon flow and nutrient cycling from standing dead *Typha angustifolia* leaf litter in a temperate freshwater marsh. Limnol. Oceanogr. **56**(2):529–539.

Kurata, A. 1981. The production of B group vitamins by epiphytic microorganisms on macrophytes in Lake Biwa. Verh. Internat. Verein. Limnol. **21**:596–599.

Kusakabe, M., T. Ohba, Y. Yoshida, H. Satake, T. Ohizumi, W. C. Evans,... and G. W. Kling. 2008. Evolution of CO_2 in Lakes Monoun and Nyos, Cameroon, before and during controlled degassing. Geochem. J. **42**(1), 93–118.

Laforsch, C., and R. Tollrian. 2004. Inducible defenses in multipredator environments: Cyclomorphosis in Daphnia cucullata. Ecology **85**(8):2302–2311.

Langelier, W. F. 1936. The analytical control of anticorrosion water treatment. J. Amer. W. W. Assoc. **28**:1500–1521.

LaRow, E. J. 1969. A persistent diurnal rhythm in *Chaoborus* larvae: II. Ecological significance. Limnol. Oceanogr. **14**:213–218.

LaRow, E. J., and D. C. McNaught. 1978. Systems and organismal aspects of phosphorus remineralization. Hydrobiologia **59**:151–154.

Larson, D. W. 2012. Runaway Devils Lake. Am. Sci. **100**:46–53.

Lathbury, A., R. Bryson, and B. Lettau. 1960. Some observations of currents in the hypolimnion of Lake Mendota. Limnol. Oceanogr. **5**:409–413.

Lavaud, J. 2007. Fast regulation of photosynthesis in diatoms: Mechanisms, evolution and ecophysiology. Funct. Plant. Sci. Biotechnol. **1**:267–287.

Laybourn-Parry, J., and D. A. Pearce. 2007. The biodiversity and ecology of Antarctic lakes: Models for evolution. Philos. T. R. Soc. B. **362**(1488):2273–2289.

Lehman, J. T. 1980. Release and cycling of nutrients between planktonic algae and herbivores. Limnol. Oceanogr. **25**:620–632.

Leopold, L. B., M. G. Wolmann, and J. P. Miller. 1964. Fluvial processes in geomorphology, San Francisco: W. H. Freeman and Co.

Lewis, W. M., Jr. 1973. The thermal regime of Lake Lanao (Philippines) and its theoretical implication for tropical lakes. Limnol. Oceanogr. **18**:200–217.

Lewis, W. M., Jr. 1978. Dynamics and succession of the phytoplankton in a tropical lake: Lake Lanao, Philippines. J. Ecol. **66**:849-880.

Lewis, W. M., Jr. 1986. Phytoplankton succession in Lake Valencia, Venezuela. Pp. 189–203 in M. Munawar and J. F. Talling, eds., Seasonality of freshwater phytoplankton. Dordrecht, Netherlands: Dr. W. Junk b.v.

Ligon, F. K., W. E. Dietrich, and W. J. Trush. 1995. Downstream ecological effects of dams. BioScience **45**:183–192.

Likens, G. E. 1965. Some chemical characteristics of meromictic lakes in North America. Pp. 19–62 in D. F. Jackson, ed., Symposium on meromictic lakes. New York: Syracuse University.

Likens, G. E., ed. 1972. Nutrients and eutrophication: The limiting-nutrient controversy. ASLO Special Symposia 1. Lawrence, KS: American Society of Limnology and Oceanography.

Likens, G. E., ed. 1981. Some perspectives of the major biogeochemical cycles. SCOPE 17. John New York: Wiley and Sons.

Likens, G. E., F. H. Bormann, R. S. Pierce. J. S. Eaton, and N. M. Johnson. 1977. Biogeochemistry of a forested ecosystem. New York: Springer-Verlag.

Likens, G. E., and A. D. Hasler. 1962. Movements of radio-sodium (Na^{24}) within an ice-covered lake. Limnol. Oceanogr. **7**:48–56.

Likens, G. E., and P. L. Johnson. 1966. A chemically stratified lake in Alaska. Science **153**:875–877.

Limpens, J., F. C. Berendse, C. Blodau, J. G. Canadell, C. Freeman, J. Holden, ... and G. Schaepman-Strub. 2008. Peatlands and the carbon cycle: From local processes to global implications—a synthesis. Biogeosciences **5**(5):1475–1491.

Lind, O. T. 1985. Handbook of common methods in limnology, 3rd ed. Chicago: Kendall-Hunt.

Lindeman, R. L. 1941. The developmental history of Cedar Creek Bog, Minnesota. Amer. Midl. Nat. **25**:101–112.

Lindeman, R. L. 1942a. Experimental simulation of winter anaerobiosis in a senescent lake. Ecology **23**:1–13.

Lindeman, R. L. 1942b. The trophic-dynamic aspect of ecology. Ecology **23**:399–418.

Lindroth, A. 1957. Abiogenic gas supersaturation of a river water. Arch. Hydrobiol. **53**:589–597.

Liu, Y., and W. B. Whitman. 2008. Metabolic, phylogenetic, and ecological diversity of the methanogenic archaea. Ann. NY Acad. Sci. **1125**(1):171–189.

Livingstone, D. A. 1963a. Alaska, Yukon, Northwest Territories, and Greenland. Pp. 559–574 in D. G. Frey, ed., Limnology in North America. Madison: University of Wisconsin Press.

Livingstone, D. A. 1963b. Chemical composition of rivers and lakes. U.S. Geol. Surv. Prof. Paper 440–G. Washington, DC: U.S. Government Printing Office.

Löffler, H. 1964. The limnology of tropical high-mountain lakes. Verh. Internat. Verein. Limnol. **15**:176–193.

Löffler, H. 1975. The onset of meromictic conditions in Goggausee, Carinthia. Verh. Internat. Verein. Limnol. **19**:2284–2289.

Lofrgen, B. M., and Y. Zhu. 1999. Seasonal climatology of surface energy fluxes on the Great Lakes. NOAA Technical Memorandum ERL GLERL-112.

Logue, J. B., H. Bürgmann, and C. T. Robinson. 2008. Progress in the ecological genetics and biodiversity of freshwater bacteria. BioScience **58**(2):103–113.

Lotka, A. J. 1922. Contribution to the energetics of evolution. P. Natl. Acad. Sci. USA **8**(6):147–151.

Löw, F., P. Navratil, K. Kotte, H. F. Schöler, and O. Bubenzer. 2013. Remote-sensing-based analysis of landscape change in the desiccated seabed of the Aral Sea—A potential tool for assessing the hazard degree of dust and salt storms. Environ. Monit. Assess. **185**(10):8303–8319.

Lowe-McConnell, R. H., R. C. M. Crul, and F. C. Roest. 1992. Symposium on resource use and conservation of the African Great Lakes. Isernhagen, Germany: Herbert Mueller.

Lozinsky, W. 1909. Über die mechanische Verwitterung der Sandsteine im germasstigen Klima. Cl. Sci., Math et Nat., Bull. Acad. Sci. Cracovie, 1–25.

Lund, J. W. G. 1964. Primary production and periodicity of phytoplankton. Verh. Internat. Verein. Limnol. **15**:37–56.

Lyautey, E., B. Lacoste, L. Ten-Hage, J. L. Rols, and F. Garabetian. 2005. Analysis of bacterial diversity in river biofilms using 16S rDNA PCR-DGGE: Methodological settings and fingerprints interpretation. Water Res. **39**(2):380–388.

Macan, T. T., and E. B. Worthington. 1951. Life in lakes and rivers (The New Naturalist Series). London: William Collins Sons.

MacIsaac, J. J., and R. C. Dugdale. 1969. The kinetics of nitrate and ammonia uptake by natural populations of marine phytoplankton. Deep-Sea-Res. **16**:45–57.

Mackereth, F. J. H. 1966. Some chemical observations on postglacial lake sediments. Proc. Roy. Soc. London (B) **250**:165–213.

Maguire, B., Jr. 1971. Phytotelmata: Biota and community structure determination in plant held waters. Pp. 439–464 in R. F. Johnston, ed., Annual review ecology and systematics, Vol. 2. Palo Alto, CA: Annual Reviews.

Malueg, K. W., J. R. Tilstra, D. W. Schults, and C. F. Powers. 1972. Limnological observations on an ultraoligotrohpic lake in Oregon, U.S.A. Verb. Internat. Verein. Limnol. 18:292–302.

Manconi, R., and R. Pronzato. 2008. Global diversity of sponges (Porifera: Spongillina) in freshwater. Pp. 27–33 in E. V. Balian, C. Lévêque, H. Segers, and K. Martens, eds., Freshwater animal diversity assessment. Dordrecht: Springer Netherlands.

Mann, K. H. 1958. Annual fluctuations in sulphate and bicarbonate hardness in ponds. Limnol. Oceanogr. 3:418–422.

Marchant, R., and H. B. N. Hynes. 1981. The distribution and production of *Gammarus pseudolimnaeus* (Crustacea: Amphipoda) along a reach of the Credit River, Ontario. Freshwat. Biol. 11:169–182.

Marie-Ève, G., T. Posch, G. Hitz, F. Pomerleau, C. Pradalier, R. Siegwart and J. Pernthaler. 2013. Short-term displacement of *Planktothrix rubescens* (cyanobacteria) in a pre-alpine lake observed using an autonomous sampling platform. Limnol. Oceanogr. 58(5)1892–1906.

Marotta, R., M. Ferraguti, C. Erséus, and L. M. Gustavsson. 2008. Combined-data phylogenetics and character evolution of Clitellata (Annelida) using 18S rDNA and morphology. Zool. J. Linn. Soc-Lond. 154(1):1–26.

Marra, J. 1978. Phytoplankton photosynthetic response to vertical movement in a mixed layer. Mar. Biol. 46:203–208.

Martin, J. W., and G. E. Davis. 2001. An updated classification of the recent Crustacea. Los Angeles: Natural History Museum of Los Angeles County.

Martin, N. V. 1955. Limnological and biological observations in the region of the Ungava or Chubb Crater, Province of Quebec. J. Fish. Res. Bd. Canada. 12:487–496.

Martin, P. S. 1960. Effect of Pleistocene climatic change on biotic zones near the equator. Pp. 265–267 in Yearbook of the American Philosophical Society, Philadelphia.

Mason, D. T. 1967. Limnology of Mono Lake, California. Univ. Cal. Pub. Zool. 83:1–102; 6 plates.

McConnell, W. J. 1963. Primary productivity and fish harvest in a small desert impoundment. Trans. Amer. Fish. Soc. 92:1–12.

McConnell, W. J., and W. F. Sigler. 1959. Chlorophyll and productivity in a mountain river. Limnol. Oceanogr. 4:335–351.

McCutchan, J. H., W. M. Lewis, Jr., and J. F. Saunders, III. 1998. Uncertainty in the estimation of stream metabolism from open-channel oxygen concentrations. J. N. Am. Benthol. Soc. 17:155–164.

McEwen, G. F. 1929. A mathematical theory of the vertical distribution of temperature and salinity in water under the action of radiation, conduction, evaporation, and mixing due to the resultant convection. Bull. Scripps Oceanogr. Tech. 2:197–306.

McGaha, Y. J. 1952. The limnological relations of insects to certain aquatic flowering plants. Trans. Amer. Microsc. Soc. 71:355–381.

McIntire, C. D., and H. F. Phinney. 1965. Laboratory studies of periphyton production and community metabolism in lotic environments. Ecol. Monogr. 35:237–258.

McKee, P. M., and G. L. Mackie. 1981. Life history adaptations of the fingernail clams *Sphaerium occidentale* and *Musculium securis* to ephemeral habitats. Canad. J. Zool. 59:2219–2229.

McKew, B. A., P. Davey, S. J. Finch, J. Hopkins, S. C. Lefebvre, M. V. Metodiev,... and R. J. Geider. 2013. The trade-off between the light-harvesting and photoprotective functions of fucoxanthin-chlorophyll proteins dominates light acclimation in *Emiliania huxleyi* (clone CCMP 1516). New Phytol. 200(1):74–85.

McLaren, I. A. 1964. Zooplankton of Lake Hazen, Ellesmere Island, and a nearby pond, with special reference to the copepod *Cyclops scutifer* Sars. Canadian J. Zool. 42:613–629.

McMullen, D. B. 1941. Methods used in the control of schistosome dermatitis in Michigan. Pp. 379–388 in J. G. Needham, P. B. Sears, and A. Leopold, eds., Symposium on hydrobiology. Madison: University of Wisconsin Press.

McNaught, D. C. 1971. Plasticity of cladoceran visual systems to environmental changes (abstract). Trans. Amer. Microsc. Soc. **90**:113–114.

Megard, R. O. 1964. Biostratigraphic history of Dead Man Lake, Chuska Mountains, New Mexico. Ecology **45**:529–546.

Megard, R. O. 1968. Planktonic photosynthesis and the environment of calcium carbonate deposition in lakes. Univ. Minnesota, Limnol. Res. Ctr., Interim Report **2**:1–47.

Megard, R. O. 1972. Phytoplankton, photosynthesis, and phosphorus in Lake Minnetonka, Minnesota. Limnol. Oceanogr. **17**:68–87.

Melack, J. M. 1978. Morphometric, physical and chemical features of the volcanic crater lakes of Western Uganda. Arch. Hydrobiol. **84**:430–453.

Melack, J. M., and P. Kilham. 1974. Photosynthetic rates of phytoplankton in East African alkaline saline lakes. Limnol. Oceanogr. **19**:743–755.

Melis, A. 2009. Solar energy conversion efficiencies in photosynthesis: minimizing the chlorophyll antennae to maximize efficiency. Plant Sci. **177**(4):272–280.

Meriläinen, J. J., J. Hynynen, A. Palomäki, H. Veijola, A. Witick, K., Mäntykoski, and K. Lehtinen. 2001. Pulp and paper mill pollution and subsequent ecosystem recovery of a large boreal lake in Finland: A paleolimnological analysis. J. Paleolimnol. **26**(1):11–35.

Meybeck, M. 2003. Global analysis of river systems: From Earth system controls to Anthropocene syndromes. Philos. T. Roy. Soc. B. **358**(1440):1935–1955.

Michel, A. E., A. S. Cohen, K. West, M. R. Johnston, and P. W. Kat. 1992. Large African lakes as natural laboratories for evolution: Examples from the endemic gastropod fauna of Lake Tanganyika. Mitt. Internat. Verein. Limnol. **23**:85–99.

Middleton, B. A. 2002. The flood pulse concept in wetland restoration. Pp. 1–10 in B. A. Middleton, ed., Flood pulsing in wetlands: Restoring the natural hydrological balance. New York: John Wiley and Sons.

Miller, R. R. 1946. *Gila cypha,* a remarkable new species of cyprinid fish from the Colorado River in Grand Canyon, Arizona. J. Washington Acad. Sci. **36**:409–415.

Minckley, W. L. 1973. Fishes of Arizona. Pub. Ariz. Game and Fish Dept. Phoenix, AZ: Sims Printing.

Minckley, W. L., and D. R. Tindall, 1963. Ecology of *Batrachospermum* sp. (Rhodophyta) in Doe Run, Meade County, Kentucky, Bull. Torrey Bot. Club **90**:391–400.

Minshall, G. W. 1978. Autotrophy in stream ecosystems. Bioscience **28**:767–771.

Minshall, G. W., R. C. Petersen, K. W. Cummins, T. L. Bott, J. R. Sedell, C. E. Cushing, and R. L. Vannote. 1983. Interbiome comparison of stream ecosystem dynamics. Ecol. Monogr. **53**(1):1–25.

Mishra, S. K., and V. P. Singh. 1999. Another look at SCS-CN Method. J. Hydrol. Eng. **4**:257–264.

Mitsch, W. J., and J. G. Gosselink. 2007. Wetlands. Hoboken, NJ: John Wiley and Sons.

Mitsch, W. J., X. Wu, R. W. Nairn, P. E. Weihe, N. Wang, R. Deal, and C. E. Boucher. 1998. Creating and restoring wetlands. BioScience **48**(12):1019–1030.

Moen, D. S., D. J. Irschick, and J. J. Wiens. 2013. Evolutionary conservatism and convergence both lead to striking similarity in ecology, morphology and performance across continents in frogs. Proc. R. Soc. B **280**(1773). doi: 10.1098/rspb.2013.2156.

Moore, E. W. 1939. Graphic determination of carbon dioxide and the three forms of alkalinity. J. Amer. Water W. Assoc. **31**:51–66.

Moore, G. M. 1939. A limnological investigation of the microscopic benthic fauna of Douglas Lake, Michigan. Ecol. Monogr. **9**:537–582.

Moore, W. G. 1942. Field studies on the oxygen requirements of certain fresh-water fishes. Ecology **23**:319–329.

Moore, W. G. 1957. Studies on the laboratory culture of Anostraca. Trans. Amer. Microsc. Soc. **76**:159–173.

Mori, K. 1976. Some limnological features of Lake Suigetsu, a typical meromictic lake in Japan. Rept. Environment. Sci. Mie University **1**:161–170.

Mortimer, C. H. 1941–1942. The exchange of dissolved substances between mud and water in lakes. J. Ecol. **29**: 280–329; **30**:147–201.

Mortimer, C. H. 1952. Water movements in lakes during summer stratification: Evidence from the distribution of temperature in Windermere. Philos. T. Roy. Soc. B. **236**:355–404.

Mortimer, C. H. 1974. Lake hydrodynamics. Mitt. Internat. Verein. Limnol. **20**:124–197.

Moskalenko, B. K. 1972. Biological productive system of Lake Baikal. Verh. Internat. Verein. Limnol. **18**:568–573.

Moss, B. 1972. The influence of environmental factors on the distribution of freshwater algae, an experimental study: I. Introduction and the influence of calcium concentration. J. Ecol. **60**:917–932.

Mossop, B., and M. J. Bradford. 2006. Using thalweg profiling to assess and monitor juvenile salmon (Oncorhynchus spp.) habitat in small streams. Can. J. Fish. Aquat. Sci. **63**:1515–1525.

Moyle, J. B. 1945. Some chemical factors influencing the distribution of aquatic plants in Minnesota. Amer. Midl. Nat. **34**:402–420.

Moyle, J. B. 1956. Relationships between the chemistry of Minnesota surface waters and wildlife management. J. Wildl. Mgt. **20**:303–320.

Muhar, S., and M. Jungwirth. 1998. Habitat integrity of running waters—Assessment criteria and their biological relevance. Hydrobiologia **386**:195–202

Müller, K. 1954. Investigations on the organic drift in North Swedish streams. Ann. Rep. Inst. Freshwat. Res. Drottningholm. **35**:133–148.

Munawar, M., and I. F. Munawar. 1975. Some observations on the growth of diatoms in Lake Ontario with emphasis on *Melosira binderana* during thermal bar conditions. Arch. Hydrobiol. **75**:490–499.

Munk, W. H., and G. A. Riley. 1952. Absorption of nutrients by aquatic plants. J. Mar. Res. **11**:215–240.

Naiman, R. J. 1976. Primary production, standing stock, and export of organic matter in a Mohave Desert thermal stream. Limnol. Oceanogr. **21**:60–73.

Naiman, R. J., and J. R. Sedell. 1979. Benthic organic matter as a function of stream order in Oregon. Arch. Hydrobiol. **87**:404–422.

National Academy of Sciences. 1969. Eutrophication: Causes, consequences, correctives. Washington, DC: Author.

Nations, J. D., R. H. Hevly, D. W. Blinn, and J. J. Landye. 1981. Paleontology, paleoecology, and depositional history of the Miocene-Pliocene Verde Formation, Yavapai County, Arizona. Arizona Geol. Survey Digest **13**:133–149.

Neal, J. T. 1975. Playas and dried lakes: Occurrence and development. Stroudsburg, PA: Dowden, Hutchinson and Ross.

Nebeker, A. V. 1976. Survival of *Daphnia*, crayfish, and stoneflies in air-supersaturated water. J. Fish. Res. Bd. Canada **33**:1208–1212.

Nelson, J. S. 2006. Fishes of the World. Hoboken, NJ: John Wiley and Sons.

Neumann, J. 1953. Energy balance and evaporation from sweet-water lakes of the Jordan Rift. Bull. Res. Coun. Israel **2**:337–357.

Neumann, J. 1959. Maximum depth and average depth of lakes. J. Fish. Res. Bd. Canada. **16**:923–927.

Newbold, J. D., J. W. Elwood, R. V. O'Neil, and W. Van Winckle. 1981. Measuring nutrient spiralling in streams. Canad. J. Fish. Aquat. Sci. **38**:860–863.

Newcombe, C. L., and J. V. Slater. 1950. Environmental factors of Sodon Lake—a dichothermic lake in southeastern Michigan. Ecol. Monogr. **20**:207–227.

Nicholls, K. H., W. Kennedy, and C. Hammett. 1980. A fish-kill in Heart Lake, Ontario, associated with the collapse of a massive population of *Ceratium hirundinella* (Dinophyceae). Freshwat. Biol. **10**:553–561.

Nielsen, A. 1950. The torrential invertebrate fauna. Oikos **2**:177–196.

Nihongi, A. 2011. Daphnia pulicaria hijacked by Vibrio cholera: Altered swimming behaviour and predation risk Implications. Pp. 181–192 in A. Nihongi, Ziarek, J. J., T. Nagai, M. Uttieri, E. Zambianchi, J. R. Strickler, and G. Kattel, eds., Zooplankton and phytoplankton: Types, characteristics and ecology. New York: Nova Science.

Nilsson, C., C. A. Reidy, M. Dynesius, and C. Revenga. 2005. Fragmentation and flow regulation of the world's large river systems. Science **308**:405–408.

Nilsson, N. A., and B. Pejler. 1973. On the relation between fish fauna and zooplankton composition in north Swedish lakes. Inst. Freshwat. Res. Drottingholm **53**:51–77.

Nininger, H. H. 1972. Find a falling star. New York: Paul S. Eriksson.

Nocentini, A. M. 1966. Struttura e dinamica della fauna macrobentonica litorale e sublitorale del Lago di Mergozzo. Mem Ist. Ital. Idrobiol. **20**:209–259.

Noe, G. B., D. L. Childers, and R. D. Jones. 2001. Phosphorus biogeochemistry and the impact of phosphorus enrichment: Why is the Everglades so unique? Ecosystems **4**(7), 603–624.

Nriagu, J. O. 1978. Dissolved silica in pore waters of Lakes Ontario, Erie, and Superior sediments. Limnol. Oceanogr. **23**:53–67.

Oana, S., and E. S. Deevey. 1960. Carbon 13 in lake waters and its possible bearing on paleolimnology. Amer. J. Sci. **258A**:253–272.

O'Brien, W. J., and D. Schmidt. 1979. Arctic *Bosmina* morphology and copepod predation. Limnol. Oceanogr. **24**:564–568.

Odén, S. 1976. The acidity problem—an outline of concepts. pages 1–36 in L. S. Dochinger and T. A. Seliga, eds., Proceedings of the first international symposium on acid precipitation and the forest ecosystem. Upper Darby, PA: USDA, Forest Service Gen. Tech. Report NE-23.

Odum, H. T. 1956a. Efficiencies, size of organisms, and community structure. Ecology **37**(3): 592–597.

Odum, H. T. 1956b. Primary production in flowing waters. Limnol. Oceanogr. **1**:102–117.

Odum, H. T. 1957. Trophic structure and productivity of Silver Springs, Florida. Ecol. Monogr. **27**:55–112.

Odum, H. T. 1995. Energy systems concepts and self-organization: A rebuttal. Oecologia **104**(4):518–522.

Ohle, W. 1952. Die hypolimnische Kohlendioxydealckumulation als productionsbiologischer Indikator. Arch. Hydrobiol. **46**:153–285.

O'Kelley, J. C. 1974. Inorganic nutrients. Pp. 610–655 in W. D. P. Stewart, ed., Algal physiology and biochemistry. Oxford, UK: Blackwell Scientific.

Olson, F. C. W. 1960. A system of morphometry. Internat. Hydrog. Rev. **37**:147–155.

O'Neill, R. V. 2001. Is it time to bury the ecosystem concept? (With full military honors, of course!). Ecology **82**(12):3275–3284.

O'Reilly, C. M., S. R. Alin, P. D. Plisnier, A. S. Cohen, and B. A. McKee. 2003. Climate change decreases aquatic ecosystem productivity of Lake Tanganyika, Africa. Nature **424**(6950):766–768.

Osmond, C. B., N. Valaane, S. M. Haslam, P. Uotila, and Z. Roksandic. 1981. Comparisons of $\delta13C$ values in leaves of aquatic macrophytes from different habitats in Britain and Finland: Some implications for photosynthetic processes in aquatic plants. Oecologia **50**(1):117–124.

Osterkamp, W. R., and W. W. Wood. 1987. Playa-lake basins on the Southern High Plains of Texas and New Mexico: Part I. Hydrologic, geomorphic, and geologic evidence for their development. Geol. Soc. Am. Bull. **99**:215–223.

Ostrofsky, M. L., and H. J. Peairs. 1981. The relative photosynthetic efficiency of *Asterionella formosa* Hass. (Bacillarophyta) in natural plankton assemblages. J. Phycol. **17**:230–233.

O'Toole, G., H. B. Kaplan, and R. Kolter. 2000. Biofilm formation as microbial development. Ann. Rev. Microbiol. **54**(1):49–79.

Otsuki, A., and R. G. Wetzel. 1972. Coprecipitation of phosphate with carbonates in a marl lake. Limnol. Oceanogr. **17**:763–767.

Owens, M. 1969. In running waters. Pp. 92–97 in R. A. Vollenweider, ed., A manual on methods for measuring primary production in aquatic environments. IBP Handbook No. 12. Oxford, UK: Blackwell Scientific.

Pace, M. L., and J. J. Cole,. 1996. Regulation of bacteria by resources and predation tested in whole-lake experiments. Limnol. Oceanogr **41**(7):1448–1460.

Pace, M. L., J. J. Cole, S. R. Carpenter, and J. F. Kitchell. 1999. Trophic cascades revealed in diverse ecosystems. Trends Ecol. Evol. **14**(12):483–488.

Pace, M. L., and Y. T. Prairie. 2005. Respiration in lakes. Ch. 7 in P. A. Del Giorgio and P. J. L. B. Williams, eds., Respiration in aquatic ecosystems. Oxford, UK: Oxford University Press.

Padial, A. A., and S. M. Thomaz 2008. Prediction of the light attenuation coefficient through the Secchi disk depth: Empirical modeling in two large neotropical ecosystems. Limnology **9**:143–151.

Paerl, H. W., R. C. Richards, R. L. Leonard, and C. R. Goldman. 1975. Seasonal nitrate cycling as evidence for complete vertical mixing in Lake Tahoe, California-Nevada. Limnol. Oceanogr. **20**:1–8.

Paerl, H. W., and L. A. Mackenzie. 1977. A comparative study of the diurnal carbon fixation patterns of nannoplankton and netplankton Limnol. Oceanogr. **22**:732–738.

Paine, R. T. 1980. Food webs: Linkage, interaction strength and community infrastructure. J. Anim. Ecol. **49**(3):667–685.

Pamatmat, M. M., and A. M. Bhagwat. 1973. Anaerobic metabolism in Lake Washington sediments. Limnol. Oceanogr. **18**:611–627.

Park, P. K., G. R. Webster, and R. Yamamoto. 1969. Alkalinity budget of the Columbia River. Limnol. Oceanogr. **14**:559–567.

Parsons, J. D. 1975. The aquatic thermal capsule. Verh. Internat. Verein. Limnol. **19**:137–143.

Pearsall, W. H. 1932. Phytoplankton in the English Lake District: II. J. Ecol. **20**:231–262.

Pearse, A. S., ed. 1936. The cenotes of Yucatán: A zoological and hydrographic survey. Washington, DC: Carnegie Inst. of Washington; 2 plates.

Pedrós-Alió, C., and T. D. Brock. 1982. Assessing biomass and production of bacteria in eutrophic Lake Mendota, Wisconsin. Appl. Environ. Microb. **44**(1):203–218.

Pennak, R. W. 1940. Ecology of the microscopic metazoa inhabiting the sandy beaches of some Wisconsin lakes. Ecol. Monogr. **10**:537–615.

Pennak, R. W. 1989. Fresh-water invertebrates of the United States: Protozoa to Mollusca, 3rd ed. New York: John Wiley and Sons.

Pennak, R. W., and D. J. Zinn. 1943. Mystacocarida, a new order of Crustacea from intertidal beaches in Massachusetts and Connecticut. Smithsonian Misc. Collect. **103**:1–11.

Peñuelas, J., J. Sardans, A. Rivas-Ubach, and I. A. Janssens. 2012. The human-induced imbalance between C, N, and P in Earth's life system. Global Change Biol. **18**:3–6.

Peter, H., I. Ylla, C. Gudasz, A. M. Romaní, S. Sabater, and L. J. Tranvik. 2011. Multifunctionality and diversity in bacterial biofilms. PLoS ONE **6**(8):e23225.

Peters, R. H., and S. MacIntyre. 1976. Orthophosphate turnover in East African lakes. Oecologia **25**:313–319.

Peterson, B. J. 1980. Aquatic primary productivity and the ^{14}C-CO_2 method: A history of the productivity problem. Ann. Rev. Ecol. Syst. **11**:359–385.

Péwé, T. L. 1969. The periglacial environment. Pp. 1–9 in T. L. Péwé, ed., The periglacial environment: Past and present. Montreal: McGill-Queens University Press.

Pickett, S. T., and M. L. Cadenasso. 2002. The ecosystem as a multidimensional concept: Meaning, model, and metaphor. Ecosystems **5**(1):1–10.

Pivovarov, A. A. 1973. Thermal conditions in freezing lakes and rivers (translation). Jerusalem: Halsted Press.

Poff, N. L., J. D. Olden, D. M. Merritt, and D. M. Pepin. 2007. Homogenization of regional river dynamics by dams and global biodiversity implications. PNAS **104**:5732–5737.

Pomeroy, L. R. 1974. The ocean's food web, a changing paradigm. BioScience **24**:499–504.

Pomeroy, L. R. 1980. Detritus and its role as a food source. Pp. 84–102 in R. K. Barnes and K. H. Mann, eds., Fundamentals of aquatic ecosystems. London: Blackwell Scientific.

Ponder, W. F. 1986. Mound springs of the Great Artesian Basin. Pp. 403–420 in P. DeDeckker and W. D. Williams, eds., Limnology in Australia. Melbourne: CSIRO; Dordrecht: Dr. W. Junk b.v.

Poole, H. H., and W. R. G. Atkins. 1929. Photo-electric measurements of submarine illumination throughout the year. J. Mar. Biol. Assoc. UK **16**:297–324.

Porter, K. G. 1976. Enhancement of algal growth and productivity by grazing zooplankton. Science **192**:1332–1334.

Potash, M. 1956. A biological test for determining the potential productivity of water. Ecology **37**:631–639.

Potts, M. 1979. Ethylene production in a hot brine environment. Arch. Hydrobiol. **87**:198–204.

Pough, F. H. 1976. Acid precipitation and embryonic mortality of spotted salamanders, *Ambystoma maculatum*. Science **192**:68–70.

Povoledo, D. 1964. Some comparative physical and chemical studies on soil and lacustrine organic matter. Mem. Ist. Ital. Idrobiol. **17**:21–32.

Povoledo, D. 1975. Closing address. Pp. 361–368 in D. Povoledo and H. L. Golterman, eds., Humic substances, their structure and function in the biosphere. Proceedings of an international meeting held at Nieuwersluis, The Netherlands, May 29–31, 1972, Centre for Agricultural Publishing and Documentation, Wageningen.

Povoledo, D., and H. L. Golterman, eds. 1975. Humic substances, their structure and function in the biosphere. Proceedings of an international meeting held at Nieuwersluis, The Netherlands, May 29–31, 1972. Centre for Agricultural Publishing and Documentation, Wageningen.

Power, M. E., and Dietrich, W. E. 2002. Food webs in river networks. Ecol. Res. **17**(4):451–471.

Pozdnyakov, D. V., A. A. Korosov, N. A. Petrova, and H. Grassl. 2013. Multi-year satellite observations of Lake Ladoga's biogeochemical dynamics in relation to the lake's trophic status. J. Great Lakes Res. **39**:34–45.

Prakash, A., A. Jensen, and M.A. Rashid. 1975. Humic substances and aquatic productivity. Pp. 259–268 in D. Povoledo and H. L. Golterman, eds., Humic substances, their structure and function in the biosphere. Proceedings of an international meeting held at Nieuwersluis, The Netherlands, May 29–31, 1972. Centre for Agricultural Publishing and Documentation, Wageningen.

Pratt, D. M., and H. Berkson. 1959. Two sources of error in the oxygen light and dark bottle method. Limnol. Oceanogr. **4**:328–334.

Proctor, V. W. 1957. Some controlling factors in the distribution of *Haematococcus pluvialis*. Ecology **38**:457–462.

Provasoli, L. 1969. Algal nutrition and eutrophication. Pp. 574–593 in Eutrophication: Causes consequences, correctives. Washington, DC: National Academy of Science.

Provasoli, L., and A. F. Carlucci. 1974. Vitamins and growth regulators. Pp. 741–787 in W. D. P. Stewart, ed., Algal physiology and biochemistry. Botan. Monogr., Vol. 10. Berkeley/Los Angeles: University of California Press.

Rabinowitch, E. I. 1948. Photosynthesis. Sci. Amer. **179**(2): 24–35.

Raburu, P. O., F. O. Masese, and C. A. Mulanda. 2009. Macroinvertebrate Index of Biotic Integrity (M-IBI) for monitoring rivers in the upper catchment of Lake Victoria Basin, Kenya. Aquat. Ecosyst. Health **12**(2):197–205.

Rader, R. 1997. A functional classification of the drift: traits that influence invertebrate availability to salmonids. Can. J. Fish. Aquat. Sci. **54**:1211–1234.

Ragotzkie, R. A. 1974. The Great Lakes rediscovered. Amer. Sci. **62**:454–464.

Ragotzkie, R. A. 1978. Heat budgets of lakes. Pp. 1–19 in A. Lerman, ed., Lakes—chemistry, geology, physics. New York: Springer-Verlag.

Ragotzkie, R. A., and G. E. Likens. 1964. The heat balance of two Antarctic lakes. Limnol. Oceanogr. **9**:412–425.

Rainwater, F. H., and L. L. Thatcher. 1960. Methods for collection and analysis of water samples. U.S. Geol. Surv. Water-supply Paper 1454. Washington, DC: U.S. Government Printing Office.

Rankin, E. T. 1995. Habitat Indices in water resource quality assessments. Chapter 13 in W. S. Davis and T. P. Simon, eds., Biological assessment and criteria: Tools for water resource planning and decision making. Boca Raton, FL: CRC Press.

Rao, Y. R., M. G. Skafel, and M. N. Charlton. 2004. Circulation and turbulent exchange characteristics during the thermal bar in Lake Ontario. Limnol. Oceanogr. **49**(6):2190–2200.

Rawson, D. S. 1951. The total mineral content of lake water. Ecology **32**:669–672.

Rawson, D. S. 1953. The standing crop of net plankton in lakes. J. Fish. Res. Bd. Canada **10**:224–237.

Rawson, D. S. 1955. Morphometry as a dominant factor in the productivity of large lakes. Verh. Internat. Verein. Limnol. **12**:164–175.

Rawson, D. S., and J. E. Moore. 1944. The saline lakes of Saskatchewan. Canadian J. Res. (D) **22**:141–201.

Redfield, A. C. 1958. The biological control of chemical factors in the environment. Amer. Sci. **46**:205–221.

Reese, E. G., and Batzer, D. P. 2007. Do invertebrate communities in floodplains change predictably along a river's length? Freshwater Biol. **52**(2):226–239.

Reid, J. W., and C. E. Williamson. 2010. Copepoda. In J. H. Thorp and A. P. Covich, eds., Ecology and classification of North American freshwater invertebrates. San Diego, CA: Academic Press.

Reid, M. A., and J. J. Brooks. 2000. Detecting effects of environmental water allocations in wetlands of the Murray–Darling Basin, Australia. Regul. River. **16**(5):479–496.

Reiswig, H. M. 2006. Classification and phylogeny of Hexactinellida (Porifera). Can. J. Zoolog. **84**(2):195–204.

Resh, V. H., and J. D. Unzicker. 1975. Water quality monitoring and aquatic organisms: The importance of species identification. J. Water Pollut. Con. F. **47**(1):9–19.

Resh, V. H., and J. O. Solem. 1978. Phylogenetic relationships and evolutionary adaptations of aquatic insects. Pp. 33–42 in R. W. Merritt and K. W. Cummins, eds., An introduction to the aquatic insects of North America. Dubuque, IA: Kendall Hunt.

Rich, P. H. 1978. Differential CO_2 and O_2 benthic community metabolism in a soft-water lake. Unpublished manuscript.

Rich, P. H., and R. G. Wetzel. 1978. Detritus in lake ecosystems. Amer. Nat. **112**:57–71.

Richards, J. H., D. N. Kuhn, and K. Bishop. 2012. Interrelationships of petiolar air canal architecture, water depth, and convective air flow in *Nymphaea odorata* (Nymphaeaceae). Am. J. Bot. **99**(12):1903–1909.

Richards, S. R., C. A. Kelly, and J. W. M. Rudd. 1991. Organic volatile sulfur in lakes of the Canadian Shield and its loss to the atmosphere. Limnol. Oceanogr. **36**:468–482.

Richardson, J. L., and M. J. Vepraskas. 2001. Wetland soils: Genesis, hydrology, landscapes, and classification. Boca Raton, FL: Lewis Publishers.

Richter, B. D., and H. E. Richter. 2000. Prescribing flood regimes to sustain riparian ecosystems along meandering rivers. Conserv. Biol. **14**(5):1467–1478.

Ricker, W. E. 1937. Physical and chemical characteristics of Cultus Lake, British Columbia. J. Biol. Bd. Canada **3**:363–402.

Rickert, D. A., and L. B. Leopold. 1972. Fremont Lake, Wyoming—Preliminary survey of a large mountain lake. U.S. Geol. Surv. Professional Research Paper 800D: 173–188.

Riemann, B. 1977. Phosphorus budget for a non-stratified Danish lake and horizontal differences in phytoplankton growth. Arch. Hydrobiol. **79**:357–381.

Riessen, H. P. 1980. Diel vertical migration of pelagic water mites. Pp. 129–137 in W. C. Kerfoot, ed., Evolution and ecology of zooplankton communities. Hanover, NH: The University Press of New England.

Rigler, F. H. 1956. A tracer study of the phosphorus cycle in lakewater. Ecology **37**:550–562.

Rigler, F. H. 1964. The phosphorus fractions and the turnover time of inorganic phosphorus in different types of lakes. Limnol. Oceanogr. **9**:511–518.

Rigler, F. H. 1973. A dynamic view of the phosphorus cycle in lakes. Pp. 539–572 in E. J. Griffith et al., eds., Environmental phosphorus handbook. New York: John Wiley and Sons.

Rigler, F. H. 1975. Nutrient kinetics and the new typology. Verh. Internat. Verein. Limnol. **19**:197–210.

Riley, G. A. 1939. Limnological studies in Connecticut. Part I. General limnological survey. Part II. The copper cycle, Ecol. Monogr. **6**:66–94.

Riley, G. A. 1963. Organic aggregates in seawater and the dynamics of their formation and utilization. Limnol. Oceanogr. **8**:372–381.

Ringelberg, J. 1991. Enhancement of the phototactic reaction in *Daphnia hyalina* by a chemical mediated by juvenile perch (*Perca fluviatilis*). J. Plankton Res. **13**:17–25.

Ristic, R., and J. Ristic. 1980. Lake Michigan: Bathymetric chart and morphometric parameters. Milwaukee: Center for Great Lakes Studies.

Rocha, A. V., and M. L. Goulden. 2009. Why is marsh productivity so high? New insights from eddy covariance and biomass measurements in a *Typha* marsh. Agr. Forest Meteorol. **149**(1):159–168.

Rodgers, G. K. 1965. The thermal bar in the Laurentian Great Lakes. University of Michigan, Great Lakes Res. Div., Pub. No. 13:358–363.

Rodgers, G. K., and D. V. Anderson. 1961. A preliminary study of the energy budget of Lake Ontario. J. Fish. Res. Bd. Canada **18**:617–636.

Rodhe, W. 1949. The ionic composition of lake waters. Verh. Internat. Verein. Limnol. **10**:377–386.

Rodhe, W. 1969. Crystallization of eutrophication concepts in northern Europe. Pp. 50–64 in Eutrophication: Causes, consequences, correctives. Proceedings of a Symposium. Washington, DC: National Academy of Sciences.

Rodhe, W., J. E. Hobbie, and R. T. Wright. 1966. Phototrophy and heterotrophy in high mountain lakes. Verh. Internat. Verein. Limnol. **16**:302–312.

Rodrigo, M. A., M. R. Miracle, and E. Vicente. 2001. The meromictic Lake La Cruz (Central Spain). Patterns of stratification. Aquat. Sci. **63**(4):406–416.

Rohden, C. V., B. Boehrer, and J. Ilmberger. 2010. Evidence for double diffusion in temperate meromictic lakes. Hydrol. Earth Syst. Sc. **14**(4):667–674.

Romani, A. M., A. Giorgi, V. Acuna, and S. Sabater. 2004. The influence of substratum type and nutrient supply on biofilm organic matter utilization in streams. Limnol. Oceanogr. **49**:1713–1721.

Rosa, E., M. Larocque, S. Pellerin, S. Gagné, and B. Fournier. 2008. Determining the number of manual measurements required to improve peat thickness estimations by ground penetrating radar. Earth Surf. Process. Landforms **34**(3):377–383. doi: 10.1002/esp.1741.

Rosenberg, G., and H. W. Paerl. 1981. Nitrogen fixation by blue-green algae associated with the siphonous green seaweed *Codium decorticatum:* Effects on ammonium uptake. Mar. Biol. **61**:151–158.

Rossi, L., and G. Vitagliano-Tadini. 1978. Role of adult faeces in the nutrition of larvae of *Asellus aquaticus* (Isopoda). Oikos **30**:109–113.

Rosswall, T. 1981. The biogeochemical nitrogen cycle. Pp. 25–49 in G. E. Likens, ed., Some perspectives of the major biogeochemical cycles. SCOPE 17. New York: John Wiley and Sons.

Roth, J. C., and S. E. Neff. 1964. Studies of physical limnology and profundal bottom fauna, Mountain Lake, Virginia. Virginia Agric. Exper. Sta., Blacksburg. Tech. Bull. **169**:1–44.

Rueter, J. G., Jr., S. W. Chisholm, and F. M. M. Morel. 1981. Effects of copper toxicity on silicic acid uptake and growth in *Thalassiosira pseudonana.* J. Phycol. **17**:270–278.

Ruttner, F. 1931. Hydrographische und hydrochemische Beobachtungen auf Java, Sumatra, und Bali. Arch. Hydrobiol. Suppl. **8**:197–454.

Ryder, R. A. 1965. A method for estimating the potential fish production of north-temperate lakes. Trans. Amer. Fish. Soc. **94**:214–218.

Ryder, R. A., S. R. Kerr, K. H. Loftus, and H. A. Regier. 1974. The morphoedaphic index, a fish yield estimator-review and evaluation. J. Fish Res. Bd. Canada **31**:663–688.

Ryding, S. O., and C. Forsberg. 1976. Six polluted lakes: A preliminary evaluation of the treatment and recovery process. Ambio **4**:151–156.

Ryther, J. H., and W. M. Dunstan. 1971. Nitrogen, phosphorus, and eutrophication in the coastal marine environment. Science **171**:1008–1013.

Ryther, J. H., and C. S. Yentsch. 1957. The estimation of phytoplankton production in the ocean from chlorophyll and light data. Limnol. Oceanogr. **2**:281–286.

Rzóska, J. 1961. Observations on tropical rainpools and general remarks on temporary waters. Hydrobiologia **17**:265–286.

Rzóska, J. 1978. On the nature of rivers, with case stories of the Nile, Zaire, and Amazon. The Hague: Dr. W. Junk N.V.

Sahoo, G. B., S. G. Schladow, J. E. Reuter, R. Coats, M. Dettinger, J. Riverson, B. Wolfe, and M. Costa-Cabral. 2012. The response of Lake Tahoe to climate change. Climatic Change **116**:71–95. doi: 10.1007/s10584-012-0600-8.

Sanders, H. L. 1955. The Cephalocarida, a new subclass of Crustacea from Long Island Sound. Proc. Nat. Acad. Sci. **41**:61–66.

Satake, K. 1980. Limnological studies on inorganic acid lakes in Japan. Japan. J. Limnol. **41**:41–50.

Saunders, G. W. 1972. Potential heterotrophy in a natural population of *Oscillatoria agardhii* var. *isothrix* Skuja. Limnol. Oceanogr. **17**:704–711.

Scavia, D., J. D. Allan, K. K. Arend, S. Bartell, D. Beletsky, N. S. Bosch, S. B. Brandt, ... and Y. Zhou. 2014. Assessing and addressing the re-eutrophication of Lake Erie: Central basin hypoxia. J. Great Lakes Res. **40**(2):226–246. doi:10.1016/j.jglr.2014.02.004.

Schagerl, M., and B. Müller, B. 2006. Acclimation of chlorophyll *a* and carotenoid levels to different irradiances in four freshwater cyanobacteria. J. Plant Physiol. **163**(7):709–716.

Schelske, C. L. 1975. Silica and nitrate depletion as related to rate of eutrophication in Lakes Michigan, Huron, and Superior. Pp. 277–298 in A. D. Hasler, ed., Coupling of land and water systems. New York: Springer-Verlag.

Schelske, C. L., and E. F. Stoermer. 1972. Phosphorus, silica, and eutrophication of Lake Michigan. Pp. 157–171 in G. E. Likens, ed., Nutrients and eutrophication: The limiting-nutrient controversy. ASLO Special Symposia 1. Lawrence, KS: American Society of Limnology and Oceanography.

Schindler, D. W. 1971a. Light, temperature, and oxygen regimes of selected lakes in the Experimental Lakes Area, northwestern Ontario. J. Fish. Res. Bd. Canada **28**:157–169.

Schindler, D. W. 1971b. A hypothesis to explain differences and similarities among lakes in the Experimental Lakes Area, northwestern Ontario. J. Fish. Res. Bd. Canada **28**:295–301.

Schindler, D. W. 1971c. Carbon, nitrogen and phosphorus and the eutrophication of freshwater lakes. J. Phycol. **7**:321–329.

Schindler, D. W. 1977. Evolution of phosphorus limitation in lakes. Science **195**:260–262.

Schindler, D. W. 2009. Lakes as sentinels and integrators for the effects of climate change on watersheds, airsheds, and landscapes. Limnol. Oceanogr. **54**:2349–2358.

Schindler, D. W., G. J. Brunskill, S. Emerson, W. S. Broecker, and T. H. Peng. 1972. Atmospheric carbon dioxide: Its role in maintaining phytoplankton standing crops. Science **177**:1192–1194.

Schindler, D. W., and G. W. Comita. 1972. The dependence of primary production upon physical and chemical factors in a small, senescing lake, including the effects of complete winter oxygen depletion. Arch. Hydrobiol. **69**:413–451.

Schmalz, R. F. 1972. Calcium carbonate: geochemistry. Pp. 104–118 in W. Fairbridge, ed., The encyclopedia of geochemistry and environmental sciences. New York: Van Nostrand Reinhold.

Schmid-Araya, J. M. 1998. Rotifers in interstitial sediments. Hydrobiologia **387**:231–240.

Schmidt, W. 1915. Über den Energie-gehalt der Seen. Mit Beispielen von Lunzer Untersee nach Messungen mit einen einfachen Temperaturlot. Internat. Rev. Hydrobiol., Suppl. 5, Leipzig.

Schmidt, W. 1928. Über Temperatur und Stabilitatsverhältnisse von Seen. Geogr. Ann. **10**:145–177.

Schönborn, W. 1962. Über Planktismus und Zyklomorphose bei *Difflugia limnetica*. Limnologica **1**:21–34.

Schrödinger, E. 1945. What is life? London: Cambridge University Press.

Schwaderer, A. S., K. Yoshiyama, P. de Tezanos Pinto, N. G. Swenson, C. A. Klausmeier, and E. Litchman. 2011. Eco-evolutionary differences in light utilization traits and distributions of freshwater phytoplankton. Limnol. Oceanogr. **56**(2):589–598.

Scott, S. L., and J. A. Osborne. 1981. Benthic macroinvertebrates of a *Hydrilla* infested central Florida lake. J. Freshwat. Ecol. **1**:41–49.

Sekar, R., K. V. K. Nair, V. N. R. Rao, and V. P. Venugopalan. 2002. Nutrient dynamics and successional changes in a lentic freshwater biofilm. Freshwater Biol. **47**(10):1893–1907.

Shannon, C. E., and W. Weaver. 1949. The mathematical theory of communication. Urbana: University of Illinois Press.

Shannon, E. E., and P. L. Brezonik. 1972. Limnological characteristics of north and central Florida lakes. Limnol. Oceanogr. **17**:97–110.

Shapiro, J. 1957. Chemical and biological studies on the yellow organic acids of lake water. Limnol. Oceanogr. **2**:161–179.

Shapiro, J. 1960. The cause of metalimnetic minimum of dissolved oxygen. Limnol. Oceanogr. **5**:216–227.

Shapiro, J. 1966. The relation of humic color to iron in natural water. Verh. Internat. Verein. Limnol. **16**:477–484.

Shapiro, J. 1990. Current beliefs regarding dominance by blue-greens: The case for the importance of CO_2 and pH. Verh. Internat. Verein. Limnol. **24**:38–54.

Shaw, S. P., and C. G. Fredine. 1956. Wetlands of the United States, their extent, and their value for water fowl and other wildlife. Washington, DC: U.S. Department of the Interior, Fish and Wildlife Circular 39, p. 67.

Sheldon, R. B., and C. W. Boylen. 1977. Maximum depth inhabited by aquatic vascular plants. Amer. Midl. Nat. **97**:248–254.

Shephard, R. B., and G. W. Minshall. 1981. Nutritional value of lotic insect feces compared with allochthonous materials. Arch. Hydrobiol. **90**:467–488.

Shepherd, A., E. R. Ivins, A. Geruo, V. R. Barletta, M. J. Bentley, S. Bettadpur, and H. J. Zwally. 2012. A reconciled estimate of ice-sheet mass balance. Science **338**(6111):1183–1189.

Short, R. A., S. P. Canton, and J. V. Ward. 1980. Detrital processing and associated macroinvertebrates in a Colorado mountain stream. Ecology **61**:727–732.

Shtarkman, Y. M., Z. A. Koçer, R. Edgar, R. S. Veerapaneni, T. D'Elia, P. F. Morris, and S. O. Rogers. 2013. Subglacial Lake Vostok (Antarctica) accretion ice contains a diverse set of sequences from aquatic, marine and sediment-inhabiting bacteria and eukarya. PloS ONE **8**(7):e67221.

Sigee, D. C. 2005. Freshwater microbiology-biodiversity and dynamic interactions of microorganisms in the aquatic environments. Hoboken, NJ: John Wiley and Sons.

Simpson, E. H. 1949. Measurement of diversity. Nature 163:688. doi:10.1038/163688a0.

Sládecková, A. 1962. Limnological investigation methods for the periphyton ("Aufwuchs") community. Bot. Rev. 28:287–350.

Šlapeta, J., D. Moreira, and P. López-García. 2005. The extent of protist diversity: Insights from molecular ecology of freshwater eukaryotes. P. Roy. Soc. B-Biol Sci. 272(1576):2073–2081.

Slobodkin, L. B. 1972. On the inconstancy of ecological efficiency and the form of ecological theories. Pp. 293–305 in E. S. Deevey, ed., Growth by intussusception. Ecological essays in honor of G. E. Hutchinson. Trans. Connecticut Acad. Arts Sci. 44:1–43.

Smith, A. J., and L. D. Delorme. 2010. Ostracoda. In J. H. Thorp, and A. P. Covich, eds., Ecology and classification of North American freshwater invertebrates. San Diego, CA: Academic Press.

Smith, D. G. 2001. Pennak's freshwater invertebrates of the United States: Porifera to Crustacea. New York: John Wiley and Sons.

Smith, F. E., and E. R. Baylor. 1953. Color responses in the Cladocera and their ecological significance. Amer. Nat. 87:49–55.

Smith, R. D., A. Ammann, C. Bartoldus, and M. M. Brinson. 1995. An approach for assessing wetland functions using hydrogeomorphic classification, reference wetlands, and functional indices No. WES/TR/WRP-DE-9). Vicksburg, MS: Army Engineer Waterways Experiment Station.

Snyder, J. M., and L. H. Wullstein. 1973. The role of desert cryptograms in nitrogen fixation. Amer. Midi. Nat. 90:257–265.

Sommer, U., Z. M. Gliwicz, W. Lampert, and A. Duncan. 1986. The PEG-model of seasonal succession of planktonic events in fresh waters. Arch. Hydrobiol. 106:433–471.

Sorokin, J. I. 1975. Sulphide formation and chemical composition of bottom sediments of some Italian lakes. Hydrobiologia 47:231–240.

Spence, D. H. N., and J. Chrystal. 1970a. Photosynthesis and zonation of freshwater macrophytes: I. Depth distribution and shade tolerance. New Phytol. 69:205–215.

Spence, D. H. N., and J. Chrystal. 1970b. Photosynthesis and zonation of freshwater macrophytes: II. Adaptability of species of deep and shallow water. New Phytol. 69:217–227.

Stanier, R. Y. 1973. Autotrophy and heterotrophy in unicellular blue-green algae. Pp. 501–518 in A. G. Carr and B. A. Whitton, eds., The biology of blue-green algae. Bot. Monogr. 9. Berkeley: University of California Press.

Stavn, R. H. 1971. The horizontal-vertical distribution hypothesis: Langmuir circulations and Daphnia distributions. Limnol. Oceanogr. 16:453–466.

Steemann Nielsen, E. 1952. Measurement of the production of organic matter in the sea by means of carbon 14. Nature 167:684–685.

Steemann Nielsen, E. 1977. The carbon-14 technique for measuring organic production by plankton algae. A report on the present knowledge. Folia Limnol. Scandinavica 17:45–48.

Steinböck, O. 1958. Grundsätzliches zum "Kryoeutrophen" See. Verh. Internat. Verein. Limnol. 13:181–190.

Stewart, K. M. 1972. Isotherms under ice. Verh. Internat. Verein. Limnol. 18:303–311.

Stewart, W. D. P. 1977. Present-day nitrogen-fixing plants. Ambio 6:166–173.

Stockner, J. G., and K. G. Porter. 1988. Microbial food webs in freshwater planktonic ecosystems. Pp. 69–83 in S. R. Carpenter, ed., Complex interactions in lake communities. New York: Springer-Verlag.

Stoermer, E. F. 1978. Phytoplankton assemblages as indicators of water quality in the Laurentian Great Lakes. Trans. Amer. Microsc. Soc. 97:2–16.

Stross, R. G. 1971. Photoperiodism and diapause in Daphnia: A strategy for all seasons (abstract). Trans. Amer. Microsc. Soc. 90:110–112.

Stuckey, R. L., and D. Moore. 1995. Return and increase in abundance of aquatic flowering plants in Put-In-Bay Harbor, Lake Erie, Ohio. Ohio J. Sci. **95**:261–266.

Stuiver, M. 1970. Oxygen and carbon isotope ratios of fresh-water carbonates as climatic indicators. J. Geophys. Res. **75**:5247–5257.

Stull, E. 1971. The contribution of individual species of algae to the primary production of Castle Lake. Ph.D. thesis, University of California, Davis.

Stumm, W., and J. J. Morgan. 1970. Aquatic chemistry. New York: John Wiley and Sons.

Swain, F. M. 1975. Biogeochemistry of humic compounds. Pp. 337–360 in D. Povoledo and H. L. Golterman, eds., Humic substances, their structure and function in the biosphere. Proceedings of an international meeting held at Nieuwersluis, The Netherlands, May 29–31, 1972. Centre for Agricultural Publishing and Documentation, Wageningen.

Sweeney, R. E., K. K. Liu, and I. R. Kaplan. 1978. Oceanic nitrogen isotopes and their uses in determining the source of sedimentary nitrogen. Pp. 9–26 in B. W. Robinson, ed., Stable isotopes in the earth sciences. DISR Bull. 220. Wellington: New Zealand Dept. Sci. Indust. Res.

Swift, B. L. 1984. Status of riparian ecosystems in the United States. Water Resources Bull. **20**:223–228.

Tagliapietra, D., M. Cornello, and G. Pessa. 2007. Indirect estimation of benthic secondary production in the Lagoon of Venice (Italy). Hydrobiologia **588**(1):205–212.

Talarico, L., and G. Maranzana. 2000. Light and adaptive responses in red macroalgae: An overview. J. Photochemistry and Photobiology **56**:1–11.

Talling, J. F. 1986. The seasonality of phytoplankton in African lakes. Hydrobiol. **138**(1):139–160.

Talling, J. F., and I. B. Talling. 1965. The chemical composition of African lake waters. Internat. Rev. Hydrobiol. **50**:421–463.

Talling, J. F., R. B. Wood, M. V. Prosser, and R. M. Baxter. 1973. The upper limit of photosynthetic productivity by phytoplankton: Evidence from Ethiopian soda lakes. Freshwat. Biol. **3**:53–76.

Tassi, F., and D. Rouwet. 2014. An overview of the structure, hazards, and methods of investigation of Nyos-type lakes from the geochemical perspective. J. Limnol **73**(1). doi: 10.4081/jlimnol.2014.836.

Taylor, W. D. 1980. Observation on the feeding and growth of the predaceous oligochaete *Chaetogaster langi* on ciliated protozoa. Trans. Amer. Microsc. Soc. **99**:360–368.

Teal, J. M. 1957. Community metabolism in a temperate cold spring. Ecol. Monogr. **27**:283–302.

Teal, J. M. 1962. Energy flow in the salt marsh ecosystem of Georgia. Ecology **43**(4):614–624.

Tejada-Martínez, A. E., I. Akkerman and Y. Bazilevs. 2011. Large-eddy simulation of shallow water Langmuir turbulence using isogeometric snalysis and the residual-based variational multiscale method. J. Appl. Mech. **79**(1):010909.

Terai, H. 1979. Taxonomic study and distribution of denitrifying bacteria in Lake Tizaki, Japan. J. Limnol. **40**:81–92.

Thienemann, A. 1925. Die Binnengewässer Mitteleuropas: eine Limnologische Einfuhrung. Die Binnengewässer **1**:1–255.

Thienemann, A. 1927. Der Bau des Seebeckens in seiner Bedeutung für den Ablauf des Lebens im See. Zool. Bot. Ges. Vienna, Verhandl. **77**:87–91.

Thomas, E. A. 1969. The process of eutrophication in Central European lakes. Pp. 29–49 in Eutrophication: Causes, consequences, correctives. Proceedings of a symposium. Washington, DC: National Academy of Sciences.

Thomas, E. A. 1973. Phosphorus and eutrophication. Pp. 585–611 in Griffith et al., eds., Environmental phosphorus handbook. New York: John Wiley and Sons.

Thoreau, H. D. 1854. Walden; or Life in the woods. Boston: Ticknor and Fields.

Thorp, J. H., and A. P. Covich, eds. 2010. Ecology and classification of North American freshwater invertebrates. San Diego, CA: Academic Press.

Tilman, D. 1982. Resource competition and community structure. Monogr. Pop. Biol. No. 17. Princeton, NJ: Princeton University Press.

Timms. B. V. 1974. Aspects of the limnology of Lake Tali Karng, Victoria. Australian J. Mar. Freshwat. Res. **25**:273–279.

Timms, B. V. 1986. The coastal dune lakes of eastern Australia. Pp. 421–432 in P. DeDeckker and W. D. Williams, eds., Limnology in Australia. Melbourne: CSIRO; Dordrecht, Netherlands: Dr. W. Junk b.v.

Tomanova, S., P. A. Tedesco, M. Campero, P. A. Van Damme, N. Moya and T. Oberdorff. 2007. Longitudinal and altitudinal changes of macroinvertebrate functional feeding groups in neotropical streams: a test of the River Continuum Concept. Fundamental and Applied Limnology Archiv für Hydrobiologie **170**:233–241.

Tonolli, L., M. Gerletti, and G. Chiaudani. 1975. Trophic conditions of Italian lakes as a consequence of human pressures. Pp. 215–225 in A.D. Hasler, ed., Coupling of land and water systems. New York: Springer-Verlag.

Torii, T., and J. Ossaka. 1965. Antarcticite: A new mineral, calcium chloride hexahydrate, discovered in Antarctica. Science **149**:975–977.

Tortell, P. D. 2000. Evolutionary and ecological perspectives on carbon acquisition in phytoplankton. Limnol. Oceanogr. **45**:744–750.

Tranvik, L. J., J. A. Downing, J. B. Cotner, S. A. Loiselle, R. G. Striegl, T. J. Ballatore, ... and G. A. Weyhenmeyer. 2009. Lakes and reservoirs as regulators of carbon cycling and climate. Limnol. Oceanogr. **54**(6):2298–2314.

Tsivoglou, E. C. 1972. Turbulence, mixing and gas transfer. Pp. 5–18 in Proceedings of a symposium on direct tracer measurement of the reaeration capacity of streams and estuaries. Environmental Protection Agency Project #16050 and Georgia Institute of Technology.

Tudorancea, C., R. H. Green, and J. Huebner. 1979. Structure, dynamics and production of the benthic fauna in Lake Manitoba. Hydrobiologia **64**:59–95.

Tuunainen, P., K. Granberg, L. Hakkari, and J. Särkkä. 1972. On the effects of eutrophication of Lake Päijänne, Central Finland. Verh. Internat. Verein. Limnol. **18**:388–402.

Twining, C. W., D. C. West, and D. M. Post. 2013. Historical changes in nutrient inputs from humans and anadromous fishes in New England's coastal watersheds. Limnol. Oceanogr. **58**:1286–1300.

Uéno, M. 1934. Yale North India Expedition. Report on amphipod Crustacea of the genus *Gammarus.* Mem. Connecticut Acad. Arts Sci. **10**:63–75; 5 plates.

U.S. Department of Agriculture, Natural Resources Conservation Service. 2010. Field indicators of hydric soils in the United States: A guide for identifying and delineating hydric soils, Version 7.0. L. M. Vasilas, G. W. Hurt, and C. V. Noble, eds., USDA, NRCS, in cooperation with the National Technical Committee for Hydric Soils.

Vagle, S., J. Hume, F. McLaughlin, E. MacIsaac, and K. Shortreed. 2010. A methane bubble curtain in meromictic Sakinaw Lake, British Columbia. Limnol. Oceanogr. **55**(3):1313–1326.

Väinölä, R., J. D. S. Witt, M. Grabowski, J. H. Bradbury, K. Jazdzewski, and B. Sket. 2008. Global diversity of amphipods (Amphipoda; Crustacea) in freshwater. Hydrobiologia **595**(1):241–255.

Vallentyne, J. R. 1974. The algal bowl: Lakes and man. U.S. Dept. of Environ., Fisheries and Marine Svc., Misc. Spec. Pub. 22.

van der Valk, A. G. 1981. Succession in wetlands: A Gleasonian approach. Ecology **62**:688–696.

van der Valk, A. G., and C. B. Davis. 1978. The role of seed banks in the vegetation dynamics of prairie glacial marshes. Ecology **59**:322–335.

van Driesche, J., and R. Van Driesche. 2000. Nature out of place: Biological invasions in the global age. Washington, DC: Island Press.

van Gorkom, H. J., and M. Donze. 1971. Localization of nitrogen fixation in Anabaena. Nature **234**:231–232.

Van Looy, K., T. Tormos, and Y. Souchon. 2014. Disentangling dam impacts in river networks. Ecol. Indic. **37**:10–20.

Vannote, R. L., G. W. Minshall, K. W. Cummins, J. R. Sedell, and C. E. Cushing. 1980. The river continuum concept. Canad. J. Fish. Aquat. Sci. **37**:130–137.

Vareschi, E. 1978. The ecology of Lake Nakuru (Kenya). I. Abundance and feeding of the lesser flamingo. Oecologia **32**:11–35.

Venäläinen, A., M. Frech, M. Heikinheimo, and A. Grelle. 1999. Comparison of latent and sensible heat fluxes over boreal lakes with concurrent fluxes over a forest: Implications for regional averaging. Agr. Forest Meteorol. **98**:535–546.

Venkiteswaran, J. J., L. I. Wassenaar, and S. L. Schiff. 2007. Dynamics of dissolved oxygen isotopic ratios: A transient model to quantify primary production, community respiration, and air-water exchange in aquatic ecosystems. Oecologia **153**(2): 385–398.

Verburg, P. M., and R. E. Hecky. 2009. The physics of the warming of Lake Tanganyika by climate change. Limnol. Oceanogr. **54**:2418–2430.

Verduin, J. 1951. Photosynthesis in naturally reared aquatic communities. Plant Physiol. **26**:45–49.

Verduin, J. 1956. Energy fixation and utilization by natural communities in western Lake Erie. Ecology **37**:40–50.

Viner, A. B. 1977. The sediments of Lake George (Uganda): IV. Vertical distribution of chemical features in relation to ecological history and nutrient cycle. Arch. Hydrobiol. **80**:40–69.

Vollenweider, R. A. 1961. Photometric studies in inland waters: I. Relations existing in the spectral extinction of light in water. Mem. Ist. Ital. Idrobiol. **13**:87–113.

Vollenweider, R. A. 1964. Über oligomiktishe Verhaltnisse des Lago Maggiore und einiger anderer insubrischer Seen. Mem. Ist. Ital. Idrobiol. **17**:191–206.

Vollenweider, R. A. 1976. Advances in defining critical loading levels for phosphorus in lake eutrohpication. Mem. Ist. Ital. Idrobiol. **33**:53–83.

von Arx, W. S. 1962. An introduction to physical oceanography. Boston: Addison-Wesley.

von Rohden, C. V., B. Boehrer, and J. Ilmberger. 2010. Evidence for double diffusion in temperate meromictic lakes. Hydrol. Earth Syst. Sc. **14**(4):667–674.

Walker, K. F. 1973. Studies on a saline lake ecosystem. Australian J. Mar. Freshwat. Res. **24**:21–71.

Walker, K. F., and G. E. Likens. 1975. Meromixis and a reconsidered typology of lake circulation patterns. Verh. Internat. Verein. Limnol. **19**:442–458.

Wallace, J. B., J. R. Webster, and W. R. Woodall. 1977. The role of filter feeders in flowing waters. Arch. Hydrobiol. **79**:506–532.

Wallace, R. L., and T. W. Snell. 2010. Rotifera. Pp. 173–235 in J. H. Thorp and A. P. Covich, eds., Ecology and classification of North American freshwater invertebrates. Oxford, UK: Elsevier.

Walsh, E. J., H. A. Smith, and R. L. Wallace. 2014. Rotifers of temporary waters. Int. Rev. Hydrobiol. **99**(1-2):3–19.

Walshe, B. M. 1947. On the function of haemoglobin in *Chironomus* after oxygen lack. J. Exper. Biol. 24:329–342.

Walshe, B. M. 1950. The function of haemoglobin in *Chironomus plumosus* under natural conditions. J. Exper. Biol. **27**:73–95.

Ward, A. K., and R. G. Wetzel. 1980. Interactions of light and nitrogen source among planktonic blue-green algae. Arch. Hydrobiol. **90**:1–25.

Ward, J. V., and J. A. Stanford. 1982a. Thermal responses in the evolutionary ecology of aquatic insects. Ann. Rev. Entomol. **27**:97–117.

Ward, J. V., and J. A. Stanford. 1982b. The serial discontinuity concept of lotic ecosystems. In T. D. Fontaine and S. M. Bartell, eds., Dynamics of lotic ecosystems. Ann Arbor, MI: Ann Arbor Science Publishers.

Warnick, S. L., and H. L. Bell. 1969. The acute toxicity of some heavy metals to different species of aquatic insects. Part I. J. Water Pollut. Con. F. **41**(2):280–284.

Waters, T. F. 1972. The drift of stream insects. Ann. Rev. Entomol. **17**:253–272.

Waters, T. F. 1977. Secondary production in inland waters. Adv. Ecol. Res. **10**:91–164.

Waters, T. F., and J. C. Hokenstrom. 1980. Annual production and drift of the stream amphipod, *Gammarus pseudolimnaeus* in Valley Creek, Minnesota. Limnol. Oceanogr. 25:700–710.

Watson, N. H. F. 1974. Zooplankton of the St. Lawrence Great Lakes-species composition, distribution, and abundance. J. Fish. Res. Bd. Canada 31:783–794.

Watt, W. D., D. Scott, and S. Ray. 1979. Acidification and other chemical changes in Halifax County lakes after 21 years. Limnol. Oceanogr. 24:1154–1161.

Way, C. M., D. J. Hornback, and A. J. Burky. 1981. Seasonal metabolism of the clam *Musculium partumeium* from a permanent and a temporary pond. Nautilus 95:55–58.

Weber, C. A. 1907. Aufbau und Vegetation der Moore Norddeutschlands. Beibl. Bot. Jahrb. 90:19–34.

Wedderburn, E. M. 1907. The temperature of freshwater lochs of Scotland, with special reference to Loch Ness. Trans. Roy. Soc. Edinburgh 45:407–489.

Weigel, B. M., L. J. Henne, and L. M. Martínez-Rivera. 2002. Macroinvertebrate-based index of biotic integrity for protection of streams in west-central Mexico. J. N. Am. Benthol. Soc. 21(4):686–700.

Wellborn, G. A., R. Cothran, and S. Bartholf. 2005. Life history and allozyme diversification in regional ecomorphs of the *Hyalella azteca* (Crustacea: Amphipoda) species complex. Biol. J. Linn. Soc. 84(2):161–175.

Wetzel, R. G. 1964. A comparative study of the primary productivity of higher aquatic plants, periphyton, and phytoplankton in a large, shallow lake. Internat. Rev. Hydrobiol. 49:1–61.

Wetzel, R. G. 1969. Excretion of dissolved organic compounds by aquatic macrophytes. Bioscience 19:539–540.

Wetzel, R. G. 1972. The role of carbon in hard-water lakes. Pp. 84–97 in G. E. Likens, ed., Nutrients and eutrophication: The limiting-nutrient controversy. ASLO Special Symposia 1. Lawrence, KS: American Society of Limnology and Oceanography.

Weyl, P. 1961. The carbonate saturometer. J. Geol. 69:32–44.

Whiteside, M. C. 1965. Paleoecological studies of Potato Lake and its environs. Ecology 46:807–816.

Whiteside, M. C., and C. Lindegaard. 1982. Summer distribution of zoobenthos in Grane Langsø, Denmark. Freshwat. Invert. Biol. 356:2–16.

Whitman, R. L., M. B. Nevers, G. C. Korinek, and M. N. Byappanahalli. 2004. Solar and temporal effects on *Escherichia coli* concentration at a Lake Michigan swimming beach. Appl. Environ. Microb. 70(7):4276–4285.

Whitney, L. V. 1938. Microstratification of inland lakes. Trans. Wisconsin Acad. Sci. Arts Lett. 31:155–173.

Whitton, B. A., and M. Potts. 2012. Introduction to the cyanobacteria. In B. A. Whitton, ed., Ecology of Cyanobacteria II (pp. 1–13). New York/London: Springer Netherlands.

Wiegert, R. G., and P. C. Fraleigh. 1972. Ecology of Yellowstone thermal effluent systems: Net primary production and species diversity of a successional blue-green algal mat. Limnol. Oceanogr. 17:215–228.

Williams, D. D., and R. R. Fulthorpe. 2003. Using invertebrate and microbial communities to assess the condition of the hyporheic zone of a river subject to 80 years of contamination by chlorobenzenes. Can. J. Zoolog. 81(5):789–802.

Williams, P. N., W. H. Mathews, and G. L. Pickard. 1961. A lake in British Columbia containing old sea-water. Nature 191:830–832.

Williams, W. D. 1964. A contribution to lake typology in Victoria, Australia. Verh. Internat. Verein. Limnol. 15:158–168.

Williams, W. D. 1966. Conductivity and the concentration of total dissolved solids in Australian lakes. Australian J. Mar. Freshwat. Res. 17:169–176.

Williams, W. D. 1969. Eutrophication of lakes. Proc. Roy. Soc. Victoria 83:17–26, 1 plate.

Williams, W. D. 1972. The uniqueness of salt lake ecosystems. Pp. 349–361 in Z. Kajak and A. Hillb Ilkowska, eds., Productivity problems of freshwaters. Warsaw-Krakow: Polish Scientific Publishers.

Williams, W. D., ed. 1981. Inland salt lakes. Hydrobiologia **81/82**:i–ix, 1–444.

Williamson, C. E., J. E. Saros, W. F. Vincent, and J. P. Smol. 2009. Lakes and reservoirs as sentinels, integrators, and regulators of climate change. Limnol. Oceanogr. **54**:2273–2282.

Wilson, D. E. 1979. The influence of humic compounds on titrimetric determinations of total inorganic carbon in freshwater. Arch. Hydrobiol. **87**:379–384.

Winberg, G. G. 1963. The primary production of bodies of water. Washington, DC: U.S. Atomic Energy Commission, Transl. AEC-tr5692, Biol. Med. Office of Tech. Serv.

Winberg, G. G. 1971. Methods for the estimation of production of aquatic animals. New York: Academic Press.

Winberg, G. G. 1972. Études sur le bilan biologique energetique et la productivité des lacs en Union Soviétique. Verh. Internat. Verein. Limnol. **18**:39–64.

Woese, C. R. 2004. A new biology for a new century. Microbiol. Mol. Biol. R. **68**(2):173–186.

Wolfe, P. E. 1953. Periglacial frost-thaw basins in New Jersey. J. Geol. **61**(2):133–141.

Wood, K. G. 1977. Chemical enhancement of CO_2 flux across the air-water interface. Arch. Hydrobiol. **79**:103–110.

Woodroffe, C. D., J. M. A. Chappell, B. G. Thom, and E. Wollensky. 1986. Geomorphological dynamics and evolution of the South Alligator tidal river and plains, Northern Territory. Darwin: Australian National University, North Australia Research Unit, Mangrove Monograph 3.

World Commission on Dams. 2000. Dams and development: A new framework for decision-making. London: Earthscan.

Wright, H. E., Jr. 1964. Origin of the lakes in the Chuska Mountains, northwestern New Mexico. Bull. Geol. Soc. Amer. **75**:589–598; 4 figs., 3 plates.

Wright, R. F., and E. T. Gjessing. 1976. Acid precipitation: Changes in the chemical composition of lakes. Ambio **5**:219–223.

Writer, J. H., L. B. Barber, G. K. Brown, H. E. Taylor, R. L. Kiesling, M. L. Ferrey, N. D. Jahns, S. E. Bartell, and H. L. Schoenfuss. 2010. Anthropogenic tracers, endocrine disrupting chemicals, and endocrine disruption in Minnesota lakes. Sci. Total Environ. **409**(1):100–111.

Wu, J., J. Huang, X. Han, X. Gao, F. He, M. Jiang, ... and Z. Shen. 2004. The Three Gorges Dam: An ecological perspective. Front. Ecol. Environ. **2**(5):241–248.

Wynne, R. H., J. J. Magnuson, M. K. Clayton, T. M. Lillesand, and D. C. Rodman. 1996. Determinants of temporal coherence in the satellite-derived 1987–1994 ice breakup dates of lakes on the Laurentian Shield. Limnol. Oceanogr. **41**:832–838.

Xenopoulos, M. A., D. M. Lodge, J. Frentress, T. A. Kreps, S. D. Bridgham, E. Grossman, and C. J. Jackson. 2003. Regional comparisons of watershed determinants of dissolved organic carbon in temperate lakes from the Upper Great Lakes region and selected regions globally. Limnol. Oceanogr. **48**(6):2321–2334.

Yang, X., and X. X. Lu. 2013. Ten years of the Three Gorges Dam: A call for policy overhaul. Environ. Res. Lett. **8**:041006. doi:10.1088/1748-9326/8/4/041006.

Yoshimura, S. 1936. A contribution to the knowledge of deep water temperatures of Japanese lakes. Part II. Winter temperatures. Jap. J. Astr. Geophys. **14**:57–83.

Yoshimura, S. 1937. Abnormal thermal stratifications in inland lakes. Proc. Imp. Acad. Japan **13**:316–319.

Yurista, P. M. 2000. Cyclomorphosis in *Daphnia lumholtzi* induced by temperature. Freshwater Biol. **43**(2):207–213.

Zaret, T. M. 1972. Predators, invisible prey, and the nature of polymorphism in the Cladocera (Class Crustacea). Limnol. Oceanogr. **17**:171–184.

Zaret, T. M. 1980. Predation and freshwater communities. New Haven, CT: Yale University Press.

Zedler J. B., and S. Kercher. 2005. Wetland resources: Status, trends, ecosystem services, and restorability. Annu. Rev. Environ. Resour. **30**:39–74.

Zobell, C. E., F. D. Sisler, and C. H. Oppenheimer. 1953. Evidence of biochemical heating in Lake Mead mud. J. Sed. Petrol. **23**:13–17.

Zosel, J., W. Oelßner, M. Decker, G. Gerlach, and U. Guth. 2011. The measurement of dissolved and gaseous carbon dioxide concentration. Measurement Science and Technology **22**(7):072001. doi:10.1088/1748-9326/8/4/041006.

Zrzavý, J., Říha, P., Piálek, L., and Janouškovec, J. 2009. Phylogeny of Annelida (Lophotrochozoa): Total-evidence analysis of morphology and six genes. BMC Evol. Biol. **9**(1):189.

Name Index

Subject Index

Ablation, 213

Absorption
  atomic absorption spectrophotometry, 318, 337
  attenuation vs., 192
  coefficient of, 191–192, 195
  of light below the water surface, 190
  by plant and bacterial pigments, 200–201
  total coefficient of, 191–192
  vertical coefficient of, 196, 191

Abudefduf troscheli, 72

Accretion, 146

Acid deposition (acid rain), 283–284, 287, 293

Acidity
  atmospheric causes of, 283–284
  edaphic factors affecting, 284–286
  high concentration in *sphagnum* bogs, 285
  humic substances and, 366–368
  hydrogen ions and pH, 283–286
  limnologic effects of, 287
  in peat bogs, 149
  total vs. free, 284

Activity coefficient, 52, 300–301

Aeschna, 174

Aggressive CO_2, 296

Agitation-aggregation, 66–67

Algae
  blue-green. *See* Cyanobacteria
  calcareous, 113
  Chara (stonewort), 328
  chelation's effect on, 347
  Desmidiaceae (calcidarious), 332
  epipelic, 63
  epiphytic, 58
  fish kills caused by oxygen-deficit algal blooms, 263
  green (Chlorophyta), 20–21, 62
  lacustrine, 20–21
  perphyton, 63
  vitamins essential to, 368

Algal succession, 46–48

Alkaline carbonate waters, biota of, 317–318

Alkaline phosphatase, 353

Alkalinity, 293–296
  diel changes in, 295–296
  factors contributing to, 294–295
  pH and, 298

Allen curve method, 87

Allochthonous material
  autochthonous material vs., 2, 163, 166
  detritus, 66
  downstream drift of, 179
  lake death from accumulation of, 95
  organic molecules, 77

Allotrophy, 78

Alluvium, 159–160

Alonopsis elongata, 204

Alosa pseudoharengus (alewife), 56

Alpine lakes, climate change in, 116

Amazon River, 156

Amiocentrus, 178

Amixis, 212–213

Ammonia, 257

Amoeba, 23

Amphipoda, 33–34, 55–56, 61, 172, 175

Anabaena, 197

Anabaenopsis, 19

Anaerobiosis, 144

Anemotrophy, 78

Angiosperms
  deepest limit for growth of, 203
  duckweeds, 57
  morphologic adaptations to fast-flowing water, 171
  stream-dwelling, 165, 171
  Zostera (eelgrass), 68

Annelida (segmented worms), 27

Anisogammarus, 56

Anthropogenic effects, 116–117

Aphanizomenon, 197

Aquatic ecosystems
  benthic community, 4
  light (solar radiation) and, 187–204. *See also* Light
  oligotrophic vs. eutrophic, 7–9